Advances in Experimental Medicine and Biology

Volume 920

Advances in Experimental Medicine and Biology presents multidisciplinary and dynamic findings in the broad fields of experimental medicine and biology. The wide variety in topics it presents offers readers multiple perspectives on a variety of disciplines including neuroscience, microbiology, immunology, biochemistry, biomedical engineering and cancer research. *Advances in Experimental Medicine and Biology* has been publishing exceptional works in the field for over 30 years and is indexed in Medline, Scopus, EMBASE, BIOSIS, Biological Abstracts, CSA, Biological Sciences and Living Resources (ASFA-1), and Biological Sciences. The series also provides scientists with up to date information on emerging topics and techniques.

More information about this series at http://www.springer.com/series/5584

Paul W. Ackermann • David A. Hart
Editors

Metabolic Influences on Risk for Tendon Disorders

Springer

Editors
Paul W. Ackermann
Department of Molecular Medicine and Surgery
Karolinska Institutet
Stockholm, Sweden

Department of Orthopedic Surgery
Karolinska University Hospital
Stockholm, Sweden

David A. Hart
McCaig Institute for Bone and Joint Health
University of Calgary
Calgary, AB, Canada

Centre for Hip Health and Mobility
University of British Columbia
Vancouver, BC, Canada

ISSN 0065-2598 ISSN 2214-8019 (eBook)
Advances in Experimental Medicine and Biology
ISBN 978-3-319-33941-2 ISBN 978-3-319-33943-6 (eBook)
DOI 10.1007/978-3-319-33943-6

Library of Congress Control Number: 2016947053

Printed on acid-free paper

This Springer imprint is published by Springer Nature
The registered company is Springer International Publishing AG Switzerland

Preface

This book on "Metabolic Influences on Risk for Tendon Disorders" was written with the aim to increase the holistic knowledge of the enigma of tendon problems. As of year 2016, we still do not have any effective interventions to treat many patients with tendon disorders. This book aspires to provide an essential understanding of how collaborative research efforts within a multidisciplinary array of basic and clinical scientists are leading to an accelerated development of more effective therapies for tendon aliments.

The quality of life for millions of people is impaired due to tendon problem such as chronic pain and impaired limb function, situations which can adversely impact their general health condition.

On the other hand, however, novel research findings show that general health factors, which control metabolic pathways, are actively involved in regulating tendon homeostasis to a much further extent than previously known. This means that metabolic influences via sedentary lifestyle, e.g. obesity, can lead to painful tendon disorders, which in turn can result in hampered mobility for the patient. The net result is a vicious circle that is currently unexplored and possibly crucial for the understanding of tendon homeostasis.

Having this said, the reader will have to bear in mind that while reading this information, the scientific details in parts of this book are not always explored in great depth. We hope that the lack of in-depth supplied knowledge in some sections, however, will inspire scientists from all corresponding disciplines to join forces to improve our understanding of tendon problems.

In writing this book, we have had a fantastic collaboration with some of the leading experts in their field of tendon research. With their help and engagement, the content of this book has widely surpassed the imagination of the editors. The editors would like to take the opportunity to sincerely thank all the contributors to the production of this book. We would also like to thank the organizers of the congress of "Metabolic diseases and tendinopathies – the missing link", who inspired many researchers in this multidiciplinary field.

At this point, we would like to give you some personal tips about how to read this book. The book starts with eight chapters to provide you the latest understanding of tendon basic biology and anatomy, which is especially recommended for those of you without any knowledge on tendons, but also of high interest for tendon researchers and clinicians.

Having the basic understanding of tendon metabolism, we suggest that you read all/some/one of the 15 chapters on how tendon disorders can be associated with altered metabolism and metabolic disorders. These 15 chapters have a common thread in the order that they are written, but can very well be read individually or in any order you like. We suggest that these chapters should be read by every clinician who deals with tendon problems.

Near the end of the book, we have included three chapters on novel and popular therapies which may affect tendon metabolism. This book does not make any attempts to cover how tendon disorders should be managed, but rather to inspire the development of integrated approaches. In the last chapter of the book, the editors have attempted to summarize the concepts and content of the book. For those of you, like ourselves, who would like to get the latest information in a short time, we recommend that you start by reading this chapter, which may further intrigue you to subsequently read most if not all of the individual chapters of the book. You may want to go back and forth from this overview chapter to individual focused chapters in order to synthesize your own perspectives.

We wish you a pleasant and engaging reading of the book!

Stockholm, Sweden — Paul W. Ackermann
Calgary, AB, Canada — David A. Hart

Contents

Part I

Basic Biology and Anatomy

Tendon Structure and Composition

1

Chavaunne T. Thorpe and Hazel R.C. Screen

Abstract

Tendons are soft, fibrous tissues that connect muscle to bone. Their main function is to transfer muscle generated force to the bony skeleton, facilitating movement around a joint, and as such they are relatively passive, inelastic structures, able to resist high forces. Tendons are predominantly composed of collagen, which is arranged in a hierarchical manner parallel to the long axis of the tendon, resulting in high tensile strength. Tendon also contains a range of non-collagenous proteins, present in low amounts, which nevertheless have important functional roles. In this chapter, we describe general tendon composition and structure, and discuss how variations in composition and structure at different levels of the tendon hierarchy confer specific mechanical properties, which are related to tendon function.

Introduction

Typical of connective tissues, tendon is predominantly composed of water, which makes up 55–70 % of total tendon weight. By far the majority of the tendon dry weight consists of collagens (60–85 %) [17]. Collagen is a stiff structural protein, providing tissues with tensile strength. As such, aligning the collagen along the long axis of the tendon results in a highly anisotropic tissue, particularly well suited to the uniaxial tensile strain transferring role of tendon.

Collagen molecules within tendon are arranged in a hierarchical manner (Fig. 1.1) and at each level of the hierarchy, the collagen is interspersed with a less fibrous, highly hydrated matrix, traditionally referred to as the ground substance [15, 32]. Materials with this arrangement of stiff fibres within a surrounding, softer matrix are generally referred to as fibre composites, and tendon provides an example of an aligned fibre composite on multiple hierarchical levels, building from the nano- to macroscale. Fibre composite materials stretch through a combination of extension of, and shearing (sliding) between, the fibrous components.

C.T. Thorpe (✉) • H.R.C. Screen
Institute of Bioengineering, School of Engineering and Materials Science, Queen Mary University of London, Mile End Road, London E1 4NS, UK
e-mail: c.thorpe@qmul.ac.uk; h.r.c.screen@qmul.ac.uk

P.W. Ackermann and D.A. Hart (eds.), *Metabolic Influences on Risk for Tendon Disorders*, Advances in Experimental Medicine and Biology 920, DOI 10.1007/978-3-319-33943-6_1

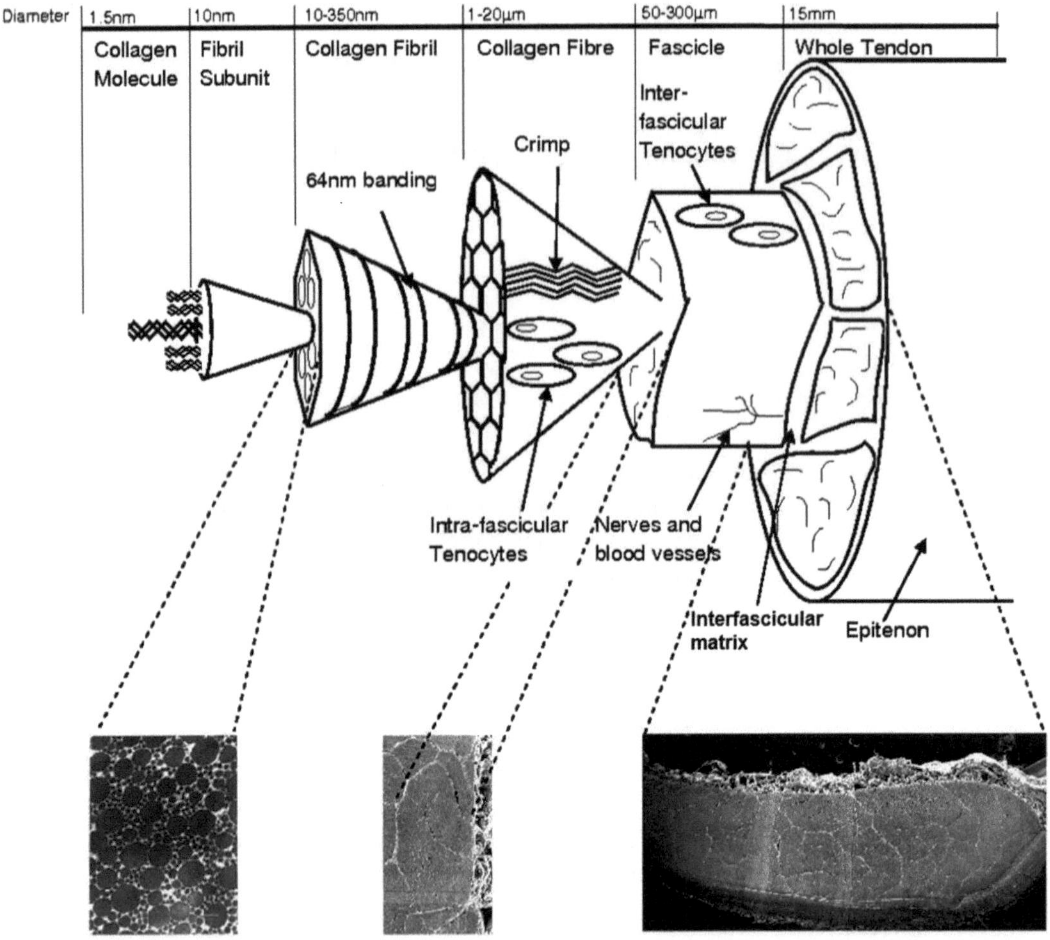

Fig. 1.1 Schematic showing the hierarchical structure of tendon, in which collagen molecules aggregate, forming subunits of increasing diameter (Adapted from Thorpe et al. [35] with kind permission from Wiley publishing)

The proportional contributions of fibre extension and shearing to total material stretch depend on the relative mechanical properties of the fibres and surrounding matrix. Alterations to local extension and shearing mechanics in a fibre composite can substantially affect the gross mechanical properties of the bulk tissue. With multiple hierarchical levels to tendon structure, there is significant scope for altering its local extension and sliding mechanics, and consequently whole tendon mechanical behaviour, through simple, often minor adjustments to composition at one or more of the tendon hierarchical levels. As such, understanding not just the overall composition of tendon, but also the organisation of matrix molecules throughout the tissue hierarchy is critical to understanding tendon function.

The smallest building blocks of tendon are collagen molecules. These are arranged longitudinally in a quarter stagger pattern, with a gap of approximately 40 nm between the ends of each molecule, resulting in the characteristic banding pattern seen in collagen fibrils. Groups of five collagen molecules are bound together by intermolecular crosslinks to form pentafibrils (also referred to as microfibrils). These pentafibrils then pack together, forming fibrils which have diameters ranging from approximately 10 to 500 nm [2, 5]. Collagen fibrils are stabilised by crosslinking, and the fibril is generally defined as the primary structural unit across many different tissue types. In tendon, fibrils aggregate to form collagen fibres, with the fibres aggregating once again to make up the largest subunit of tendon, the fascicle. Fascicles are visible to the naked eye, with diameters ranging from 150 to 500 μm. Each fascicle is surrounded by a connective tissue compartment called the interfascicular matrix (IFM). The IFM (sometimes also referred to as the endotenon) binds the fascicles to make the complete tendon unit, and is particularly important for tendons which function as energy stores [39].

The tendon surface is covered by the epitenon, which is a connective tissue sheath continuous with the IFM. An additional loose connective tissue layer, the paratenon, surrounds the tendons in regions away from joints, to facilitate movement of tendons below the skin. However, where a tendon passes around a joint, it is contained within a synovial sheath in order to ensure smooth gliding past surrounding structures.

Collagens

Type I collagen is the predominant collagen in tendon, making up approximately 90 % of the total collagen content [4, 18]. Other fibrillar collagens are also present, including types III, V and XI. Collagen type III is the second most abundant tendon collagen, comprising up to 10 % of the collagen content. It plays an important role in collagen fibrillogenesis by regulating the size of type I collagen fibrils [14]. It is often localised to the IFM [27], but the role it plays in this tendon compartment is yet to be established. Type V collagen is found in the centre of collagen-I fibrils, where it is thought to provide a template for fibrillogenesis [24]. Several non-fibrillar collagens are also present in tendon at low amounts; type VI is often localised to the pericellular matrix. The fibril associated collagens, types XII and XIV provide a molecular bridge between type I collagen and other matrix molecules, and play important roles during tendon development [1].

Proteoglycans

Proteoglycans are the most abundant non-fibrous proteins in tendons, making up 1–5 % of tendon dry weight. Proteoglycans are a class of glycoproteins, consisting of a core protein attached to one or more polysaccharide chains. These side chains are termed glycosaminoglycan (GAG) side chains and are negatively charged, which attracts water into the tendon. Proteoglycans are a primary component of the matrix interspersing the collagenous units at the fibril, fibre and fascicle levels of the tendon hierarchy [32] (Fig. 1.2).

The most abundant and well characterised proteoglycan in tendon is decorin, which makes up 80 % of the total proteoglycan content. Decorin is a small leucine rich proteoglycan (SLRP), which consists of a horseshoe shaped core protein, attached to which is a single chondroitin or dermatan sulphate side chain. The core protein has a specific binding site on the collagen fibril, and is able to interact with other decorin molecules bound to adjacent collagen fibrils via its side chain, forming an interfibrillar bridge. Decorin is found both within fascicles and the IFM [32]. Decorin shares its fibril binding site with the SLRP biglycan, which contains several dermatan or chondroitin sulphate side chains. The class II SLRPS lumican and fibromodulin contain keratan sulphate side chains, and share a fibril binding site, which is distinct to that used by decorin and biglycan [32].

It is well established that the SLRPs have a critical role in tendon development, where they regulate collagen fibrillogenesis. Specific SLRPs are thought to have a distinct, time-dependent role in this process, with biglycan and lumican modulating fibrillogenesis during early development, whereas decorin and fibromodulin play a primary role at later stages of development and maturation [8, 41].

In addition, it has been suggested that the SLRPs (decorin in particular) may contribute directly to the mechanical properties of mature tendon by transferring load between discontinuous collagen fibrils via interfibrillar bridges.

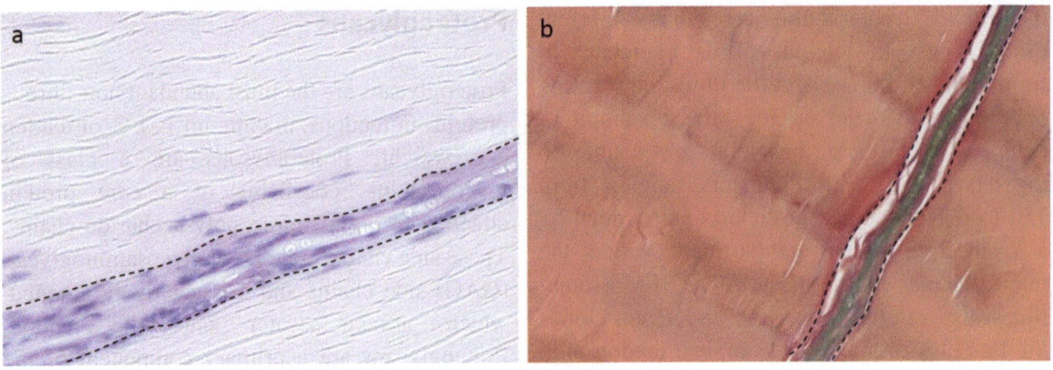

Fig. 1.2 (**a**) Histological image of longitudinal tendon section showing Alcian Blue/Periodic Acid Schiff staining of proteoglycans (*purple*), which are enriched in the IFM (enclosed by *dotted lines*). Cell nuclei are counterstained *blue*. (**b**) Histological image showing elastic Van Gieson's staining of elastin (*black*), which is predominantly localised to the IFM

Individually, these interfibrillar bonds are weak, but when combined they may reach sufficient magnitudes to enable force transfer. However, this is controversial, as it is uncertain whether collagen fibrils are discontinuous, and recent work suggests that decorin may actually promote sliding between fibrils rather than transferring force between them [23, 32].

Other less well studies proteoglycans in tendon include lubricin and versican. Lubricin (also referred to as superficial zone proteoglycan or PRG4) is a large proteoglycan, which is localised to the tendon surface, the IFM and compressive regions of tendon [9, 20, 29]. It provides lubrication, allowing gliding at the tendon surface, and may also facilitate sliding between adjacent fascicles. Versican is a large aggregating proteoglycan which is preferentially expressed in the IFM, particularly in the pericellular region [25]. It interacts with elastic fibres, and so may contribute to the structural properties of the IFM [12], but its precise role in tendon is yet to be determined.

It has been demonstrated that proteoglycan distribution varies along the length of a tendon, with the SLRPs present at highest abundance in the tensional mid-substance region, and a predominance of cartilage associated proteoglycans (aggrecan, biglycan, lubricin) in areas of tendon subjected to compression [22]. It is likely that the predominance of SLRPs in tensional regions is related to their organisational roles in collagen fibrillogenesis, while the high abundance of cartilage-associated proteoglycans in compressive regions attract water into the tendon which enables resistance to compression.

Glycoproteins and Other Molecules

There are several other glycoproteins that form part of the tendon structure. Glycoproteins have carbohydrate groups attached to their polypeptide chains. In tendon, the most abundant glycoprotein is cartilage oligomeric matrix protein (COMP), which consists of a cylindrical core surrounded by five subunits [26]. Each subunit is able to bind to type I collagen, and therefore one COMP molecule can provide a link between 5 collagen molecules. COMP is present at high concentrations in the inter-fibrillar matrix, but absent from the IFM [27]. However, the precise function of COMP in tendon is uncertain, as COMP-null knock-out mice do not demonstrate tendon abnormalities [30].

The glycoprotein tenascin-C is also found at low levels in tendon. It is composed of six subunits, bound together by N-terminal interchain cross-linking domains [13]. Its function in tendon is yet to be established, but it predominates in regions where tendons experience high forces, and its levels are modulated by mechanical loading [13], so it has been suggested it contributes to tissue elasticity.

Elastic fibres are also found in varying amounts throughout the tendon matrix, with reported concentrations from 1 to 10 % of tendon dry weight. These fibres are predominantly localised to the IFM (Fig. 1.2) but are also present within fascicles, particularly around cells [10]. Elastic fibres have a central core of elastin, which is surrounded by a sheath of fibrillins 1 and 2, as well as other associated proteins [16]. Elastic fibres are highly elastic, fatigue resistant and able to store and return energy [32]. In ligament, elastin resists transverse and shear deformation [11], but it is yet to be established whether it has a similar role in tendon.

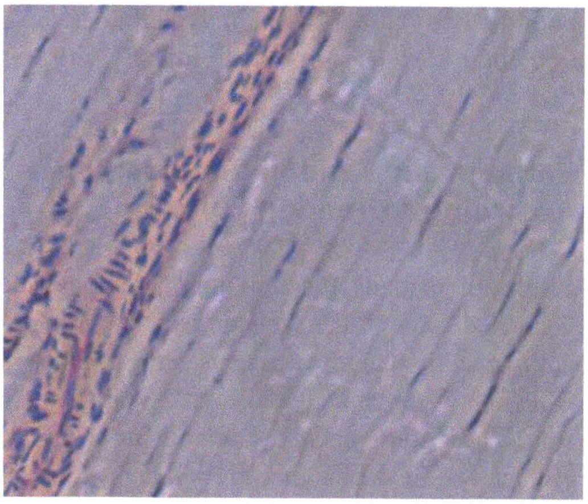

Fig. 1.3 Longitudinal tendon section stained with haematoxylin and eosin, demonstrating differences in cell morphology between tendon compartments, with elongated cells within the fascicles, and an abundance of rounded cells within the IFM

Cells

The tendon extracellular matrix is synthesised and maintained by the resident cell populations. There are two distinct populations found within tendon; those that reside between the collagen fibres within the fascicles (tenocytes) or those that are found in the IFM between the fascicles (interfascicular cells). Tenocytes have an elongated morphology and complex network of cytoplasmic processes that link adjacent cells via gap junctions [21]. Interfascicular cells are present at higher density, and have a rounded morphology (Fig. 1.3). It is likely that the interfascicular space contains a heterogeneous population of fibroblasts, progenitor and vascular cells [7]. In general, tendon cell phenotype is poorly understood and defined. However, recent data do indicate that the IFM cell population may be more metabolically active than the intrafascicular tenocytes (see Chap 5). [28, 34].

Bone Insertion

The tendon to bone insertion consists of a specialised interface, known as the enthesis. This is a highly complex structure which forms a connection between elastic tendon and rigid bone, which is approximately 100 times stiffer than tendon [6]. Entheses can be fibrous or fibrocartilaginous in nature. At fibrous entheses, the tendon inserts directly onto the metaphysis or epiphysis of a long bone. Fibrocartilaginous entheses are more complex, consisting of four distinct zones, with a gradual transition between each region [3]. Zone 1 is the tendon proper, with highly aligned collagen fibres, zone 2 is made of unmineralised fibrocartilage and is rich in cartilage associated proteins, including type II collagen and proteoglycans. Zone 3 consists of mineralised fibrocartilage, and is rich in collagen type X, which is associated with endochondral ossification, as well as varying amounts of bone mineral. Zone 4 is analogous to bone.

Myotendinous Junction

Unlike the tendon to bone insertion, the transition between tendon and muscle, known as the myotendinous junction, is abrupt rather than gradual. To reduce stresses at the myotendinous junction, the tendon collagen fibres and muscle cell membrane (sarcolemma) interdigitate via finger-like projections, which increases the interface area between tendon and muscle [19]. Despite the specialisations in structure seen at both the bony insertion and myotendinous junction, these regions are prone to injury [3, 19].

Variations in Tendon Composition According to Tendon Function

While all tendons function to transfer muscle generated force to the skeleton, specific tendons have an additional role acting as energy stores, in which they stretch and recoil with each stride to store and return energy, maximising exercise efficiency. In man, the predominant energy storing tendons are the Achilles and patellar tendons. In order to store and return energy effectively, energy storing tendons require specific mechanical properties, including increased extensibility, elasticity and fatigue resistance. These specialised properties are conferred by alterations in matrix structure and composition, targeting different levels of the tendon fibre composite hierarchy.

Variation in Tendon Collagen

Variations in collagen content and organisation between energy storing and other (positional) tendons are seen throughout the hierarchy. Energy storing tendons have a lower total collagen content than positional tendons [31]. However, type III collagen levels are elevated in energy storing tendons [33]. Type III collagen is abundant in tissues that require a high degree of compliance, such as skin and blood vessels, and therefore the higher levels in energy storing tendons may provide the greater extensibility and improved recoil needed in this tendon type. Collagen crosslink profile also differs between tendon types. Such differences in crosslinking are likely to influence fibril stiffness and subsequent local mechanics throughout the tendon hierarchy. However, the influence of crosslink type on tendon mechanical properties is yet to be determined [4]. Differences are also seen at the fibril level; fibrils in energy storing tendons have a bimodal distribution, with a lower mass average fibril diameter, which may contribute to the increased compliance seen in this tendon type. In addition, collagen crimp angles are greater in energy storing tendons [37], which may provide greater energy storing capacity. Differences in collagen packing are also seen at the fascicular level; fascicles from energy storing tendons have a smaller diameter than those from positional tendons. Collagen turnover rate also varies with tendon type, with lower levels of collagen turnover in energy storing tendons resulting in an extremely long half-life of approximately 200 years [38].

Variation in Non-collagenous Components

There are also several variations seen in the abundance of non-collagenous matrix components between tendons with different functions. Levels of sulphated glycosaminoglycans are significantly higher in energy storing tendons than in positional tendons [31], indicative of increased total proteoglycan levels in these tendons. Correspondingly, energy storing tendons have an elevated water content, which is negatively correlated with tissue stiffness [4]. Few studies have assessed how the content and distribution of specific proteins differs between tendon types and this remains an important area for future research, in order to understand how compositional differences can lead to altered mechanical behaviour. However, recent advances have been made by correlating local matrix composition with local tendon mechanical properties, with particular focus on the IFM. Mechanical testing studies indicate that sliding between adjacent fascicles is the primary mechanism by which energy storing tendons are able to stretch and recoil efficiently [39]. To achieve this, the IFM in this tendon type must be highly specialised to facilitate low stiffness inter-fascicle sliding and elastic recoil. Recent studies have started to elucidate the structural and compositional IFM specialisations in energy storing tendons that confer these mechanical properties. It is apparent that there are differences in IFM content between tendon types, with a larger volume of interfascicular matrix in energy storing tendons [40]. In addition a very recent study has compared the distribution of proteins within the IFM and fascicular regions of energy storing and positional tendons, demonstrating enrichment of lubricin and elastin in the IFM, which is particularly pronounced in energy storing tendons [36]. Based on the properties of these proteins, it is likely that lubricin facilitates inter-fascicular sliding, whereas elastin enables efficient recoil of the IFM.

Conclusions

It is evident that tendon structure is optimised to fulfil its functional role, with highly aligned and abundant collagen providing the high tensile strength required for efficient force transfer. However, in order to fully understand tendon structure function relationships, the specific functional roles of each component of the tendon matrix must be determined. Further, there is much still to be elucidated regarding how small variations in structure and composition throughout the levels of the tendon hierarchy results in the altered mechanical properties evident in tendons which function as energy stores.

References

1. Banos CC, Thomas AH, Kuo CK (2008) Collagen fibrillogenesis in tendon development: current models and regulation of fibril assembly. Birth Defects Res Part C 84:228–244
2. Barnard K, Light ND, Sims TJ, Bailey AJ (1987) Chemistry of the collagen cross-links. Origin and partial characterization of a putative mature cross-link of collagen. Biochem J 244:303–309
3. Benjamin M, McGonagle D (2009) Entheses: tendon and ligament attachment sites. Scand J Med Sci Sports 19:520–527
4. Birch HL (2007) Tendon matrix composition and turnover in relation to functional requirements. Int J Exp Pathol 88:241–248
5. Canty EG, Kadler KE (2005) Procollagen trafficking, processing and fibrillogenesis. J Cell Sci 118:1341–1353
6. Connizzo BK, Yannascoli SM, Soslowsky LJ (2013) Structure–function relationships of postnatal tendon development: a parallel to healing. Matrix Biol 32:106–116
7. Dyment N, Galloway J (2015) Regenerative biology of tendon: mechanisms for renewal and repair. Curr Mol Biol Rep 1:124–131
8. Ezura Y, Chakravarti S, Oldberg Å, Chervoneva I, Birk DE (2000) Differential expression of lumican and fibromodulin regulate collagen fibrillogenesis in developing mouse tendons. J Cell Biol 151:779–788
9. Funakoshi T, Schmid T, Hsu H-P, Spector M (2008) Lubricin distribution in the goat infraspinatus tendon: a basis for interfascicular lubrication. J Bone Joint Surg 90:803–814
10. Grant TM, Thompson MS, Urban J, Yu J (2013) Elastic fibres are broadly distributed in tendon and highly localized around tenocytes. J Anat 222:573–579
11. Henninger HB, Valdez WR, Scott SA, Weiss JA (2015) Elastin governs the mechanical response of medial collateral ligament under shear and transverse tensile loading. Acta Biomater 25:304–312
12. Isogai Z, Aspberg A, Keene DR, Ono RN, Reinhardt DP, Sakai LY (2002) Versican interacts with fibrillin-1 and links extracellular microfibrils to other connective tissue networks. J Biol Chem 277:4565–4572
13. Järvinen TAH, Józsa L, Kannus P, Järvinen TLN, Hurme T, Kvist M, Pelto-Huikko M, Kalimo H, Järvinen M (2003) Mechanical loading regulates the expression of tenascin-C in the myotendinous junction and tendon but does not induce de novo synthesis in the skeletal muscle. J Cell Sci 116:857–866
14. Kadler KE, Hojima Y, Prockop DJ (1990) Collagen fibrils in vitro grow from pointed tips in the C- to N-terminal direction. Biochem J 268:339–343
15. Kastelic J, Galeski A, Baer E (1978) The multicomposite structure of tendon. Connect Tissue Res 6:11–23
16. Kielty CM, Sherratt MJ, Shuttleworth CA (2002) Elastic fibres. J Cell Sci 115:2817–2828
17. Kjaer M (2004) Role of extracellular matrix in adaptation of tendon and skeletal muscle to mechanical loading. Physiol Rev 84:649–698
18. Kjaer M, Langberg H, Heinemeier K, Bayer ML, Hansen M, Holm L, Doessing S, Kongsgaard M, Krogsgaard MR, Magnusson SP (2009) From mechanical loading to collagen synthesis, structural changes and function in human tendon. Scand J Med Sci Sports 19:500–510
19. Knudsen AB, Larsen M, Mackey AL, Hjort M, Hansen KK, Qvortrup K, Kjær M, Krogsgaard MR (2015) The human myotendinous junction: an ultrastructural and 3D analysis study. Scand J Med Sci Sports 25:e116–e123
20. Kohrs RT, Zhao C, Sun Y-L, Jay GD, Zhang L, Warman ML, An K-N, Amadio PC (2011) Tendon fascicle gliding in wild type, heterozygous, and lubricin knockout mice. J Orthop Res 29:384–389
21. McNeilly CM, Banes AJ, Benjamin M, Ralphs JR (1996) Tendon cells in vivo form a three dimensional network of cell processes linked by gap junctions. J Anat 189:593–600
22. Rees SG, Dent CM, Caterson B (2009) Metabolism of proteoglycans in tendon. Scand J Med Sci Sports 19:470–478

23. Rigozzi S, Müller R, Stemmer A, Snedeker JG (2013) Tendon glycosaminoglycan proteoglycan sidechains promote collagen fibril sliding—AFM observations at the nanoscale. J Biomech 46:813–818
24. Riley G (2004) The pathogenesis of tendinopathy. A molecular perspective. Rheumatology 43:131–142
25. Ritty TM, Roth R, Heuser JE (2003) Tendon cell array isolation reveals a previously unknown fibrillin-2-containing macromolecular assembly. Structure 11:1179–1188
26. Smith RKW, Zunino L, Webbon PM, Heinegård D (1997) The distribution of Cartilage Oligomeric Matrix Protein (COMP) in tendon and its variation with tendon site, age and load. Matrix Biol 16:255–271
27. Södersten F, Hultenby K, Heinegård D, Johnston C, Ekman S (2013) Immunolocalization of collagens (I and III) and cartilage oligomeric matrix protein in the normal and injured equine superficial digital flexor tendon. Connect Tissue Res 54:62–69
28. Spiesz EM, Thorpe CT, Chaudhry S, Riley GP, Birch HL, Clegg PD, Screen HRC (2015) Tendon extracellular matrix damage, degradation and inflammation in response to in vitro overload exercise. J Orthop Res 33:889–897
29. Sun Y-L, Wei Z, Zhao C, Jay GD, Schmid TM, Amadio PC, An K-N (2015) Lubricin in human achilles tendon: the evidence of intratendinous sliding motion and shear force in achilles tendon. J Orthop Res 33:932–937
30. Svensson L, Aszódi A, Heinegård D, Hunziker EB, Reinholt FP, Fässler R, Oldberg Å (2002) Cartilage oligomeric matrix protein-deficient mice have normal skeletal development. Mol Cell Biol 22:4366–4371
31. Thorpe CT (2010) Extracellular matrix synthesis and degradation in functionally distinct tendons. In: Institute of Orthopaedics and Musculoskeletal Science (ed). University College London, London, p 267
32. Thorpe CT, Birch HL, Clegg PD, Screen HRC (2013) The role of the non-collagenous matrix in tendon function. Int J Exp Pathol 94:248–259
33. Thorpe CT, Birch HL, Clegg PD, Screen HRC (2015) Chapter 1 – Tendon Physiology and Mechanical Behavior: Structure–Function Relationships. In: Gomes ME, Reis RL, Rodrigues MT (eds) Tendon regeneration. Academic, Boston, pp 3–39
34. Thorpe CT, Chaudhry S, Lei II, Varone A, Riley GP, Birch HL, Clegg PD, Screen HRC (2015) Tendon overload results in alterations in cell shape and increased markers of inflammation and matrix degradation. Scand J Med Sci Sports 25:e381–e391
35. Thorpe CT, Clegg PD, Birch HL (2010) A review of tendon injury: why is the equine superficial digital flexor tendon most at risk? Equine Vet J 42:174–180
36. Thorpe CT, Karunaseelan KJ, Ng Chieng Hin J, Riley GP, Birch HL, Clegg PD, Screen HR (2016) Distribution of proteins within different compartments of tendon varies according to tendon type. J Anat doi:10.1111/joa.12485 [Epub ahead of print]
37. Thorpe CT, Klemt C, Riley GP, Birch HL, Clegg PD, Screen HR (2013) Helical sub-structures in energy-storing tendons provide a possible mechanism for efficient energy storage and return. Acta Biomater 9:7948–7956
38. Thorpe CT, Streeter I, Pinchbeck GL, Goodship AE, Clegg PD, Birch HL (2010) Aspartic acid racemization and collagen degradation markers reveal an accumulation of damage in tendon collagen that is enhanced with aging. J Biol Chem 285:15674–15681
39. Thorpe CT, Udeze CP, Birch HL, Clegg PD, Screen HRC (2012) Specialization of tendon mechanical properties results from interfascicular differences. J R Soc Interface 9:3108–3117
40. Thorpe CT, Udeze CP, Birch HL, Clegg PD, Screen HRC (2013) Capacity for sliding between tendon fascicles decreases with ageing in injury prone equine tendons: a possible mechanism for age-related tendinopathy? Eur Cells Mater 25:48–60
41. Zhang G, Chen S, Goldoni S, Calder BW, Simpson HC, Owens RT, McQuillan DJ, Young MF, Iozzo RV, Birk DE (2009) Genetic evidence for the coordinated regulation of collagen fibrillogenesis in the cornea by decorin and biglycan. J Biol Chem 284:8888–8897

2 Collagen Homeostasis and Metabolism

S. Peter Magnusson, Katja M. Heinemeier, and Michael Kjaer

Abstract

The musculoskeletal system and its collagen rich tissue is important for ensuring architecture of skeletal muscle, energy storage in tendon and ligaments, joint surface protection, and for ensuring the transfer of muscular forces into resulting limb movement. Structure of tendon is stable and the metabolic activity is low, but mechanical loading and subsequent mechanotransduction and molecular anabolic signaling can result in some adaptation of the tendon especially during youth and adolescence. Within short time, tendon will get stiffer with training and lack of mechanical tissue loading through inactivity or immobilization of the human body will conversely result in a dramatic loss in tendon stiffness and collagen synthesis. This illustrates the importance of regular mechanical load in order to preserve the stabilizing role of the connective tissue for the overall function of the musculoskeletal system in both daily activity and exercise. Adaptive responses may vary along the tendon, and differ between mid-substance and insertional areas of the tendon.

S.P. Magnusson (✉)
Institute of Sports Medicine, Department of Orthopedic Surgery M, Bispebjerg Hospital, Center for Healthy Aging, Faculty of Health and Medical Sciences, University of Copenhagen, Copenhagen, Denmark

Department of Physical Therapy, Musculoskeletal Rehabilitation Research Unit, Bispebjerg Hospital, Copenhagen, Denmark
e-mail: P.Magnusson@sund.ku.dk

K.M. Heinemeier
Institute of Sports Medicine, Department of Orthopedic Surgery M, Bispebjerg Hospital, Copenhagen, Denmark

Department of Biomedical Sciences, Centre for Healthy Ageing, Faculty of Health and Medical Sciences, University of Copenhagen, Copenhagen, Denmark
e-mail: KH@sund.ku.dk

M. Kjaer
Institute of Sports Medicine, Department of Orthopedic Surgery M, Bispebjerg Hospital, Copenhagen, Denmark

Center for Healthy Aging, Faculty of Health and Medical Sciences, University of Copenhagen, Copenhagen, Denmark
e-mail: michaelkjaer@sund.ku.dk

P.W. Ackermann and D.A. Hart (eds.), *Metabolic Influences on Risk for Tendon Disorders*, Advances in Experimental Medicine and Biology 920, DOI 10.1007/978-3-319-33943-6_2

Keywords
Tendon • Connective tissue • Physical training • Inactivity

Introduction

Homeostasis is the maintenance of internal stability of a physiological system in response to a perturbation in an attempt to preserve normal function. While this intuitively seems like a natural and optimal response for a physiological system to function, it was only decades ago that the connective tissue of tendon was considered exclusively inert and unable to readily adapt both metabolically and structurally to mechanical loading. However, today it is well recognized that tendon can respond to different perturbations typical of mechanical type, and also of hormonal nature.

Tendon tissue plays an essential role in transmitting contractile forces to bone to produce movement and is therefore uniquely designed to withstand quite considerable loads that may at times reach up to ~8 times body weight during human locomotion [1–3]. It has a surprisingly aligned and architectural well sculptured structure. An acute ultimate high load can cause rupture of the tendon structure, and repetitive use of tendon often results in overuse injuries such as tendinopathies, which is a common clinical condition characterized by pain during activity, localized tenderness upon palpation, swelling of the tendon and impaired performance [4, 5]. This situation represents an example of loss of tissue homeostasis. Tendinopathy is a problem in both elite and recreational athletes as well as in the workplace [6–8]. In some elite sports, the prevalence can reach as high as 45 % [6, 9–11], and the symptoms as well as performance reduction may be very long lasting (years) [12]. To date the injury mechanism is poorly understood. Although tendons have traditionally been considered as these largely inert structures, recent studies show that human tendons are metabolically quite active in responsive to mechanical loading [13, 14]. Understanding specifically how tendon tissue adapts to mechanical loading (training) and unloading (immobilization) – and at what time of life – will be the key to understand homeostasis of tendon the pathogenesis of tendinopathy, and thus provide the basis for prevention of these overuse injuries.

Tendon Composition

Tendon is organized in a strict hierarchical manner [15]. Collagen molecules dominated the structure, and they are organized in a precise pattern to yield the characteristic 67-nm *D*-periodization that forms collagen fibrils. The collagen molecule is ~ 300 nm in length and 1.5 nm in diameter [16] and aggregated molecules of the fibril are stabilized by covalent intramolecular cross-links [17]. The cross-links bind the collagen molecules to one another and thereby confer integrity on the fibril. Groups of fibrils then form fibers known as fascicle bundles, which finally comprise the tendon proper. There are at least 28 different collagen proteins, but tendon and ligament are predominantly made up of type I collagen [18]. The fibrillar collagen is embedded in a hydrophilic extracellular matrix consisting of proteoglycans, glycoproteins and glycosaminoglycans, which are involved in the development, organization and growth control of tendon [19]. The large proteoglycan molecules are dominated by versican and aggrecan, whereas smaller proteoglycans, the small leucine rich proteoglycan (SLRP) are dominated by decorin, and has some amounts of biglycan, lumican and

Table 2.1 Molecular components of tendon tissue

Mechanical loading of tendon
Collagen in fibrils/fascicles – no turnover
(95 % stable structures after adolescence)
High turnover Collagen
(5 % – fibril fragments inter-fascicular?)
Non-collagenous matrix molecules
(Proteoglycans, (a) large: aggrecan, versican (b) SLRP: decorin (80 %), biglycan, lumican, fibromodulin)
COMP, Lubrican, Elastin, Tenascin-C, Fibronectin
Cross-links
(a) enzymatic: lysyl oxydase initiated, (b) advanced glycation end products (AGE's)

fibromodulin (Table 2.1). In addition molecules like cartilage oligometric protein (COMP), lubricant, elastin, tenascin -C and fibronectin are represents important components of the tendon, and each have roles to play in mechanical properties such as maximal tendon strength, tendon stiffness and tissue elasticity (see Chap. 1). Finally, cross-link molecules of both enzymatic and non-enzymatic nature are a part of tendon.

The chief cell in tendon is the fibroblast, and compared to other tissue there are relatively few of them and they are evenly distributed within the tendon. The cells are elongated along the collagen fibril direction, and they have processes that spread in between collagen fibers, which allows for cell-cell signaling. As there currently are no definitive markers for fibroblasts in tendon, it cannot fully be ruled out that other types of cells may also be present in the tendon.

Tendon Metabolism

Strictly speaking, the oxidative metabolism is low in tendon, which only contains few cell, and content of oxidative enzymes and of mitochondria is relatively low because of the low number of cells. Although mechanical loading of tendon can stimulate increased interstitial tissue concentrations of both glucose and free fatty acids, the changes are small [20]. With the use of near-infrared spectroscopy (NIRS) it has been demonstrated that exercise resulted in only a moderate drop in oxygen saturation and thus estimated oxygen consumption within the human tendon [21], and the hypoxia degree during even intense exercise upon the Achilles tendon is still much less pronounced compared to the hypoxia observed in the tendon region when blood flow is occluded with a pneumatic cuff [22]. This does not exclude hypoxic phenomenon in relation to tendon and the development of pathology, but does not prove it either.

Collagen Synthesis and Turnover

Tendon mechanical properties and its cross sectional area have been shown to increase in stiffness and size, respectively, in response to training in humans and animals, which indicates that mechanical stimuli can lead to adaptive responses of the tendon cells, resulting in changes of the extracellular matrix. However, it is yet to be determined when and to what extent these changes can occur, and the mechanisms responsible for these adjustments are still debated, and a relatively large discrepancy exists between results from animal and human studies. Mechanical loading of tendon tissue during exercise or training can initiate a signaling cascade that stimulates the cells located in the tissue to increase their production of matrix proteins, ultimately leading to tendon hypertrophy. Such is "mechanotransduction" is well established and described *in vitro* [23]. Conversely, when an artificial tendon made from human tendon fibroblast is detensioned (cut), the expression of tendon phenotypical important molecules like collagen and tenomodulin is reduced [24]. Cell culture studies on tendon and ligament fibroblasts show that fibroblasts respond to mechanical stretch by increasing their production and secretion of certain growth factors that in turn act on the fibroblasts to induce expression and synthesis of collagen [23]. Growth factors involved in this signaling cascade include transforming growth factor b1 (TGFb1) and connective tissue growth factor (CTGF) [25, 26]. In addition, more indirect evidence suggests that insulin like growth factor-I (IGF-I) could act as a link between mechanical load and collagen synthesis in tendon tissue [27, 28]. In small animal models such mode of mechanotransduction is involved in tendon adaptation to loading. As an example, intense tendon loading in rats, generated by electrically induced muscle training or synergist ablation, leads to substantial increases in mRNA expression of collagen-inducing growth factors, IGF-I and TGFb1, in parallel with increased mRNA expression of collagen type I and III in the tendon tissue [29–31]. In addition, repeated bouts of treadmill running have been shown to elevate levels of IGF-I protein in rat tendons [28]. Thus it seems likely that the tendon cells respond to loading by increasing growth factor production, and that the action of these growth factors leads to induction of collagen expression. However, a causal link between the increased expression of these growth factors and the increase in collagen expression have not been proven, and the exact molecular signaling pathways involved in transforming mechanical signals into biochemical signals in tendons are still largely unknown [32]. More recently, the administration of growth hormone (GH) to individuals that were subjected to lower limb immobilization for 2 weeks, counteracted the lowering of expression of collagen and og lysyl oxidase (LOX) that is the enzyme responsible for cross link formation [33]. Interestingly this improved expression of LOX during immobilization with GH treatment was associated to less loss in tendon stiffness during inactivity.

To support a loading-induced tendon collagen synthesis in adult humans, microdialysis studies have shown increased levels of markers for collagen synthesis in the peritendinous tissue surrounding the Achilles tendon tissue, in response to both acute exercise and long-term training [14, 34]. However, these microdialysis data are only determining the peritendinous area and are likely to reflect the collagen synthesis at the very periphery of the tendon, or even outside the tendon, rather than that of the actual tendon tissue. A more direct way of measuring collagen synthesis in human tendon tissue is to trace incorporation of labeled amino acids in tendon tissue. With this approach, an increased rate of collagen synthesis was observed in patellar tendons of young men in response to acute kicking exercise [35]. However, several other studies using the same technique, and the same exercise model, could not confirm this loading-induced collagen synthesis in adult human tendon [36–39].

Gene-expression of growth factors and collagen in response to exercise do not show similar trends in adult humans vs rodents in response to tendon loading. Three studies have investigated gene expression in human patellar tendon tissue in response to acute exercise. Two of these investigations found decreased or unchanged growth factor and collagen mRNA expression in tendon biopsies from the mid-portion of the tendon [40, 41], while one study found modest increases in collagen and CTGF mRNA expression in tissue from the proximal part of the patellar tendon in response to exercise [36]. In other words, the response of adult human tendon tissue to acute loading does not seem to mimic that of rodent tendon tissue. This suggests that adult human tendon tissue is far less responsive than that of small animals, and such differences may relate to the fact that rats and mice are still in a growth phase at the point when they are typically used in experiments [29–31], and consequently their tendons may have more potential for adaptation than adult human tendons. As diverging results exist in tendon research, it is currently debated to what degree acute exercise influences rates of collagen production in the adult human tendon and at what basal rate tendon tissue is actually metabolized. Using the carbon-14 (^{14}C) bomb pulse method, large amounts of ^{14}C were retained in healthy Achilles tendon core from adults born during the peak of the ^{14}C bomb pulse, and this gives clear evidence of an extremely slow rate of Achilles tendon tissue turnover in adult life [42]. In comparison, no remnants of ^{14}C from the bomb pulse were found in skeletal muscle tissue from the same persons, indicating rapid turnover of this tissue [42]. The evidence of slow tendon turnover is supported by two earlier studies, in which accumulation of pentosidine (non-enzymatic cross-link) and D-aspartate were measured to estimate long-term tissue turnover in adult human biceps tendon [43] and mature horse tendon [44] respectively.

The slow rate of tendon tissue renewal fits poorly with data from studies investigating acute tendon collagen synthesis rates with use of either microdialysis sampled pro-collagen propeptides or analyzing the incorporation of labeled amino acids into tendon tissue. These studies indicate relatively high basal rates of tendon collagen synthesis, almost resembling the synthesis rate of myofibrillar protein in skeletal muscle [35]. However, the measure of acute incorporation of labeled amino acids suffers from the complication that all new synthesis of collagen is detected, even though this newly synthesized collagen may often be rapidly degraded, and thus never incorporated into the tissue matrix [45]. Hypothetically, a potentially large production of excess collagen, which is broken down relatively quickly and never incorporated into the more permanent tissue structures (i.e., collagen fibrils), will contribute to the measured synthesis rate. In fact if it is assumed that 90–95 % of collagen is contained in stable structures of the tendon that do not turnover, whereas 5–10 % of collagen is contained in a rapid turnover pool of molecules, the different finding in studied with different design can in fact fit together and not be mutually contradictory. If this hypothesis holds true, the findings of high basal levels of tendon protein synthesis (measured with amino acid tracers) could be reconciled with a relatively permanent tendon matrix in adult human and horse tendon [42–44] (Table 2.1). It does, however, remain difficult to reconcile a very slow tissue turnover with the fact that human tendons can hypertrophy in response to long-term loading (e.g. [46]), as the hypertrophic response indicates some degree of synthetic activity. One possible explanation is that loading-induced tendon growth takes place at the very periphery of the tendon. This could be reconciled both with the ^{14}C bomb pulse data, showing very low turnover rates in the tendon core, and with the fact that microdialysis experiments consistently indicate a loading-induced collagen synthesis in peritendinous tissue [34, 42]. This hypothesis is supported by recent data on 6-month-old mice, which showed that overload-induced plantaris tendon hypertrophy was based on growth and cell proliferation only in the most superficial layers of tendon tissue, while the "original" core tendon remained relatively constant [47]. A greater potential for growth at the tendon periphery is further supported by an early study that showed greater levels of IGF-I protein expression in cells located in the rat Achilles tendon periphery compared to those located in the deeper part of the tendon [28]. Also more recent studies suggest greater potential for growth and cell proliferation in the superficial parts of the tendon in rodents [48, 49]. In other words, it may be speculated that a new layer of collagenous matrix is added, comparable to a tree ring, when the tendon grows in response to loading.

Another explanation for the diverging results in tendon research, with regard to the adaptability of tendon tissue to loading and the overall metabolic activity, could be that large differences exist between different types of tendons. Data from horses show that high-load tendons have slower turnover than tendons subjected to more moderate loads [44]. Though perhaps counterintuitive, it may be speculated that high-load tendons simply cannot "afford" to have a constant remodeling going on as this may reduce strength. Therefore the high-load Achilles tendon may well have slower turnover than tendons that are loaded less, such as the patellar tendon. This could explain why patellar tendon hypertrophy is seen relatively consistently in response to long-term loading (e.g. [46]), while data on Achilles tendon hypertrophy in relation to loading are far less consistent [50]. In addition there may well be regional differences within individual tendons. As an example, training induced hypertrophy of the patellar tendon is in most cases seen exclusively at the proximal and distal parts, and not the mid-tendon (e.g. [46]). It remains to be investigated, if the apparent differences in responsiveness between different tendons, and between regions within tendons, are connected to different levels of tissue turnover. Finally, a potential explanation for exercise-induced hypertrophy could be that it is merely a consequence of increased water content and not an actual accrual of collagen matrix. This leaves the training induced increase in tendon stiffness to be explained, and potentially an increase in stiffness could happen in the absence of collagen accrual if cross-linking of the existing matrix was augmented by loading. In favor of this hypothesis, one study in rats showed a substantial increases in mRNA expression of the collagen cross-linking enzyme lysyl oxidase (LOX) in response to 4 days of strength training [29].

Whether dynamic changes in collagen and thus most likely also in fibril turnover in the human body takes place (1) in the surface of the tendon, (2) in the inter-fascular space with more loosely arranged tissue or (3) represents dynamic changes in single fibril thickness is yet to be determined.

Force Transmission Within the Tendon

Information on the mechanical properties and function of the tendon has historically been based on measurements on whole tendon on animal and human cadaver studies. However, recent findings suggest that both the Achilles and patellar tendon, which are some of the strongest tendons in the human body, should not necessarily be regarded as one single force-transmitting structure. Data on cadaver show that muscle activation of the separate triceps surae muscles can influence medial and lateral forces in the Achilles tendon [51]. It has also been demonstrated *in vivo* that during isometric contractions with the plantar flexor muscles there is a differential displacement between the soleus and gastrocnemius aponeuroses just proximal to the junction with the Achilles tendon [52]. During passive dorsiflexion in vivo it can be seen that there is greater displacement in the deep layer of the tendon compared to superficial aspects [53]. These results indicate that the Achilles tendon is exposed to intratendinous shear and stress gradients during human loading depending on muscle activation. While it may be intuitive that three separate muscles may produce uneven forces and shear within a commonly shared tendon, there is evidence suggesting that other tendon also should not be regarded as one single force-transmitting structure [54, 55].

Patellar tendinopathy mostly involves the proximal and posterior portion of the patellar tendon, suggesting that perhaps this part of the tendon has different mechanical properties. It has been shown that if individual collagen fascicles from the anterior and posterior portion of the patellar tendon are mechanically tested, the tendon fascicles from the anterior portion display far greater peak and yield stress and tangent modulus compared to that of the posterior portion of the tendon [56, 57]. In other words, the anterior fascicles are stronger than the posterior fascicles. Moreover, it appears that there are also region specific biochemical and structural difference within the patellar tendon [57]. These findings demonstrate that there are region-specific material properties within a tendon. To what extent these regional differences contribute to the etiology of tendinopathy, or how they relate to adaptation in response to load remains unknown.

The fact that fascicles from the anterior and posterior portion of the human patellar tendon displayed substantially different mechanical properties [56] also implies that they may operate largely independently. In fact, it has been shown that lateral force transmission between adjacent fascicles is rather small, and therefore the fascicles may be considered functionally independent structures and they extend the entire length of the whole tendon [58]. The fact that sliding between fascicles can take place may be advantageous as tendons wrap around bones, for example.

The fibroblast is located between fibrils and in the inter-fascicular space. During tensile loading the fibroblast and its cell nucleus undergo deformation that may play role in the mechanical signal transduction pathway of this tissue [59, 60]. Loading can potentially place strain on several components of the tendon that contribute to an 'injury' or material fatigue that requires repair. Such a repair process may be a fine balance between synthesis and degradation of the various components of the extracellular matrix. However, it should also be noted that too little stimulation (relative inactivity) may also off-set such anabolic-catabolic balance [61].

The collagen fibril is the smallest functional tensile bearing unit of the tendon. The fibril are embedded in the matrix consisting of many different molecules, including water, proteoglycans, glycosaminoglycans, elastin, and glycoproteins, which resembles a fiber-reinforced composite material in which forces are transmitted laterally to adjacent fibers through the matrix. The proteoglycan decorin and its associated glycosaminoglycan have been suggested as a main force-transmitting complex [62, 63]. However, it has been shown in both ligament and tendon that when the majority of these proteins are removed, the elastic and viscous properties of the tendon are left unaffected [64, 65]. This suggests that either different linker molecules exists or that collagen fibrils are continuous throughout the length of the tendon [66], but this has not yet to be confirmed.

The tropocollagen molecule is comprised of three polypeptides arranged as a triple helical structure stabilized by H-bonds [67]. The molecules are organized in a precise pattern to form collagen fibrils, which are the principle tensile bearing structure within the tendon. An important contributor to the mechanical properties of the tendon is the intermolecular cross-links [17, 68]. During loading, the triple helix itself may elongate, the gap region between longitudinally adjoining molecules of the fibril may increases, or there may be a relative slippage between laterally adjacent molecules [69–72], or a combination thereof. However, individual collagen molecules have a fracture modulus that far exceeds that of the tendon fibril [67, 73], and is therefore unlikely the 'weak link' in terms of fracture/injury. Cross-linking in collagen is believed to be a major source of mechanical strength in connective tissues, and improved mechanical properties during development is likely related to maturation of enzymatic cross-links [74]. It was recently shown that isolated fibrils display an initial rise in modulus followed by a plateau with reduced modulus, which was finally followed by an even greater increase in stress and modulus before failure [75]. This mechanical behavior may relate to cross-link maturity of the tissues.

Chronic Loading of Tendon – Physical Training

Knowledge of whether and how tendons adapt to increased loading, has until recently been somewhat limited. The average tensile stress (force/area) exerted on the tendon will depend on its cross-sectional area. Therefore, for a given force a larger cross-sectional area will yield a reduction in stress, and possibly confer a reduction in injury risk. Human tendons, including the patellar and Achilles tendons, typically have a fracture stress of ~100 MPa. However, most tendons are only subjected to stresses of up to 30 MPa [76], which gives tendons a reasonable safety margin, although the Achilles tendon might experience stresses of up to ~70 MPa [1].

In humans, there is some evidence suggesting that habitual long distance running (>5 years) is associated with a markedly greater cross-sectional area (22 %) of the Achilles tendon compared to that of non-runners [77]. However, a total training stimulus of approximately 9 months of running in previously untrained subjects did not result in hypertrophy of the Achilles tendon [50]. This lack of tissue adaptation may relate to the relative unresponsiveness of tendon tissue.

The lack of response to 9 months of endurance training may also relate the absolute load or the time of exposure: It has been shown that resistance training can result in tendon hypertrophy [46, 78], and it has been shown that athletes that expose the patellar tendon to intermittent high loads on one side but not the other for years also have a greater tendon cross-sectional tendon [79] on the loaded side. The data from determining collagen turnover with 14C-bomb pulse technique, indicate that larger structural changes in tendon towards physical training requires that these occur in childhood or in adolescence [42]. A greater tendon cross-section will yield a lower average stress on the tendon when loaded. Curiously, it appears that the tendon hypertrophy is largely regional, such that the increase occurs primarily in the ends of the tendon [79]. In humans it has been shown that resistance training with large loads may also increase the material stiffness (modulus) of the tendon [78], although the structural site for the change remains unknown.

Tendon collagen fibrils are the basic force-transducing units of the tendon, and the morphology of the fibrils is commonly believed to largely influence the mechanical properties and function of the tendon. However, currently the picture is not coherent: in animal models it has been shown that an increase in physical activity may decrease, increase, or leave the fibrils size unchanged [80–82]. In humans, it has been shown that fibril morphology is unaltered by lifelong endurance training [83], while it is influenced by heavy resistance training in patients with tendinopathy [84]. The intermolecular cross-links may also be an important contributor to the properties of tendon tissue. The cross-links can be divided into two broad groups, the enzymatic and the non-enzymatic. Enzymatic cross-links are formed as a result of the lysyl oxidase (LOX) enzyme acting on the type I collagen molecule [85]. In tendons, the enzymatic cross-link composition changes with maturation, and it has been shown that it can be modulated with heavy resistance training [86]. Non-enzymatic cross-links are formed when reducing sugars bind to amino acids and are referred to as advanced glycation end-product (AGE). AGE accumulation is dependent on collagen turnover rates in the specific tissue and therefore naturally collects during the aging process to a higher extent in tissue with low collagen turnover [87]. In tendon tissue it has been shown that both endurance training and resistance training may reduce AGE accumulation [83, 86].

Unloading of Tendon

It is well known that immobilization is associated with a decline in maximal muscle strength and muscle mass and that both contribute to a reduced muscle function [88, 89]. Interestingly, a recent study report that unloading affects tendon properties to a greater extent than it does muscle loss [90]. Some animal immobilization studies show that tendon stiffness is reduced without a change in tendon size [91, 92], which would imply an alteration in the material composition, although some have shown opposite results [93]. Data on young subjects also show that mechanical properties of the tendon decrease without tendon atrophy [90, 94, 95]. Short-term immobilization has been reported to reduce tendon modulus by up to 30 % in 23 days [90], and up to 58 % after 90 days [96]. Overall, it appears that the tendon responds quite robustly to even brief periods of immobilization and that it results in a reduced modulus, i.e. material stiffness. Along with the reduction in stiffness with immobilization, it has also been demonstrated that the synthesis of collagen is markedly reduced with immobilization [97]. These two events may however not be related, and rapid adaptations in other matrix proteins or tendinous structures may take place and explain the change in mechanical properties. Both in young and in elderly individuals, 2 weeks of immobilization has demonstrated a marked loss in tendon stiffness, and interestingly, administration of growth hormone (GH) over the 14 days, counteracted this loss somewhat [33, 98]. Even more interesting was the finding that the effect of GH also had an inhibiting effect upon the decline in lysyl oxidase (LOX) mRNA that was observed with inactivity [33]. Although no proof of causality, its interesting that changes in stiffness and in LOX mRNA were accompanied, and at least not speaking against a role for enzymatic cross link formation (or lack thereof) in the determination of tendon stiffness.

With respect to the fibril morphology it has been shown that there is a decrease in the number and average diameter of collagen fibrils with immobilization [99, 100]. In addition, the interfibrillar spacing has been reported to increases with immobilization [101]. This reduced fibril density may explain the loss of tendon material properties while tendon CSA remains unchanged, which could be due to increased interstitial water and/or other ECM components [90, 102]. It was recently shown in humans that a 2-week immobilization period resulted in reduced tendon stiffness, although the fibril morphology was unaffected. Such rapid changes in mechanical properties may be related to changes in enzymatically driven cross-links [33]. It has been proposed that tension on the tendon is a key factor in maintaining homeostasis [103], and this rapid change in mechanical properties in response to inactivity underscores this notion.

The Aging Tendon

With aging muscle mass, strength and overall activity level declines. The fact that activity level declines along with muscle mass has made it difficult to delineate these separate effects and therefore, to what extent aging, *per se*, affects tendon mechanics has not been entirely clear. For quite some time only animal data was available in terms of tendon mechanics and these have not provided a coherent picture. Some of these studies reported increased tendon modulus and strength with age [104–106], while others have found the opposite [107–111], or no change at all [100, 112–114]. In vitro results on human tendons generally suggest that there are either no changes or a reduction in the modulus and strength of the tendon with aging [115–117]. More recently it has been possible with the use of ultrasound technology to examine tendon properties, in vivo [118]. It seems that these data corroborates the in vitro data: when whole tendon extensibility is compared in young and old at a common force (not maximal force), the extensibility remains unchanged [119, 120] or is actually increased [121–125]. A recent study examined trained and untrained young and older persons and could demonstrate that the strain and stiffness was unaltered with age and or with life-long habitual endurance training [83]. However, older trained persons had a greater (patellar) tendon cross/sectional area and therefore the average stress (force/area) was lower in the old trained individual. The latter finding may be of importance in terms of risk of overuse injury.

From a structural standpoint it seems that aging alters the tendon composition although it largely maintains its mechanical properties. It has been shown that the density of pentosidine, which is an advanced glycation end product, will increase sevenfold tendon in old compared to young [126]. Presumably an increase in advanced glycation end products should augment tendon stiffness. At the same time, the collagen content of the is reduced substantially (34 %) [126]. It is possible that the elevated nonenzymatic cross-link density in old serves to maintain tendon stiffness despite the diminished collagen concentration

Tendon Connections with Bone and Skeletal Muscle

Most of the information on tendon homeostasis and metabolism comes from studies of mid-tendon, and the insertional parts of the tendon whether it be to bone or to the skeletal muscle are much less studied. It is known that many bony insertional parts of the tendon (e.g. Achilles tendon) is more rich of collagen type II content as well as of aggrecan, both substances dominating cartilage like tissue [127]. This fits with the finding of compressive forces in addition to tensile forces in that insertional part of the tendon. Further, it is notable that in studies investigating the structural adaptation to chronic physical loading (training), it is found that the increased cross sectional area of the Patella tendon observed in trained vs untrained individuals is most pronounced in areas close to the bony insertion of the tendon whether it is close to the Patella bone, or to the tibial bone [79].

The connection between muscle and tendon tissue is often subjected to tissue failure when subjected to excessive strain ("muscle strain injury") [128]. Animal studies have demonstrated the interphase between tendon and muscle – the myotendinous junction – to be tendon finger-like processes that merge into skeletal muscle [129, 130], and more lately human studies have confirmed tendon made ridge-like (rather than finger-like) protrusions that interdigitates with groove-like evaginations in the muscle cell [131]. The myotendinous junction is known to consist of both actin microfilaments, actin binding proteins, proteins that link actin bundles to the sarcolemma, trans-membrane proteins that link cytoskeletal elements to extracellular components, and proteins that link external lamina to the collagen-fibril rich matrix. Further, a high content of collagen dominated by collagen type I and III is present, and interestingly the myotendinous junction is the only location in the muscle tendon unit where collagen type XXII is present [131, 132]. Physical training is known – at least in animals – to cause an increase in the tissue interphase "folding" and thus extent of the area of the myotendinous junction, which results in a larger contact surface area between muscle and tendon, and thus results in a higher tolerable mechanical load [133]. Apart from increasing the tolerable load due to contact area expansion, it might also be that more shear-like rather than simple tensile stress due to the folding may improve tolerable tissue load even further, and the ridge-like tendon form into the muscle cell may contribute to a "grab- or trap-effect" when muscle contractions shorten but thicken the muscle cell and thus "anchoring" the tendon parts into the muscle [131]. To what extent physical training influence the myotendinous junction in humans has not been documented.

Conclusion

The collagen rich tissues of the musculoskeletal system are important for maintaining the architecture of the skeletal muscle, ensuring force transmission, storing energy, protecting joint surface and stability, and ensuring the transfer of muscular forces into resulting limb movement. The tendon structure is relatively stable and its metabolic activity is low, but mechanical loading and subsequent mechanotransduction and molecular anabolic signaling can result in some adaptation of the tendon especially during youth and adolescence can result in increased tendon size. Within short time, tendon will get stiffer with training and lack of mechanical tissue loading through inactivity or immobilization of the human body will result in a dramatic loss tendon stiffness and collagen synthesis. This illustrates the importance of regular mechanical load in order to preserve the stabilizing role of the connective tissue for the overall function of the musculoskeletal system in both daily activity and exercise. Adaptive responses may vary along the tendon, and differ between mid-substance and insertional areas of the tendon.

References

1. Finni T, Komi PV, Lepola V (2000) In vivo human triceps surae and quadriceps femoris muscle function in a squat jump and counter movement jump. Eur J Appl Physiol 83(4–5):416–426
2. Giddings VL, Beaupre GS, Whalen RT, Carter DR (2000) Calcaneal loading during walking and running. Med Sci Sports Exerc 32(3):627–634
3. Magnusson SP, Aagaard P, Dyhre-Poulsen P, Kjaer M (2001) Load–displacement properties of the human triceps surae aponeurosis in vivo. J Physiol 531(Pt 1):277–288
4. Khan K, Cook J (2003) The painful nonruptured tendon: clinical aspects. Clin Sports Med 22 (4):711–725
5. Maffulli N, Khan KM, Puddu G (1998) Overuse tendon conditions: time to change a confusing terminology. Arthroscopy 14(8):840–843
6. Ferretti A (1986) Epidemiology of jumper's knee. Sports Med 3(4):289–295
7. Frost P, Bonde JP, Mikkelsen S, Andersen JH, Fallentin N, Kaergaard A et al (2002) Risk of shoulder tendinitis in relation to shoulder loads in monotonous repetitive work. Am J Ind Med 41(1):11–18
8. Tanaka S, Petersen M, Cameron L (2001) Prevalence and risk factors of tendinitis and related disorders of the distal upper extremity among U.S. workers: comparison to carpal tunnel syndrome. Am J Ind Med 39(3):328–335
9. Gruchow HW, Pelletier D (1979) An epidemiologic study of tennis elbow. Incidence, recurrence, and effectiveness of prevention strategies. Am J Sports Med 7(4):234–238
10. Lian OB, Engebretsen L, Bahr R (2005) Prevalence of jumper's knee among elite athletes from different sports: a cross-sectional study. Am J Sports Med 33(4):561–567
11. Knobloch K, Yoon U, Vogt PM (2008) Acute and overuse injuries correlated to hours of training in master running athletes. Foot Ankle Int 29(7):671–676
12. Kettunen JA, Kvist M, Alanen E, Kujala UM (2002) Long-term prognosis for jumper's knee in male athletes. A prospective follow-up study. Am J Sports Med 30(5):689–692
13. Bojsen-Moller J, Kalliokoski KK, Seppanen M, Kjaer M, Magnusson SP (2006) Low-intensity tensile loading increases intratendinous glucose uptake

in the Achilles tendon. J Appl Physiol 101 (1):196–201
14. Langberg H, Rosendal L, Kjaer M (2001) Training-induced changes in peritendinous type I collagen turnover determined by microdialysis in humans. J Physiol 534(Pt 1):297–302
15. Kastelic J, Galeski A, Baer E (1978) The multicomposite structure of tendon. Connect Tissue Res 6(1):11–23
16. Kadler KE, Holmes DF, Trotter JA, Chapman JA (1996) Collagen fibril formation. Biochem J 316(Pt 1):1–11
17. Bailey AJ (2001) Molecular mechanisms of ageing in connective tissues. Mech Ageing Dev 122(7):735–755
18. Riley GP (2005) Gene expression and matrix turnover in overused and damaged tendons. Scand J Med Sci Sports 15(4):241–251
19. Kjaer M (2004) Role of extracellular matrix in adaptation of tendon and skeletal muscle to mechanical loading. Physiol Rev 84(2):649–698
20. Langberg H, Skovgaard D, Karamouzis M, Bulow J, Kjaer M (1999) Metabolism and inflammatory mediators in the peritendinous space measured by microdialysis during intermittent isometric exercise in humans. J Physiol 515(Pt 3):919–927
21. Boushel R, Langberg H, Green S, Skovgaard D, Bulow J, Kjaer M (2000) Blood flow and oxygenation in peritendinous tissue and calf muscle during dynamic exercise in humans. J Physiol 524(Pt 1): 305–313
22. Boushel R, Langberg H, Olesen J, Nowak M, Simonsen L, Bulow J et al (2000) Regional blood flow during exercise in humans measured by near-infrared spectroscopy and indocyanine green. J Appl Physiol 89(5):1868–1878
23. Chiquet M, Gelman L, Lutz R, Maier S (2009) From mechanotransduction to extracellular matrix gene expression in fibroblasts. Biochim Biophys Acta 1793(5):911–920
24. Bayer ML, Schjerling P, Herchenhan A, Zeltz C, Heinemeier KM, Christensen L et al (2014) Release of tensile strain on engineered human tendon tissue disturbs cell adhesions, changes matrix architecture, and induces an inflammatory phenotype. PLoS One 9(1), e86078
25. Yang G, Crawford RC, Wang JH (2004) Proliferation and collagen production of human patellar tendon fibroblasts in response to cyclic uniaxial stretching in serum-free conditions. J Biomech 37 (10): 1543–1550
26. Schild C, Trueb B (2002) Mechanical stress is required for high-level expression of connective tissue growth factor. Exp Cell Res 274(1):83–91
27. Abrahamsson SO, Lohmander S (1996) Differential effects of insulin-like growth factor-I on matrix and DNA synthesis in various regions and types of rabbit tendons. J Orthop Res 14(3):370–376
28. Hansson HA, Engstrom AM, Holm S, Rosenqvist AL (1988) Somatomedin C immunoreactivity in the Achilles tendon varies in a dynamic manner with the mechanical load. Acta Physiol Scand 134 (2):199–208
29. Heinemeier KM, Olesen JL, Haddad F, Langberg H, Kjaer M, Baldwin KM et al (2007) Expression of collagen and related growth factors in rat tendon and skeletal muscle in response to specific contraction types. J Physiol 582(Pt 3):1303–1316
30. Heinemeier KM, Olesen JL, Schjerling P, Haddad F, Langberg H, Baldwin KM et al (2007) Short-term strength training and the expression of myostatin and IGF-I isoforms in rat muscle and tendon: differential effects of specific contraction types. J Appl Physiol 102(2):573–581
31. Olesen JL, Heinemeier KM, Haddad F, Langberg H, Flyvbjerg A, Kjaer M et al (2006) Expression of insulin-like growth factor I, insulin-like growth factor binding proteins, and collagen mRNA in mechanically loaded plantaris tendon. J Appl Physiol 101 (1):183–188
32. Gumucio JP, Sugg KB, Mendias CL (2015) TGF-beta superfamily signaling in muscle and tendon adaptation to resistance exercise. Exerc Sport Sci Rev 43:93–99
33. Boesen AP, Dideriksen K, Couppe C, Magnusson SP, Schjerling P, Boesen M et al (2013) Tendon and skeletal muscle matrix gene expression and functional responses to immobilisation and rehabilitation in young males: effect of growth hormone administration. J Physiol 591(Pt 23):6039–6052
34. Langberg H, Skovgaard D, Petersen LJ, Bulow J, Kjaer M (1999) Type I collagen synthesis and degradation in peritendinous tissue after exercise determined by microdialysis in humans. J Physiol 521(Pt 1):299–306
35. Miller BF, Olesen JL, Hansen M, Dossing S, Crameri RM, Welling RJ et al (2005) Coordinated collagen and muscle protein synthesis in human patella tendon and quadriceps muscle after exercise. J Physiol 567(Pt 3):1021–1033
36. Dideriksen K, Sindby AK, Krogsgaard M, Schjerling P, Holm L, Langberg H (2013) Effect of acute exercise on patella tendon protein synthesis and gene expression. Springer Plus 2(1):109
37. Hansen M, Kongsgaard M, Holm L, Skovgaard D, Magnusson SP, Qvortrup K et al (2009) Effect of estrogen on tendon collagen synthesis, tendon structural characteristics, and biomechanical properties in postmenopausal women. J Appl Physiol 106(4):1385–1393
38. Hansen M, Miller BF, Holm L, Doessing S, Petersen SG, Skovgaard D et al (2009) Effect of administration of oral contraceptives in vivo on collagen synthesis in tendon and muscle connective tissue in young women. J Appl Physiol 106(4):1435–1443
39. Petersen SG, Miller BF, Hansen M, Trappe TA, Kjaer M, Holm L (2010) Exercise & NSAID: effect

on muscle protein synthesis in knee osteoarthritis patients? Med Sci Sports Exerc 43:425–431
40. Heinemeier KM, Bjerrum SS, Schjerling P, Kjaer M (2013) Expression of extracellular matrix components and related growth factors in human tendon and muscle after acute exercise. Scand J Med Sci Sports 23(3):e150–e161
41. Sullivan BE, Carroll CC, Jemiolo B, Trappe SW, Magnusson SP, Dossing S et al (2009) Effect of acute resistance exercise and sex on human patellar tendon structural and regulatory mRNA expression. J Appl Physiol 106(2):468–475
42. Heinemeier KM, Schjerling P, Heinemeier J, Magnusson SP, Kjaer M (2013) Lack of tissue renewal in human adult Achilles tendon is revealed by nuclear bomb (14)C. FASEB J 27(5):2074–2079
43. Bank RA, TeKoppele JM, Oostingh G, Hazleman BL, Riley GP (1999) Lysylhydroxylation and non-reducible crosslinking of human supraspinatus tendon collagen: changes with age and in chronic rotator cuff tendinitis. Ann Rheum Dis 58(1):35–41
44. Thorpe CT, Streeter I, Pinchbeck GL, Goodship AE, Clegg PD, Birch HL (2010) Aspartic acid racemization and collagen degradation markers reveal an accumulation of damage in tendon collagen that is enhanced with aging. J Biol Chem 285(21):15674–15681
45. McAnulty RJ, Laurent GJ (1987) Collagen synthesis and degradation in vivo. Evidence for rapid rates of collagen turnover with extensive degradation of newly synthesized collagen in tissues of the adult rat. Coll Relat Res 7(2):93–104
46. Kongsgaard M, Reitelseder S, Pedersen TG, Holm L, Aagaard P, Kjaer M et al (2007) Region specific patellar tendon hypertrophy in humans following resistance training. Acta Physiol (Oxf) 191 (2):111–121
47. Gumucio JP, Phan AC, Ruehlmann DG, Noah AC, Mendias CL (2014) Synergist ablation induces rapid tendon growth through the synthesis of a neotendon matrix. J Appl Physiol 2014:jap 00720
48. Tan Q, Lui PP, Lee YW (2013) In vivo identity of tendon stem cells and the roles of stem cells in tendon healing. Stem Cells Dev
49. Mendias CL, Gumucio JP, Bakhurin KI, Lynch EB, Brooks SV (2012) Physiological loading of tendons induces scleraxis expression in epitenon fibroblasts. J Orthop Res 30(4):606–612
50. Hansen P, Aagaard P, Kjaer M, Larsson B, Magnusson SP (2003) Effect of habitual running on human Achilles tendon load-deformation properties and cross-sectional area. J Appl Physiol 95(6):2375–2380
51. Arndt A, Bruggemann GP, Koebke J, Segesser B (1999) Asymmetrical loading of the human triceps surae: I. Mediolateral force differences in the Achilles tendon. Foot Ankle Int 20(7):444–449
52. Bojsen-Moller J, Hansen P, Aagaard P, Svantesson U, Kjaer M, Magnusson SP (2004) Differential displacement of the human soleus and medial gastrocnemius aponeuroses during isometric plantar flexor contractions in vivo. J Appl Physiol 97 (5):1908–1914
53. Arndt A, Bengtsson AS, Peolsson M, Thorstensson A, Movin T (2012) Non-uniform displacement within the Achilles tendon during passive ankle joint motion. Knee Surg Sports Traumatol Arthrosc 20(9):1868–1874
54. Kim YS, Kim JM, Bigliani LU, Kim HJ, Jung HW (2011) In vivo strain analysis of the intact supraspinatus tendon by ultrasound speckles tracking imaging. J Orthop Res 29(12):1931–1937
55. Stegman KJ, Djurickovic S, Dechev N (2014) In Vivo estimation of flexor digitorum superficialis tendon displacement with speckle tracking on 2-d ultrasound images using laplacian, gaussian and rayleigh techniques. Ultrasound Med Biol 40(3):568–582
56. Haraldsson BT, Aagaard P, Krogsgaard M, Alkjaer T, Kjaer M, Magnusson SP (2005) Region-specific mechanical properties of the human patella tendon. J Appl Physiol 98(3):1006–1012
57. Hansen P, Haraldsson BT, Aagaard P, Kovanen V, Avery NC, Qvortrup K et al (2010) Lower strength of the human posterior patellar tendon seems unrelated to mature collagen cross-linking and fibril morphology. J Appl Physiol 108(1):47–52
58. Haraldsson BT, Aagaard P, Qvortrup K, Bojsen-Moller J, Krogsgaard M, Koskinen S et al (2008) Lateral force transmission between human tendon fascicles. Matrix Biol 27(2):86–95
59. Screen HR, Lee DA, Bader DL, Shelton JC (2004) An investigation into the effects of the hierarchical structure of tendon fascicles on micromechanical properties. Proc Inst Mech Eng H 218(2):109–119
60. Arnoczky SP, Lavagnino M, Whallon JH, Hoonjan A (2002) In situ cell nucleus deformation in tendons under tensile load; a morphological analysis using confocal laser microscopy. J Orthop Res 20 (1):29–35
61. Arnoczky SP, Lavagnino M, Egerbacher M (2007) The mechanobiological aetiopathogenesis of tendinopathy: is it the over-stimulation or the under-stimulation of tendon cells? Int J Exp Pathol 88(4):217–226
62. Scott JE (1991) Proteoglycan: collagen interactions in connective tissues. Ultrastructural, biochemical, functional and evolutionary aspects. Int J Biol Macromol 13(3):157–161
63. Scott JE (2003) Elasticity in extracellular matrix 'shape modules' of tendon, cartilage, etc. A sliding proteoglycan-filament model. J Physiol 553(Pt 2): 335–343
64. Svensson RB, Hassenkam T, Hansen P, Kjaer M, Magnusson SP (2011) Tensile force transmission in human patellar tendon fascicles is not mediated by glycosaminoglycans. Connect Tissue Res 52:415–421

65. Lujan TJ, Underwood CJ, Jacobs NT, Weiss JA (2009) Contribution of glycosaminoglycans to viscoelastic tensile behavior of human ligament. J Appl Physiol 106(2):423–431
66. Provenzano PP, Vanderby R Jr (2006) Collagen fibril morphology and organization: implications for force transmission in ligament and tendon. Matrix Biol 25(2):71–84
67. Buehler MJ (2008) Nanomechanics of collagen fibrils under varying cross-link densities: atomistic and continuum studies. J Mech Behav Biomed Mater 1(1):59–67
68. Barnard K, Light ND, Sims TJ, Bailey AJ (1987) Chemistry of the collagen cross-links. Origin and partial characterization of a putative mature cross-link of collagen. Biochem J 244(2):303–309
69. Fratzl P, Misof K, Zizak I, Rapp G, Amenitsch H, Bernstorff S (1998) Fibrillar structure and mechanical properties of collagen. J Struct Biol 122(1–2):119–122
70. Mosler E, Folkhard W, Knorzer E, Nemetschek-Gansler H, Nemetschek T, Koch MH (1985) Stress-induced molecular rearrangement in tendon collagen. J Mol Biol 182(4):589–596
71. Sasaki N, Odajima S (1996) Elongation mechanism of collagen fibrils and force-strain relations of tendon at each level of structural hierarchy. J Biomech 29(9):1131–1136
72. Sasaki N, Odajima S (1996) Stress–strain curve and Young's modulus of a collagen molecule as determined by the X-ray diffraction technique. J Biomech 29(5):655–658
73. Lorenzo AC, Caffarena ER (2005) Elastic properties, Young's modulus determination and structural stability of the tropocollagen molecule: a computational study by steered molecular dynamics. J Biomech 38(7):1527–1533
74. Bailey AJ, Paul RG, Knott L (1998) Mechanisms of maturation and ageing of collagen. Mech Ageing Dev 106(1–2):1–56
75. Svensson RB, Mulder H, Kovanen V, Magnusson SP (2013) Fracture mechanics of collagen fibrils: influence of natural cross-links. Biophys J 104(11):2476–2484
76. Ker RF, Alexander RM, Bennett MB (1988) Why are mammalian tendons so thick? J Zoo Lond 216:309–324
77. Rosager S, Aagaard P, Dyhre-Poulsen P, Neergaard K, Kjaer M, Magnusson SP (2002) Load–displacement properties of the human triceps surae aponeurosis and tendon in runners and non-runners. Scand J Med Sci Sports 12(2):90–98
78. Arampatzis A, Karamanidis K, Albracht K (2007) Adaptational responses of the human Achilles tendon by modulation of the applied cyclic strain magnitude. J Exp Biol 210(Pt 15):2743–2753
79. Couppe C, Kongsgaard M, Aagaard P, Hansen P, Bojsen-Moller J, Kjaer M et al (2008) Habitual loading results in tendon hypertrophy and increased stiffness of the human patellar tendon. J Appl Physiol 105(3):805–810
80. Michna H (1984) Morphometric analysis of loading-induced changes in collagen-fibril populations in young tendons. Cell Tissue Res 236(2):465–470
81. Patterson-Kane JC, Parry DA, Birch HL, Goodship AE, Firth EC (1997) An age-related study of morphology and cross-link composition of collagen fibrils in the digital flexor tendons of young thoroughbred horses. Connect Tissue Res 36(3):253–260
82. Patterson-Kane JC, Firth EC, Parry DA, Wilson AM, Goodship AE (1998) Effects of training on collagen fibril populations in the suspensory ligament and deep digital flexor tendon of young thoroughbreds. Am J Vet Res 59(1):64–68
83. Couppe C, Svensson RB, Grosset JF, Kovanen V, Nielsen RH, Olsen MR et al (2014) Life-long endurance running is associated with reduced glycation and mechanical stress in connective tissue. Age (Dordr) 36(4):9665
84. Kongsgaard M, Qvortrup K, Larsen J, Aagaard P, Doessing S, Hansen P et al (2010) Fibril morphology and tendon mechanical properties in patellar tendinopathy: effects of heavy slow resistance training. Am J Sports Med 38(4):749–756
85. Bailey AJ, Paul RG, Knott L (1998) Mechanisms of maturation and ageing of collagen. Mech Ageing Dev 106(1–2):1–56
86. Kongsgaard M, Kovanen V, Aagaard P, Doessing S, Hansen P, Laursen AH et al (2009) Corticosteroid injections, eccentric decline squat training and heavy slow resistance training in patellar tendinopathy. Scand J Med Sci Sports 19(6):790–802
87. Verzijl N, DeGroot J, Oldehinkel E, Bank RA, Thorpe SR, Baynes JW et al (2000) Age-related accumulation of Maillard reaction products in human articular cartilage collagen. Biochem J 350(Pt 2):381–387
88. Narici MV, Maganaris CN (2007) Plasticity of the muscle-tendon complex with disuse and aging. Exerc Sport Sci Rev 35(3):126–134
89. Suetta C, Hvid LG, Justesen L, Christensen U, Neergaard K, Simonsen L et al (2009) Effects of ageing on human skeletal muscle after immobilisation and re-training. J Appl Physiol
90. de Boer MD, Maganaris CN, Seynnes OR, Rennie MJ, Narici MV (2007) Time course of muscular, neural and tendinous adaptations to 23 day unilateral lower-limb suspension in young men. J Physiol Lond 583(Pt 3):1079–1091
91. Matsumoto F, Trudel G, Uhthoff HK, Backman DS (2003) Mechanical effects of immobilization on the Achilles' tendon. Arch Phys Med Rehabil 84(5):662–667
92. Rumian AP, Draper ER, Wallace AL, Goodship AE (2009) The influence of the mechanical environment on remodelling of the patellar tendon. J Bone Joint Surg (Br) 91(4):557–564

93. Eliasson P, Fahlgren A, Pasternak B, Aspenberg P (2007) Unloaded rat Achilles tendons continue to grow, but lose viscoelasticity. J Appl Physiol 103 (2):459–463
94. Kinugasa R, Hodgson JA, Edgerton VR, Shin DD, Sinha S (2010) Reduction in tendon elasticity from unloading is unrelated to its hypertrophy. J Appl Physiol 109(3):870–877
95. Shin D, Finni T, Ahn S, Hodgson JA, Lee HD, Edgerton VR et al (2008) Effect of chronic unloading and rehabilitation on human Achilles tendon properties: a velocity-encoded phase-contrast MRI study. J Appl Physiol 105(4):1179–1186
96. Reeves ND, Maganaris CN, Ferretti G, Narici MV (2005) Influence of 90-day simulated microgravity on human tendon mechanical properties and the effect of resistive countermeasures. J Appl Physiol 98(6):2278–2286
97. de Boer MD, Maganaris CN, Seynnes OR, Rennie MJ, Narici MV (2007) Time course of muscular, neural and tendinous adaptations to 23 day unilateral lower-limb suspension in young men. J Physiol 583(Pt 3):1079–1091
98. Boesen AP, Dideriksen K, Couppe C, Magnusson SP, Schjerling P, Boesen M et al (2014) Effect of growth hormone on aging connective tissue in muscle and tendon: gene expression, morphology, and function following immobilization and rehabilitation. J Appl Physiol 116(2):192–203
99. Majima T, Yasuda K, Tsuchida T, Tanaka K, Miyakawa K, Minami A et al (2003) Stress shielding of patellar tendon: effect on small-diameter collagen fibrils in a rabbit model. J Orthop Sci 8(6):836–841
100. Nakagawa Y, Hayashi K, Yamamoto N, Nagashima K (1996) Age-related changes in biomechanical properties of the Achilles tendon in rabbits. Eur J Appl Physiol Occup Physiol 73(1–2):7–10
101. Nakagawa Y, Totsuka M, Sato T, Fukuda Y, Hirota K (1989) Effect of disuse on the ultrastructure of the achilles tendon in rats. Eur J Appl Physiol Occup Physiol 59(3):239–242
102. Tsuchida T, Yasuda K, Kaneda K, Hayashi K, Yamamoto N, Miyakawa K et al (1997) Effects of in situ freezing and stress-shielding on the ultrastructure of rabbit patellar tendons. J Orthop Res 15 (6):904–910
103. Majima T, Yasuda K, Fujii T, Yamamoto N, Hayashi K, Kaneda K (1996) Biomechanical effects of stress shielding of the rabbit patellar tendon depend on the degree of stress reduction. J Orthop Res 14(3):377–383
104. Nielsen HM, Skalicky M, Viidik A (1998) Influence of physical exercise on aging rats. III. Life-long exercise modifies the aging changes of the mechanical properties of limb muscle tendons. Mech Ageing Dev 100(3):243–260
105. Wood LK, Arruda EM, Brooks SV (2011) Regional stiffening with aging in tibialis anterior tendons of mice occurs independent of changes in collagen fibril morphology. J Appl Physiol 111(4):999–1006
106. Viidik A, Nielsen HM, Skalicky M (1996) Influence of physical exercise on aging rats: II. Life-long exercise delays aging of tail tendon collagen. Mech Ageing Dev 88(3):139–148
107. Dressler MR, Butler DL, Wenstrup R, Awad HA, Smith F, Boivin GP (2002) A potential mechanism for age-related declines in patellar tendon biomechanics. J Orthop Res 20(6):1315–1322
108. Vogel HG (1978) Influence of maturation and age on mechanical and biochemical parameters of connective tissue of various organs in the rat. Connect Tissue Res 6(3):161–166
109. Dunkman AA, Buckley MR, Mienaltowski MJ, Adams SM, Thomas SJ, Satchell L et al (2013) Decorin expression is important for age-related changes in tendon structure and mechanical properties. Matrix Biol 32(1):3–13
110. LaCroix AS, Duenwald-Kuehl SE, Brickson S, Akins TL, Diffee G, Aiken J et al (2013) Effect of age and exercise on the viscoelastic properties of rat tail tendon. Ann Biomed Eng 41(6):1120–1128
111. Simonsen EB, Klitgaard H, Bojsen-Moller F (1995) The influence of strength training, swim training and ageing on the Achilles tendon and m. soleus of the rat. J Sports Sci 13(4):291–295
112. Haut RC, Lancaster RL, DeCamp CE (1992) Mechanical properties of the canine patellar tendon: some correlations with age and the content of collagen. J Biomech 25(2):163–173
113. Haut RC (1983) Age-dependent influence of strain rate on the tensile failure of rat-tail tendon. J Biomech Eng 105(3):296–299
114. Connizzo BK, Sarver JJ, Birk DE, Soslowsky LJ, Iozzo RV (2013) Effect of age and proteoglycan deficiency on collagen fiber re-alignment and mechanical properties in mouse supraspinatus tendon. J Biomech Eng 135(2):021019
115. Flahiff CM, Brooks AT, Hollis JM, Vander Schilden JL, Nicholas RW (1995) Biomechanical analysis of patellar tendon allografts as a function of donor age. Am J Sports Med 23(3):354–358
116. Hubbard RP, Soutas-Little RW (1984) Mechanical properties of human tendon and their age dependence. J Biomech Eng 106(2):144–150
117. Johnson GA, Tramaglini DM, Levine RE, Ohno K, Choi NY, Woo SLY (1994) Tensile and viscoelastic properties of human patellar tendon. J Orthop Res 12(6):796–803
118. Seynnes OR, Bojsen-Moller J, Albracht K, Arndt A, Cronin NJ, Finni T et al (2015) Ultrasound-based testing of tendon mechanical properties: a critical evaluation. J Appl Physiol 118(2):133–141
119. Carroll CC, Dickinson JM, Haus JM, Lee GA, Hollon CJ, Aagaard P et al (2008) Influence of aging on the in vivo properties of human patellar tendon. J Appl Physiol 105(6):1907–1915

120. Couppe C, Hansen P, Kongsgaard M, Kovanen V, Suetta C, Aagaard P et al (2009) Mechanical properties and collagen cross-linking of the patellar tendon in old and young men. J Appl Physiol 107 (3):880–886
121. Karamanidis K, Arampatzis A (2006) Mechanical and morphological properties of human quadriceps femoris and triceps surae muscle-tendon unit in relation to aging and running. J Biomech 39(3):406–417
122. Mian OS, Thom JM, Ardigo LP, Minetti AE, Narici MV (2007) Gastrocnemius muscle-tendon behaviour during walking in young and older adults. Acta Physiol (Oxford) 189(1):57–65
123. Morse CI, Thom JM, Birch KM, Narici MV (2005) Tendon elongation influences the amplitude of interpolated doublets in the assessment of activation in elderly men. J Appl Physiol 98(1):221–226
124. Onambele GL, Narici MV, Maganaris CN (2006) Calf muscle-tendon properties and postural balance in old age. J Appl Physiol 100(6):2048–2056
125. Stenroth L, Peltonen J, Cronin NJ, Sipila S, Finni T (2012) Age-related differences in Achilles tendon properties and triceps surae muscle architecture in vivo. J Appl Physiol 113(10):1537–1544
126. Couppe C, Hansen P, Kongsgaard M, Kovanen V, Suetta C, Aagaard P et al (2009) Mechanical properties and collagen cross-linking of the patellar tendon in old and young men. J Appl Physiol 107 (3):880–886
127. Riley G (2005) Chronic tendon pathology: molecular basis and therapeutic implications. Exp Rev Mol Med 7(5):1–25
128. Tidball JG, Salem G, Zernicke R (1993) Site and mechanical conditions for failure of skeletal muscle in experimental strain injuries. J Appl Physiol 74(3):1280–1286
129. Roffino S, Carnino A, Chopard A, Mutin M, Marini JF (2006) Structural remodeling of unweighted soleus myotendinous junction in monkey. Comptes Rendus Biologies 329(3):172–179
130. Ciena AP, Luques IU, Dias FJ, Yokomizo de Almeida SR, Iyomasa MM, Watanabe IS (2010) Ultrastructure of the myotendinous junction of the medial pterygoid muscle of adult and aged Wistar rats. Micron 41(8):1011–1014
131. Knudsen AB, Larsen M, Mackey AL, Hjort M, Hansen KK, Qvortrup K et al (2014) The human myotendinous junction: an ultrastructural and 3D analysis study. Scand J Med Sci Sports 25: e116–e123
132. Trotter JA (2002) Structure-function considerations of muscle-tendon junctions. Comp Biochem Physiol A Mol Integr Physiol 133(4):1127–1133
133. Kojima H, Sakuma E, Mabuchi Y, Mizutani J, Horiuchi O, Wada I et al (2008) Ultrastructural changes at the myotendinous junction induced by exercise. J Orthop Sci 13(3):233–239

Blood Supply

3

Keitaro Kubo

Abstract

It has been suggested that blood circulation within the tendons contributes to repair of the tendon after the exercises. Recently, blood circulation of human tendons could be measured using red laser lights (Kubo et al. 2008b). Using this technique, we were able to measure changes in blood volume and oxygen saturation of human tendons by various treatments. During a 60-min heating, the blood volume and oxygen saturation of the tendon increased significantly from the resting level, and continued to increase by 35 min. These changes in blood circulation of tendon were considerably different from the temperatures of muscle and skin. Furthermore, when the needle tip was moved up and down from the targeted depth (up-and-down manipulation) at approximately 1 mm amplitude, the blood volume and oxygen saturation of the treated tendon increased significantly. After the removal of the acupuncture needle, the blood volume and oxygen saturation of the tendon increased gradually for the non-treated side. These results suggested that the change in blood circulation of the tendon during acupuncture with up-and-down manipulation was caused by axon reflex, and increase in blood flow in the tendons after the needle removal might be caused through the central nervous system. It is well known that heating and acupuncture treatments were quite effective in the management of tendon injuries. Therefore, these phenomena would be related to the changes in blood circulation of tendons due to heating and acupuncture treatments.

Keywords

Blood volume • Oxygen saturation • Heating • Acupuncture • Tendon injury

K. Kubo (✉)
Department of Life Science (Sports Sciences),
The University of Tokyo, Komaba 3-8-1, Meguro-ku,
Tokyo 153-8902, Japan
e-mail: kubo@idaten.c.u-tokyo.ac.jp

P.W. Ackermann and D.A. Hart (eds.), *Metabolic Influences on Risk for Tendon Disorders*,
Advances in Experimental Medicine and Biology 920, DOI 10.1007/978-3-319-33943-6_3

Introduction

Blood circulation of human tendon has so far been investigated using radioisotope scanning and laser Doppler flowmetry [1, 2]. However, the radioisotope scanning technique exposes subjects to radiation, and laser Doppler flowmetry requires the invasive insertion of a probe. Therefore, these methodologies seem to be difficult for applying to investigate the changes in blood circulation of human tendon in the athletic and medical fields. Recently, we claimed that the blood volume and oxygen saturation of human tendon could be determined using red laser lights [3]. The present chapter will address the recent findings on blood circulation of human tendon using this technique. It should acknowledge that the anatomical point of view concerning the blood vessels and vascular system of tendons are beyond the scope of this chapter [4].

Measurement of Blood Circulation of Human Tendon Using Red Laser Lights

In the late 1980s, near infrared spectroscopy has made it possible to evaluate changes in the oxygenation of tissues non-invasively [5]. This technique has been used to assess changes in blood volume and the balance between oxygen delivery and its utilization in the human muscle *in vivo*. However, this technique requires a large volume of measured samples, because the difference in absorption by oxy- and deoxy-hemoglobin in the near-infrared region is small. Kashima [6] developed a method for measuring the blood volume and oxygen saturation in a small volume of tissue (<1 cm^3) using three red laser lights based on the Beer-Lambert law.

We used red laser lights (BOM-L1TRSF, Omega Wave, Tokyo, Japan) to measure the blood volume and oxygen saturation of human Achilles tendon *in vivo* [3]. A probe (SF-DS, Omega Wave, Tokyo, Japan) was positioned 30 mm proximal to the calcaneus. This instrument used three red laser lights (635, 650, and 690 nm), and calculated the relative tissue levels of oxyhemoglobin, deoxyhemoglobin, blood volume, and oxygen saturation. The distance between the light source and photodetector was 5 mm. According to the findings of Kashima [6], the measurement depth was estimated at 3–5 mm when the distance between the light source and photodetector was 5 mm. The details of this technique and principles of this instrument have been described elsewhere [3, 6]. Briefly, two-point detection and the differential calculation method were used for measuring the blood volume and oxygen saturation only in the deep region of the tissue (measurement depth of 3–5 mm). The blood volume and oxygen saturation at specific depths of tissue could be measured by changing the location of the two detectors. The offset value of the blood volume was reduced, and highly sensitive measurements were achieved using the two-point detection method.

During contraction at 50 % of maximal voluntary contraction, the blood volume and oxygen saturation of the Achilles tendon as well as the medial gastrocnemius muscle (using near infrared spectroscopy) decreased significantly from the resting level (Fig. 3.1). However, the extents of decrease in blood volume and oxygen saturation were considerably lower in tendon than in muscle. We also observed that there was a distinct difference in the changes of blood circulation between the tendon and the skin measured by laser Doppler flowmetry. Using ultrasonography, we confirmed that the depth of the Achilles tendon from the skin was 2.2 mm at its superficial surface and 7.8 mm at its deep surface. Considering these findings, the present methodology was appropriate for investigating blood circulation of human tendon *in vivo*.

Effect of Heating on Blood Circulation of Human Tendon

Heat treatment has been used as a physical therapy for muscle and tendon injuries [7]. The healing of injuries induced by heat treatment

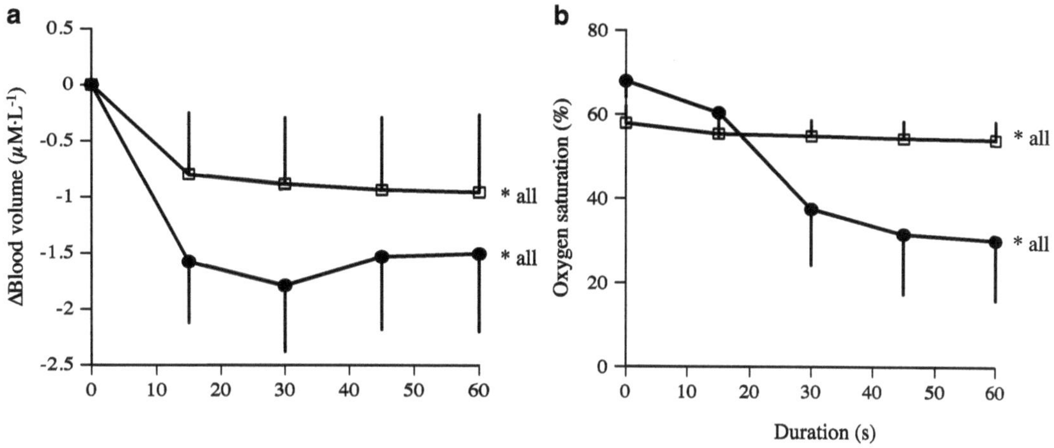

Fig. 3.1 Changes in blood volume (**a**) and oxygen saturation (**b**) of the Achilles tendon (*open*) and medial gastrocnemius muscle (*closed*) during 50 % of the maximum voluntary contraction (Modified from Kubo et al. 2008b). * significantly different from the resting level

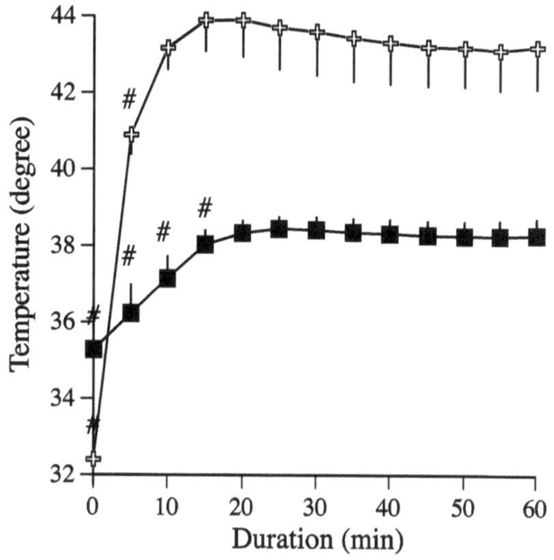

Fig. 3.2 The time course change in the temperatures of muscle (*closed square*) and skin (*open cross*) during 60 min of heating (Modified from Kubo et al. 2012). # significantly different from the point of 60 min

was considered to be related to the increase in the blood flow of each tissue. According to the previous findings [8, 9], the blood flow of muscle and skin increased considerably when the temperature of each tissue elevated. However, no study has investigated the effects of heat treatment on blood circulation of human tendon *in vivo*. In athletic and medical fields, the application time of heat treatment was around 20 min. The duration of heating would be selected, because the heat treatment (e.g., hot pack application, hot water immersion) to the human limbs produced peak temperatures of skin and muscle at approximately 10 min [10, 11]. However, we should decide the duration of heat treatment according to the changes in blood circulation of each tissue during heating. Therefore, we aimed to investigate changes in blood circulation of tendon during heating, and thus to suggest an appropriate duration of heat treatment for the injured tendon based on the present findings [12].

During a 60-min heating, the blood volume and oxygen saturation of Achilles tendon were measured using red laser lights. A hot pack provided the specific localized temperature control. Throughout the experiment, the temperatures of muscle and skin were measured by a deep body thermometer. During heating, the temperature of each muscle and skin continued to increase by 20 min and 10 min, respectively (Fig. 3.2). From the point of 15 min, the blood volume and oxygen saturation of the heated tendon increased significantly from the resting level, and continued to increase by 35 min (Fig. 3.3a, b).

The application of heat treatment for the injured muscle and tendon has so far been used during 15–30 min, because the hot pack application and hot water immersion produced peak

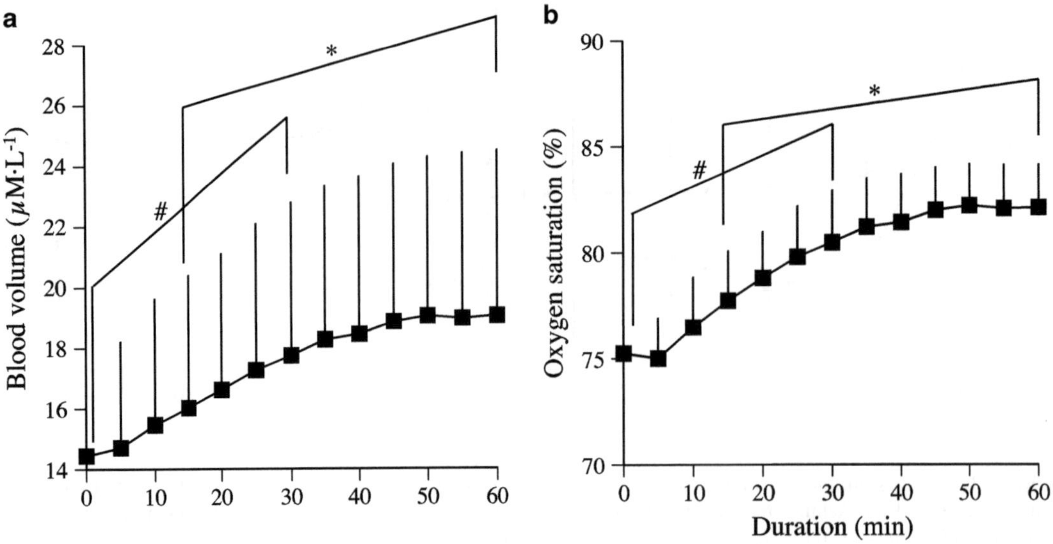

Fig. 3.3 The time course change in the blood volume (**a**) and oxygen saturation (**b**) of the Achilles tendon during 60 min of heating (Modified from Kubo et al. 2012). * significantly different from the resting level. # significantly different from the point of 60 min

temperature of surface and muscle at approximately 10 min [10, 11]. In the present study, the increases in temperatures of each tissue leveled out at 10–20 min, and were agreed with these previous studies [10, 11]. However, the present study showed that changes in blood circulation of tendon were different from the changes in their temperatures. During a 60-min heating, the blood volume and oxygen saturation of tendon continued to increase by 35 min (Fig. 3.3a, b). Although the mechanisms, which resulted in the increase of blood volume of tendon during heating, are unknown, it seems reasonable to suppose that increase in blood volume of tendon would result from opening of vessels and increase in capillary permeability. In addition, increased oxygen saturation during heating would be related to lower oxygen consumption of tendon [13, 14]. Giombini et al. [7] suggested that increases in oxygen and nutrients are necessary to effect tissue repair. Consequently, it is likely that the duration of heating more than 35 min would be able to supply sufficient blood and oxygen to treat the injured tendons. According to previous finding [7], heat treatment was quite effective in the management of tendon injuries, especially overuse tendinopathies. Indeed, previous study indicated that synthesis and proliferation of collagen were closely related to oxygen availability [15]. Considering the present and these previous findings, we may say that the duration of heating more than 35 min would be necessary to treat the injured tendons.

Effect of Acupuncture on Blood Circulation of Human Tendon

In recent years, acupuncture has been increasingly used to treat acute and chronic pain conditions [16]. This treatment is expected to improve blood flow to the injured tissues. According to the previous studies using animals and humans [17, 18], the blood flow in skin and muscle increased considerably during needle stimulation at specific sites. Sandberg et al. [17] reported that needle insertion into the anterior tibial muscle in healthy female subjects increased both skin and muscle blood flow. On the other hand, the blood flow of the skin and the muscle of the non-treated site and/or contralateral limb increased when acupuncture treatment was applied [19, 20]. For example, Ernst and Lee [19] demonstrated that the temperature of the

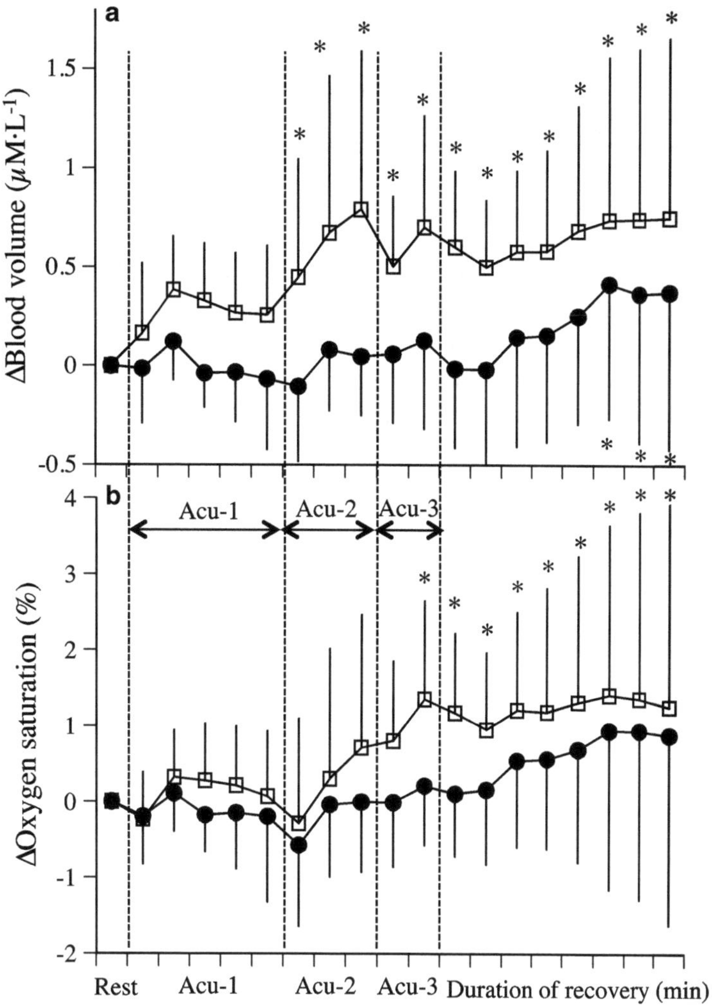

Fig. 3.4 The time course changes in the blood volume (**a**) and oxygen saturation (**b**) of the treated tendon (*open*) and non-treated tendon (*closed*) during acupuncture treatment and recovery periods (Modified from Kubo et al. 2011). * significantly different from the resting level

non-treated sites as well as the treated site increased during and after acupuncture. They suggested that these interesting phenomena were brought about by a central effect rather than a peripheral effect. Therefore, it is expected that the blood circulation of both the treated and non-treated tendons change during acupuncture treatment.

During the acupuncture treatment (10 min) and recovery period (40 min), the blood volume and oxygen saturation of the treated and the non-treated tendons were measured using red laser lights [21]. After the needle insertion, the needle was left in place for 5-min without any manipulations (Acu-1). Then, the needle tip was moved up and down from the targeted depth (up-and-down manipulation) at approximately 1 mm amplitude and 2 Hz for 3 min (Acu-2). After this technique, the needle was left in place for 2-min without manipulations (Acu-3). The time course changes in the blood volume and oxygen saturation of the treated and the non-treated tendons during acupuncture treatment and recovery period are shown in Figure 3.4. The blood volume and oxygen saturation did not change during acupuncture without manipulation (Acu-1). During up-and-down manipulation (Acu-2), all measured variables for the treated side increased significantly, while those for the non-treated side did not. During the recovery period, the blood volume and oxygen saturation of the tendon remained high values for the treated side. Considering these results, we may say that the up-and-down manipulation is necessary when we apply acupuncture treatment to the injured tendon. Some previous studies showed that the manual acupuncture (like Acu-2) caused decrease of heart rate and decrease of muscle sympathetic activity [22, 23]. If the changes in blood circulation would be caused by central effects [22, 23], the blood volume and oxygen saturation of the non-treated tendon may increase during acupuncture. However, the expected increases in these variables were not observed in this study. Accordingly, it is likely that the change in blood circulation of the tendon during acupuncture with up-and-down manipulation (Acu-2) is caused by axon reflex rather than somato-viceral reflex (see Chap. 4).

Other interesting finding was that the blood volume and oxygen saturation of the tendon increased gradually for the non-treated side. Ernst and Lee [19] also showed that the skin temperature (corresponding to the blood flow) in the non-treated face and foot as well as the treated hand increased for a long time after the needle removal. These present and previous results suggested that the increase in blood flow in the tendons after the needle removal might be caused through the central nervous system. Although it was unknown that the exact mechanism for the changes in the blood circulation of the non-treated tendon after removal of the needle, further investigations are needed to clarify the mechanism of this interesting phenomenon. Furthermore, blood circulation in the injured tendon in a plaster cast may be improved by applying acupuncture treatment to the contralateral healthy limb.

Conclusion

Recent studies demonstrated that blood circulation of human tendon was able to measure using three red laser lights based on the Beer-Lambert [3]. In addition, this technique might be appropriate for investigating the physiology and pathology of the human tendon *in vivo*. However, these results as quoted above were obtained from the healthy subjects [12, 21]. In the future study, we need to investigate the effects of the changes in blood circulation of the tendon on the healing of actual injured tendons.

References

1. Langberg H, Bulow J, Kjaer M (1998) Blood flow in the peritendinous space of the human Achilles tendon during exercise. Acta Physiol Scand 163:149–153
2. Astrom M, Westlin N (1994) Blood flow in the human Achilles tendon assessed by laser Doppler flowmetry. J Orthop Res 12:246–252
3. Kubo K, Ikebukuro T, Tsunoda N et al (2008) Noninvasive measures of blood volume and oxygen saturation of human Achilles tendon by red laser lights. Acta Physiol 193:257–264
4. Jozsa LG, Kannus P (1997) Structure and metabolism of normal tendons. In: Jozsa LG, Kannus P (eds) Human tendons: anatomy, physiology and pathology. Human Kinetics, Champaign, pp 46–95
5. Chance BS, Nioka S, Kent J et al (1988) Time resolved spectroscopy of hemoglobin and myoglobin in resting and ischemic muscle. Anal Biochem 174:698–707
6. Kashima S (2003) Spectroscopic measurement of blood volume and its oxygenation in a small volume of tissue using red laser lights and differential calculation between two point detections. Opt Laser Technol 35:485–489
7. Giombini A, Giovannini V, Di Cesare A et al (2007) Hyperthermia induced by microwave diathermy in the management of muscle and tendon injuries. Br Med Bull 83:379–396
8. Akyurekli D, Gerig LH, Raaphorst GP (1997) Changes in muscle blood flow distribution during hyperthermia. Int J Hyper 13:481–496
9. Davis SL, Fadel PJ, Cui J et al (2006) Skin blood flow influences near-infrared spectroscopy-derived measurements of tissue oxygenation during heat stress. J Appl Physiol 100:221–224
10. Draper DO, Harris ST, Schulthies S et al (1998) Hot-pack and 1-MHz ultrasound treatments have an additive effect on muscle temperature increase. J Athl Train 33:21–24
11. Ichinoseki-Sekine N, Naito H, Saga N et al (2007) Changes in muscle temperature induced by 434 MHz microwave hyperthermia. Br J Sports Med 41:425–429
12. Kubo K, Ikebukuro T (2012) Effects of duration of heating on blood circulation of human muscle and tendon *in vivo*. Gazzetta Medica Italiana 171:731–737
13. Boushel R, Langberg H, Green S et al (2000) Blood flow and oxygenation in peritendinous tissue and calf muscle during dynamic exercise in humans. J Physiol 524:305–313

14. Kubo K, Ikebukuro T, Tsunoda N et al (2008) Changes in oxygen consumption of human muscle and tendon following repeat muscle contractions. Eur J Appl Physiol 104:859–866
15. LaVan FB, Myers B (1990) Oxygen and wound healing. Clin Plast Surg 17:463–472
16. Kleinhenz J, Streitberger K, Windeler J et al (1999) Randomised clinical trial comparing the effects of acupuncture and a newly designed placebo needle in rotator cuff tendonitis. Pain 83:235–241
17. Sandberg M, Lundeberg T, Lindberg LG et al (2003) Effects of acupuncture on skin and muscle blood flow in healthy subjects. Eur J Appl Physiol 90:114–119
18. Sato A, Sato Y, Shimura M et al (2000) Calcitonin generated peptide produces skeletal muscle vasodilation following antidromic stimulation of unmyelinated afferents in the dorsal roots in rats. Neurosci Lett 283:137–140
19. Ernst M, Lee MHW (1985) Sympathetic vasomotor changes induced by manual and electrical acupuncture of the Hoku point visualized by thermography. Pain 21:25–33
20. Maeda M, Kachi H, Ishibashi N et al (1998) The effect of electrical acupuncture-stimulation therapy using thermography and plasma endothelin (ET-1) levels in patients with progressive systemic sclerosis (PSS). J Dermatol Sci 17:151–155
21. Kubo K, Yajima H, Takayama M et al (2011) Changes in blood circulation of the contralateral Achilles tendon during and after acupuncture and heating. Int J Sports Med 32:807–813
22. Moriyama T (1987) Microneurographic analysis of the effects of acupuncture stimulation on muscle nerve sympathetic activity in humans: excitation followed by inhibition. Nippon Seirigaku Zassi 49:711–721 (in Japanese)
23. Nishijo K, Mori H, Yoshikawa K et al (1997) Decreased heart rate by acupuncture stimulation in human via facilitation of cardiac vagal activity and suppression of cardiac sympathetic nerve. Neurosci Lett 227:165–168

Tendon Innervation

4

Paul W. Ackermann, Paul Salo, and David A. Hart

Abstract

The regulation of tendon metabolism including the responses to loading is far from being well understood. During the last decade, however, accumulating data show that tendon innervation in addition to afferent functions, via efferent pathways has a regulatory role in tendon homeostasis via a wide range of neuromediators, which coordinate metabolic and neuro-inflammatory pathways.

Innervation of intact healthy tendons is localized in the surrounding structures, i.e paratenon, endotenon and epitenon, whereas the tendon proper is practically devoid of neuronal supply. This anatomical finding reflects that the tendon metabolism is regulated from the tendon envelope, i.e. interfascicular matrix (see Chap. 1).

Tendon innervation after injury and during repair, however, is found as extensive nerve ingrowth into the tendon proper, followed by a time-dependent emergence of different neuronal mediators, which amplify and fine-tune inflammatory and metabolic pathways in tendon regeneration. After healing nerve fibers retract to the tendon envelope.

In tendinopathy innervation has been identified to consist of excessive and protracted nerve ingrowth in the tendon proper, suggesting pro-inflammatory, nociceptive and hypertrophic (degenerative) tissue responses.

In metabolic disorders such as eg. diabetes impaired tendon healing has been established to be related to dysregulation of neuronal growth factors.

P.W. Ackermann (✉)
Department of Molecular Medicine and Surgery, Karolinska Institutet, SE-17176, Stockholm, Sweden

Department of Orthopedic Surgery, Karolinska University Hospital, SE-17176, Stockholm, Sweden
e-mail: paul.ackermann@karolinska.se

P. Salo
McCaig Institute for Bone & Joint Health, University of Calgary, Calgary, AB, Canada

D.A. Hart
McCaig Institute for Bone and Joint Health, University of Calgary, Calgary, AB, Canada

Centre for Hip Health and Mobility, University of British Columbia, Vancouver, BC, Canada

P.W. Ackermann and D.A. Hart (eds.), *Metabolic Influences on Risk for Tendon Disorders*, Advances in Experimental Medicine and Biology 920, DOI 10.1007/978-3-319-33943-6_4

Targeted approaches to the peripheral nervous system including neuronal mediators and their receptors may prove to be effective therapies for painful, degenerative and traumatic tendon disorders.

Keywords
Tendon • Innervation • Peripheral nervous system • Homeostasis • Tendinopathy • Neuropeptides

Introduction

Tendon homeostasis is the balance between formation and resorption [1]. Mechanical loading is the most powerful and well-known extrinsic factor to regulate tendon protein synthesis and degradation [1]. While loading over time leads to a net gain in collagen repeated loading, however, exceeding the tendons capacity of new collagen formation is harmful and may trigger the development of tendinopathy.

Prolonged unloading may additionally be detrimental for the human tendon. Disuse leads to reduced tendon mechanical stiffness [2]. Prolonged unloading post injury also demonstrated negative effects on tendon mechanical properties and production of extracellular matrix molecules [3–5].

Although it is well known both clinically and experimentally that loading improves while unloading impairs tendon protein synthesis, the exact mechanisms and regulatory factors responsible for this mechano-biological transduction are still not fully known. In spite of this gap in knowledge, accumulating data do suggest that the peripheral nervous system including specific mediators and their receptors play an important role in tendon homeostasis and repair as well as in tendinopathy [6].

Innervation of Tendons

Generally, tendons exhibit a low degree of innervation, which may partly explain the slow adaptation to repetitive loading, prolonged healing and vulnerability to chronic injuries [7, 8] (Fig. 4.1). The innervation of tendons originates from neighbouring muscular, cutaneous and peritendinous nerve trunks [9]. From the myotendinous junction nerve fibres cross and enter the endotenon septa. In the paratenon, nerve fibres form rich plexuses and send small branches that penetrate the epitenon. Nerve fibres do not under normal conditions enter the tendon proper, but terminate as nerve endings on the different surfaces of the tendon (paratenon, epitenon, endotenon) [10].

The nerves innervating tendons are composed of a low degree of myelinated, fast transmitting Aα- and Aβ-fibres and a higher degree of unmyelinated, slow transmitting Aγ-, Aδ-, B- and C-fibres [8, 11].

The nerve endings of (1) Aα- and Aβ-fibres are of types I-III and mediate mechanoception. These include Type I or Ruffini corpuscles (pressure and stretching sensors), II or Vater-Pacini corpuscles (pressure sensors, reacting to acceleration and deceleration of movement), type III or Golgi tendon organs (tension receptors) [12, 13]. Tension receptors (Type III) have been found mostly in the myotendinous junction and insertion areas [13].

The nerve endings of (2) Aγ-, Aδ-, and C-fibres are of type IVa, so called nociceptors, mediating deep tissue pain and hyperalgesia that are characteristic features of pain in tendinopathy. The nerve endings of B-fibres, which are autonomic, consist of type IVb fibres that are mainly localised in the walls of small arteries, arterioles, capillaries, and postcapillary veins exerting vasomotor actions [13].

In addition to these classical afferent functions it is now well established that the peripheral nervous system also participates in the efferent regulation of a wide variety of efferent physiological responses, including actions on

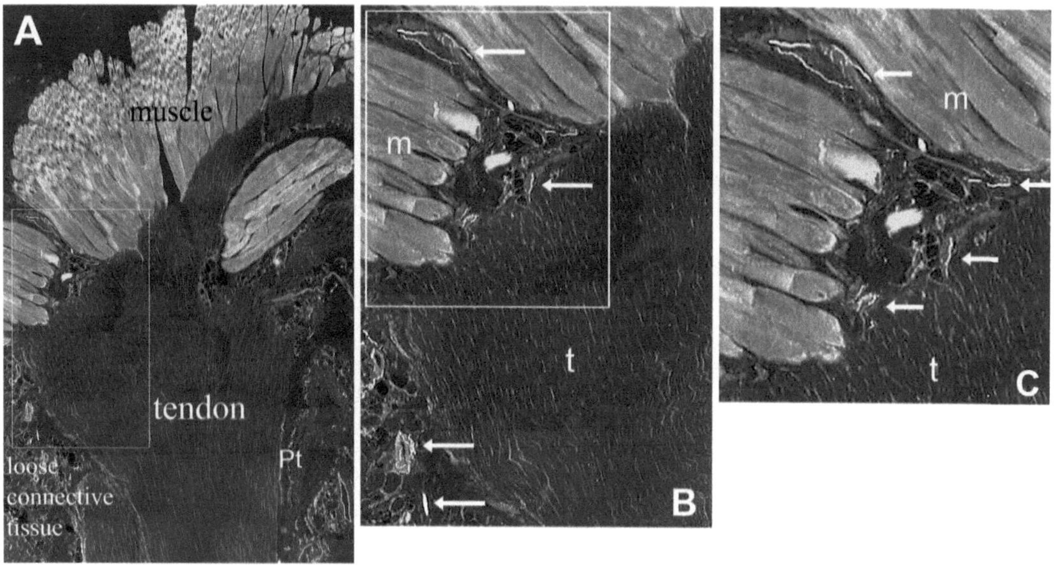

Fig. 4.1 (**a–c**). Overview of several arranged micrographs of longitudinal sections through the Achilles tendon after incubation with antisera to general nerve marker PGP 9.5. Micrographs depict the proximal half of the Achilles tendon at increasing magnification in figures (**a–c**). *Arrows* denote varicosities and nerve terminals. The typical vascular localisation of NPY is depicted in lower *left* (**b**), whereas the free nerve endings are typical localisation of SP (**c**). The immunoreactivity is seen in the paratenon and surrounding loose connective tissue, whereas the proper tensinous tissue, notably, is almost devoid of nerve fibers pt = paratenon (Reproduced with permission from Ackermann et al. [32])

cell proliferation, expression of cytokines and growth factors, inflammation, immune responses and hormone release. In the last decade, further characterization of tendon innervation related to mediator phenotypes and neuronal regulation of tendon homeostasis has received increasing attention.

In addition to classical neurotransmitters (monoamines, acetylcholine, amino acids) several neuropeptides, which act as chemical messengers in the central and peripheral nervous system, have been identified in tendons [14]. Neuropeptides differ from classical neurotransmitters in several respects [8, 15] since they generally exhibit more long-term effects than the classical neurotransmitters.

The effects of neuropeptides and classical transmitters are also elicited by different receptor mechanisms. While classical transmitters act on ligand-gated ion channels, neuropeptides act by binding to specific plasma membrane receptors, called G-protein coupled receptors [16]. Several G-protein coupled receptors have been identified in tendons and when stimulated they generate various second messengers, which can trigger a wide range of effector mechanisms regulating cellular excitability and function [14] (Table 4.1).

Neuromediators in Healthy Tendon

In general, tendons seem to exhibit neuronal mediators of a similar variety as observed in other organs of the body including those of various sensory including opioid, autonomic, and excitatory neuromediators (Table 4.1). In contrast to most other tissues, however, the tendon proper during normal conditions is devoid of nerve fibers. Innervation is found in the tendon envelope, i.e. the paratenon, endotenon and surrounding loose connective tissue (Fig. 4.1). Another vital feature of tendon neuroanatomy is the presence of counteracting mediators, i.e. pro- and anti-inflammatory peptides [17]. These

Table 4.1 Neuromediators in tendons

Type	Sub-Type	Mediator	Receptor	Actions
Sensory (Type IVa)	Sensory	SP	NK1	Pro-inflammatory
		CGRP	CRLR, RAMP-1	
		NKA	NK2*	
		NKB*, NPK*, NPG*	NK3*	
	Opioid and opioid like	Enkephalins: LE, ME, MEAP	δ-opiod receptor	Anti-inflammatory
		Dynorphins: DYN B[ND]	κ-opiod receptor[ND]	
		Endomorphins[ND]	μ-opiod receptor[ND]	
		Nociceptin[ND]	N/OFQ receptor*	
		Opioid like: GAL, SOM	GalR1-3*, SSTR1-5*	
Autonomic (Type IVb)	Sympathetic	Noradrenaline	α-,β- adrenoceptors	Pro-(anti)-inflammatory
		NPY	Y1-3*	
	Para-sympathetic	Acetylcholine	Nicotinic, muscarinic	Anti-inflammatory
		VIP	VPAC1-2*, PAC1*	
Excitatory	Glutamatergic amino acid	Glutamate	NMDA, mGlu, AMPA*, Kainate*,	Sensitization

ND Not detected in tendon, * Not yet assessed in tendon

observations would suggest that the homeostatic regulation of healthy tendon tissue is highly dependent on the balance of neuromediator modulation occurring in the tendon envelope.

Sensory Neuromediators

The sensory nerves Type IVa act principally through release of slowly acting mediators, i.e. neuropeptides and opiods. In tendons, sensory neuropeptides with nociceptive and pro-inflammatory effects (substance P (SP), calcitonin gene related peptide (CGRP) and neurokinin A (NKA)), sensory modulatory neuropeptides with anti-inflammatory actions (galanin (GAL), somatostatin (SOM)), as well as opioid neuropeptides with anti-nociceptive and anti-inflammatory effects (Leu-enkephalin (LE), Met-enkephalin (ME), Met-enkephalin-Arg-Phe (MEAP), Met-enkephalin-Arg-Gly-Leu (MEAGL), nociceptin) have been identified (Table 4.1) (Fig. 4.2a–d) [14].

Abundant amounts of sensory neuromediators have been detected in peri-vascular nerve fibers in the surrounding loose connective tissues, which may reflect an important role in the regulation of blood flow to the tendon structures. Both SP and CGRP have been reported to be potent vasodilators [18] and exert pro-inflammatory effects [19, 20]. The occurrence of sensory neuromediators in free nerve endings not associated with vessels, predominantly seen in the paratenon, may suggest nociceptive, trophic and immune regulatory roles.

In tendons receptors for SP (neurokinin 1, NK1) and CGRP (calcitonin receptor-like receptor, CRLR, and receptor activity-modifying proteins, RAMP-1) have been identified [21] (Table 4.1). These receptors have been found localized on tendon cells, immune cells, blood vessels and on free nerve endings unrelated to vessels. These localizations corroborate the suggested trophic, immune regulatory, vasoregulatory and nociceptive effects of the sensory neuromediators on tendon metabolism.

Autonomic Neuromediators

The sympathetic nervous system regulates inflammatory processes at local and systemic levels through the balanced release of sympathetic and parasympathetic mediators. The sympathetic mediator norepinephrine (noradrenaline, NA) together with neuropeptide Y (NPY) are released upon injury or nociceptive input, while parasympathetic mediators acetylcholine (ACh)

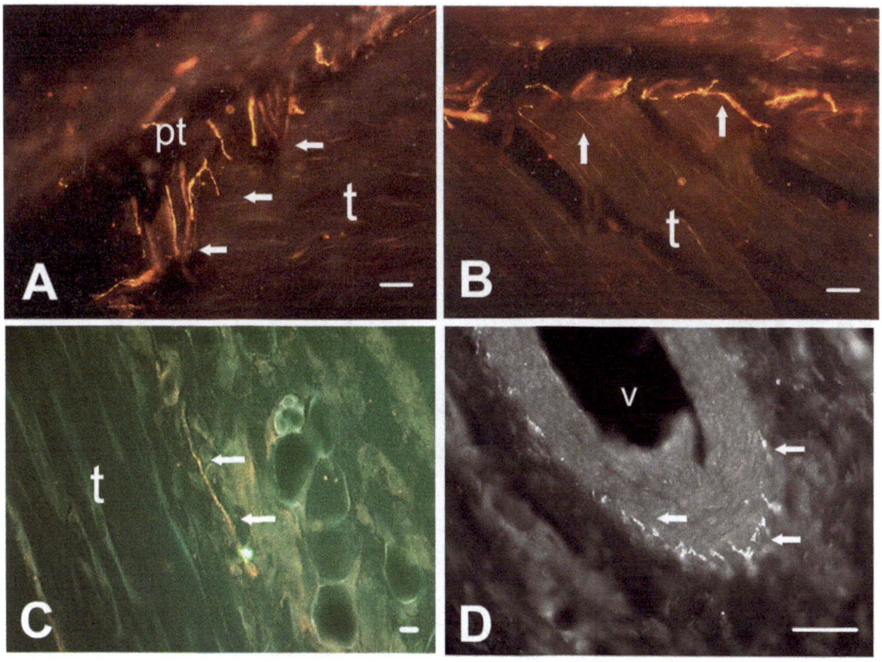

Fig. 4.2 (**a–d**) Immunofluorescence micrographs of longitudinal sections through the Achilles tendon after double staining with antisera to SP and CGRP (**a**), SP and GAL (**b**) LE and DOR (**c**) and incubation with antisera to ME (**d**). A co-existence of SP and CGRP is seen in nerve fibers localised in the paratenon (**a**), suggesting possible pro-inflammatory actions. Moreover, SP is also co-localised with GAL (**b**), reflecting anti-inflammatory actions. The immunoreactivity displaying co-existence of LE and DOR is seen as free nerve endings in the paratenon (**c**), indicating a potential peripheral anti-nociceptive system. ME immunoreaction is localised in a vessel wall (**d**). *t* tendon tissue, *Pt* paratenon; Bar = 50 μm (Reproduced with permission from Ackermann et al. [66, 67])

and vasoactive intestinal polypeptide (VIP) are released by vagus nerve activation called the 'cholinergic anti-inflammatory pathway' [17] (Fig. 4.3a–c).

The occurrence of sympathetic NA and NPY as well as parasympathetic mediators ACh and VIP has been demonstrated in tendon [22–27]. The observations of NA and NPY around blood vessels in the loose connective tissue around the main body of the tendon suggest that the sympathetic tendon vasoregulation occurs predominantly in the tendon envelope (Fig. 4.3a–b).

Adrenergic receptors responding to NA (A--adrenoceptors) and to NPY (Y1) have been identified on tendon cells, blood vessel walls and on nerve fibers [24, 27]. These localizations suggest that adrenergic stimulation of tendons may be involved in proliferation of tenocytes, endothelial cells and possibly nerve cells.

Excitatory Neuromediators

Accumulating data suggest that modulation of glutamate signalling by inhibition of its receptors, ionotropic (NMDA, AMPA, Kainate) and metabotropic (mGlu), may have potential for targeted therapy in several persistent pain conditions [28, 29]. Moreover, glutamate signaling is implicated in programmed cell death, apoptosis.

Recently, glutamate and several of its receptors have been identified in tendon on nerve fibers, blood vessels and cells (Fig. 4.4) [29–31]. These localizations of glutamate have

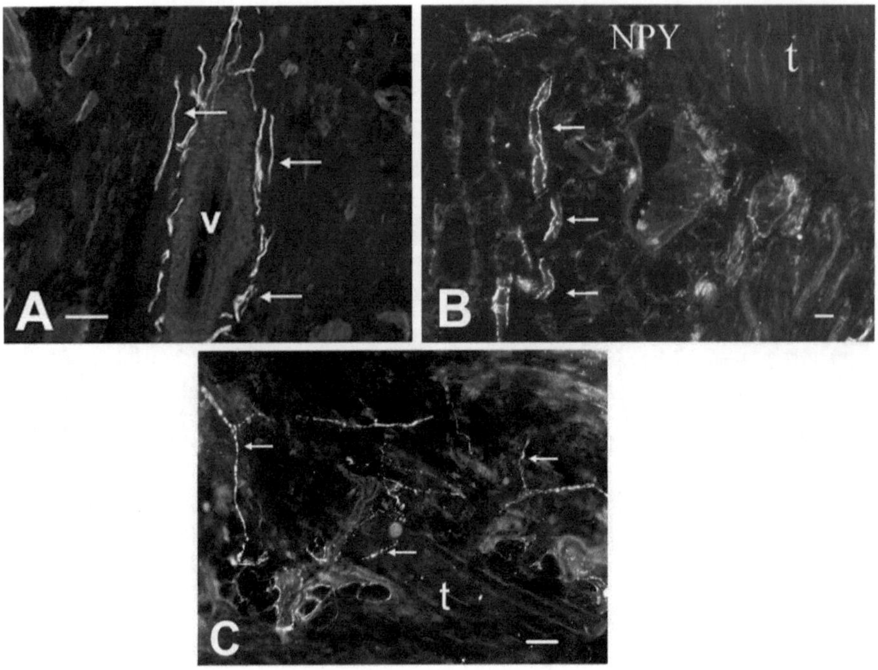

Fig. 4.3 (**a–c**) Immunofluorescence micrographs of longitudinal sections through the Achilles tendon after incubation with antisera to Noradrenaline (NA) (**a**), NPY (**b**) and VIP (**c**). NA-positive fibers are mainly found as nerve terminals in outer layers of the blood vessel walls. The NPY-positive fibers are arranged as nerve terminals in the vessel walls. VIP-positive nerves are arranged as a "fence", surrounding the proper tendon, of small varicosities in the paratenon. *t* tendon tissue, *Pt* paratenon; Bar = 50 μm (Reproduced with permission from Ackermann et al. [22])

been verified by identification of several glutamate receptors, eg. NMDAR1. The localization of glutamate receptors suggest that glutamate signaling may also be involved in regulating tendon homeostasis [14].

Nerve Ingrowth After Tendon Injury

After tendon injury and during healing the peripheral nervous system responds by nerve ingrowth into the tendon proper, which during healthy conditions is more or less aneuronal and hypovascular (Figs. 4.5–4.6) [32, 33]. Nerve sprouting and growth within the tendon proper is followed by a time dependent expression of neuropeptides during the tendon healing process. After the healing process is finished sprouting nerve fibers within the tendon proper retract to the surrounding structures, ie. the paratenon and loose connective tissue. These observations of early nerve regeneration are in line with observations on bone, ligament and skin healing indicating that nerve ingrowth and subsequent retraction are fundamental aspects of tissue repair [34–38].

Inflammatory Healing Phase

At 1 week after tendon injury increased occurrence of SP- and CGRP-positive nerve fibers has been demonstrated to be predominantly located in blood vessel walls surrounded by inflammatory cells in the loose connective tissue [33] (Fig. 4.7a). The findings comply with the nociceptive role of sensory neuropeptides, but also with a pro-inflammatory role [39].

During the first week post tendon injury, increased levels of glutamatergic signaling molecules have been found both experimentally [40] and in vivo on patients with Achilles tendon ruptures [41]. Increased glutamate levels during healing [41] seem to persist until at least week

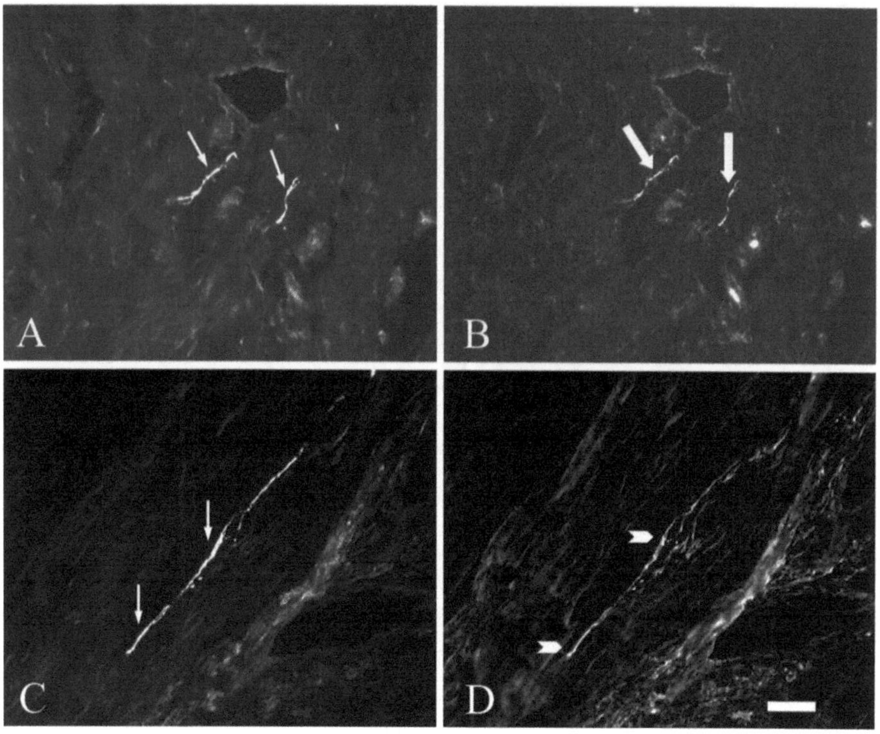

Fig. 4.4 (**a–d**) Immunofluorescence double-staining micrographs of longitudinal sections through tendinopathic patellar tendons focusing on the tendon proper, after incubation with antisera for PGP9.5 (**a**, **c**, *thin arrows*), N-methyl-D-aspartate receptor type 1 (NMDAR1) (**b**) and glutamate (**d**). *Thick arrows* denote neuronal NMDAR1 and within the tendon proper arrowheads denote neuronal glutamate, which is totally absent in the controls (Scale bar = 25 um) (Reproduced with permission from Schizas et al. [31])

6 [unpublished data], suggesting essential actions in tendon healing as has also been shown in maintenance of bone tissue [40].

Proliferative Healing Phase

From 1 to 6 weeks post rupture, a striking shift in neuronal occurrence has been demonstrated to occur from the surrounding loose connective tissue into the proper tendinous tissue [33]. This suggests the transition of a predominantly inflammatory into a proliferative repair phase [42].

During weeks 2 to 6 post injury, the expression of SP and CGRP peaked at the rupture site of the proper tendon (Fig. 4.6). SP and CGRP was observed in sprouting free nerve endings among fibroblasts in the healing tendinous tissue, suggesting a stimulatory role of sensory neuropeptides on cell proliferation and stem cell recruitment (Fig. 4.7b) (see Chap. 5) [43, 44]. SP and CGRP are also known to stimulate proliferation of endothelial cells [45, 46], indicating that SP- and CGRP fibers around newly formed blood vessels in the rupture site would comply with a role in angiogenesis (Fig. 4.7).

Additionally the expression of sensory neuropeptide receptors, SP (NK1) and CGRP (CRLR and RAMP-1), in the healing tendon is significantly up-regulated at 2 weeks, but not at 1 week, post tendon rupture [21].

Remodeling Healing Phase

During weeks 6 to 16 post tendon rupture it has been demonstrated that the nerve fibers retract from the proper tendon tissue to their normal location in the paratenon and surrounding loose

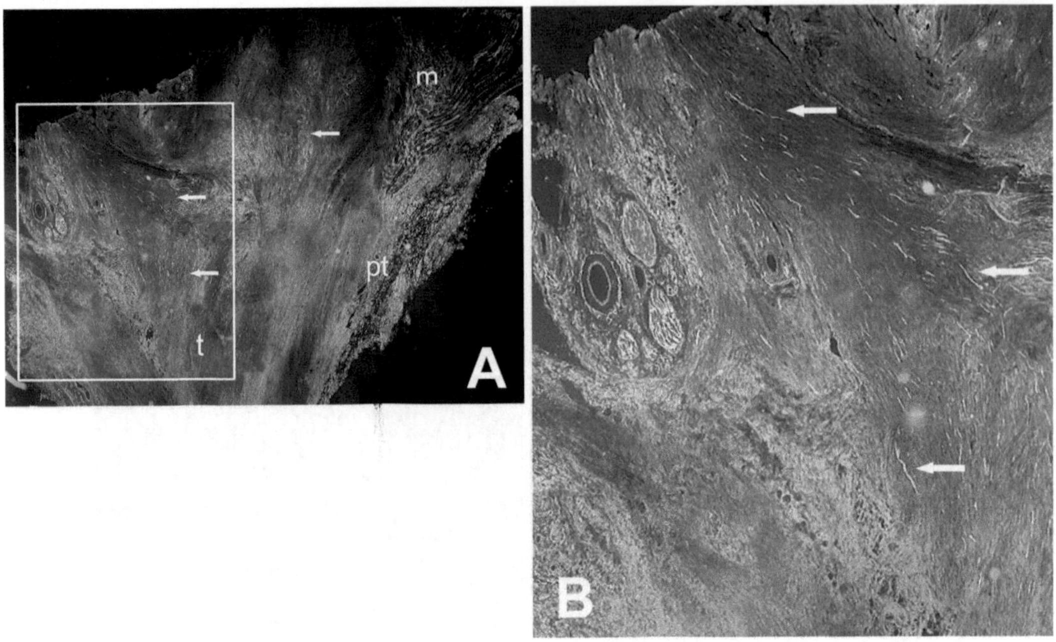

Fig. 4.5 (**a–b**) Overview micrographs of longitudinal sections through the Achilles tendon at 2 weeks post rupture. Incubation with antisera to a nerve growth marker, GAP-43. Micrographs depict the proximal half of the Achilles tendon at increasing magnification in figures (**a–b**). *Arrows* denote varicosities and nerve terminals. The GAP-positive fibers, indicating new nerve fiber ingrowth, are abundantly observed in the healing proper tendon tissue(Reproduced with permission from Ackermann et al. [32])

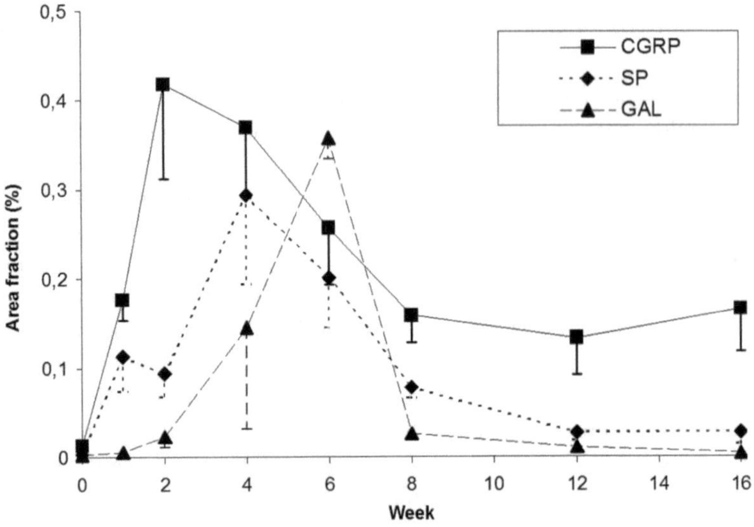

Fig. 4.6 Area occupied by nerve fibers (%) immunoreactive to SP, CGRP and GAL in relation to total area, in the mid third of the tendon, over 16 weeks post rupture (mean ± s.e.m.) (Reproduced with permission from Ackermann et al. [33])

connective tissue [33] (Fig. 4.6). This process appears to end simultaneously with the completion of paratenon repair.

Interestingly, between weeks 4 and 6, corresponding to the transition of the proliferative into the remodeling phase, a dramatic increase in the expression of the autonomic neuropeptides VIP and NPY has been demonstrated [33]. Subsequent to the elevated expression of autonomic neuropeptides, a

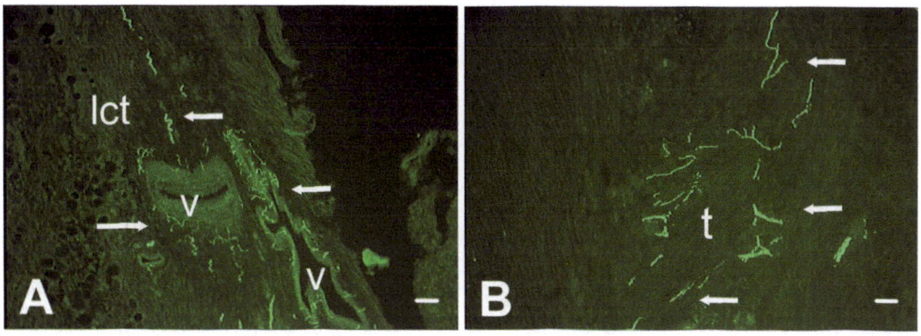

Fig. 4.7 (**a–b**) Immunofluorescence micrograph of longitudinal sections through healing Achilles tendon 1- (**a**) and 2- (**b**) weeks post rupture after incubation with antisera to CGRP. Nerve fibers immunoreactive to CGRP at week 1 are seen as vascular and free nerve endings in the loose connective tissue (**a**). At week 2, CGRP- immunoreactivity occurs mainly in the healing tendinous tissue as sprouting free nerve fibers (**b**). v = blood vessel; lct = loose connective tissue; t = proper tendon tissue; Bar = 50 µm (Reproduced with permission from Ackermann et al. [33])

Table 4.2 Neuromediators in tendinopathy

Type	Sub-type	Mediator	Receptor
Autonomic	Sympathetic	Noradrenaline ↓	A-adrenoceptors ↑
		NPY	Y1 ↑
	Parasympathetic	Acetylcholine ↑	Nicotinic*
		VIP*	Muscarinic ↑
Sensory	Sensory	SP ↑	NK1 ↑
	Opioid and opioid like	Cannabioids*	CB1 ↑
Excitatory	Glutamatergic amino acid	Glutamate ↑	NMDA1 ↑
			Phosfo- NMDA1 ↑
			mGluR1 ↑
			mGluR5 ↑
			mGluR6-7 →

significantly decreased expression of SP and CGRP was observed in the healing tendon [33] (Fig. 4.6). Hypothetically, an autonomic modulation may be required to end the inflammatory and reparative processes, thereby facilitating entry to and maintenance of the remodelling phase.

Protracted Nerve Ingrowth in Tendinopathy

The underlying histology in tendinopathies with chronic pain is often characterized as reflecting a failed healing response. The innervation pattern in tendinopathic tissue resembles that observed during the proliferative phase of healing after tendon injury (Table 4.2).

Sensory Neuromediators

Chronic painful tendons with tendinopathy exhibit new ingrowth of sensory nerve fibers (Fig. 4.8) [47–49] (49, 50, 98), correspondingly to what is also observed during tissue proliferation in healing tendons [32]. The observation of increased ingrowth of sensory nerves into the painful tendon proper, seen as sprouting free nerve endings, possibly represents nociceptors responding to mechanical stimuli by initiating pain signalling.

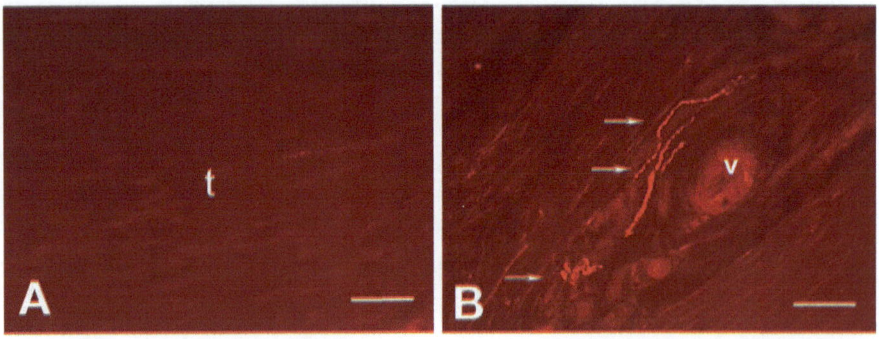

Fig. 4.8 (**a–b**) Immunofluorescence micrographs of longitudinal sections from biopsies of healthy Achilles tendon (**a**) and tendinosis tissue (**b**) after immunostaining for SP. *Arrows* denote varicosities and nerve terminals. The micrograph illustrates SP-positive nerve fibres in close vicinity to a proliferated vessel (**b**). v = blood vessel. Bar = 50 μm (Reproduced with permission from Lian et al. [47])

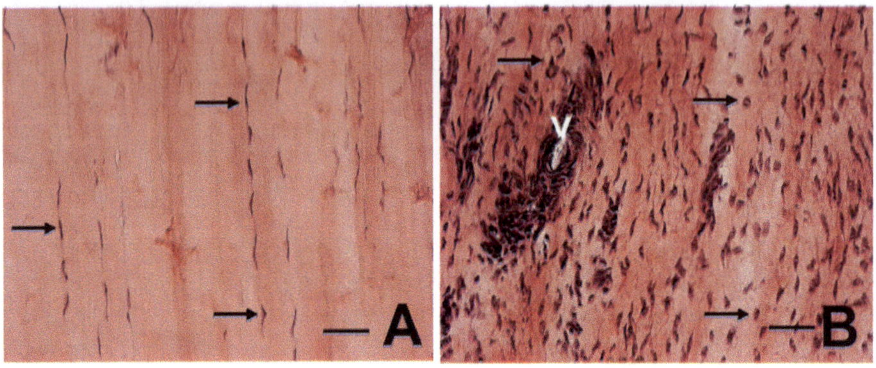

Fig. 4.9 (**a–b**) Hematoxylin and eosin micrographs of longitudinal sections from biopsies of healthy patellar tendon (**a**) and painful tendinopathy (**b**). *Arrows* denote tenocytes. The healthy tendon is homogeneous, with organized parallel collagen structure and thin, elongated tenocytes (**a**). The tendinopathy, on the other hand, is marked by collagen disorganization, increased cell count, activated tenocytes, and vascular ingrowth in the tendon proper (**b**). V = blood vessel. Bar = 50 μm (Reproduced with permission from Lian et al. [47])

The neuronal dysregulation in tendinopathy, characterized by aberrant increase of sprouting sensory nerves and increased expression of SP, possibly triggers pain signalling and also the hyperproliferative/degenerative changes associated with tendinopathy [47, 48, 50, 51] (Fig. 4.9).

Autonomic Neuromediators

Interestingly, tendinopathic patients exhibit a decreased occurrence of sympathetic nerve fibers, immunoreactive to noradrenaline (Fig. 4.10) (see Chap. 7) [47]. The reduction in vasoregulatory noradrenaline would seem to comply with an altered blood flow and a suppressed anti-nociceptive function [52, 53].

Excitatory Neuromediators

Elevated levels of glutamate have, similarly to the findings of early healing, been detected in patients with tendinopathy by microdialysis and by immunohisthochemistry [31, 54]. The specific localization for the increased glutamate levels in tendinopathic patients is observed in morphologically altered tenocytes, in the endothelial and adventitial layers of blood vessel walls and in nerve fibers (Fig. 4.11) [29–31].

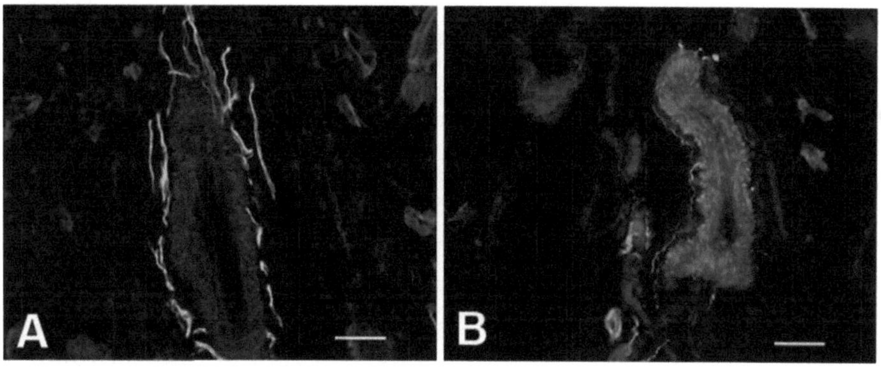

Fig. 4.10 (**a–b**) Immunofluorescence micrographs of longitudinal sections through the patellar tendon of healthy control (**a**) and painful tendinopathy (**b**) stained for TH (a marker for noradrenaline). *Arrows* denotes nerve fibres. In the healthy tendon, a strong relation is seen between blood vessels and TH positive nerves (**a**). In painful tendinopathy, a decreased number of TH positive nerves, which are blood vessel related is seen. V = blood vessel. Bar = 50 μm (Reproduced with permission from Lian et al. [47])

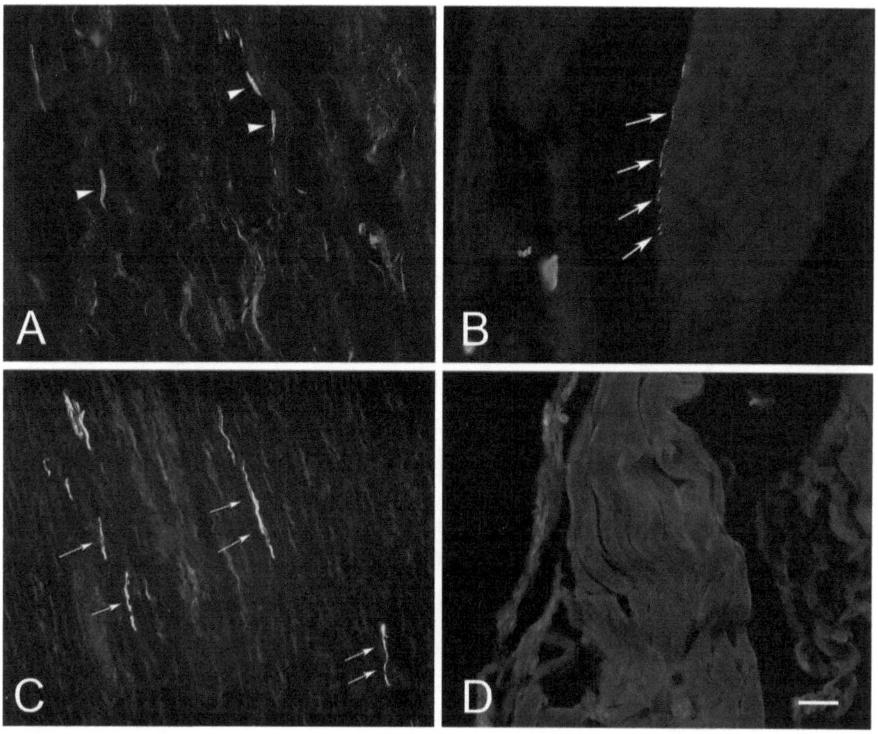

Fig. 4.11 (**a–d**) Micrographs of longitudinal sections through patellar tendon biopsies after incubation with antisera to phospho-NMDAR1 (activated NMDA) in patients with patellar tendinosis (**a, b**) and controls (**d**). Micrographs of longitudinal sections through patellar tendon biopsies stained for PGP 9.5 (a general nerve marker) in patients with patellar tendinosis are shown in (**c**). Phospho-NMDAR1 immunoreactivity in the tendon proper was exclusively observed in tendinosis as immunoreactive cells (**a**, *arrowheads*) as well as penetrating nerve fibres (**b**, *arrows*). The controls did not exhibit phospho-NMDAR1 immunoreactivity within the tendon proper (**d**). The occurrence of PGP 9.5 within the tendon proper depicted extensive nerve ingrowth in tendinopathy not seen in the controls (**c**, *thin arrows*; bar = 100 mm) (Reproduced with permission from Aubdool and Brain [39])

Several glutamate receptors have also been identified in tendinopathy patients, eg. ionotropic (NMDAR1 and activated NMDAR (Phospho-NMDAR)), metabotropic mGluR 5, which increases NMDA excitability and mGluR 6,7 that decrease NMDA excitability. Of the identified glutamate receptors, a significant up-regulation has been demonstrated of most receptors (Table 4.2).

Closer analysis has demonstrated the specific localization of the different glutamate receptors. NMDAR-1 and phospho-NMDAR1 were detected on sprouting nerve fibers, newly formed blood vessels and transformed tenocytes [29]. These localizations suggest involvement of glutamate receptors in tendinopathy regulating excitability of tenocytes, endothelial cells and nerves.

Recent reports on glutamatergic signaling in tendinopathic patients demonstrated that elevated glutamate co-existed with its up-regulated receptor NMDA1 in nerve fibers, morphologically altered tenocytes and blood vessels, which may reflect cell hyperexcitation involved in cell proliferation/differentiation. None of the controls, however, exhibited neuronal coexistence of glutamate and NMDAR1 [29, 31].

One of the recent findings established a possible mechanism responsible for activating NMDAR-1 in tendinopathy. It was demonstrated that the elevated occurrence of NMDAR-1 was correlated to that of SP ($r^2 = 0.54$, $p = 0.03$) in tendinopathic tendons, while their occurrence in controls exhibited no correlation [29]. These data suggest that SP may be involved in the up-regulation of NMDAR1. In fact, SP is known to activate NMDAR1 by removing the magnesium block [55].

Effect of Hampered Neuronal Supply

Accumulating experimental evidence and clinical observations suggest that reduced neuronal supply leads to impairment of the mechanical properties of connective tendon tissue [14]. Thus, several studies have indicated that denervation impairs the mechanical properties of both normal and injured tendons.

Sensory Neuromediators

A selective denervation of sensory neuromediators by the use of Spanish pepper (capsaicin) has in experimental tendon healing reduced the concentrations of SP by approximately 60 % [56] (Fig. 4.12). The study demonstrated that higher residual SP levels correlated with increased mechanical properties

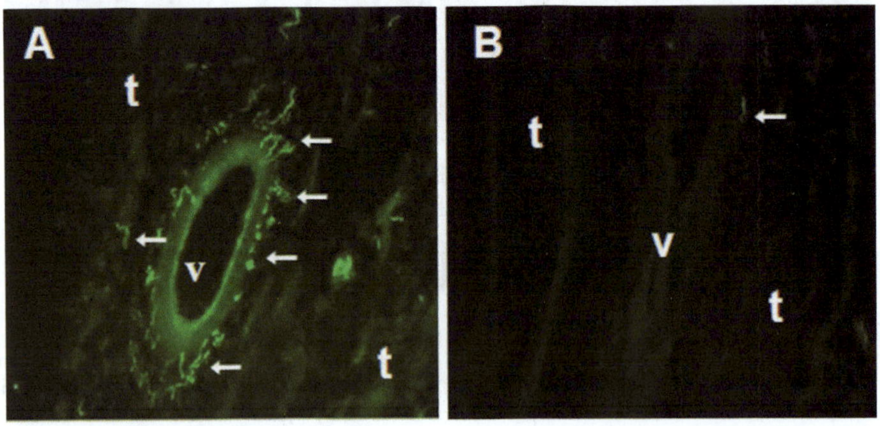

Fig. 4.12 (**a–b**) Immunofluorescence micrographs of sections of tissue from the right hind paw of denervated (**b**) and controls (**a**) after incubation with antisera to SP. *Arrows* denotes varicosities and nerve terminals. A marked reduction of SP immunoreactivity is seen in the denervated group (**a–b**) (Reproduced with permission from Bring et al. [56])

of the transverse area, ultimate tensile strength, and stress at failure ($r = 0.39$, $p = 0.036$; $r = 0.53$, $p = 0.005$; and $r = 0.43$, $p = 0.023$, respectively).

Moreover, femoral nerve transection has been demonstrated to impair healing of the medial collateral ligament in rabbits [57]. In one study, blood flow, angiogenesis and mechanical strength of the ligament scar were all significantly decreased in denervated limbs compared to normally innervated limbs, 6 weeks after injury [57].

Autonomic Neuromediators

Surgical sympathectomy reduced failure loads of healing MCLs by 50 % compared to normally innervated healing MCLs, at 2 weeks after injury [58].

Chemical sympathectomy by systemic administration of guanethidine leads to degradation of the mechanical properties of the intact medial collateral ligament (MCL) of the knee joint in rats after only 10 days of treatment [59]. Ligaments from guanethidine treated animals exhibited a larger cross-sectional area, a higher wet weight, a decreased modulus of elasticity and a decreased stress at failure. These structural changes may to some extent be explained by the significantly increased mRNA levels for the matrix degrading enzymes MMP-13 and cathepsin K, and increased ligament blood flow induced by chemical sympathectomy.

Observations in Diabetes Mellitus

It has recently been demonstrated that patients with metabolic disorders such as diabetes mellitus are at greater risk of developing various musculoskeletal disorders [60]. Thus, diabetics often exhibit neuropathy and also decreased levels of sensory neuropeptides, which may be associated with defective tissue healing [61]. - Diabetes is associated with impaired connective tissue healing and reduced biomechanical properties, correlated to down-regulated extracellular matrix proteins (see Chaps. 16, 17, 18, and 19) [62].

In injured tendons of diabetic rats there are lower mRNA and protein levels of nerve growth factor (NGF) as well as of its receptor (TrkA) as compared to injured healthy controls. Hence, neuronal mediators as well as neurotrophic factors such as NGF may be potential targets for novel regenerative approaches in tendon disorders, eg. tendinopathy.

Taken together these above mentioned studies strongly support the idea that nerve derived factors have a powerful metabolic control on the structure, function and healing capacity of tendon tissue.

Neuronal Effects of Tendon Loading

Physical activity has been demonstrated to accelerate the neuronal plasticity in tendon repair [63]. It has moreover been demonstrated that exercise leads to increased levels of various neuromediators and their receptors, including SP and CGRP, which may be involved in regulating the healing response [21, 64, 65].

Maybe the most fascinating example of mechano-neuro transduction is the upregulation by exercise of neuromediator receptors (Fig. 4.13). Thus, mRNA-levels of the SP- and CGRP-receptors in mobilized tendons are significantly increased compared to immobilized controls at 17 days post tendon injury [21]. It may prove that enhanced tendon repair after early loading is related to an increased peripheral sensitivity to nerve factor stimulation, as a result of a local up-regulation of neuronal receptors.

Conclusion

Aggregated knowledge acquired during the last decade demonstrates that neuronal regulation plays an essential role in tendon metabolism, repair, and also pathology. This conception supports the idea that neuronal mediators

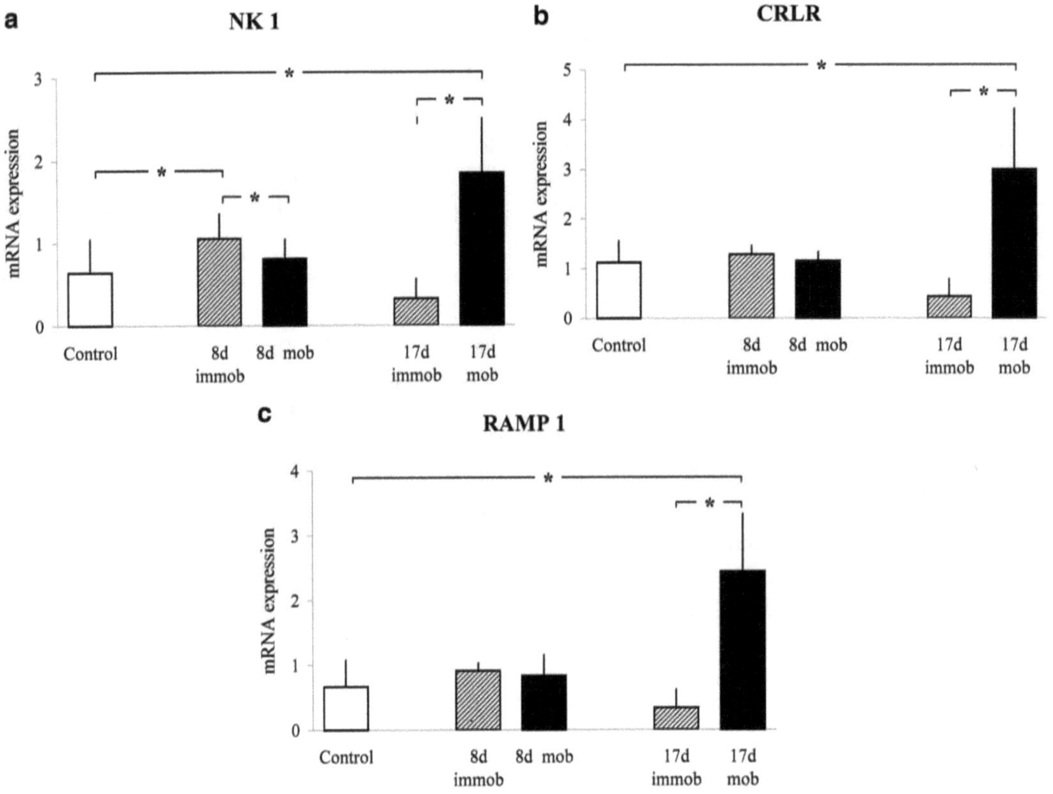

Fig. 4.13 (**a–b**) Normalized expression of mRNA for the SP- (NK1) (**a**) and CGRP- (CRLR (**b**) and RAMP-1 (**c**) receptors in the healing area, in the Achilles tendon of rats subjected to two different levels of physical activity (freely mobilized versus plaster immobilized) at 8 and 17 days postrupture (mean + SD). * = $p < 0.05$; n.s. = $p > 0.05$. Between 8 and 17 days, there was an immense increase of the receptor expression in the mobile healing group, while the expression in the immobilized healing group fell back to levels comparable to the intact tendon control group (Reproduced with permission from Bring et al. [5])

effectuate crucial, but as yet incompletely defined roles in mechanically active connective tissues such as tendons.

Dysregulation of various neuromediators and their receptors as has been observed experimentally and clinically in eg. diabetes and tendinopathy, leads to a loss of normal tendon homeostasis resulting in chronic pain and gradual degeneration.

These novel findings of neuronal plasticity modulating tendon homeostasis and capable of responding to dysregulation in pathology should stimulate the development of targeted pharmacological and tissue engineering approaches to improve healing and treat painful tendon disorders.

Funding Source Studies from the author's laboratories were supported by the regional agreement on medical training and clinical research (ALF) between Stockholm County Council and Karolinska Institutet (project nr. SLL 20110177), and the Swedish National Centre for Research in Sports, as well as the Swedish Medical Research Council (2012–3510) (PWA).

References

1. Magnusson SP, Langberg H, Kjaer M (2010) The pathogenesis of tendinopathy: balancing the response to loading. Nat Rev Rheumatol 6(5):262–268
2. Reeves ND, Maganaris CN, Narici MV (2003) Effect of strength training on human patella tendon mechanical properties of older individuals. J Physiol 548(Pt 3):971–981

3. Schizas N, Andersson T, Fahlgren A, Aspenberg P, Ahmed M, Ackermann P (2010) Compression therapy promotes proliferative repair during rat Achilles tendon immobilization. J Orthop Res; Jan 7. [Epub ahead of print]
4. Bring D, Reno C, Renstrom P, Salo P, Hart D, Ackermann P (2009) Prolonged immobilization compromises up-regulation of repair genes after tendon rupture in a rat model. Scand J Med Sci Sports 20 (3):411–417
5. Bring DKI, Reno C, Renstrom P, Salo P, Hart DA, Ackermann PW (2009) Joint immobilization reduces the expression of sensory neuropeptide receptors and impairs healing after tendon rupture in a rat model. J Orthop Res 27(2):274–280
6. Ackermann PW (2014) Healing and repair mechanism. In: Karlsson J, Calder J, van Diek N (eds) Achilles tendon disorders. Current conceps, 2nd edn. DJO Publications, pp 17–26
7. Ackermann PW, Salo PT, Hart DA (2009) Neuronal pathways in tendon healing. Front Biosci 14:5165–5187
8. Ackermann PW (2001) Peptidergic innervation of periarticular tissue
9. Stilwell DL Jr (1957) The innervation of tendons and aponeuroses. Am J Anat 100(3):289–317
10. Ackermann PW (2014) Tendinopathies in sports: from basic research to the field. In: Doral MN, Karlsson J (ed) Sports injuries. Springer, Berlin/Heidelberg, pp 1–15
11. Hogervorst T, Brand RA (1998) Mechanoreceptors in joint function. J Bone Joint Surg Am 80 (9):1365–1378
12. Strasmann T, Weihe E, Halata Z (1990) CGRP-like immunoreactivity in sensory nerve endings of the Golgi tendon organ. A light- and electron-microscopic study in the grey short-tailed opossum (Monodelphis domestica). Acta Anat 137(3):278–281
13. Jozsa L, Kannus P (1997) Human tendons. Anatomy, physiology, and pathology. Human Kinetics, Champaign
14. Ackermann PW, Franklin SL, Dean BJ, Carr AJ, Salo PT, Hart DA (2014) Neuronal pathways in tendon healing and tendinopathy–update. Front Biosci 19:1251–1278
15. Hokfelt T, Johansson O, Ljungdahl A, Lundberg JM, Schultzberg M (1980) Peptidergic neurones. Nature 284(5756):515–521
16. Audet M, Bouvier M (2012) Restructuring G-protein-coupled receptor activation. Cell 151(1):14–23
17. Tracey KJ (2002) The inflammatory reflex. Nature 420:853–859
18. Brain SD, Williams TJ, Tippins JR, Morris HR, MacIntyre I (1985) Calcitonin gene-related peptide is a potent vasodilator. Nature 313(5997):54–56
19. Brain SD, Williams TJ (1985) Inflammatory oedema induced by synergism between calcitonin gene-related peptide (CGRP) and mediators of increased vascular permeability. Br J Pharmacol 86(4):855–860
20. Maggi CA (1995) Tachykinins and calcitonin gene-related peptide (CGRP) as co-transmitters released from peripheral endings of sensory nerves. Prog Neurobiol 45(1):1–98
21. Bring D, Reno C, Renstrom P, Salo P, Hart D, Ackermann P (2010) Prolonged immobilization compromises up-regulation of repair genes after tendon rupture in a rat model. Scand J Med Sci Sports 20 (3):411–417
22. Ackermann PW, Li J, Finn A, Ahmed M, Kreicbergs A (2001) Autonomic innervation of tendons, ligaments and joint capsules. A morphologic and quantitative study in the rat. J Orthop Res 19(3):372–378
23. Danielson P, Alfredson H, Forsgren S (2006) Immunohistochemical and histochemical findings favoring the occurrence of autocrine/paracrine as well as nerve-related cholinergic effects in chronic painful patellar tendon tendinosis. Microsc Res Tech 69(10):808–819
24. Danielson P, Alfredson H, Forsgren S (2007) In situ hybridization studies confirming recent findings of the existence of a local nonneuronal catecholamine production in human patellar tendinosis. Microsc Res Tech 70(10):908–911
25. Danielson P, Alfredson H, Forsgren S (2007) Studies on the importance of sympathetic innervation, adrenergic receptors, and a possible local catecholamine production in the development of patellar tendinopathy (tendinosis) in man. Microsc Res Tech 70(4):310–324
26. Ljung BO, Forsgren S, Friden J (1999) Sympathetic and sensory innervations are heterogeneously distributed in relation to the blood vessels at the extensor carpi radialis brevis muscle origin of man. Cells Tissues Organs 165(1):45–54
27. Wall ME, Faber JE, Yang X, Tsuzaki M, Banes AJ (2004) Norepinephrine-induced calcium signaling and expression of adrenoceptors in avian tendon cells. Am J Physiol 287(4):C912–C918
28. Wozniak KM, Rojas C, Wu Y, Slusher BS (2012) The role of glutamate signaling in pain processes and its regulation by GCP II inhibition. Curr Med Chem 19(9):1323–1334
29. Schizas N, Weiss R, Lian O, Frihagen F, Bahr R, Ackermann PW (2012) Glutamate receptors in tendinopathic patients. J Orthop Res 30(9):1447–1452
30. Scott A, Alfredson H, Forsgren S (2007) VGluT2 expression in painful Achilles and patellar tendinosis: Evidence of local glutamate release by tenocytes. J Orthop Res
31. Schizas N, Lian O, Frihagen F, Engebretsen L, Bahr R, Ackermann PW (2010) Coexistence of up-regulated NMDA receptor 1 and glutamate on nerves, vessels and transformed tenocytes in tendinopathy. Scand J Med Sci Sports 20(2):208–215
32. Ackermann PW, Ahmed M, Kreicbergs A (2002) Early nerve regeneration after achilles tendon rupture–a prerequisite for healing? A study in the rat. J Orthop Res 20(4):849–856

33. Ackermann PW, Li J, Lundeberg T, Kreicbergs A (2003) Neuronal plasticity in relation to nociception and healing of rat achilles tendon. J Orthop Res 21 (3):432–441
34. Hukkanen M, Konttinen YT, Santavirta S, Paavolainen P, Gu XH, Terenghi G et al (1993) Rapid proliferation of calcitonin gene-related peptide-immunoreactive nerves during healing of rat tibial fracture suggests neural involvement in bone growth and remodelling. Neuroscience 54(4):969–979
35. Kishimoto S (1984) The regeneration of substance P-containing nerve fibers in the process of burn wound healing in the guinea pig skin. J Investig Dermatol 83(3):219–223
36. Li J, Ahmad T, Spetea M, Ahmed M, Kreicbergs A (2001) Bone reinnervation after fracture: a study in the rat. J Bone Miner Res 16(8):1505–1510
37. Martin P (1997) Wound healing–aiming for perfect skin regeneration. Science (New York) 276(5309):75–81
38. Salo PT, Beye JA, Seerattan RA, Leonard CA, Ivie TJ, Bray RC (2008) Plasticity of peptidergic innervation in healing rabbit medial collateral ligament. Can J Surg 51(3):167–172
39. Aubdool AA, Brain SD (2011) Neurovascular aspects of skin neurogenic inflammation. The journal of investigative dermatology Symposium proceedings/ The Society for Investigative Dermatology, Inc [and] European Society for Dermatological Research 15(1):33–39
40. Molloy TJ, Wang Y, Horner A, Skerry TM, Murrell GA (2006) Microarray analysis of healing rat Achilles tendon: evidence for glutamate signaling mechanisms and embryonic gene expression in healing tendon tissue. J Orthop Res 24(4):842–855
41. Greve K, Domeij-Arverud E, Labruto F, Edman G, Bring D, Nilsson G et al (2012) Metabolic activity in early tendon repair can be enhanced by intermittent pneumatic compression. Scand J Med Sci Sports 22 (4):e55–e63
42. Ackermann PW, Domeij-Arverud E, Leclerc P, Amoudrouz P, Nader GA (2012) Anti-inflammatory cytokine profile in early human tendon repair. Knee Surg Sports Traumatol Arthrosc
43. Nilsson J, von Euler AM, Dalsgaard CJ (1985) Stimulation of connective tissue cell growth by substance P and substance K. Nature 315(6014):61–63
44. Hong HS, Lee J, Lee E, Kwon YS, Lee E, Ahn W et al (2009) A new role of substance P as an injury-inducible messenger for mobilization of CD29(+) stromal-like cells. Nat Med 15(4):425–435
45. Haegerstrand A, Dalsgaard CJ, Jonzon B, Larsson O, Nilsson J (1990) Calcitonin gene-related peptide stimulates proliferation of human endothelial cells. Proc Natl Acad Sci U S A 87(9):3299–3303
46. Ziche M, Morbidelli L, Pacini M, Geppetti P, Alessandri G, Maggi CA (1990) Substance P stimulates neovascularization in vivo and proliferation of cultured endothelial cells. Microvasc Res 40(2):264–278
47. Lian O, Dahl J, Ackermann PW, Frihagen F, Engebretsen L, Bahr R (2006) Pronociceptive and antinociceptive neuromediators in patellar tendinopathy. Am J Sports Med 34(11):1801–1808
48. Schubert TE, Weidler C, Lerch K, Hofstadter F, Straub RH (2005) Achilles tendinosis is associated with sprouting of substance P positive nerve fibres. Ann Rheum Dis 64(7):1083–1086
49. Sanchis-Alfonso V, Rosello-Sastre E, Subias-Lopez A (2001) Neuroanatomic basis for pain in patellar tendinosis ("jumper's knee"): a neuroimmunohistochemical study. Am J Knee Surg 14(3):174–177
50. Andersson G, Backman LJ, Scott A, Lorentzon R, Forsgren S, Danielson P (2011) Substance P accelerates hypercellularity and angiogenesis in tendon tissue and enhances paratendinitis in response to Achilles tendon overuse in a tendinopathy model. Br J Sports Med 45(13):1017–1022
51. Backman LJ, Eriksson DE, Danielson P (2014) Substance P reduces TNF-alpha-induced apoptosis in human tenocytes through NK-1 receptor stimulation. Br J Sports Med 48(19):1414–1420
52. Kager I, Mousa SA, Sieper J, Stein C, Pipam W, Likar R (2011) Blockade of intra-articular adrenergic receptors increases analgesic demands for pain relief after knee surgery. Rheumatol Int 31(10):1299–1306
53. Klatt S, Fassold A, Straub RH (2012) Sympathetic nerve fiber repulsion: testing norepinephrine, dopamine, and 17beta-estradiol in a primary murine sympathetic neurite outgrowth assay. Ann N Y Acad Sci 1261:26–33
54. Alfredson H, Forsgren S, Thorsen K, Lorentzon R (2001) In vivo microdialysis and immunohistochemical analyses of tendon tissue demonstrated high amounts of free glutamate and glutamate NMDAR1 receptors, but no signs of inflammation, in Jumper's knee. J Orthop Res 19(5):881–886
55. Madden DR (2002) The structure and function of glutamate receptor ion channels. Nat Rev Neurosci 3 (2):91–101
56. Bring DK, Paulson K, Renstrom P, Salo P, Hart DA, Ackermann PW (2012) Residual substance P levels after capsaicin treatment correlate with tendon repair. Wound repair and regeneration : official publication of the Wound Healing Society [and] the European Tissue Repair. Society 20(1):50–60
57. Ivie TJ, Bray RC, Salo PT (2002) Denervation impairs healing of the rabbit medial collateral ligament. J Orthop Res 20(5):990–995
58. Grorud KW, Jensen KT, Provenzano PP, Vanderby R Jr (2007) Adjuvant neuropeptides can improve neuropathic ligament healing in a rat model. J Orthop Res 25(6):703–712
59. Dwyer KW, Provenzano PP, Muir P, Valhmu WB, Vanderby R Jr (2004) Blockade of the sympathetic nervous system degrades ligament in a rat MCL model. J Appl Physiol 96(2):711–718

60. Ramchurn N, Mashamba C, Leitch E, Arutchelvam V, Narayanan K, Weaver J et al (2009) Upper limb musculoskeletal abnormalities and poor metabolic control in diabetes. Eur J Int Med 20(7):718–721
61. Pradhan L, Nabzdyk C, Andersen ND, LoGerfo FW, Veves A (2009) Inflammation and neuropeptides: the connection in diabetic wound healing. Exp Rev Mol Med 11, e2
62. Ahmed AS, Schizas N, Li J, Ahmed M, Ostenson CG, Salo P et al (2012) Type 2 diabetes impairs tendon repair after injury in a rat model. J Appl Physiol 113(11):1784–1791
63. Bring DK, Kreicbergs A, Renstrom PA, Ackermann PW (2007) Physical activity modulates nerve plasticity and stimulates repair after Achilles tendon rupture. J Orthop Res 25(2):164–172
64. Ytteborg E, Torgersen JS, Pedersen ME, Helland SJ, Grisdale-Helland B, Takle H (2013) Exercise induced mechano-sensing and Substance P mediated bone modeling in Atlantic salmon. Bone 53(1):259–268
65. Jonsdottir IH (2000) Special feature for the Olympics: effects of exercise on the immune system: neuropeptides and their interaction with exercise and immune function. Immunol Cell Biol 78(5):562–570
66. Ackermann PW, Finn A, Ahmed M (1999) Sensory neuropeptidergic pattern in tendon, ligament and joint capsule. A study in the rat. Neuroreport 10(10):2055–2060
67. Ackermann PW, Spetea M, Nylander I, Ploj K, Ahmed M, Kreicbergs A (2001) An opioid system in connective tissue: a study of achilles tendon in the rat. J Histochem Cytochem 49(11):1387–1395

Tendon Stem Cells: Mechanobiology and Development of Tendinopathy

5

James H-C. Wang and Issei Komatsu

Abstract

Millions of people suffer from tendon injuries in both occupational and athletic settings. However, the restoration of normal structure and function to injured tendons still remains as one of the greatest challenges in orthopaedics and sports medicine. In recent years, a remarkable advancement in tendon research field has been the discovery of tendon stem/progenitor cells (TSCs). Unlike tenocytes, the predominant resident cell in tendons, TSCs have the ability to self-renew and multi-differentiate. Because of these distinct properties, TSCs may play a critical role in tendon physiology as well as pathology such as tendinopathy, which is a prevalent chronic tendon injury. Additionally, because TSCs are tendon-specific stem cells, they could potentially be used in tendon tissue engineering *in vitro*, and serve as a promising cell source for cell-based therapy to effectively repair or even regenerate injured tendons in clinical settings.

Keywords

Tendon stem cells • Self-renewal • Multi-differentiation • Mechanobiology

Abbreviations

ATSC	Achilles tendon stem cell
Col. I	Collagen type I
Col. II	Collagen type II
FEM	Finite element method
ITR	Intensive treadmill running
MSCs	Mesenchymal stem cells
MTR	Moderate treadmill running
NS	Nucleostemin
Scx	Scleraxis
Tenom	Tenomodulin
TSCs	Tendon stem/progenitor cells

J.H.-C. Wang (✉) • I. Komatsu
MechanoBiology Laboratory, Department of Orthopaedic Surgery, University of Pittsburgh School of Medicine, 210 Lothrop Street, BST, E1640, Pittsburgh, PA 15213, USA
e-mail: wanghc@pitt.edu

Introduction

Tendons are bands of connective tissues that are particularly rich in collagens. The most abundant tendon component is type I collagen, which

P.W. Ackermann and D.A. Hart (eds.), *Metabolic Influences on Risk for Tendon Disorders*, Advances in Experimental Medicine and Biology 920, DOI 10.1007/978-3-319-33943-6_5

constitutes about 70–80 % of the dry tendon mass and about 95 % of the total collagen in tendons [1–3]. The remaining 5 % consists of type III and V collagens. Within the tendon matrix, collagens are cross-linked [4, 5], which increases the tendon's mechanical strength [6]. Besides collagens, tendons also contain proteoglycans in small quantities, but their amounts vary in tendons in different anatomical locations and also depend on the mechanical loading conditions (e.g., tension vs. compression) on a tendon [7, 8].

Mechanically, tendons are responsible for transmitting muscular forces to bone, and because they are live tissues, tendons also respond to mechanical loads by changing their metabolism as well as their structural and mechanical properties. While appropriate mechanical loading benefits tendons, chronic mechanical loading placed on tendons plays a major role in the development of tendinopathy [9, 10]. This major tendon disorder is often manifested by degenerative changes in the tendon, including lipid deposition, proteoglycan accumulation, and calcification either alone or in combination [11]. Despite years of extensive research, restoration of tendon structure and function after injury still remains as one of the greatest challenges in orthopaedic surgery and sports medicine.

In pursuit of new strategies to promote healing of tendon injuries, significant progress has been made in tendon research in recent years. We and others have identified a new population of tendon cells termed tendon stem/progenitor cells (TSCs) in humans, mice and rabbits [12–15]. Much like other tendon cells, TSCs also respond to the various mechanical loads placed on tendons. This chapter will provide an overview of the recent advancements in TSC mechanobiology. In particular, the role of TSCs in the mechanical loading-induced development of tendinopathy is highlighted. Future research developments in the field are also suggested and discussed.

Discovery of TSCs

Until recently, tenocytes were considered to be the only cell type in tendons. However, about 5 % of tendon cells were recently discovered to be TSCs in the tendons of humans, mice, rats and rabbits [12–15]. TSCs can self-renew and differentiate into tenocytes to maintain tendon homeostasis. Like other adult stem cells, TSCs also have multi-differentiation potential, which allows them to differentiate not only into tenocytes, but also into non-tenocytes including chondrocytes, osteocytes and adipocytes, but only under certain pathophysiological conditions [12, 13].

Unlike TSCs, tenocytes are specialized tendon cells that lack the capacity to differentiate into other cell types. In culture, TSCs proliferate more quickly than their counterparts (tenocytes) and appear as small, cobblestone-shaped cells with large nuclei, while tenocytes grow as highly

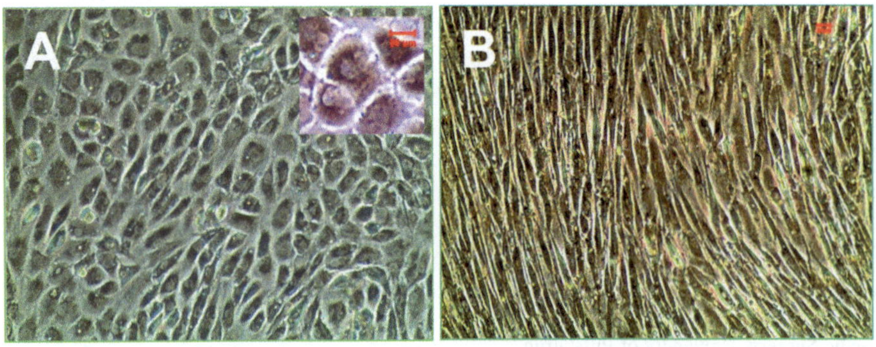

Fig. 5.1 Tendon stem/progenitor cells (TSCs) and tenocytes in culture exhibit a striking difference in morphology. (A) TSCs. The cells are cobblestone-like in culture when grown to confluence. They also have large nuclei (inset). (B) Tenocytes in culture. These cells are fibroblast-like and assume a highly elongated shape in confluent conditions

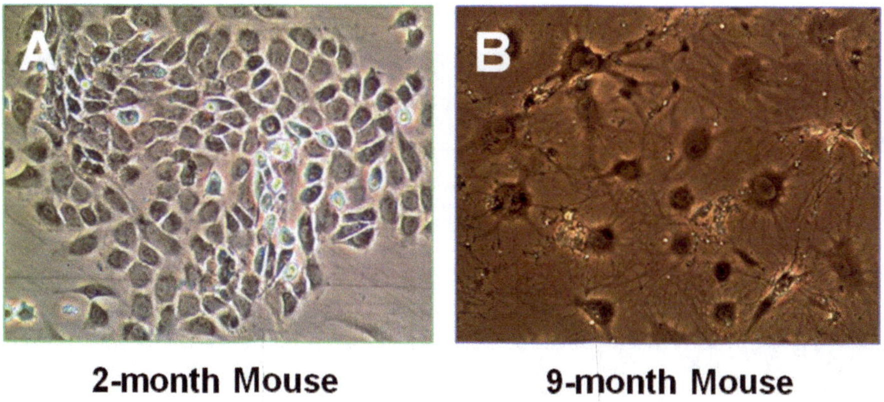

Fig. 5.2 Age affects TSC morphology and growth in culture. (**A**) Shown is a colony of young TSCs derived from 2-months old mouse. The cells are large in numbers and cobble-stone shaped in culture. (**B**) Aging TSCs from 9-months old mouse are lower in numbers, and are flat and spread-out in culture. Their growth is also slower in culture

elongated, fibroblast-like cells with smaller nuclei [13] (Fig. 5.1). Tenocytes also lack a characteristic feature of TSCs, which is the expression of stem cell markers such as Oct-4, SSEA-1/4 and nucleostemin (NS).

The typical features of TSCs in tendons, however, change during aging. In a recent study, we determined the effects of aging on TSCs *in vitro*. TSCs derived from aging mice (9 and 24 months) proliferated significantly slower than TSCs obtained from young mice (2.5 and 5 months) (Fig. 5.2). A likely reason for this was revealed during the analysis of cellular protein expression pattern, i.e., expression of the stem cell markers Oct-4, NS, Sca-1 and SSEA-1 in TSCs decreased in an age-dependent manner, which in part, may contribute to the lowered proliferation rate of TSCs in aging mice. These findings indicate that aging impairs the proliferative ability of TSCs and reduces their stemness [16] (Fig. 5.3).

A Mechanical Loading System to Study TSC Mechanobiology *In Vitro*

To study the mechanobiology of TSCs under well controlled conditions, we developed a novel *in vitro* model system (Fig. 5.4). This *in vitro* system enabled us to examine mechanobiological responses (e.g. proliferation, self-renewal, and differentiation) of TSCs under well controlled, *in vivo*-like multiple mechanical loading conditions [17], which are not possible in an animal model. Therefore, such a system is essential to define the loading-induced, TSC-based mechanisms of degenerative tendinopathy.

Use of Mouse Treadmill Running to Study of Mechanobiology of Tendon *In Vivo*

To study the mechanobiology of tendons *in vivo*, we developed a mouse treadmill running model (Fig. 5.5) and optimized two different running regimens: (1) moderate treadmill running (MTR); and (2) intensive treadmill running (ITR). In the MTR regimen, mice were trained to run at the speed of 13 m/min, 15 min/day and 5 days/week. Then, they ran at the same speed for 50 min/day, 5 days/week for 3 weeks. In the ITR regimen, mice were trained as in MTR in the first week. They then ran at the same speed for 3 h/day, 4 h/day, and 5 h/day for 5 days in the second, third, and fourth weeks, respectively. The mice in the control group did not run on treadmills, but freely moved in cages during treadmill running experiments.

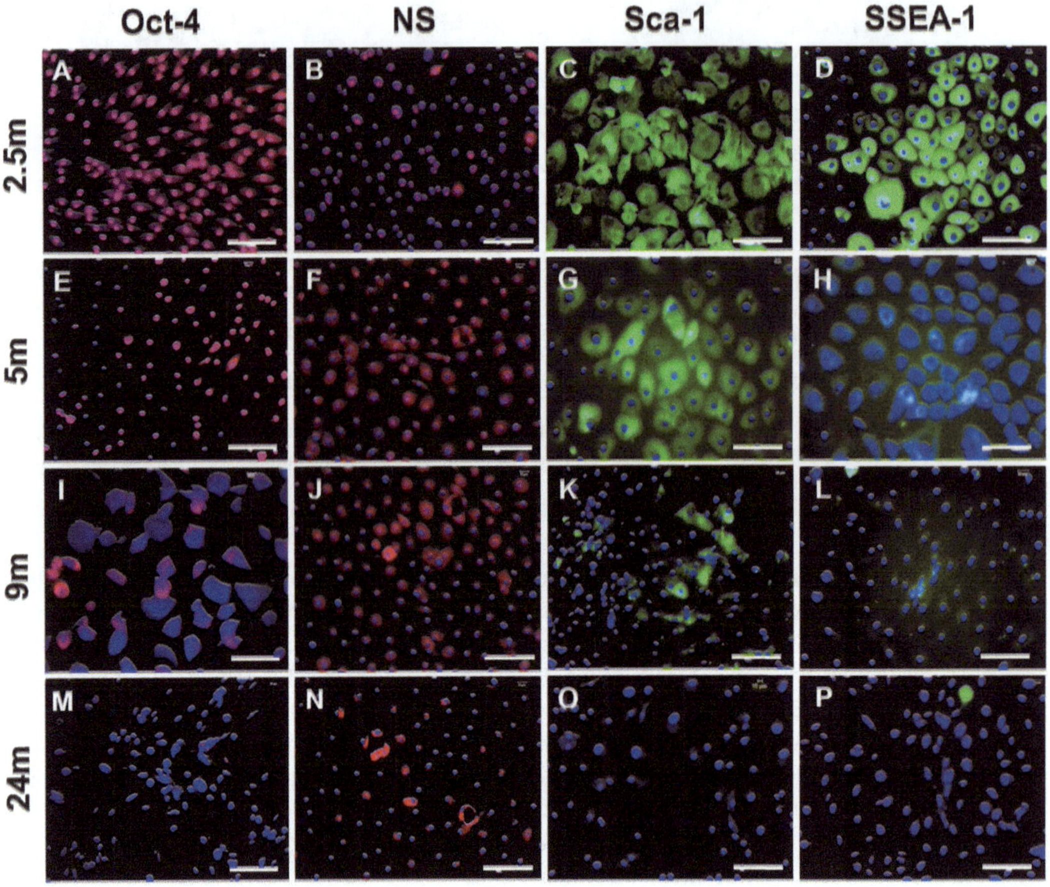

Fig. 5.3 Age reduces TSC stemness. Expression of the stem cell markers, Oct-4, NS, Sca-1, and SSEA-1 in TSCs from 2.5 months (2.5 m), 5 months (5 m), 9 months (9 m) and 24 months (24 m) old mice are shown. Increase in the mouse age decreased the number of TSCs expressing the stem cell markers, which are abundant in TSCs from 2.5 months old mice but scant in TSCs from 24 months old mice. Bar −100 μm

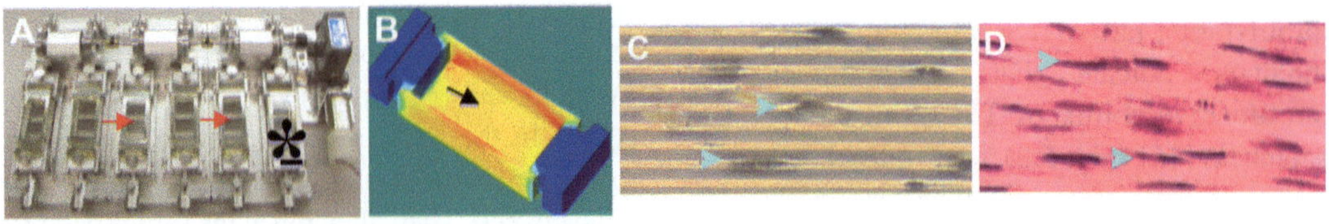

Fig. 5.4 A novel *in vitro* model system to study mechanobiology of TSCs. (**A**) The stretching apparatus. Silicone dishes (*red arrows*) are used to grow and stretch tendon cells. (**B**) Strain distribution in the silicone dish by FEM (finite element method). The surface strains are uniform in the central region of the culture surface (*black arrow*), where cells are plated and stretched. (**C**) Tendon cells (TSCs, tenocytes, or both) on microgrooved culture surfaces (*arrows* point to two tendon cells residing on microgrooves with a width of 10 μm). (**D**) Tendon cells in the tendon section (arrows point to tendon cells). As seen in **C** and **D**, tendon cell density, cell shape and organization can be controlled in this model system to mimic *in vivo* conditions in the tendon

The Role of TSCs in the Development of Tendinopathy

Tendinopathy is a serious healthcare problem for people in both occupational and athletic settings [18–20]. Many occupational activities result in tendinopathy and present with clinical symptoms including pain and signs of inflammation such as redness and swelling; all of these may impair one's ability to work [19]. When tendoninopathy further progresses, one or more of the following degenerative changes are often noticed: lipid deposition, proteoglycan accumulation and calcification [11] (Fig. 5.6). Often times, the etiology for this tendon disorder is recognized as the chronic mechanical loading on tendons [18, 21–28]. However, the precise cellular mechanisms leading to the pathology of tendinopathy remain unclear. As a result, tendinopathy continues to be a nemesis for both physicians and patients alike [10]. Currently, there is no protocol in place that can effectively prevent tendinopathy or restore tendon structure and function after injury [22, 29].

In our quest to address this lacuna, we focused our research on TSCs that play a major role in the maintenance of tendon homeostasis [12, 13, 15]. To explore the possibility that TSCs may also play a critical role in the development of degenerative tendinopathy, we used both *in vitro* and *in vivo* model systems described above to examine the mechanobiological responses of TSCs to various mechanical loading conditions.

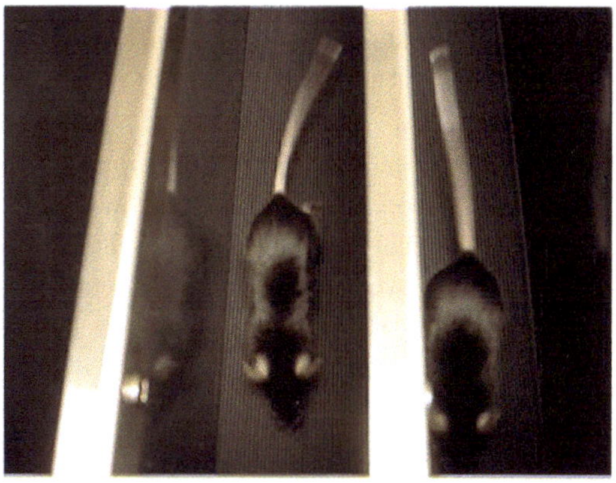

Fig. 5.5 Mouse treadmill running is used to study TSC mechanobiology. Mice run on the treadmill with 6 running lanes, according to a specified running regimen; moderate treadmill running (MTR) or intensive treadmill running (ITR)

Mechanobiological Response of TSCs *In Vitro*

Applying small mechanical loading on TSCs using our *in vitro* model system induces tenocyte differentiation, while large mechanical loading induces both tenocyte and non-tenocyte differentiation [30] (Fig. 5.7). Specifically, small mechanical stretching (4 %) significantly increased only the expression of the tenocyte related gene, collagen type I (Col. I), but not the expression of non-tenocyte related genes, PPARγ, collagen type II (Col. II), Sox-9 and Runx-2, which are markers of adipocytes, chondrocytes and osteocytes, respectively. However, large mechanical stretching of 8 %

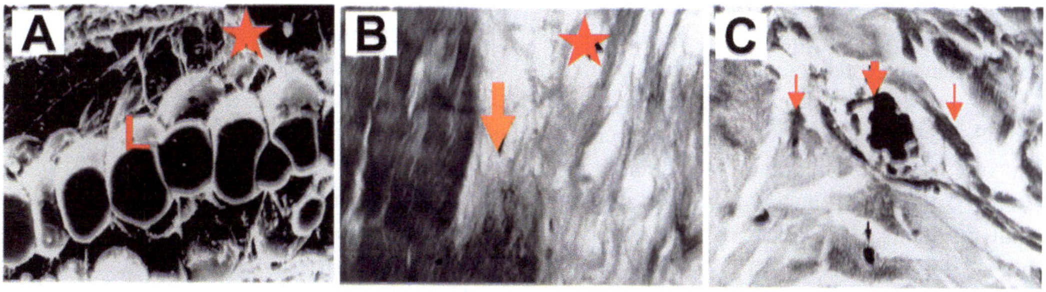

Fig. 5.6 Degenerative tendinopathy in human tendons. Three forms of degenerative changes were observed either alone or in combination in affected tendons: Lipid formation (**A**, L and star), accumulation of proteoglycans (**B**, *arrow* and *star*) and calcified tissue (**C**, *arrows*) [11]

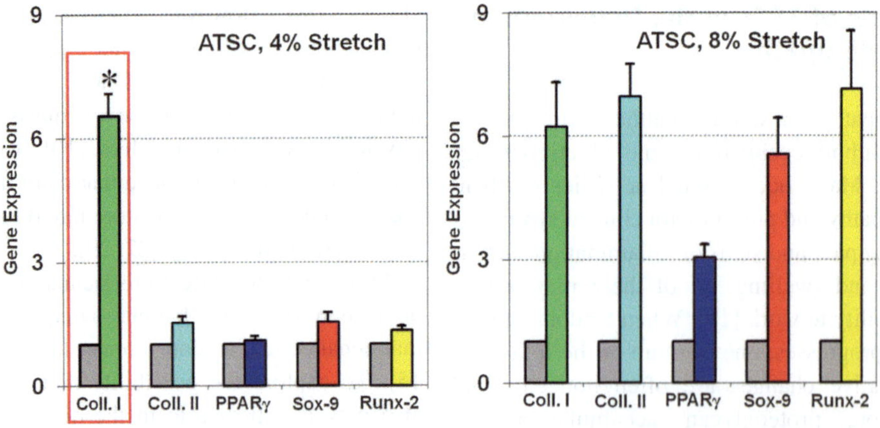

Fig. 5.7 Gene expression profile in TSCs derived from mouse Achilles tendons depends on mechanical loading conditions *in vitro*. TSCs stretched to 4 % (**a**) showed specific up-regulation of collagen type I (Coll. I), which is a tenocyte related gene, but not the non-tenocyte related genes, collagen type II (Coll. II), Sox-9, PPARγ or Runx-2. However, 8 % stretching up-regulated both tenocyte and non-tenocyte genes to varying degrees. *Gray bars* are controls without loading

significantly increased the expression of both tenocyte and non-tenocyte related genes.

Moreover, by performing *in vitro* stretching experiments on isolated tenocytes [30], we further showed that tenocytes do not express non-tenocyte related genes, even under overloading (8 % stretching) conditions [31]. Thus, these findings indicate that TSCs are likely the primary cells responsible for degenerative tendinopathy by undergoing aberrant non-tenocyte differentiation into fatty-, cartilage-like and bone-like tissues in tendons that compromise tendon structure.

In addition to studying the effects of various mechanical loading conditions on TSCs, several studies have investigated the molecular signaling pathways involved in the mechanobiological responses of TSCs. It was reported that Wnt5a/Wnt5b/JNK signaling pathway is involved in the osteogenic differentiation of rat TSCs in response to mechanical loads [32–34]. Particularly, JNK activity was decreased and osteogenic differentiation was inhibited in Wnt5a/Wnt5b/JNK knockdown TSCs after uniaxial mechanical loading. This inhibition was rescued by the JNK activator, anisomycin, or JNK1-cDNA, suggesting that Wnt5a/Wnt5b is located upstream of JNK and regulates the function of JNK in the Wnt5a/Wnt5b/JNK signaling pathway [32]. Wnt3 also promoted osteogenic differentiation of TSCs *in vitro* [35]. Moreover, increased amounts of Wnt3a were observed in human patellar tendons with tendinopathy and animal models of tendinopathy [36]. In another study, mechanical loading of TSCs increased Wnt5a protein levels, but not other Wnt proteins [37]. The discrepancy between these two studies may be due to differences in the mechanical loading variables for TSCs including different amounts of mechanical loads, frequency and duration (see Chap. 7).

To date, only a few studies have investigated the mechanotransduction mechanisms of TSCs. Since TSCs are adult stem/progenitor cells, it is conceivable that the mechano-signaling pathways that have been established in mesenchymal stem cells (MSCs) and tenocytes may also be applied to TSCs. It has been demonstrated that focal adhesion kinase, cytoskeleton and RhoA/ROCK play important roles in mechanical loading-induced tenogenic differentiation of MSCs [38, 39]. Mechanical forces are also known to regulate Scleraxis (Scx) expression through the activation of SMAD2/3-mediated TGF-β signaling in adult mouse tendons [40]. Stretching of tendon constructs *in vitro* also increased ERK/MAPK phosphorylation [41]. Moreover, EGR1 is a known

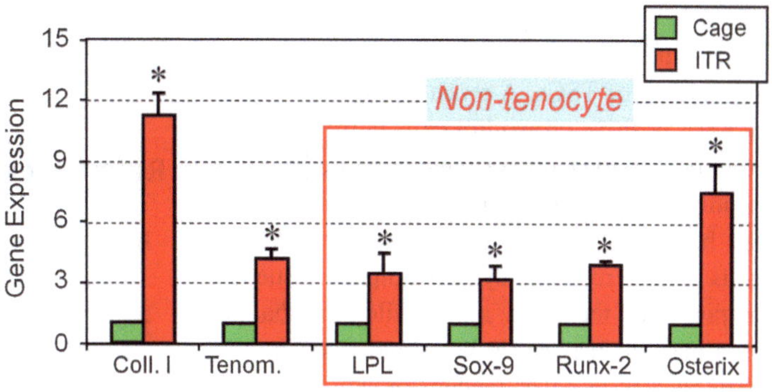

Fig. 5.8 Intensive treadmill running (ITR) up-regulates both tenocyte-related genes, collagen type I (Coll. I) and tenomodulin (Tenom) and non-tenocyte related genes, LPL, Sox-9, Runx-2 and Osterix in mouse Achilles tendons

mechanosensitive transcription factor involved in the tendon cell response [42]. EGR1 is expressed in response to mechanical stimuli, mainly via the MAPK pathways. However, further investigations regarding its role in the mechanotransduction of TSCs are needed.

The Mechanobioological Responses of TSCs *In Vivo*

The effects of mechanical loading on TSCs *in vivo* were also determined. Our results revealed that a moderate treadmill running regimen (MTR, see above for definition) nearly doubled the proliferation rate of TSCs when compared to control mice in cages. Moreover, cellular production of total collagen in the treadmill running group increased by 70 and 200 % in patellar and Achilles tendon cells, respectively, compared to cells from the control group [17].

A continuation of our study to determine the effects of intensive treadmill running (ITR, see above) on mouse tendons [31] revealed that after ITR, all genes related to fatty, cartilage and bony tissues (LPL, Sox-9, Runx-2 and Osterix), and tendon-related genes (Coll. I, and tenomodulin or Tenom) were highly expressed in the mouse patellar tendons (Fig. 5.8). This is consistent with the findings that excessive mechanical loading via treadmill running results in elevated expression of cartilage-related genes (Col. II, aggrecan and Sox-9) in the rat supraspinatus tendon [43] and that long-term treadmill running (up to 16 weeks) leads to degenerative tendinopathy as evidenced by accumulated proteoglycan and chondrocytes present in the rat tendon [26].

Concluding Remarks

Tendon injuries frequently occur among people in all walks of life. Current treatments are largely ineffective and, as a result, the restoration of normal tendon structure and function continues to be a challenge in the clinical treatment of tendons. Therefore, basic scientific studies on tendons are needed so that new treatment strategies can be devised. The recent discovery of TSCs has offered an opportunity for tendon researchers to look into the stem cell mechanisms responsible for chronic tendon injury or tendinopathy. Studies have, indeed, suggested that TSCs may be responsible for the development of degenerative tendinopathy due to excessive mechanical loading constantly placed on tendons. Therefore, effective treatment options for such conditions should not only be anti-inflammatory, but should also include methods to deter degeneration of tendons. This could likely be achieved by blocking the aberrant differentiation of TSCs, even under excessive mechanical loading conditions.

There are several research directions to explore in TSC mechanobiology. First, the *in vivo* characterization of TSCs should be pursued. Currently, TSCs are characterized *in vitro*. The *in vivo* identities and niche of TSCs with respect to anatomical locations and regulatory factors in tendons remain to be investigated [35, 44]. Second, the exact origins and identities of "TSCs" *in vivo* after a tendon injury are unknown. This requires lineage specific marking/tracing of TSCs. A recent study showed that α-SMA positive progenitor cells in the paratenon migrate into the tendon wound to participate in its healing [45]. Other potential sources of "TSCs" in the healing of tendon injury may include: stem cells in the circulatory system, adipose tissue near tendons, and the bursa, a fluid-filled sac that acts as a cushion between a bone and other moving parts such as muscles, tendons or skin [46, 47].

Third, the effects of aging on TSC mechanobiology should be thoroughly investigated. Although it remains to be demonstrated definitively, aging is presumed to cause tendon cell senescence, which refers to the cell status in which permanent cell cycle arrest cannot be reactivated [48]. Tendon cell senescence may lead to tendon degeneration and impair tendon healing, which are often seen in aging patients. But the relationship between aging and reduced tendon potential to heal needs to be demonstrated. These detrimental effects caused by aging can be reversed by applying moderate physiological loads on tendons that induces anabolic changes by regulating TSCs [49]. Thus, exercise in the form of mechanical loading can improve aging-associated impairment in healing tendons by reactivating senescent tendon cells, especially TSCs. Therefore, this line of research warrants future study.

Finally, it has been well established that stem cells are excellent sources for the engineering of injured tissues. Indeed, implantation of engineered tendon tissue containing TSCs in de-cellularized fibroblast matrix effectively improved the mechanical properties in a patellar tendon injury model [50] and promoted tendon-like tissue formation in rat patellar tendons [51]. However, the role played by moderate mechanical loading in the healing of TSC-treated injured tendons remains to be investigated; devising optimal exercise regimens for tendon injury patients is of great clinical importance.

Acknowledgements We gratefully acknowledge the funding support received from NIH grants AR061395 and AR065949 (JHW). We also thank Dr. Nirmala Xavier for her assistance in the preparation of this review.

References

1. Evans JH, Barbenel JC (1975) Structural and mechanical properties of tendon related to function. Equine Vet J 7(1):1–8
2. Riley GP, Harrall RL, Constant CR, Chard MD, Cawston TE, Hazleman BL (1994) Tendon degeneration and chronic shoulder pain: changes in the collagen composition of the human rotator cuff tendons in rotator cuff tendinitis. Ann Rheum Dis 53(6):359–366
3. Wang JH, Guo Q, Li B (2012) Tendon biomechanics and mechanobiology-a minireview of basic concepts and recent advancements. J Hand Ther 25(2):133–140
4. Eyre DR, Koob TJ, Van Ness KP (1984) Quantitation of hydroxypyridinium crosslinks in collagen by high-performance liquid chromatography. Anal Biochem 137(2):380–388
5. Bailey AJ, Light ND (1985) Intermolecular cross-linking in fibrotic collagen. Ciba Found Symp 114:80–96
6. Thompson JI, Czernuszka JT (1995) The effect of two types of cross-linking on some mechanical properties of collagen. Biomed Mater Eng 5(1):37–48
7. Riley GP, Harrall RL, Constant CR, Chard MD, Cawston TE, Hazleman BL (1994) Glycosaminoglycans of human rotator cuff tendons: changes with age and in chronic rotator cuff tendinitis. Ann Rheum Dis 53(6):367–376
8. Berenson MC, Blevins FT, Plaas AH, Vogel KG (1996) Proteoglycans of human rotator cuff tendons. J Orthop Res 14(4):518–525
9. Maffulli N, Khan KM, Puddu G (1998) Overuse tendon conditions: time to change a confusing terminology. Arthroscopy 14(8):840–843
10. Khan KM, Maffuli N (1998) Tendinopathy: an Achilles' heel for athletes and clinicians. Clin J Sport Med 8:151–154
11. Kannus P, Jozsa L (1991) Histopathological changes preceding spontaneous rupture of a tendon. A controlled study of 891 patients. J Bone Joint Surg Am 73(10):1507–1525
12. Bi Y, Ehirchiou D, Kilts TM, Inkson CA, Embree MC, Sonoyama W, Li L, Leet AI, Seo BM, Zhang L et al (2007) Identification of tendon stem/progenitor cells and the role of the extracellular matrix in their niche. Nat Med 13(10):1219–1227
13. Zhang J, Wang JH (2010) Characterization of differential properties of rabbit tendon stem cells and tenocytes. BMC Musculoskelet Disord 11:10
14. Zhou Z, Akinbiyi T, Xu L, Ramcharan M, Leong DJ, Ros SJ, Colvin AC, Schaffler MB, Majeska RJ, Flatow EL et al (2010) Tendon-derived stem/progenitor cell aging: defective self-renewal and altered fate. Aging Cell 9(5):911–915

15. Rui YF, Lui PP, Li G, Fu SC, Lee YW, Chan KM (2010) Isolation and characterization of multipotent rat tendon-derived stem cells. Tissue Eng Part A 16(5):1549–1558
16. Zhang J, Wang JH (2015) Moderate exercise mitigates the detrimental effects of aging on tendon stem cells. PLoS One
17. Zhang J, Pan T, Liu Y, Wang JH (2010) Mouse treadmill running enhances tendons by expanding the pool of tendon stem cells (TSCs) and TSC-related cellular production of collagen. J Orthop Res 28(9):1178–1183
18. Almekinders LC (1998) Tendinitis and other chronic tendinopathies. J Am Acad Orthop Surg 6(3):157–164
19. Paavola M, Kannus P, Jarvinen TA, Khan K, Jozsa L, Jarvinen M (2002) Achilles tendinopathy. J Bone Joint Surg Am 84-A(11):2062–2076
20. Maffulli N, Wong J, Almekinders LC (2003) Types and epidemiology of tendinopathy. Clin Sports Med 22(4):675–692
21. Archambault JM, Wiley JP, Bray RC (1995) Exercise loading of tendons and the development of overuse injuries. A review of current literature. Sports Med 20(2):77–89
22. Wang JH (2006) Mechanobiology of tendon. J Biomech 39(9):1563–1582
23. Flick J, Devkota A, Tsuzaki M, Almekinders L, Weinhold P (2006) Cyclic loading alters biomechanical properties and secretion of PGE2 and NO from tendon explants. Clin Biomech (Bristol, Avon) 21(1):99–106
24. Backman C, Boquist L, Friden J, Lorentzon R, Toolanen G (1990) Chronic achilles paratenonitis with tendinosis: an experimental model in the rabbit. J Orthop Res 8(4):541–547
25. Nakama LH, King KB, Abrahamsson S, Rempel DM (2005) Evidence of tendon microtears due to cyclical loading in an in vivo tendinopathy model. J Orthop Res 23:1199–1205
26. Scott A, Cook JL, Hart DA, Walker DC, Duronio V, Khan KM (2007) Tenocyte responses to mechanical loading in vivo: a role for local insulin-like growth factor 1 signaling in early tendinosis in rats. Arthritis Rheum 56(3):871–881
27. Soslowsky LJ, Thomopoulos S, Tun S, Flanagan CL, Keefer CC, Mastaw J, Carpenter JE (1999) Neer Award. Overuse activity injures the supraspinatus tendon in an animal model: a histologic and biomechanical study. J Shoulder Elbow Surg 2000 9 (2):79–84
28. Fung DT, Wang VM, Andarawis-Puri N, Basta-Pljakic J, Li Y, Laudier DM, Sun HB, Jepsen KJ, Schaffler MB, Flatow EL (2010) Early response to tendon fatigue damage accumulation in a novel in vivo model. J Biomech 43(2):274–279
29. Stone D, Green C, Rao U, Aizawa H, Yamaji T, Niyibizi C, Carlin G, Woo SL (1999) Cytokine-induced tendinitis: a preliminary study in rabbits. J Orthop Res 17(2):168–177
30. Zhang J, Wang JH (2010) Mechanobiological response of tendon stem cells: implications of tendon homeostasis and pathogenesis of tendinopathy. J Orthop Res 28(5):639–643
31. Zhang J, Li B, Wang JH (2011) The role of engineered tendon matrix in the stemness of tendon stem cells in vitro and the promotion of tendon-like tissue formation in vivo. Biomaterials 32(29):6972–6981
32. Liu X, Chen W, Zhou Y, Tang K, Zhang J (2015) Mechanical tension promotes the osteogenic differentiation of rat tendon-derived stem cells through the wnt5a/wnt5b/jnk signaling pathway. Cell Physiol Biochem 36(2):517–530
33. Park KH, Kang JW, Lee EM, Kim JS, Rhee YH, Kim M, Jeong SJ, Park YG, Kim SH (2011) Melatonin promotes osteoblastic differentiation through the BMP/ERK/Wnt signaling pathways. J Pineal Res 51 (2):187–194
34. Charoenpanich A, Wall ME, Tucker CJ, Andrews DM, Lalush DS, Dirschl DR, Loboa EG (2014) Cyclic tensile strain enhances osteogenesis and angiogenesis in mesenchymal stem cells from osteoporotic donors. Tissue Eng Part A 20(1–2):67–78
35. Lui PPY, Cheuk YC, Lee YW, Chan KM (2012) Ectopic chondro-ossification and erroneous extracellular matrix deposition in a tendon window injury model. J Orthop Res 30(1):37–46
36. Lui PP, Lee YW, Wong YM, Zhang X, Dai K, Rolf CG (2013) Expression of Wnt pathway mediators in metaplasic tissue in animal model and clinical samples of tendinopathy. Rheumatology 52(9):1609–1618
37. Shi Y, Fu Y, Tong W, Geng Y, Lui PPY, Tang T, Zhang X, Dai K (2012) Uniaxial mechanical tension promoted osteogenic differentiation of rat tendon-derived stem cells (rTDSCs) via the Wnt5a-RhoA pathway. J Cell Biochem 113(10):3133–3142
38. Chen JC, Jacobs CR (2013) Mechanically induced osteogenic lineage commitment of stem cells. Stem Cell Res Therapy 4(5):107–107
39. Xu B, Song G, Ju Y, Li X, Song Y, Watanabe S (2012) RhoA/ROCK, cytoskeletal dynamics, and focal adhesion kinase are required for mechanical stretch-induced tenogenic differentiation of human mesenchymal stem cells. J Cell Physiol 227(6):2722–2729
40. Maeda T, Sakabe T, Sunaga A, Sakai K, Rivera AL, Keene DR, Sasaki T, Stavnezer E, Iannotti J, Schweitzer R et al (2011) Conversion of mechanical force into TGF-β-mediated biochemical signals. CurrBiol 21(11):933–941
41. Paxton JZ, Hagerty P, Andrick JJ, Baar K (2012) Optimizing an intermittent stretch paradigm using erk1/2 phosphorylation results in increased collagen synthesis in engineered ligaments. Tissue Eng A 18 (3–4):277–284
42. Guerquin M-J, Charvet B, Nourissat G, Havis E, Ronsin O, Bonnin M-A, Ruggiu M, Olivera-Martinez-I, Robert N, Lu Y et al (2013) Transcription factor EGR1 directs tendon differentiation and promotes tendon repair. J Clin Invest 123(8):3564–3576

43. Archambault JM, Jelinsky SA, Lake SP, Hill AA, Glaser DL, Soslowsky LJ (2007) Rat supraspinatus tendon expresses cartilage markers with overuse. J Orthop Res 25(5):617–624
44. Lui PP, Kong SK, Lau PM, Wong YM, Lee YW, Tan CL, Wong OT (2014) Allogeneic tendon-derived stem cells (tdscs) promote tendon healing and suppress immunoreactions in hosts – in vivo model. Tissue Eng Part A
45. Dyment NA, Hagiwara Y, Matthews BG, Li Y, Kalajzic I, Rowe DW (2014) Lineage tracing of resident tendon progenitor cells during growth and natural healing. PLoS ONE 9(4), e96113
46. Utsunomiya H, Uchida S, Sekiya I, Sakai A, Moridera K, Nakamura T (2013) Isolation and characterization of human mesenchymal stem cells derived from shoulder tissues involved in rotator cuff tears. Am J Sports Med 41(3):657–668
47. Song N, Armstrong AD, Li F, Ouyang H, Niyibizi C (2014) Multipotent mesenchymal stem cells from human subacromial bursa: potential for cell based tendon tissue engineering. Tissue Eng Part A 20 (1–2):239–249
48. Campisi J, d'Adda di Fagagna F (2007) Cellular senescence: when bad things happen to good cells. Nat Rev Mol Cell Biol 8(9):729–740
49. Zhang J, Wang JH (2015) Moderate exercise mitigates the detrimental effects of aging on tendon stem cells. PLoS ONE
50. Jiang D, Xu B, Yang M, Zhao Z, Zhang Y, Li Z (2014) Efficacy of tendon stem cells in fibroblast-derived matrix for tendon tissue engineering. Cytotherapy 16 (5):662–673
51. Zhang J, Li B, Wang JH (2011) The role of engineered tendon matrix in the stemness of tendon stem cells in vitro and the promotion of tendon-like tissue formation in vivo. Biomaterials 32(29): 6972–6981

Informing Stem Cell-Based Tendon Tissue Engineering Approaches with Embryonic Tendon Development

6

William Okech and Catherine K. Kuo

Abstract

Adult tendons fail to regenerate normal tissue after injury, and instead form dysfunctional scar tissue with abnormal mechanical properties. Surgical repair with grafts is the current standard to treat injuries, but faces significant limitations including pain and high rates of re-injury. To address this, we aim to regenerate new, normal tendons to replace dysfunctional tendons. A common approach to tendon tissue engineering is to design scaffolds and bioreactors based on adult tendon properties that can direct adult stem cell tenogenesis. Despite significant progress, advances have been limited due, in part, to a need for markers and potent induction cues. Our goal is to develop novel tendon tissue engineering approaches informed by embryonic tendon development. We are characterizing structure–property relationships of embryonic tendon to identify design parameters for three-dimensional scaffolds and bioreactor mechanical loading systems to direct adult stem cell tenogenesis. We will review studies in which we quantified changes in the mechanical and biochemical properties of tendon during embryonic development and elucidated specific mechanisms of functional property elaboration. We then examined the effects of these mechanical and biochemical factors on embryonic tendon cell behavior. Using custom-designed bioreactors, we also examined the effects of dynamic mechanical loading and growth factor treatment on embryonic tendon cells. Our findings have established cues to induce tenogenesis as well as metrics to evaluate differentiation. We finish by discussing how we have evaluated the

W. Okech
Department of Biomedical Engineering, University of Rochester, 215 Robert B. Goergen Hall, PO Box 270168, Rochester, NY 14627-0168, USA

C.K. Kuo (✉)
Department of Biomedical Engineering, University of Rochester, 215 Robert B. Goergen Hall, PO Box 270168, Rochester, NY 14627-0168, USA

Department of Orthopaedics, University of Rochester, 215 Robert B. Goergen Hall, PO Box 270168, Rochester, NY 14627-0168, USA

Center for Musculoskeletal Research, University of Rochester, 215 Robert B. Goergen Hall, PO Box 270168, Rochester, NY 14627-0168, USA
e-mail: Catherine.K.Kuo@rochester.edu

P.W. Ackermann and D.A. Hart (eds.), *Metabolic Influences on Risk for Tendon Disorders*, Advances in Experimental Medicine and Biology 920, DOI 10.1007/978-3-319-33943-6_6

tenogenic differentiation potential of adult stem cells by comparing their responses to that of embryonic tendon cells in these culture systems.

Keywords
Embryonic tendon development • Tenogenesis • Tissue engineering • Stem cell • Mechanoregulation • Crosslinking

List of Acronyms and Abbreviations

3D	Three-dimensional
BAPN	Beta-aminopropionitrile
Col I	Collagen I
Col III	Collagen III
Col XII	Collagen XII
DAPI	4′,6-Diamidino-2-Phenylindole, Dihydrochloride
DMAB	p-Dimethylaminobenzaldehyde
DNA	Deoxyribonucleic acid
E	Embryonic day
ECM	Extracellular matrix
Egr-1	Early growth response-1
FGF4	Fibroblast growth factor 4
FGF8	Fibroblast growth factor 8
FV-AFM	Force volume-atomic force microscopy
GFP	Green fluorescent protein
HH	Hamilton and Hamburger
HP	Hydroxylysyl pyridinoline
LP	Lysyl pyridinoline
LC-MS/MS	Liquid chromatography tandem-mass spectrometry
LOX	Lysyl oxidase
Mkx	Mohawk
MSC	Mesenchymal stem cell
P	Postnatal day
PDGF	Platelet-derived growth factor
Scx	Scleraxis
SHG	Second Harmonic Generation
$TGF\beta_1$	Transforming growth factor beta 1
$TGF\beta_2$	Transforming growth factor beta 2
Tnmd	Tenomodulin
TPC	Tendon progenitor cell

Introduction

Tendons are dense, fibrous connective tissues that possess a well organized and hierarchical structure of collagen type I fibrils that are aligned parallel to the axis of mechanical tension. Their main function is to connect muscle to bone, which allows for the transmission of muscle-derived forces to the skeleton and regulates musculoskeletal function by guiding movement and stabilizing the skeletal system. When adult tendon tissue is injured, the native wound healing process promotes the formation of scar tissue associated with abnormal mechanical properties, matrix content, and organization. Surgical repair with grafts is the current standard for treating both acute injuries (e.g., ruptures) and chronic pathological conditions (e.g., tendinopathies). Unfortunately, this approach faces significant limitations, including autologous donor site morbidity, allogeneic tissue disease transmission, and high failure rates. To overcome this, we are interested in regenerating new tendon during healing or fully replacing dysfunctional tendon with new, normal tissue.

To do so, there is a need to identify exogenous cues (both mechanical and chemical) that can direct cell function in this process. Growth factors are potent cues that influence cell function. Considering adult tissues are unable to recapitulate normal tendon composition and functional properties, it is not surprising that the growth factor profiles identified in embryonic tendon development are distinct from that of adult tendon in homeostasis and during healing

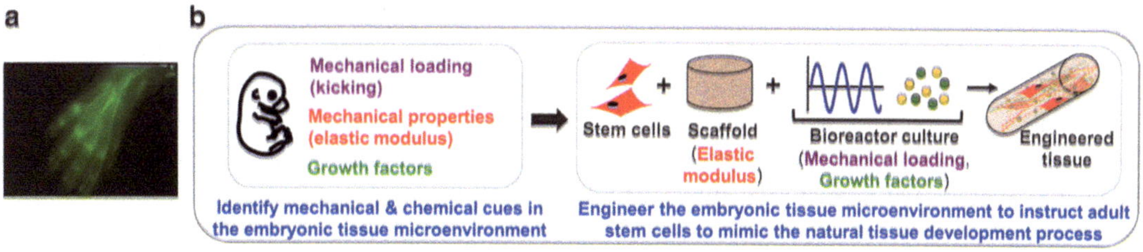

Fig. 6.1 (**a**) Tendons (*green*) in the forelimb of a mouse embryo during development. (**b**) Overview of our strategy to inform tissue engineering with embryonic development: Characterize the mechanical (elastic modulus, dynamic loading) and chemical (growth factors, extracellular matrix components) properties of the embryonic tendon microenvironment and their effects on embryonic tendon progenitor cells (TPCs). These mechanical and chemical factors are then engineered into scaffold and bioreactor culture systems to present stem cells with select embryonic cues and test the hypothesis that embryonic factors will guide adult stem cell tenogenesis in the formation of new tendon tissue

[17, 36]. For example, transforming growth factor beta 2 (TGFβ2) and members of the fibroblast growth factor (FGF4 and FGF8) family are implicated in embryonic tendon development, whereas TGFβ1 and platelet-derived growth factor (PDGF) have been reported in healthy and healing adult tendons [6, 7, 12, 14, 25, 44]. Less well understood are the differences in the mechanical environment between embryonic and adult tendons [17, 36]. In this chapter we discuss a series of studies in which we have demonstrated that the embryonic and adult tendon mechanical microenvironments are significantly different. By quantitatively characterizing the embryonic mechanical microenvironment, we have identified physical signals that may guide tenogenic cell function, and also new markers to evaluate functional tendon formation.

Our goal is to develop novel tendon tissue engineering approaches informed by embryonic tendon development. To do so, we are characterizing structure–property relationships of embryonic tendon to identify design parameters for three-dimensional scaffolds and bioreactor mechanical loading systems that can direct adult stem cells toward a tendon cell fate (Fig. 6.1). We will review studies in which we quantified changes in the mechanical and biochemical properties of tendon during embryonic development and elucidated specific mechanisms of functional property elaboration. We then discuss our work examining the effects of these mechanical and biochemical factors on embryonic tendon cell behavior. Using custom-designed bioreactors, we also examined the effects of dynamic mechanical loading and growth factor treatment on embryonic tendon cells. Our findings have established cues to induce tenogenesis as well as metrics to evaluate differentiation. We finish by discussing how we have evaluated the tenogenic differentiation potential of adult stem cells by comparing their responses to that of embryonic tendon cells in these culture systems.

Embryonic Tendon Mechanical Properties Elaborate During Development

Multiple cell types of various tissue systems have been shown to interact with and respond to the elastic modulus and ECM composition of the tissue. In particular, stem cell differentiation can be influenced by the tissue microenvironment, including its chemical composition and mechanical properties, as well as dynamic physical forces that are imposed on the tissue and cells (see Chap. 5). An important question in designing a strategy to direct cell behavior in the formation of new tendon tissue is the selection of cues that should be presented to the cells in a 3D scaffold (e.g., synthetic matrix) and during culture (e.g., in a bioreactor). Our approach is driven by the hypothesis that stem cells will generate normal tendon in response to embryonic tendon microenvironmental cues that are associated with normal tendon development, rather than when subjected to adult tendon microenvironmental cues that are associated with scarred healing (abnormal tendon formation). In quantitatively characterizing the mechanical and biochemical properties of the embryonic tendon, we identified potentially tenogenic cues. Additionally, we examined and identified mechanisms in the elaboration of tendon mechanical properties to provide functional markers for the evaluation of the progression and quality of newly forming tendon (either native or engineered). In this section, we will describe the novel application of materials science tools and approaches to characterizing these embryonic tendon mechanical properties and the contributors to these properties during tissue formation.

Embryonic Tendon Elastic Modulus Is Dependent on the Developmental Stage

Tensile Testing Measures the Bulk Mechanical Properties of Embryonic Tendon

Traditional testing methods, such as tensile testing, focus on bulk mechanical properties in which a tissue is characterized as a composite material. Few studies have characterized embryonic tendon mechanical properties. McBride et al. (1988) used tensile testing to characterize the elastic modulus of chick embryonic tendon at the latest stages of development from Hamilton Hamburger (HH) 40–43 (approximately days 14–17; the chick embryo hatches between days 19 and 21) [32]. They demonstrated significant linear increases from 0.216 ± 0.06 to 1.02 ± 0.27 MPa during these stages [32], which is substantially lower than that of the adult chicken calcaneal tendon (210 MPa) as measured using tensile testing [34]. These studies highlight the significant differences in embryonic and adult tendon mechanical properties and the rapid pace of functional tendon formation during embryonic development.

Atomic Force Microscopy (AFM) Measures Cell Length-Scale Embryonic Tendon Mechanical Properties

While tensile testing is useful for measuring the functional properties of the tendon, AFM is a powerful tool for quantifying the cell length-scale mechanical properties of the embryonic tendon that the cells experience locally. Cells sense the elastic modulus of their local environment via force-sensitive membrane-bound structures such as integrins and cadherins that are approximately 35–40 nm [21] and 25–28 nm in length [2], respectively. Integrins bind the extracellular matrix (ECM) at focal adhesions while cadherins interact with other cadherins in neighboring cells at adherens junctions. Both structures perform the complex task of sensing and converting both mechanical and chemical stimuli from the extracellular matrix and surrounding cells, respectively, into intracellular biological signals that result in cellular responses. Thus, characterization of the local (i.e., cell length-scale) mechanical properties of the tissue would be useful information in studying tissue-mediated cell behavior.

In Marturano et al. [30], we utilized force volume (FV)-AFM to characterize the nanoscale and microscale elastic moduli of embryonic chick calcaneous tendons during development from stages HH 28 to HH 43. Nanoscale (20-nm tip radius) and microscale (5-μm sphere radius) tips were used to perform indentation testing, and the elastic modulus was calculated from the indentation measurements using either standard curves based on agarose gel measurements or Hertzian theory equations. Results showed that average nanoscale and microscale elastic moduli both increased non-linearly with developmental stage, from HH 28 to HH 43 (Fig. 6.2a). We also examined the topography and regional elastic modulus values obtained with FV-AFM for embryonic tendons between HH 28 and HH 43. Both topography and elastic moduli increased in spatial heterogeneity across the tissue with developmental stage (Fig. 6.2b), which explained the increase in standard deviations in elastic moduli

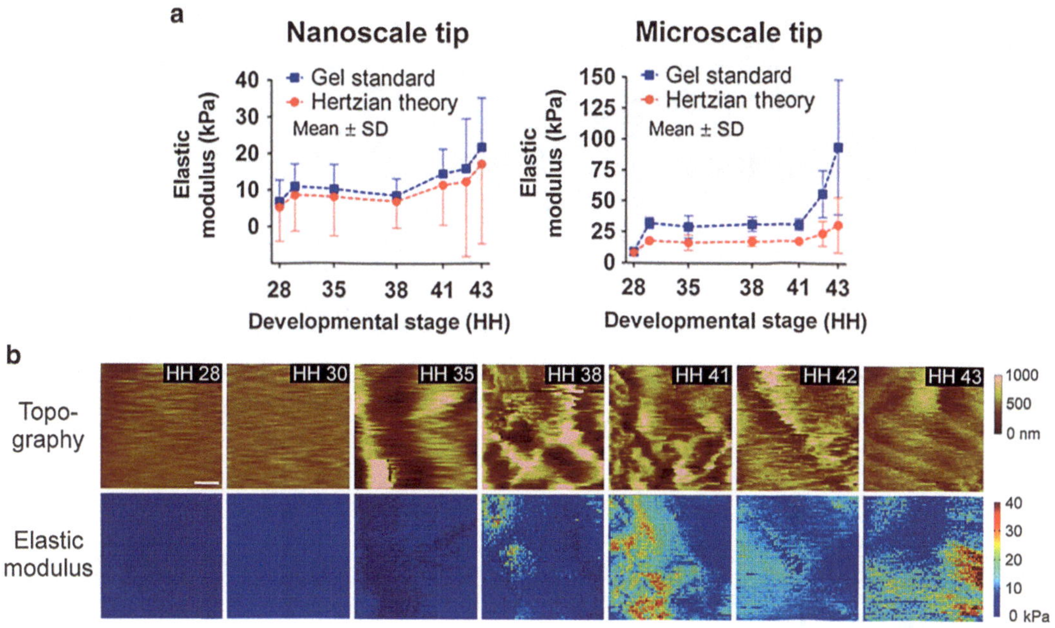

Fig. 6.2 (**a**) Nano- and microscale tendon elastic moduli measured by FV-AFM as a function of the developmental stage (HH 28 to HH 43) and calculated using both agarose gel standards and the Hertzian theory. (**b**) FV-AFM maps of topography and nanoscale elastic modulus of embryonic tendon from HH 28 to HH 43. Scale bar, 2 μm. *Figure reproduced and adapted from Marturano et al.* [30]

during development (Fig. 6.2a). These data demonstrate that cell length-scale mechanical properties elaborate throughout development as a function of embryonic stage, and that they vary regionally within the tissue.

Notably, the microscale elastic modulus was nearly three times greater than the nanoscale when calculated based on agarose gel standards (5 to 108 kPa vs. 7 to 21 kPa, respectively, $p < 0.05$; Fig. 6.2a), and two times larger than the nanoscale modulus (9 to 31 kPa vs. 5 to 17 kPa, $p < 0.05$) when calculated based on the Hertzian model. Similar differences between microscale and nanoscale elastic moduli have also been observed in mechanical property characterization of articular cartilage [42]. A separate study using gel-fiber composite materials proposed that nanoscale tips provided information about the mechanical properties of individual collagen fibrils and proteoglycans within cartilage, and that microscale tips measured the cross-linked network of collagen fibrils [29]. Similarly, in Marturano et al. [30] the nanoscale tip may have been measuring individual ECM molecules and cell contributions, whereas the microscale tip measured a network of ECM molecules and cells. These findings provided new insights into the dynamic elaboration of mechanical properties and tissue structure at the cell length-scales during embryonic tendon development.

ECM Content Changes During Embryonic Development But Does Not Correlate Directly with Mechanical Properties

In Marturano et al. (2013), we also characterized the cell and ECM content of embryonic tendons in relation to elastic modulus. Between HH 28 and HH 43 there was a qualitative increase in collagen type I deposition (Fig. 6.3a) and a simultaneous decrease in cell density (Fig. 6.3b). Biochemical assays to quantify hydroxyproline (representative of collagen) content and DNA content (not shown) corroborated the qualitative observations (Fig. 6.3) [30]. We then analyzed spatial and quantitative correlations between

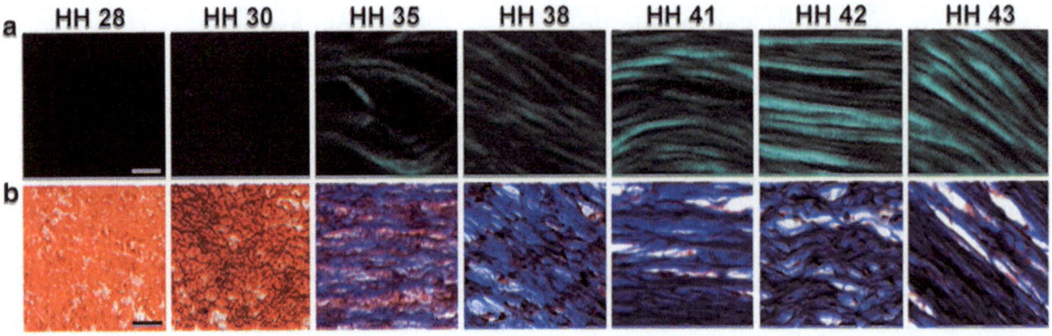

Fig. 6.3 (**a**) Collagen distribution (scale bar, 2 μm) and (**b**) cellularity (scale bar, 20 μm) during the development of embryonic tendon (HH 28 to HH 43) as detected using SHG imaging and trichrome staining, respectively. *Figure reproduced and adapted from Marturano et al.* [30]

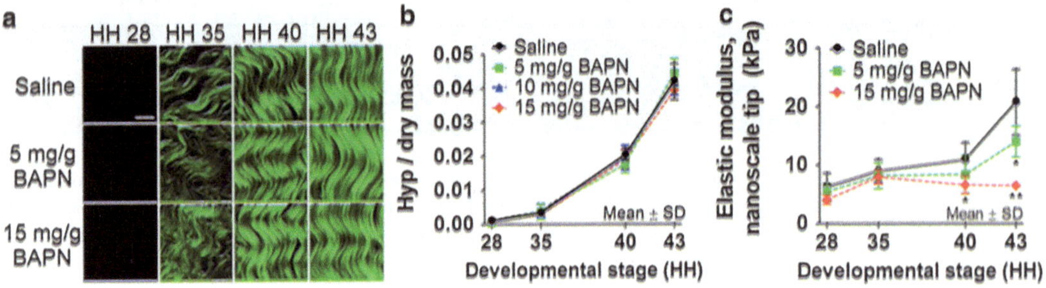

Fig. 6.4 (**a**) The collagen microstructure imaged using SHG imaging after BAPN or saline (control) treatment. Scale bar, 10 μm. (**b**) Hydroxyproline (representative of collagen) content quantified using p-Dimethylaminobenzaldehyde (DMAB) absorbance assay. (**c**) Nanoscale elastic modulus of BAPN- or saline (control)-treated tendons, measured using FV-AFM. *Figure reproduced and adapted from Marturano et al.* [30]

nanoscale elastic modulus and collagen fibers (second harmonic generation, SHG), nuclei (DAPI staining), and glycosaminoglycans (GAGs; Alcian blue staining). Statistically, there was a weak, positive relationship between the elastic modulus and collagen fibers ($r = 0.13$, $p < 0.05$), and a negative relationship between the elastic modulus and cell nuclei ($r = -0.07$, $p = 0.23$) and with total GAGs ($r = -0.04$, $p = 0.50$). These data indicate that there is no meaningful correlation between elastic modulus and the cell and ECM components of interest. Importantly, this significant finding reveals that tendon mechanical properties should not be evaluated only on the basis of ECM protein content and/or organization.

Lysyl Oxidase (LOX)-Mediated Collagen Crosslink Density Contributes to the Mechanical Properties of Embryonic Tendon

To investigate collagen crosslinks as contributors to the elastic modulus of the embryonic tendon tissue, β-aminopropionitrile (BAPN), or vehicle control, was injected into the eggs of developing chick embryos of various stages between HH 28 and HH 43 and allowed to continue developing for 24 h. BAPN binds the active site of lysyl oxidase (LOX) to inhibit its activity, thereby halting its ability to activate crosslinking of collagens and elastin [43]. After 24 h, chick embryos were sacrificed and tendons were harvested for SHG imaging, hydroxyproline quantification, and FV-AFM [30]. BAPN treatment had no effect on collagen fiber organization (Fig. 6.4a) or hydroxyproline content (Fig. 6.4b). However, despite no change in the collagen density or organization, BAPN treatment significantly reduced the elastic modulus of tendons in a dose-dependent manner. At its highest dosage administered (15 mg/g), BAPN reduced the elastic modulus of HH 40 tendons by 38.8 % ($p < 0.05$) and of HH 43 tendons by 68.4 % ($p < 0.001$) (Fig. 6.4c) without affecting cell viability [30]. These findings demonstrate LOX-mediated collagen crosslinks play a critical role in the elaboration of mechanical properties of tendon during embryonic development.

LOX-Mediated Collagen Crosslinks Are Potential Markers of Tendon Development

In previous studies, several markers of tenogenic differentiation and embryonic tendon development were identified, including scleraxis (Scx), mohawk (Mkx), tenomodulin (Tnmd), tenascin-C, collagen type I (Col I), and early growth response-1 (Egr-1) (see Chap. 7) [4, 8, 9, 11, 18, 20, 22, 27, 28, 30, 39, 46]. These molecular markers identify tendon cells and tissue, but do not provide information about the functional (mechanical) properties of the tendon. Based on our findings in Marturano et al. [30], we were interested in examining the potential to use LOX-mediated crosslinks as markers of functional tendon development.

Liquid chromatography tandem-mass spectrometry (LC-MS/MS) was employed to quantify LOX-mediated crosslinks as a function of developmental stage [31]. Specifically, hydroxylysyl pyridinoline (HP) and lysyl pyridinoline (LP) were quantified in tendons of HH 28 to HH 43 chick embryos. In adult tendon, the HP crosslinks, which link three hydroxylysines, are more prevalent than the LP crosslinks, which link two hydroxylysines with one lysine [13]. In embryonic tendon, we found that both the HP-to-dry mass ratio (Fig. 6.5a) and the HP + LP-

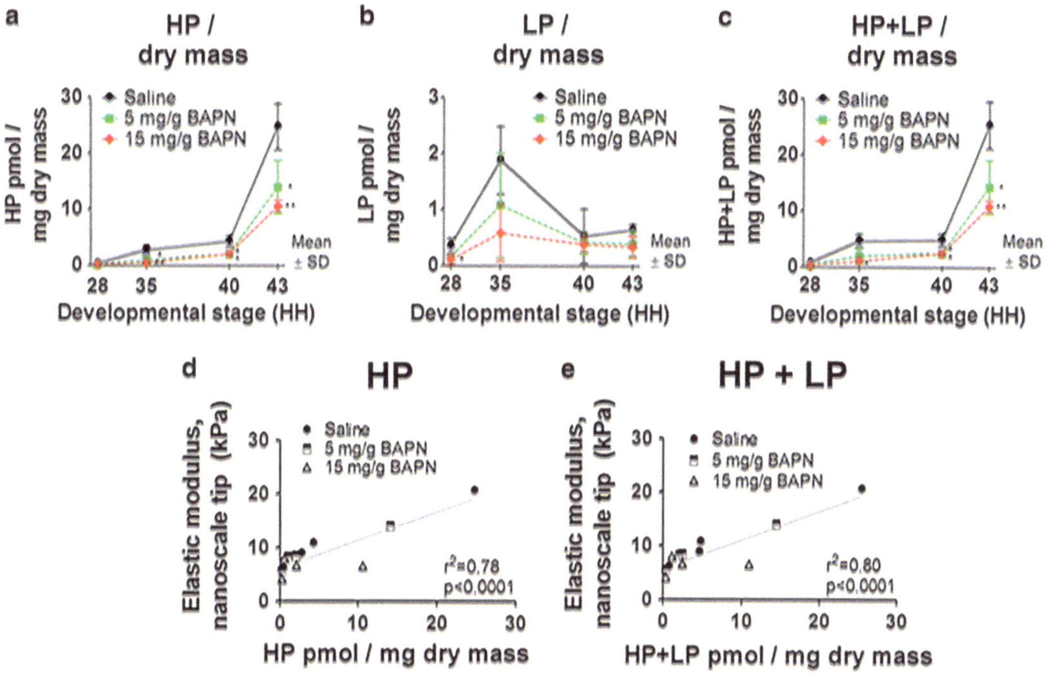

Fig. 6.5 LC-MS/MS measurements of (**a**) HP, (**b**) LP, and (**c**) total HP + LP density of BAPN- and saline (control)-treated HH 28 to 43 embryonic chick tendons. Correlation between elastic modulus and (**d**) HP or (E) HP + LP density of BAPN- and saline (control)-treated HH 28 to 43 embryonic chick tendons. *Figure reproduced and adapted from Marturano et al.* [31] *with permission from Acta Biomaterialia*

to-dry mass ratio (Fig. 6.5c) increased with developmental stage, with smaller increases at early stages and larger changes at later stages (Fig. 6.5a,c) [31]. These results demonstrated a similar trend to that obtained for nanoscale elastic modulus changes during development [30]. In contrast, the LP-to-dry mass ratio (Fig. 6.5b) did not follow this trend, with higher levels associated with earlier embryonic stages [31]. Our data showed that the HP crosslinks were present at higher levels than LP crosslinks in embryonic tendon, consistent with adult tendon [13]. Statistical analysis demonstrated that nanoscale elastic modulus correlated highly with the total HP + LP-to-dry mass ratio (Fig. 6.5e; $r^2 = 0.8$; $p < 0.0001$) and the HP-to-dry mass ratio (Fig. 6.5d; $r^2 = 0.78$; $p < 0.0001$) [31]. These findings demonstrate the potential to assess LOX-mediated crosslinkers as functional markers of embryonic tendon tissue formation.

A major advantage to using LC-MS/MS to evaluate the functional properties of tendon is that it requires very small sample sizes and numbers. This is in contrast to the need for larger sample sizes for gripping in a mechanical tester, and a higher sample number required for statistical rigor with mechanical testing. In the case of LC-MS/MS, only a small biopsy would be needed. The findings of this study are significant because they demonstrate that LOX-mediated crosslink density could be useful as a functional marker of tendon tissue formation, offering significant advantages over mechanical testing to obtain similar information.

The Actin Cytoskeleton Contributes to Embryonic Tendon Mechanical Properties and Is a Biomarker of Development

As seen in a number of our studies, embryonic tendon has a high density of tendon progenitor cells (TPCs) that gradually decrease in density as ECM (e.g., collagen) is deposited and new tissue is formed (Fig. 6.3) [25, 30, 37]. To characterize these cells further, we used DAPI and

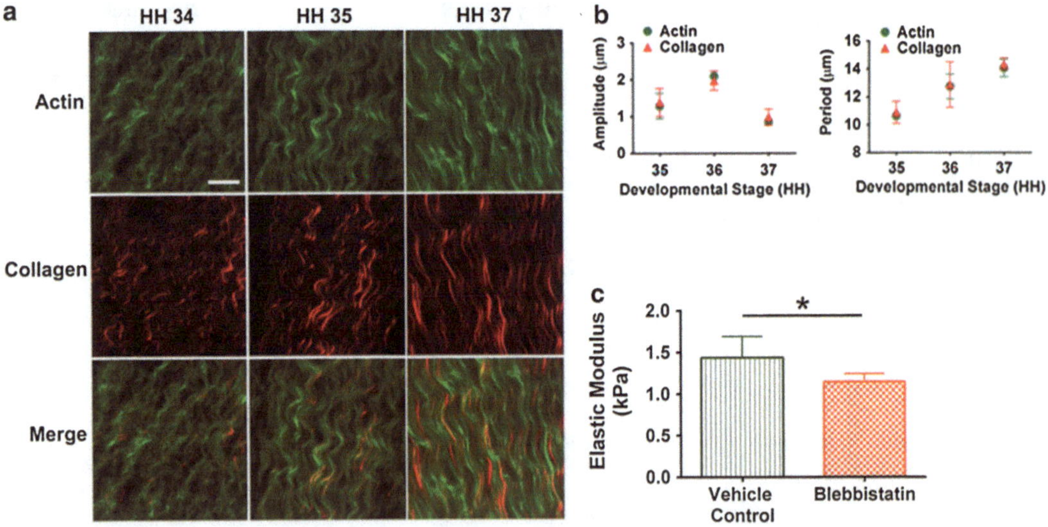

Fig. 6.6 (a) Representative confocal images of actin and collagen crimp patterns in HH 34, 35, and 37 embryonic tendons, as detected by phalloidin staining for actin (*green, top panel*) and SHG for collagen (*red, middle panel*). *Bottom panel* showed phalloidin staining and SHG images merged. Scale bar, 10 μm. (**b**) Amplitude and period of the crimp pattern of the actin cytoskeletal network and extracellular collagen of HH 35, 36, and 37 embryonic tendons. (**c**) Elastic modulus of HH 36 embryonic limb explants after 24 h of blebbistatin or vehicle control treatment. *Figure reproduced and adapted from Schiele et al.* [37] *with permission from Journal of Orthopaedic Research*

phalloidin staining to visualize cell nuclei and actin cytoskeleton filaments, respectively, in HH 34 to HH 37 embryonic tendons (not shown). The TPCs appeared densely packed, and image analysis found no significant differences in cell number (1.58×10^6 cells/mm^3) between developmental stages. A particularly interesting finding was that the actin filaments appeared to form a contiguous network between adjacent cells and spanned multiple cells (Fig. 6.6a) [37]. We also found that a crimp pattern emerged in the actin cytoskeletal network across cells, and that this overlapped with the crimping extracellular collagen (Fig. 6.6a). This was confirmed with image analysis of the crimp pattern, finding identical amplitude and period in the crimp of the actin filaments and collagen fibers (Fig. 6.6b). Furthermore, this actin filament crimp pattern was unique to each of the stages analyzed, demonstrating actin crimp pattern to be a stage-specific biomarker for tendon development.

Based on these results, we proposed the apparent intercellular actin cytoskeletal network contributes to the mechanical properties of embryonic tendon at the earlier stages of development [37]. We tested this hypothesis by culturing HH 36 chick embryo calcaneal tendon explants in the presence of blebbistatin (25 μM) or vehicle control. Blebbistatin inhibits rigid crosslinking between non-muscle myosin II and the actin filaments to disrupt the actin cytoskeleton [24, 38]. Blebbistatin treatment did not alter the collagen fiber organization but disrupted the actin stress fibers, and significantly reduced the elastic modulus by 21.4 %, from 1.4 ± 0.3 kPa to 1.1 ± 0.2 kPa ($p = 0.02$) (Fig. 6.6c). This finding demonstrated that the actin cytoskeleton is not only a unique biomarker of tendon cell morphology during development, but that it is also a significant contributor to tendon mechanical properties during embryonic tissue formation.

Embryonic Tendon Development May be Influenced by Muscle-Generated Mechanical Loading

Functional muscle, which attaches to tendon at the myotendinous junction, puts tendon under tensile load with each contraction, as occurs during leg kicking. Thus, during muscle-induced activities, such as repeated kicking, the tendon experiences cyclic tensile loading. It is possible that the resident cells of the tendon experience these mechanical forces with each muscle contraction, transducing the mechanical strains that they sense into biological signals. It would be exceedingly challenging to investigate such effects of tensile loading on tendon cells *in vivo*, as this would require separating the mechanical from the soluble influences of muscle during development. To circumvent this challenge, our group has implemented a novel approach that utilizes bioreactor culture systems to impose mechanical loading on embryonic tendon progenitor cells and tissues, in the absence of unknown factors (e.g., growth factors, cytokines, etc.). These benchtop culture systems provide highly controlled microenvironments in which individual or specific combinations of physical and biochemical cues can be administered to study their role in the regulation of cell function. This section will focus on the potential role of skeletal muscle-induced mechanical loading in the development of embryonic tendon *in vivo*, and the effect of mechanical stimulation and growth factor treatment on the behavior of tenogenically differentiating cells.

Muscle Paralysis Affects Embryonic Tendon Development

During embryonic development, the formation of the myotendinous junction (where muscle inserts into tendon) enables tendons to transmit muscle contraction-induced mechanical forces to bones. In the complete absence of muscle, chick limb tendon formation will commence, but subsequently will atrophy [22, 23]. The reasons for this are not completely understood, but loss of muscle contraction-induced mechanical loading and loss of muscle-secreted biological factors have both been implicated.

In two different studies, immobilization of embryos using the neuromuscular blocking agent decamethonium bromide [15, 33] resulted in a reduction in the tendon cross-sectional area of chick embryos [16]. The loss of differentiated craniofacial muscles in Myf5-Myod1-deficient zebrafish has been reported to decrease tendon progenitor formation in both the head and fin, though not in the axial region [10]. Furthermore, mouse embryos with muscular dysgenesis (*mdg*), a spontaneous mutation that results in the loss of excitation-contraction coupling and therefore causes muscle paralysis, have thinner flexor digitorum superficialis tendons, as compared to wildtype [19]. Taken together, it could be inferred from these muscle paralysis studies that skeletal muscle contraction is important for the normal development of embryonic limb tendon. However, these studies did not account for alterations in muscle-secreted soluble cues and how this may have affected the development of the embryonic tendon.

Scx is a basic helix-loop-helix transcription factor that is expressed in tendon progenitors in the embryo and in differentiated tenocytes of the adult [6, 39]. It has been reported that in the absence of muscle, there is diminished Scx expression. Multiple studies have also demonstrated that expression of this tendon-specific marker can be rescued by the ectopic application of FGF4 [5, 12] or FGF8 [6], both of which are secreted by the muscle adjacent to tendon. In all of the above studies, it is not clear how muscle contraction-induced mechanical loading and muscle-secreted soluble factors each contribute to embryonic tendon development because they cannot be studied in isolation *in vivo*. However, while not definitive, these studies collectively suggest muscle-generated mechanical and chemical cues both play a role in the regulation of embryonic tendon development.

Novel Mechanical Loading Bioreactor Systems to Investigate Mechanical and Chemical Influences of Tendon Progenitor Cell Function

To examine the individual effects of muscle-generated mechanical and biochemical stimuli on embryonic tendon progenitor cell (TPC) behavior, we performed novel *in vitro* studies with custom-designed bioreactors. Using bioreactor culture systems, we subjected TPCs to individual and combinations of specific mechanical forces and growth factor treatments under highly controlled conditions [8, 9]. This novel approach enabled simulations to study the specific effects of muscle-generated mechanical loading or growth factors, without the influence of unknown factors that would additionally affect the cells *in vivo*.

TPCs, the primary resident cells within the embryonic tendon, can be identified via Scx expression. To obtain mouse embryonic TPCs for our studies, we harvested Scx-GFP positive axial and limb cells from Scx-GFP transgenic mouse embryos from embryonic day (E) 13 to postnatal day (P) 7 [8]. Axial and limb TPCs arise from different locations in the embryo. In particular, axial tendon progenitors arise from the syndetome, a compartment within the somites [6], whereas limb tendon progenitors originate from the superficial limb mesenchyme [39]. We were interested in understanding whether TPCs of different anatomical origins would have different responses to the same factors implicated in embryonic tendon development.

In the bioreactor cultures, TPCs were subjected to either mechanical (cyclic tensile) loading (1 % strain, 0.5 Hz, 1 h/day) or exogenous growth factor treatments (FGF4, TGFβ2) [8]. FGF4 and TGFβ2 were selected because they have been implicated in embryonic tendon development [5–7, 12, 25, 35]. In this study, we tested the hypotheses that putative tendon

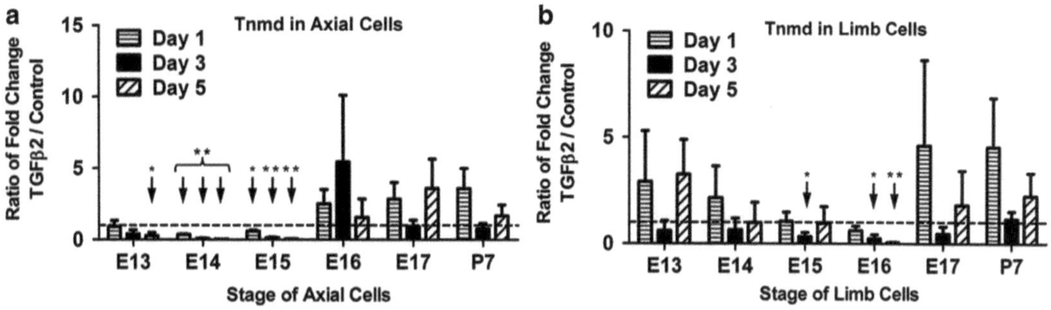

Fig. 6.7 (**a**) Effects of TGFβ2 on Tnmd gene expression in limb cells (E13 to P7) in TPCs after 1, 3, and 5 days of treatment. (**b**) Effects of TGFβ2 on Tnmd gene expression in axial cells (E13 to P7) in TPCs after 1, 3, and 5 days of treatment. *p < 0.05, **p < 0.01, and ***p < 0.001. *Figure reproduced and adapted from Brown et al.* [8] *with permission from Journal of Biomechanics*

developmental factors (mechanical loading, FGF4, TGFβ2) are tenogenic *in vitro*, and that the TPC response to mechanical loading or growth factor treatment vary with anatomical origin and developmental stage.

TGFβ2 significantly increased Scx gene expression at all stages (E13 to P7), except E16, in both limb and axial TPCs after just 1 day of treatment [8]. In contrast, FGF4 failed to upregulate Scx gene expression, despite *in vivo* findings by others in which FGF4 rescues Scx expression [7, 12]. Additionally, FGF4 significantly downregulated Tnmd and TGFβ2 gene expression in limb TPCs after 3 days. Collectively, these data suggest that TGFβ2 is a potent factor for tenogenesis *in vitro*, but that FGF4 may not be.

Interestingly, limb and axial TPCs responded differently to TGFβ2 treatment. Tnmd (late stage marker) was not regulated in limb TPCs during the early stages of development (E13 to E15) (Fig. 6.7b), but was significantly downregulated in axial TPCs at the same developmental stages (Fig. 6.7a) [8]. Under control conditions (no growth factor or mechanical loading), both limb and axial TPCs upregulated elastin expression over culture time (0 to 5 days). When treated with FGF4, this trend was abrogated in limb TPCs, whereas axial TPCs continued to increase elastin expression over time. Collectively, these results demonstrate the importance of considering the anatomical origin of the tendon in a tissue engineering approach.

Developmental stage-dependent TPC responses to mechanical loading and growth factor treatments were also analyzed for both limb and axial TPCs as part of the study [8]. Mechanical loading did not regulate TGFβ2 gene expression in the earlier stages of development (E13 and E14). However, during the later stages of development (E15 to P7) we saw significant downregulation of TGFβ2 gene expression in response to mechanical loading. In axial TPCs, TGFβ2 treatment significantly decreased Tnmd expression from E13 to E15 but did not regulate the expression between E16 to P7. Interestingly, loading did not affect Scx expression in either cell type. Future studies to further examine the role of mechanical loading in tenogenesis will be worthwhile, as we discuss below. Taken together, these results indicate that the stage of tenogenic differentiation of a progenitor cell is important to consider when designing a tissue engineering treatment strategy.

Overall, this study demonstrated a novel approach using bioreactor culture systems to investigate embryonic TPC responses to individual mechanical and biochemical cues. The results of this study provided evidence for TGFβ2 to be a potent tenogenic factor *in vitro*, and suggested that the role of FGF4 in tenogenesis may be more complex than *in vivo* studies indicate. Additionally, our results demonstrate that TPC responses are dependent on anatomical origin and developmental stage. These findings have brought new insights to the role of muscle-generated mechanical and chemical stimuli on embryonic tendon development.

MSCs Demonstrate Tenogenic Differentiation Potential Based on Their Ability to Mimic Embryonic Tendon Progenitor Cell Behavior

Our goal is to learn from embryonic tendon development to inform tendon tissue engineering strategies with adult stem cells. An attractive adult stem cell type for tissue engineering and healing strategies is the MSC. Adult MSCs have the potential to differentiate towards the osteogenic, chondrogenic, and adipogenic lineages. They can be harvested from various tissues including bone marrow, adipose, tendon, muscle, and more [3, 9, 26, 27, 38, 47]. The advantages of MSCs include high expansion potential, multilineage differentiation capacity, easy accessibility, and low immunogenicity, the latter being useful for allogeneic transplants.

Chemical stimuli, such as growth factors, are commonly utilized to direct MSCs towards various mesenchymal lineages *in vitro*. To induce MSCs to tenogenically differentiate into tendon fibroblasts, investigators have used 3D constructs subjected to uniaxial static tension [1, 45]. We have demonstrated that uniaxial dynamic tension and a 3D culture environment have the potential to increase the expression of embryonic tenogenic markers in human and murine MSCs [9, 27]. Recently, investigators have also examined the effect of substrate stiffness [40, 41] on tendon marker gene expression in MSCs. Collectively, these studies demonstrate that the mechanical microenvironment can play an important role in the regulation of tenogenic cell behavior. To further accelerate progress of stem cell-based tendon tissue engineering, tenogenic cues and markers will need to be established. In this section, we review studies that examine the influence of exogenous stimuli on MSC behavior in comparison with embryonic TPCs (model of tenogenically differentiating cells). These studies demonstrate the potential for select tendon developmental factors to induce tenogenic differentiation of MSCs when exogenously applied. These findings also show that MSCs have tenogenic potential by directly demonstrating their capacity to respond to these cues similarly as tenogenically differentiating cells of the embryonic tendon (TPCs).

MSCs Respond Similarly as Embryonic TPCs to Exogenously Applied Embryonic Tendon Factors

After establishing embryonic TPCs as a model system to study the influence of mechanical and soluble cues on tenogenic cell behavior [8], we tested the hypothesis that the cues involved in regulating TPCs *in vivo* can induce similar responses in both MSCs and TPCs *in vitro* [9]. Our goal was to assess the potential for MSCs to mimic TPC behavior as a method to evaluate whether the MSC responses to applied cues are tenogenic. MSCs were harvested from the bone marrow of 4-month old male Scx-GFP mice, selected by plastic adherence, seeded on 2-dimensional substrates, and then mechanically loaded under cyclic tension, or treated with TGFβ2 or FGF4. Gene expression responses of the MSCs were compared with that of embryonic TPCs that were subjected to the same conditions.

Mechanical loading did not regulate the gene expression of tendon markers Scx, TGFβ2, Tnmd, and Col I in MSCs or TPCs [9]. This finding was similar to our earlier findings with axial and limb TPCs [8], which could suggest that mechanical loading is not tenogenic. However, in another previous study in which human MSCs were mechanically loaded under tension, we found that while loading did not regulate Col I, III, or XII gene expression levels, it significantly enhanced collagen protein deposition within the 3D constructs over static conditions (Fig. 6.8) [27]. Specifically, cyclically loaded samples had 50 % more collagen than statically loaded samples after 7 days of culture, as determined by histomorphometric analysis of staining intensity [27]. Importantly, these findings

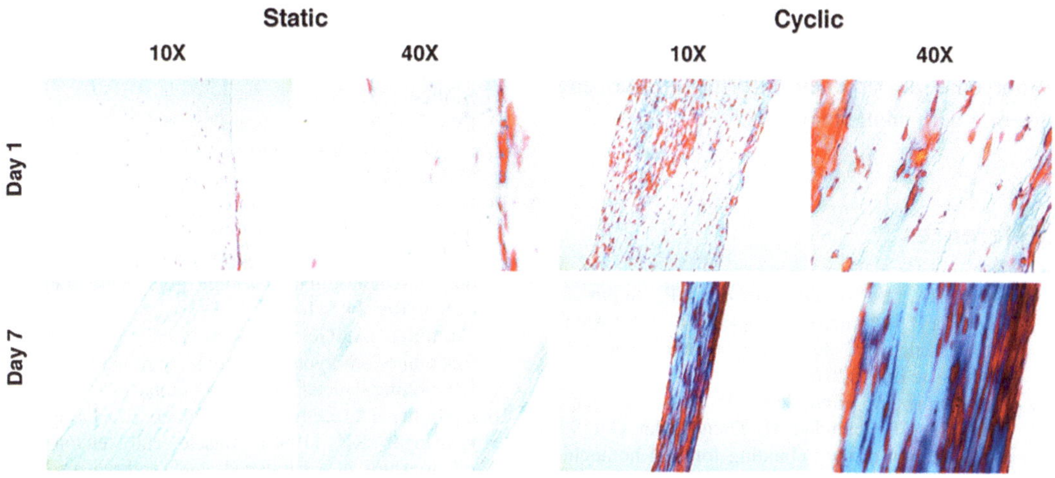

Fig. 6.8 Longitudinal 5-μm-thick sections of MSCs cultured in 3D collagen type I engineered constructs, harvested after 1 and 7 days of static and cyclic loading, and then histologically stained with Mallory's Trichrome stain. *Figure reproduced and adapted from Kuo et al.* [27]

demonstrated mechanical loading may indeed be tenogenic, and that its effects are post-transcriptional.

With regards to chemical stimulation, treatment with the growth factor FGF4 elicited dissimilar responses in MSCs and TPCs. In MSCs, FGF4 treatment significantly downregulated the gene expression levels of all the tendon markers (Scx, TGFβ2, Tnmd, Col I, and elastin) assessed in the study. In contrast, FGF4 treatment of TPCs downregulated only the mRNA levels of elastin, and not the other markers. Notably, the TPCs in this study were harvested from the limb, and demonstrated the same response to FGF4 as we previously reported [8]. TGFβ2 significantly upregulated Scx expression in both cell types, but had no effect on TGFβ2 or Col I gene expression, and significantly downregulated Tnmd and elastin mRNA levels. Similarly, in TPCs, TGFβ2 had no effect on TGFβ2 or Col I mRNA levels, but in contrast to MSCs, it had no effect on elastin and Tnmd expression. Together, these data suggest that TGFβ2 may induce tenogenic differentiation of progenitor cells *in vitro*. These data also support that MSCs have the ability to tenogenically differentiate, based on their similar responses to TGFβ2 treatment when compared to TPCs. Future studies to further understand the inherent differences between MSCs and TPCs may lead to novel approaches to enhance tenogenic differentiation of adult stem cells.

Taken together, these studies demonstrate the importance of thoroughly evaluating cell responses at the gene, protein, and functional levels to understand how a mechanical or biochemical cue is affecting a cell.

Future Directions and Conclusions

Tissue engineering and regenerative medicine aim to restore normal cellular and tissue function. Progress with approaches to engineer or regenerate new tendon has been limited, due in part to a need for effective methods to evaluate tenogenic differentiation and neotissue formation. In this chapter, we reviewed a series of studies in which we characterized structure–property relationships of embryonic tendon as well as tendon progenitor cell function during development. Our findings identified potential cues and markers to induce and assess tendon formation, respectively. We also demonstrated the potential to guide tenogenic differentiation of adult MSCs with factors that play integral roles in tenogenic differentiation of embryonic TPCs during normal development. Future studies to reveal the mechanistic effects of these factors on MSCs will be important in establishing embryonic development-informed tissue engineering with adult stem cells.

References

1. Awad HA, Butler DL, Boivin GP, Smith FN, Malaviya P, Huibregtse B, Caplan AI (1999) Autologous mesenchymal stem cell-mediated repair of tendon. Tissue Eng 5:267–277
2. Baumgartner W, Hinterdorfer P, Ness W, Raab A, Vestweber D, Schindler H, Drenckhahn D (1999) Determination of the unbinding force of homophilic interaction of vascular endothelial cadherin by atomic force microscopy. Biophys J 76:A351–A351
3. Bi Y, Ehirchiou D, Kilts TM, Inkson CA, Embree MC, Sonoyama W, Li L, Leet AI, Seo BM, Zhang L, Shi S, Young MF (2007) Identification of tendon stem/progenitor cells and the role of the extracellular matrix in their niche. Nat Med 13:1219–27
4. Birk DE, Nurminskaya MV, Zycband EI (1995) Collagen fibrillogenesis in situ: fibril segments undergo post-depositional modifications resulting in linear and lateral growth during matrix development. Dev Dyn 202:229–243
5. Brent AE, Braun T, Tabin CJ (2005) Genetic analysis of interactions between the somitic muscle, cartilage and tendon cell lineages during mouse development. Development 132:515–528
6. Brent AE, Schweitzer R, Tabin CJ (2003) A somitic compartment of tendon progenitors. Cell 113:235–248
7. Brent AE, Tabin CJ (2004) FGF acts directly on the somitic tendon progenitors through the Ets transcription factors Pea3 and Erm to regulate scleraxis expression. Development 131:3885–3896
8. Brown JP, Finley VG, Kuo CK (2014) Embryonic mechanical and soluble cues regulate tendon progenitor cell gene expression as a function of developmental stage and anatomical origin. J Biomech 47:214–222
9. Brown JP, Galassi TV, Stoppato M, Schiele NR, Kuo CK (2015) Comparative analysis of mesenchymal stem cell and embryonic tendon progenitor cell response to embryonic tendon biochemical and mechanical factors. Stem Cell Res Ther 6:89
10. Chen JW, Galloway JL (2014) The development of zebrafish tendon and ligament progenitors. Development 141:2035–2045
11. Docheva D, Hunziker EB, Fassler R, Brandau O (2005) Tenomodulin is necessary for tenocyte proliferation and tendon maturation. Mol Cell Biol 25:699–705
12. Edom-Vovard F, Schuler B, Bonnin MA, Teillet MA, Duprez D (2002) Fgf4 positively regulates scleraxis and tenascin expression in chick limb tendons. Dev Biol 247:351–366
13. Eyre DR, Paz MA, Gallop PM (1984) Cross-linking in collagen and elastin. Annu Rev Biochem 53:717–748
14. Galatz LM, Sandell LJ, Rothermich SY, Das R, Mastny A, Havlioglu N, Silva MJ, Thomopoulos S (2006) Characteristics of the rat supraspinatus tendon during tendon-to-bone healing after acute injury. J Orthop Res 24:541–550
15. Germiller JA, Goldstein SA (1997) Structure and function of embryonic growth plate in the absence of functioning skeletal muscle. J Orthop Res 15:362–370
16. Germiller JA, Lerner AL, Pacifico RJ, Loder RT, Hensinger RN (1998) Muscle and tendon size relationships in a paralyzed chick embryo model of clubfoot. J Pediatr Orthop 18:314–318
17. Glass ZA, Schiele NR, Kuo CK (2014) Informing tendon tissue engineering with embryonic development. J Biomech 47:1964–1968
18. Guerquin MJ, Charvet B, Nourissat G, Havis E, Ronsin O, Bonnin MA, Ruggiu M, Olivera-Martinez I, Robert N, Lu Y, Kadler KE, Baumberger T, Doursounian L, Berenbaum F, Duprez D (2013) Transcription factor EGR1 directs tendon differentiation and promotes tendon repair. J Clin Invest 123:3564–76
19. Huang AH, Riordan TJ, Pryce B, Weibel JL, Watson SS, Long F, Lefebvre V, Harfe BD, Stadler HS, Akiyama H, Tufa SF, Keene DR, Schweitzer R (2015) Musculoskeletal integration at the wrist underlies the modular development of limb tendons. Development 142:2431–2441
20. Ito Y, Toriuchi N, Yoshitaka T, Ueno-Kudoh H, Sato T, Yokoyama S, Nishida K, Akimoto T, Takahashi M, Miyaki S, Asahara H (2010) The Mohawk homeobox gene is a critical regulator of tendon differentiation. Proc Natl Acad Sci U S A 107:10538–10542
21. Kanchanawong P, Shtengel G, Pasapera AM, Ramko EB, Davidson MW, Hess HF, Waterman CM (2010) Nanoscale architecture of integrin-based cell adhesions. Nature 468:580–584
22. Kardon G (1998) Muscle and tendon morphogenesis in the avian hind limb. Development 125:4019–4032
23. Kieny M, Chevallier A (1979) Autonomy of tendon development in the embryonic chick wing. J Embryol Exp Morphol 49:153–165
24. Kovacs M, Toth J, Hetenyi C, Malnasi-Csizmadia A, Sellers JR (2004) Mechanism of blebbistatin inhibition of myosin II. J Biol Chem 279:35557–35563
25. Kuo CK, Petersen BC, Tuan RS (2008) Spatiotemporal protein distribution of TGF-betas, their receptors, and extracellular matrix molecules during embryonic tendon development. Dev Dyn 237:1477–1489

26. Kuo CK, Tuan RS (2003) Tissue engineering with mesenchymal stem cells. IEEE Eng Med Biol Mag 22:51–56
27. Kuo CK, Tuan RS (2008) Mechanoactive tenogenic differentiation of human mesenchymal stem cells. Tissue Eng Part A 14:1615–1627
28. Lejard V, Blais F, Guerquin MJ, Bonnet A, Bonnin MA, Havis E, Malbouyres M, Bidaud CB, Maro G, Gilardi-Hebenstreit P, Rossert J, Ruggiero F, DupreZ D (2011) EGR1 AND EGR2 INVOLVEMENT IN VERTEBRATE TENDON DIFFERENTIATION. J BIOL CHEM 286:5855–5867
29. Loparic M, Wirz D, Daniels AU, Raiteri R, Vanlandingham MR, Guex G, Martin I, Aebi U, STOLZ M (2010) Micro- and nanomechanical analysis of articular cartilage by indentation-type atomic force microscopy: validation with a gel-microfiber composite. Biophys J 98:2731–2740
30. Marturano JE, Arena JD, Schiller ZA, Georgakoudi I, Kuo CK (2013) Characterization of mechanical and biochemical properties of developing embryonic tendon. Proc Natl Acad Sci U S A 110:6370–6375
31. Marturano JE, Xylas JF, Sridharan GV, Georgakoudi I, Kuo CK (2014) Lysyl oxidase-mediated collagen crosslinks may be assessed as markers of functional properties of tendon tissue formation. Acta Biomater 10:1370–1379
32. Mcbride DJ, Trelstad RL, Silver FH (1988) Structural and mechanical assessment of developing chick tendon. Int J Biol Macromol 10:194–200
33. Mikic B, Wong M, Chiquet M, Hunziker EB (2000) Mechanical modulation of tenascin-C and collagen-XII expression during avian synovial joint formation. J Orthop Res 18:406–415
34. Nakagaki WR, Biancalana A, Benevides GP, Gomes L (2007) Biomechanical and biochemical properties of chicken calcaneal tendon under effect of age and nonforced active exercise. Connect Tissue Res 48:219–228
35. Pryce BA, Watson SS, Murchison ND, Staverosky JA, Dunker N, Schweitzer R (2009) Recruitment and maintenance of tendon progenitors by TGFbeta signaling are essential for tendon formation. Development 136:1351–1361
36. Schiele NR, Marturano JE, Kuo CK (2013) Mechanical factors in embryonic tendon development: potential cues for stem cell tenogenesis. Curr Opin Biotechnol 24:834–840
37. Schiele NR, von Flotow F, Tochka ZL, Hockaday LA, Marturano JE, Thibodeau JJ, Kuo CK (2015) Actin cytoskeleton contributes to the elastic modulus of embryonic tendon during early development. J Orthop Res 33:874–881
38. Schiller ZA, Schiele NR, Sims JK, Lee K, Kuo CK (2013) Adipogenesis of adipose-derived stem cells may be regulated via the cytoskeleton at physiological oxygen levels in vitro. Stem Cell Res Ther 4:79
39. Schweitzer R, Chyung JH, Murtaugh LC, Brent AE, Rosen V, Olson EN, Lassar A, Tabin CJ (2001) Analysis of the tendon cell fate using Scleraxis, a specific marker for tendons and ligaments. Development 128:3855–3866
40. Sharma RI, Snedeker JG (2010) Biochemical and biomechanical gradients for directed bone marrow stromal cell differentiation toward tendon and bone. Biomaterials 31:7695–7704
41. Sharma RI, Snedeker JG (2012) Paracrine interactions between mesenchymal stem cells affect substrate driven differentiation toward tendon and bone phenotypes. PLoS One 7, e31504
42. Stolz M, Raiteri R, Daniels AU, Vanlandingham MR, Baschong W, Aebi U (2004) Dynamic elastic modulus of porcine articular cartilage determined at two different levels of tissue organization by indentation-type atomic force microscopy. Biophys J 86:3269–3283
43. Tang SS, Trackman PC, Kagan HM (1983) Reaction of aortic lysyl oxidase with beta-aminopropionitrile. J Biol Chem 258:4331–4338
44. Wurgler-Hauri CC, Dourte LM, Baradet TC, Williams GR, Soslowsky LJ (2007) Temporal expression of 8 growth factors in tendon-to-bone healing in a rat supraspinatus model. J Shoulder Elbow Surg 16: S198–S203
45. Young RG, Butler DL, Weber W, Caplan AI, Gordon SL, Fink DJ (1998) Use of mesenchymal stem cells in a collagen matrix for Achilles tendon repair. J Orthop Res 16:406–413
46. Zhang G, Young BB, Ezura Y, Favata M, Soslowsky LJ, Chakravarti S, Birk DE (2005) Development of tendon structure and function: regulation of collagen fibrillogenesis. J Musculoskelet Neuronal Interact 5:5–21
47. Zhang J, Wang JH (2010) Characterization of differential properties of rabbit tendon stem cells and tenocytes. BMC Musculoskelet Disord 11:10

Cell Signaling in Tenocytes: Response to Load and Ligands in Health and Disease

7

Michelle E. Wall, Nathaniel A. Dyment, Josie Bodle, Jon Volmer, Elizabeth Loboa, Anna Cederlund, Ann M. Fox, and Albert J. Banes

Abstract

Signaling in tenocytes during development, homeostasis and injury involves multiple and redundant pathways. Given that tendons transmit mechanical forces from muscle to bone to effect movement, a key function for tenocytes is the detection of and response to mechanical stimulation. Mechanotransduction involves matrix-integrin-cytoskeleton to nucleus signaling, gap junction intercellular communication, changes in intracellular calcium (Ca^{2+}), activation of receptors and their pathways, and responses to biochemical factors such as hormones, growth factors, adenosine triphosphate (ATP) and its derivatives, and neuromodulators. The primary cilium also plays a key role in the detection of mechanical signals. During development, transforming growth factor-β (TGF-β), bone morphogenetic protein (BMP), and hedgehog (Hh) signaling modulate tendon differentiation and formation. The response to injury is complex and varied involving not only inflammatory mediators such as interleukin-1β but also mechanosensing. This chapter reviews the signaling pathways tenocytes use during mechanotransduction, development and in response to injury.

M.E. Wall (✉) • J. Volmer
Flexcell International Corp., 2730 Tucker St., Suite 200, Burlington 27215, NC, USA
e-mail: info@flexcellint.com

N.A. Dyment
Department of Reconstructive Sciences, School of Dental Medicine, University of Connecticut Health Center, Farmington, CT, USA

J. Bodle • E. Loboa • A.M. Fox
Joint Department of Biomedical Engineering, North Carolina State University & University of North Carolina at Chapel Hill, Raleigh, NC, USA

A. Cederlund
Musculoskeletal Research Programme, University of Aberdeen, Foresterhill, Aberdeen, Scotland, UK

A.J. Banes
Flexcell International Corp., 2730 Tucker St., Suite 200, Burlington 27215, NC, USA

Joint Department of Biomedical Engineering, North Carolina State University & University of North Carolina at Chapel Hill, Raleigh, NC, USA

P.W. Ackermann and D.A. Hart (eds.), *Metabolic Influences on Risk for Tendon Disorders*, Advances in Experimental Medicine and Biology 920, DOI 10.1007/978-3-319-33943-6_7

Introduction

Mechanical stimulation is important for maintaining tendon structure and function and may influence cellular responses to external factors, including hormones, growth factors, nucleotides, and neurotransmitters, possibly released from surrounding blood vessels and nerves during trauma to the tendon (Fig. 7.1 [13, 49]). Specific signaling pathways may be activated in response to mechanical stimulation that drive matrigenesis, mitogenesis (MEK/MAPK), a stress response (JAK/STAT and JNK/SAPK), apoptosis, or other responses [4, 7, 11, 14, 15, 62–64]. The extent of the response can be altered by the microenvironment itself. Tenocytes have a homeostatic set point provided by cytoskeletal tension and connections to the extracellular matrix, that when altered, changes tenocyte shape, increases matrix metalloproteinase-13 (MMP-13) mRNA expression, and alters cilia length [5, 44]. Tenocyte responses to both physical and chemical changes in their environment may involve second messengers (such as cyclic adenosine monophosphate (cAMP), cyclic guanosine monophosphate, ATP, guanosine triphosphate, nitric oxide, prostaglandin E2 (PGE_2), inositol triphosphate (IP_3), and diacylglycerol) that can act in autocrine and/or paracrine fashions, ion channels (such as Ca^{2+}-dependent, Ca^{2+}-independent, stretch-activated and voltage-gated), changes in RNA and protein expression, cilia deformation, cytoskeletal interactions (involving integrins, focal adhesion kinase, paxillin, filamin, integrin-linked kinase, vinculin, and talin), changes in intracellular calcium concentration ($[Ca^{2+}]_{ic}$), and intercellular communication, among other mechanisms (Figs. 7.1 and 7.4).

The detection of mechanical signals is key for tendon function given that their primary role is to transmit mechanical forces from muscle to bone to effect movement. Under normal physiologic conditions, tendons are subjected to low strains. At 1 % strain in whole tendon *in vivo*, the matrix crimp pattern is unaffected and cells are nominally deformed [6]. However, an applied 3 % strain deforms collagen fibrils to a straightened position. The tendon is taut and can undergo reversible deformation thus, the cells are subjected to a level of deformation above a nominal threshold. At 5 % strain, the tendon is subjected to the upper limit of elastic deformation and suffers some plastic deformation. Plastic deformation could include cell-cell contact disruption or alterations in protein arrangement within the tenocyte plasma membrane, such as connexin 43 (Cx43), the primary constituent of gap junction channels in tenocytes [72] and initiate an inflammatory response that involves interleukins, MMPs, and insulin like growth factor [88, 89, 93, 95].

This chapter will review some of the key mechanisms tenocytes use in mechanotransduction, including Ca^{2+} signaling, intercellular communication via gap junctions, norepinephrine (NE) activation of adrenoceptors, ATP and purinoceptors, and primary cilium. In addition, this chapter will review signaling pathways used during tendon development and in response to injury.

Tenocyte Mechanotransduction

Calcium Signaling and Gap Junction Intercellular Communication

One method by which tenocytes can detect and respond to mechanical stimulation is through intercellular communication pathways whereby $[Ca^{2+}]_{ic}$ increases in a coordinated fashion among interconnected cells [17]. Calcium wave propagation has been demonstrated in many cell types and can occur through direct intercellular signaling via gap junctions [25, 87], paracrine signaling mechanisms [32, 56, 57], or both [28, 56, 57]. Cells can propagate Ca^{2+} waves to neighboring cells through the passage of IP_3, or a signal that produces IP_3 in neighboring cells, through gap junctions and activation of IP_3 receptors on the endoplasmic reticulum [17]. However, it has become apparent that cells can use multiple mechanisms to propagate intercellular Ca^{2+} waves, including the release of nucleotides (e.g., ATP, uridine triphosphate (UTP)) and

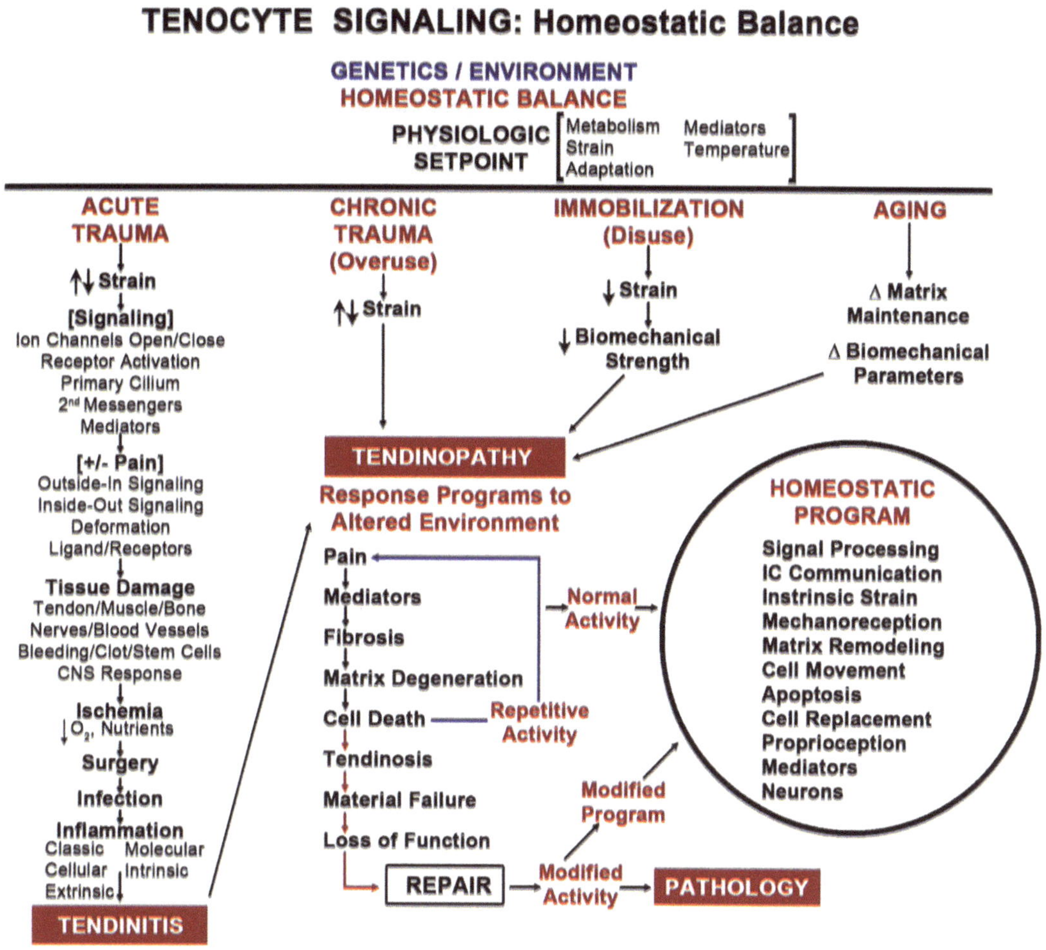

Fig. 7.1 A synopsis of normal and pathologic processes that results in homeostatic balance or pathologic process in tenocyte signaling pathways (Altered with permission from Banes et al. [13])

subsequent activation of purinoceptors on neighboring cells [28, 55–57].

Tenocytes *in vivo* have immunohistochemically detectable gap junctions [72]. *In vitro*, human tenocytes form functional gap junctions in both monolayer and three-dimensional collagen gels [60]. Gap junction channels form between cells when hexameric gap junction protein (connexins) structures on neighboring cells dock to create a pathway for direct intercellular exchange of ions and molecules [39, 40, 86, 107]. Connexins have been shown to be associated with the actin cytoskeleton, which may help stabilize gap junctions during periods of prolonged mechanical loading [103]. Additionally, connexin hemichannels can contribute to intercellular communication via autocrine and paracrine pathways (see [47] for review). *In vitro*, Cx32 and Cx43 are expressed in human, avian, murine and equine tenocytes, with Cx43 being the most prevalent species [17, 60, 107]. Avian tenocytes also express Cx26 [17]. Cx43 connects tenocytes in a syncytium, whereas Cx32 connects tenocytes between syncytial layers (Fig. 7.2 [72, 107]).

In tendon, gap junctions are involved in mechanotransduction pathways [101]. The magnitude of the applied mechanical load can alter gap junction intercellular communication in tenocytes, where low levels of strain (4 %)

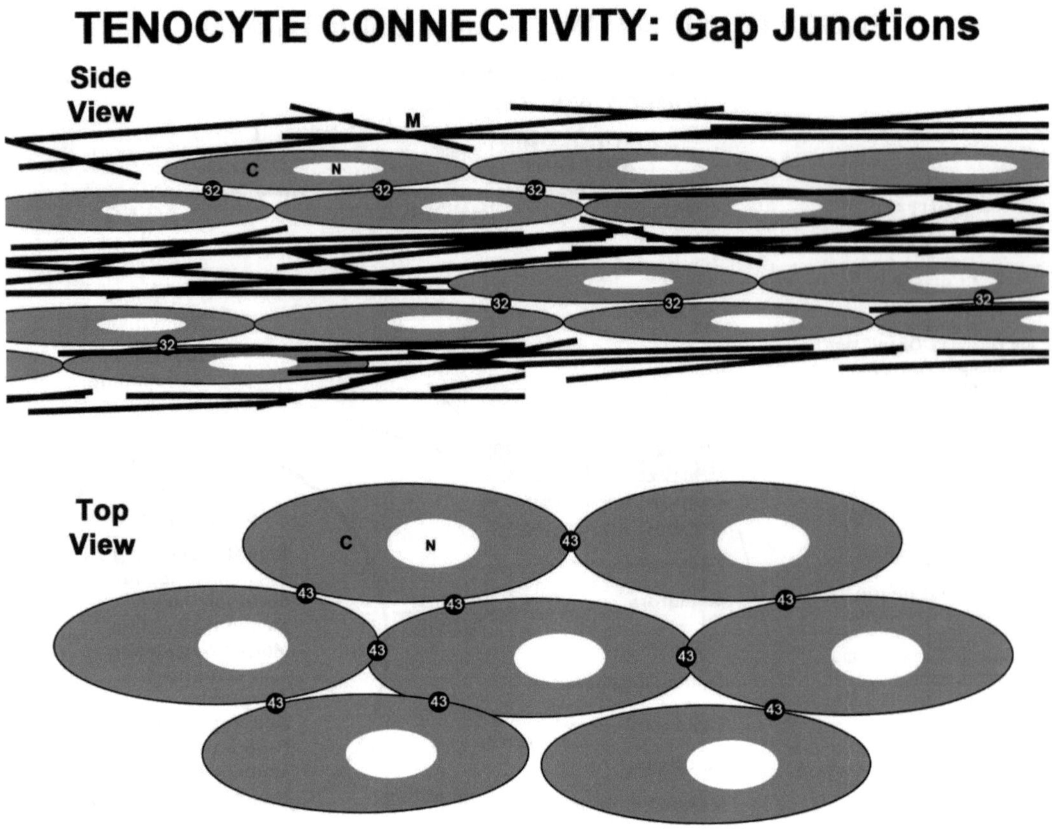

Fig. 7.2 *Side view* is a view of multiple tenocyte syncytial layers connected by connexin (Cx) 32; *Top view* is an in face view of a single tenocyte syncytial layer connected by Cx43. 32, Cx32 gap junctions connecting over and underlying tenocyte syncytia; 43, Cx43 gap junctions in adjacent cells in a syncytium; *C* cell body, *N* nucleus, *M* extracellular matrix

increased communication but high levels of strain (8 %) decreased communication [69]. Gap junctions also modulate load-induced DNA and collagen synthesis, which may contribute to matrix remodeling in response to load [12]. Results of a study by Waggett et al. [100] indicated that signaling mediated by Cx43 gap junctions may inhibit load-induced collagen secretion, whereas Cx32 signaling may stimulate load-induced collagen secretion in tenocytes. Thus, a change in connexin expression or signaling could significantly alter the molecular signals transferred between cells [46] and potentially the physiological response of the tenocyte.

Gap junctions are gated by several mechanisms including phosphorylation of serine residues on connexins. The full promoter sequence for Cx43 contains two activator protein-1 (AP-1) and four cAMP response elements (CRE)-like sites that are reported to mediate cellular response to cAMP [108]. As a second messenger, cAMP can modulate changes in gene expression via activation of transcription factors (e.g., CREB) that bind to specific promoter sequences that drive gene transcription. In avian tenocytes, cyclic equibiaxial strain activated cAMP response element binding (CREB) protein and AP-1 transcription factors [16]. Cyclic AMP as well as Ca^{2+} can increase protein kinase A and C activation, respectively, which ultimately results in Cx43 phosphorylation [27, 86]. Norepinephrine activation of adrenoceptors can also increase cAMP and $[Ca^{2+}]_{ic}$ in connective tissues [21, 59, 102]. Thus, strain and NE may activate common or additive site(s) in the Cx43 promoter to drive transcription. Other ligands may also act to alter Cx43 expression during injury or disease. IL-1β was found to upregulate Cx43 expression in tenocytes [83]. Furthermore, strain-induced cell death is gap junction-dependent [83].

Norepinephrine and Adrenoceptors

Norepinephrine is a neurotransmitter that binds adrenoceptors, G-protein coupled receptors important in the regulation of many functions including vasoconstriction (see Chap. 4) [34, 48, 99]. Cells from avian tendons express the α_{1A} and α_{1B}-adrenoreceptor subtypes [102] and respond to NE by increasing $[Ca^{2+}]_{ic}$ primarily via activation of α_{1A}-adrenoreceptors [102]. Human and rabbit tenocytes have been found to express α_{2A}-adrenoceptors [8]. Normal human tendons show positive immunoreactivity for adrenoceptors in both the tendon proper and in the tendon blood vessel walls [33]. Human tenocytes also express tyrosine hydroxylase which suggests that tenocytes may be capable of producing endogenous catecholamines [8]. Additionally, tendons that present with tendinosis have increased immunoreactivity for α_1-adrenoreceptors in the blood vessel walls suggesting a role of catecholamines in tendinopathy [33]. Thus, adrenoceptors may provide a mechanism for neurohormonal modulation of tenocyte function under both normal and diseased states.

Purinoceptors and ATP

Purinoceptors are metabotropic, G protein coupled (P2Y class) or ionotropic, ligand-gated ion channels (P2X class; [31]). $P2Y_2$ reacts with ATP or UTP, and $P2Y_1$ reacts primarily with adenosine diphosphate (ADP) but partially with ATP [30, 31]. In tendon, ATP activates $P2Y_2$ purinoceptors [97]. Tenocytes secrete ATP particularly in response to fluid shear or stretch [97]. The effect of secreted ATP is modulated by ecto-NTPases, which are expressed by tenocytes and appear to act principally at the cell surface in tendon [96]. Addition of 1 μM ATP to tenocytes or ligament cells *in vitro* and in whole tendons *ex vivo* increased $[Ca^{2+}]_{ic}$ [44, 55]. ATP and UTP may also act to amplify responses to mechanical stimulation because they activate a common pathway through an increase in $[Ca^{2+}]_{ic}$ [97]. Tenocytes from $P2Y_2$ knockout mice that lack the receptor for ATP do not respond to substrate strain by increasing $[Ca^{2+}]_{ic}$ [54]. ATP can also modulate the contraction of a linear, three-dimensional, collagen gel seeded with tenocytes or MC3T3-E1 cells [84] indicating a role of ATP in matrix remodeling. Increased ATP secretion may activate an inhibitory pathway, which may dampen a response to load [96, 97]. Additionally, ATP or a breakdown product, such as ADP or adenosine, may act as a stop, or modulating signal(s) for some genes impacted by mechanical load.

In addition to modulating responses to mechanical load, ATP is an important modulator of inflammatory gene expression in tenocytes. ATP can inhibit IL-1β-induced MMP mRNA and protein expression, cyclooxygenase-2 (COX2) expression, and PGE_2 secretion [97].

Deformation Sensing

Tenocytes can detect mechanical signals and, in turn, impose a mechanical signal via the matrix and/or substrate upon which they are cultured through both autobaric and parabaric effects (Fig. 7.3). Tenocytes detect substrate strain [104] likely via integrin connections to matrix through tensegrity [51]. More recently, tenocytes have been shown to also utilize the primary cilium to detect strain [61]. The illustration in Fig. 7.3 depicts how tenocytes respond to substrate strain, individually and in a syncytium. In the case of an autobaric effect, a tenocyte may respond to strain or a ligand by contracting, hence loading the cell itself and the matrix to which it is attached. The cell may then spread and flatten, altering its substrate even further. In a second set of responses, a tenocyte may alter

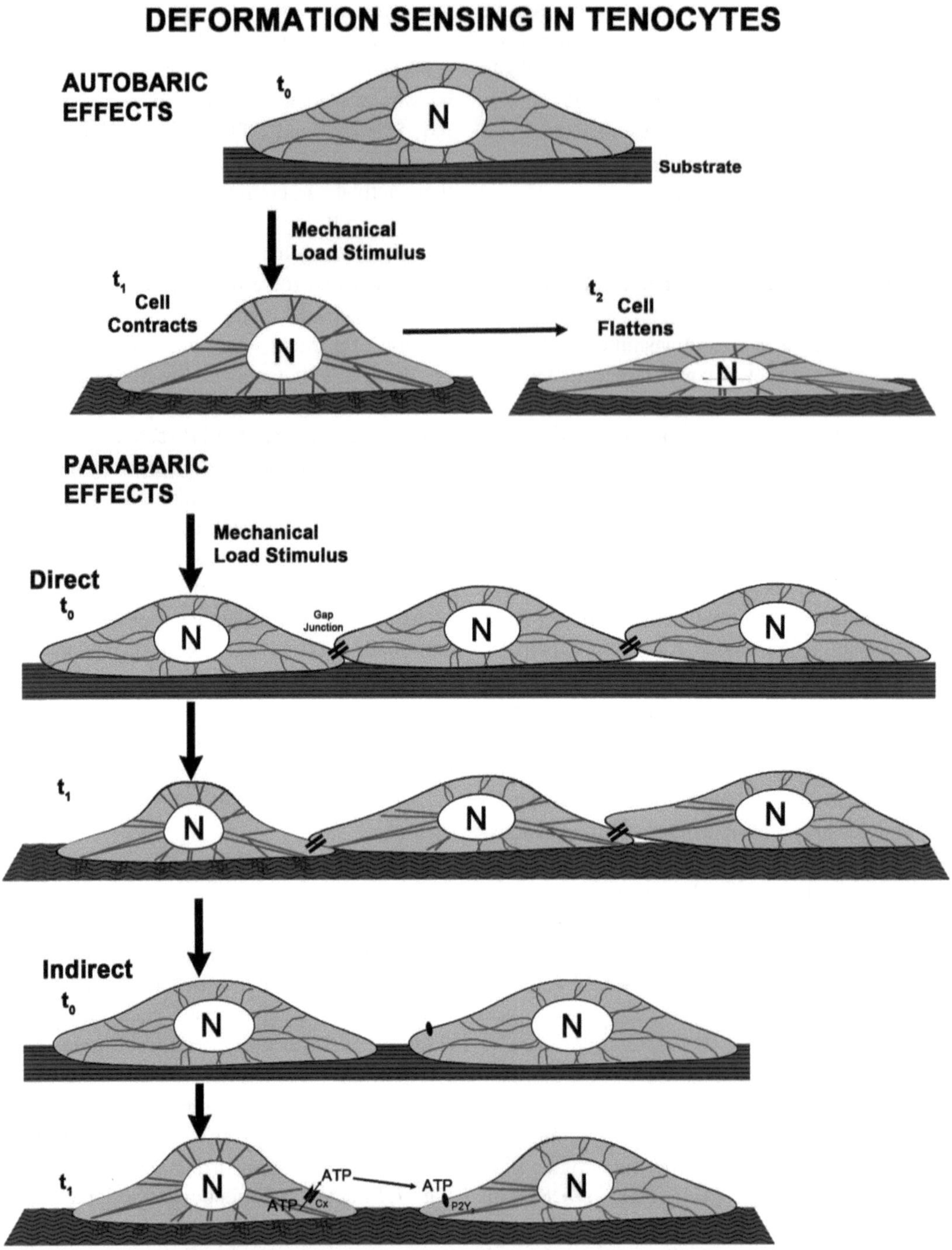

Fig. 7.3 Autobaric effects – At t_0, a cell receives a mechanical signal by deformation, and the response is a cell contraction event. The cell contracts and flattens (t_1) thus applying a secondary deformation to itself. Parabaric effects – Similar to the autobaric response to a mechanical stimulus, a cell or series of cells receive a signal then applies deformation via contraction to themselves and adjacent cells. The cell-driven deformation can be direct when cells are directly attached or indirect if cells are adjacent and deform the matrix, thereby transmitting load to an adjacent cell. In addition, in direct effects, the signal can be transmitted to adjacent cells via gap junction intercellular communication. In indirect effects, the signal can also be transmitted to neighboring cells via second messengers such as the release of adenosine triphosphate (ATP) through connexin (Cx) hemichannels from the mechanically loaded cell. The ATP can then activate purinoceptors ($P2Y_2$) on neighboring non-stimulated cells. (Altered with permission from Banes et al. [13])

shape by flattening, spreading and if attached to other tenocytes in a syncytium, will also apply load to attached cells in a direct parabaric effect. If a responding tenocyte is mechanically loaded but not directly attached to a neighboring cell, it may apply its load to the substrate and indirectly to the neighboring tenocyte in an indirect parabaric effect. In each case, the tenocyte can signal to cells via gap junctions or purinoceptors.

Role of Primary Cilia in Tenocytes

Primary cilia were discovered on mammalian cells in 1898, though they have been largely regarded as vestigial organelles for a majority of that time since their initial discovery [109]. Research over the last 20 years has revealed that primary cilia are prevalent on nearly all somatic cells. A typical (non-motile) primary cilium is composed of: (1) a tubular axoneme, approximately 0.2 microns in diameter and up to 10 microns in length, delimited by the ciliary membrane (contiguous with the cell's plasma membrane) and typically projecting outward from the cell body, (2) the axoneme containing nine circumferential microtubule doublets, devoid of a central pair of microtubules arranged in a 9 + 0 configuration, (3) a cylindrical basal body, of nine microtubular triplets derived from the mother centriole, which nucleate the microtubules of the axoneme [92], and (4) transition fibers and a basal foot that anchor the primary cilium to the cell membrane and/or ciliary pocket and the actin cytoskeleton [73]. The cilium is typically observed in a juxtanuclear position and often colocalizes with the Golgi apparatus [78]. The expression of the primary cilium is intimately associated with the phases of the cell cycle; however, there is some evidence that the details of this characteristic may be cell-type specific. Tenocytes *in vivo* and *in vitro* have primary cilium that face each other and vary in length [85].

Primary cilia generally present characteristics of chemo and mechanosensitivity and are thought to, in part, coordinate mechanotransduction pathways, particularly in mechano-active connective tissues [74, 75, 105]. Additionally, a variety of important signaling pathways localize their signaling activity to the base and axoneme of the primary cilium, including proteins of the Hh, Wnt, TGF-β, and platelet-derived growth factor pathways [81]. Though several important signaling mechanisms localize their activity to the primary cilium, many of the underlying mechanisms behind cilia expression and function remain elusive.

In the context of connective tissue, primary cilia were initially observed in cartilage [78, 79] and bone [105]. Imaging primary cilia in dense 3-dimensional (3D) connective tissue for quantitative analysis requires a sophisticated approach to generate accurate data on the length, shape and dimensions of the primary cilium. Groups have analyzed ciliary structure with transmission electron microscopy, epi-fluorescence microscopy with deconvolution, standard confocal fluorescence microscopy and multi-photon microscopy. Farnum and coworkers developed a method of morphometric analysis through using mulitphoton microscopy in such a way that allows them breakdown planar ciliary angle as well as elevation angle. Additionally, their approach allowed them to identify the prominence of primary cilium in the dense extracellular matrix of tendon and cartilage tissue [1, 41]. Farnum and Wilsman [42] further pioneered a more extensive analysis of the primary cilia in dense connective tissue, such as cartilage and tendon, to generate a more thorough understanding of the relationship between cilia orientation and mechanosensitivity.

Ciliary mechanosensitivity has been largely demonstrated in changes in cilia-associated proteins in response to fluid shear stress. These studies empirically demonstrated that primary cilia mediate mechanotransduction of shear forces in osteoblast cell types [70]. Other work in cartilage has shown that healthy and diseased cartilage tissue differentially express primary cilia, likely dependent on the mechanical environment that develops during osteoarthritis. Following these studies, cilia mechanosensitivity has been demonstrated in tendon explants cultured under cyclic tensile strain

[45, 61]. Gardner et al. [45] demonstrated that stress deprivation of tendon explants induced cilia elongation in tenocytes within 24 h, whereas cyclic tensile loading at 3 % strain induced a shortening of ciliary length. The authors suggest that in the absence of loading typically experienced by tendon *in situ*, tenocytes extend their cilia to increase their sensitivity to detect mechanical changes in their environment. In the study by Lavagnino et al. [61], rat tail tendon fascicles were incrementally exposed to strain levels up to 8 % and the ciliary deflection angle was measured at each strain increment. Detection of changes in ciliary deflection under this range of strains indicated scaling of deflection concordant with physiologic loading *in vivo*, supporting the hypothesis that cilia play a role in tendon mechanosensory mechanisms [61].

Other studies in mesenchymal (MSCs) and adipose stem cells (ASCs) have shown that the primary cilium may play a role in lineage specification [24, 50, 98]. Disrupting the primary cilium structure and/or specific cilia associated proteins results in downregulation of gene expression and end-product markers of osteogenic, chondrogenic and adipogenic differentiation [24, 50, 98]. Further, evidence suggests that culture under chemical induction towards osteogenic, chondrogenic and adipogenic lineages confers differential cilia expression linked to lineage specification. Cell morphology is a cursory indicator for cell phenotype and there is evidence that the primary cilium's length and orientation is in part modulated by cell shape and thus cytoskeletal organization. McMurray et al. [71] reported that when MSCs were cultured on grooved substrates, MSCs tended to orient along the direction of the grooves yielding an elongated cell morphology as well as elongated cilia which oriented with the long axis of the cell. This study also reported modulations of ciliary localized Wnt signaling in response to changes in culture substrate architecture. Similar observations have been reported on tenocytes in rat extensor tendons, which orient their cell body and cilia along the direction of the collagen fibrils and the long axis of the tendon tissue [35]. It follows that this principle likely extends to other connective tissue cells with mechanosensitive cilia, including tenocytes and that cilia expression may be involved in healthy and or disease tenocyte physiology. These data taken together suggest that the mechanosensitivity of primary cilia is, in part, indicative of the physiological state of connective tissue cell types and may also be involved in maintenance of tenocytes and/or tendon tissue physiology (Fig. 7.4).

Signaling in Tendon Development

The embryological origin of tendons is dependent on their anatomical position: axial tendons are derived from cells at the interface of the sclerotome and myotome known as the syndetome, limb tendons are derived from the lateral plate mesoderm, and cranial tendons, like other cranial mesenchyme, are neural crest derived. These embryological origins were clearly defined following the discovery of the basic helix-loop-helix transcription factor scleraxis (Scx), which is one of the earliest markers that specifies tendon primordia [91]. The progenitor cell populations that contribute to the initiation and differentiation of embryonic tendon are becoming clearer. Fate mapping studies indicate that two distinct progenitor pools exist at the time of tendon condensation: one pool that gives rise to the midsubstance and another pool that gives rise to the enthesis (i.e., tendon-to-bone insertion site) [22, 36, 90, 94]. While our understanding is not complete, there are known signaling pathways and transcription factors that regulate the specification and differentiation of these progenitor pools into their respective regions.

Around the time of condensation, the progenitor pool that gives rise to the enthesis expresses both Scx and the SRY-related transcription factor Sox9, while unlike the Scx-only population of the midsubstance and Sox9-only population of the underlying boney eminence [22]. Enthesis cells can also be traced back to a Gli1 and Gdf5 origin unlike the tendon midsubstance [36, 90]. Tight regulation of TGF-β and BMP

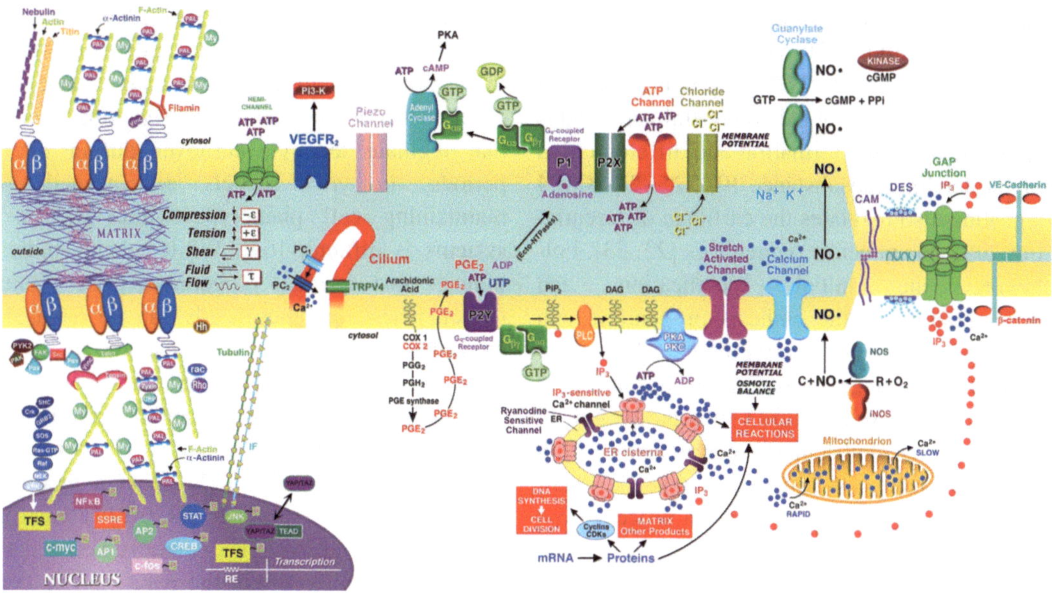

Fig. 7.4 The detection of and response to external mechanical stimuli (i.e., compression, tension, shear, fluid flow) involves multiple pathways and signaling mediators. A matrix-integrin-mechanosensory protein complex-cytoskeleton machinery is linked to a kinase cascade (tyrosine or nontyrosine kinase cascade or the JACSTAT kinase cascade) system. A mechanosensory protein complex contains talin, vinculin (Vinc), tensin, paxillin (PAX), Src, and focal adhesion kinase (FAK). In this model, a load deformation displaces matrix molecules tethered to clustered integrins at focal adhesions. The displacement is transduced to an integrin (β), to an integrin-binding protein, and then to associated proteins. Matrix-integrin-cytoskeletal interactions may also involve actin, myosin (My), nebulin, titin, α-actinin, filamin, palladin (PAL), tublin, and intermediate filaments (IF). Activated extracellular signal-regulated protein kinases (ERK) enter the nucleus and up-regulate transcription factor expression (TFS, AP1, AP2, SSRE, CREB, c-fos, c-myc,STAT, JNK) and activate nuclear binding proteins, such as nuclear factor κB (NF-κB; P for phosphorylation). Polycistin-1 (PC_1) is co-localized with the primary cilium and activated when the cilium is deformed by fluid shear stress. The shear stress signal is transferred from PC_1 to polycistin-2 (PC_2) and induces the influx of calcium (Ca^{2+}) though PC_2, which in turn activates ryanodine receptors in the endoplasmic reticulum (ER) to release Ca^{2+}, resulting in Ca^{2+-} induced Ca^{2+} release. Changes in intracellular Ca^{2+} through the release of intracellular Ca^{2+} stores from the ER through or entry of extracellular Ca^{2+} through channels such as the store-operated, stretch-activated, mechanosensitive Ca^{2+} channels, and voltage independent or dependent Ca^{2+} channels. The release of adenosine triphosphate (ATP) and, at lower levels, uridine triphosphate (UTP), following the activation of ionotropic P2X and metabotropic, G protein-coupled P2Y receptors in an autocrine/paracrine fashion. ATP acts on $P2Y_2$ receptors, the primary ATP/UTP responsive receptor in tenocytes, activating the Gαq-protein, driving phospholipase C (PLC) and producing inositol trisphosphate (IP_3) and diacylglycerol (DAG). IP_3 acts on IP_3-sensitive Ca^{2+} channels in the ER to mobilize intracellular Ca^{2+}, and DAG activates a protein kinase C (PKC) pathway. Activation of adenyl cyclase activity yields cyclic adenosine monophosphate (cAMP), which stimulates cAMP-dependent protein kinase A (PKA), which may act at Raf in the kinase cascade. Initial action of ATP is terminated quickly by membrane-bound ecto-NTPases to its metabolites: ADP, AMP, and adenosine. Adenosine activates G protein-coupled P1 receptors, activating stimulatory (Gs) or inhibitory (Gi) signaling. Phosphoinositide 3-kinase (PI_3K) are activated by vascular endothelial growth factor receptor ($VEGFR_2$). Gap junctions pass IP_3, which propagates a Ca^{2+} wave from cell to cell after a mechanical signal is detected. Connexin hemichannels can pass ATP outside the cell. CAM, cell adhesion molecule; DES, desmosome; PPi, pyrophosphate; AP-1, activator protein-1; AP-2, activator protein-2; CREB, cAMP response element binding protein; MEK, MAPK/ERK kinase; NO, nitric oxide; PKB, protein kinase B; STAT, signal transducer and activator of transcription; SHC, Src homology protein complex; Crk, Src homology adaptor protein that binds paxillin and C3G; GRB_2, growth factor receptor binding adaptor protein linking receptors to the Ras pathway through FAK and SOS (Son of Sevenless), a guanine nucleotide exchange factor; Ras, GTPase that regulates activation of Raf; IF, intermediate filament; YAP/ TAZ, Yki transcription co-activators; TEAD, transcription factor; PYK2, a nonreceptor tyrosine kinase of the FAK family; PAK, p21-actived kinase; SSRE, shear stress response element; JNK, c-Jun N-terminal kinase; Hh, hedgehog; TRPV4, transient receptor potential vanilloid 4 channel; COX 1, cyclooxygenase 1; COX 2, cyclooxygenase. (Used with permission from Flexcell International Corp.)

signaling within this region controls the specification and differentiation of these progenitor pools [18, 22, 23, 82]. TGF-β signaling is crucial for the specification of the boney eminence and is also critical for the formation and differentiation of the tendon midsubstance [9, 22, 82]. BMP4 signaling also regulates the cartilage differentiation of bone eminence progenitors [22, 23]. Following the initial specification and differentiation of the enthesis progenitors, Hh signaling becomes prominent (Fig. 7.4 [68, 90]). Overexpression of Hh signaling in Scx-expressing progenitors yields production of enthesis extracellular matrix components within the midsubstance [68], suggesting that Hh signaling is important in the maturation process from enthesis progenitors to fibrocartilage cells. Hh signaling also regulates the mineralization of enthesis cells from unmineralized fibrochondrocytes to mineralized fibrochondrocytes [26, 36, 90]. Conditional deletion of Hh signaling in these cells leads to a severe reduction in mineralized fibrocartilage production. Therefore, Hh signaling is critical to the formation of the mineralized fibrocartilage zone of the enthesis.

The specification of the midsubstance progenitor pool is regulated via TGF-β signaling as removal of TGF-β signaling in limb mesenchyme leads to complete loss of limb tendon formation [82]. While the role of TGF-β signaling in the later differentiation and maturation events is less clear because a lack of inducible knockout models, it is likely that TGF-β signaling continues to play a prominent role because of its function in collagen transcription. TGF-β signaling has been shown to stimulate collagen transcription in a variety of cell types including tendon fibroblasts [19, 29, 66]. It stimulates the expression of both Scx and the Mohawk (Mkx) homeobox transcription factor, another factor important in tenogenesis, via Smad3 interactions [19]. In fact, Scx has known binding sites to both Cola1 and Col1a2 promoters [10, 66]. Therefore, TGF-β-mediated regulation of these transcription factors likely plays a role in collagen transcription during tendon development.

While the majority of *in vivo* cell signaling data related to tenogenic differentiation has come from embryological studies, less is known about the cell signaling events that regulate cell turnover during growth and tissue maintenance. In order to better understand the signaling events that regulate tenocyte turnover during these periods, *in vivo* models are needed as maintaining cell phenotype within *in vitro* systems is difficult [64]. While recent studies indicate that resident progenitors exist within tendon tissue and these cells can be isolated, expanded in culture, and display *in vitro* multipotent potential [20, 76], fewer studies have demonstrated expansion and differentiation of resident progenitors in an *in vivo* system. Using an alpha smooth muscle actin Cre reporter system (αSMA-CreERT2; R26R-tdTomato), Dyment et al. [37] demonstrated that SMA-labeled cells within the paratenon and/or perivasculature contribute to ScxGFP+ cells during tendon healing. This model system also labels a proliferative internal population within growing tendon [37] that may be a resident progenitor population. *In vivo* fate mapping models such as these are needed to characterize resident progenitor populations as they give rise to mature tenocytes. Unfortunately, improved markers are needed to better classify cells as they progress through the lineage. Hopefully with improved transgenic models and sensitive techniques such as single cell analyses, the markers that define cells at multiple stages of the lineage and the signaling pathways that regulate this process will be elucidated in the not so distant future.

Inflammation and Response to Injury

Inflammation in response to injury is complex, which is no less true in the tendon than in any other tissue (see Chap. 20). Tenocytes show an expected variety of reactions to inflammatory signals typically associated with the activation of resident immune cells and the recruitment of inflammatory cells. However, inflammation in tendon is complicated by the fact that injuries often take the form of mechanical injuries to the extracellular matrix. Therefore, the inflammatory

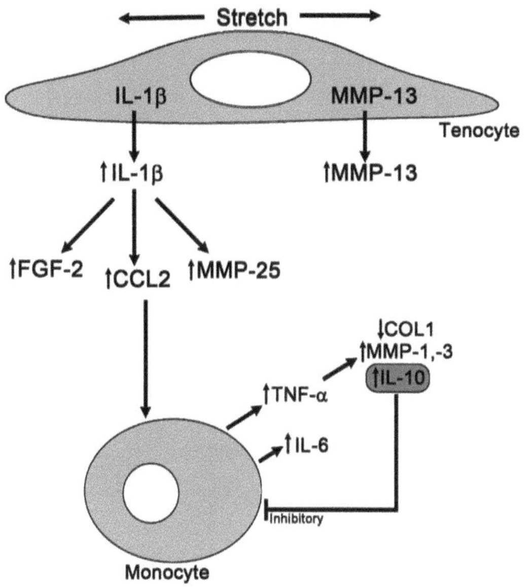

Fig. 7.5 Stretch induces tenocytes to produce interleukin-1β (IL-1β) and matrix metalloproteinase-13 (MMP-13). IL-1β can act in an autocrine and paracrine fashion to generate repair signals (such as fibroblast growth factor-2 (FGF-2)), matrix remodeling effectors (such as MMP-25) and inflammatory signals (such as chemokine (C-C motif) ligand 2 (CCL2)). CCL2 can act on tissue resident monocytes to induce differentiation and production of inflammatory cytokines such as IL-6 and tumor necrosis factor-α (TNF-α). TNF-α can then act on tenocytes to inhibit collagen deposition and release matrix remodeling enzymes, promoting inflammatory debridement of the injured area. TNF-α and IL-6 can induce the production of IL-10 [2], which can initiate a negative feedback loop, suppressing the damaged induced inflammatory response

response in tendon is likely integrated into a mechanosensing apparatus capable of detecting injurious mechanical forces. This field of study is underrepresented in the literature [77], but studies have elucidated some of the major players in the system (Fig. 7.5).

Tenocytes in isolated tendons exposed to forces capable of producing overt microstructural damage expressed IL-1β and MMP-13. Results of IL-1β siRNA transfection experiments demonstrated that MMP-13 expression was at least partially dependent on expression with IL-1β, suggesting an autocrine/paracrine role for IL-1β in tendon [95]. In a separate study, in an *ex-vivo* bovine model, stretch induced a coordinated pro-inflammatory response across multiple classes of distinct, yet unidentified, cells in tendon. Upon a stretch challenge, the matrix degrading enzymes MMP-1 and C1,2C were found expressed by putative tenocytes near microtears. At the same time, immunohistochemical analysis revealed two separate unidentified cell populations located near damaged tissue expressed IL-6 or COX-2 [93]. These mediators have the potential to not only initiate an inflammatory response, but also have the capacity to initiate tendon repair. IL-6 was required for tendon repair in a knockout mouse model [67], and PGE_2, a product of the COX-2 arachidonic acid pathway, increased tendon strength in a rat treatment model [43].

Part of the link between IL-1β and the observed downstream inflammatory effects may be TNF-α. TNF-α has an array of effects on tenocytes in culture including blocking the production of collagen I, stimulating the production of MMP-1, and (in conjunction with IL-6) causing the production of immunoregulatory IL-10 [53]. Increased matrix metalloprotease expression combined with reduced collagen expression may promote inflammatory infiltration of an injured tissue. Coordinated activity of IL-10 and IL-6 in an inflammatory environment can potentially suppress runaway inflammatory responses due to matrix damage caused by infiltrating inflammatory cells. Continued suppression, possibly combined with an influx of T-cells due to IL-15 production (see Table 7.1), can potentially lead to chronic tendon injury observations, such as expression of insulin like growth factor and suppressors of cytokine signaling [3], and eventually lead to repair responses [52].

Taken together, these findings describe an initiation of inflammatory debridement of injured tissue, complete with regulatory steps that can suppress damage-induced inflammatory insult in response to TNF-α. However, although some of the downstream inflammatory actors have been identified and associated with IL-1β, the link between IL-1β production and TNF-α production in tendon is not well understood. As part of an effort to dissect this link in tendon, our lab exposed primary human tenocytes in culture to IL-1β and performed microarray analysis. GO analysis [38]

Table 7.1 Primary human tenocytes isolated from three human flexor carpi radialis tendons were exposed to 100 pM interleukin-1β for 24 h *in vitro*

Gene	Fold increase	Function
		Inflammation
CCL2	6.5	Chemotactic factor for monocytes and basophils
IL1β	6.0	Activated by caspase 1. Wide array of activities including proliferation and differentiation.
CXCL3	5.5	Chemokine for neutrophils
CCL7	5.4	Chemokine for macrophages.
CCL8	5.3	Chemokine for multiple inflammatory cells
IFNA8	4.5	Interferon in the TGFβ pathway.
TSLP	4.1	Induces monocyte-mediated recruitment of T-cells
BCL2A1	4.0	Inhibits release of cytochrome c. Apoptosis-protective
BDKRB1	3.8	Receptor. Responds to inflammatory responses to tissue damage
VNN2	3.1	Promotes neutrophils migration
CXCL10	2.9	Chemokine for monocytes. Promotes cell adhesion.
TNFRSF11β	2.9	Decoy receptor. Blocks bone resorption
IL15	2.5	T-cell regulator. Promotes survival via BCL2
		Proliferation & Repair
EREG	5.1	Epiregulin. Ligand for epidermal growth factor receptor.
TNFAIP6	5.1	Matric binding factor that promotes matrix stability
LAMA4	4.2	Laminin 4, matrix protein
FGF2	3.6	Fibroblast mitogenic factor. Promotes migration and proliferation.
PLK2	3.0	Important in proliferation
CDK6	2.8	Allows G1 progression
PLA2G3	2.8	Secreted phospholipase. Generates arachidonic acid.
LOXL4	2.5	Metabolic enzyme essential in matrix crosslinking
LAMA3	2.1	Laminin 3, matrix protein
PLA2G4	2.0	Cytosolic phospholipase. Generates arachidonic acid.
		Remodeling
HABP2	3.8	Serine protease that cleaves fibrinogen. Binds to matrix and is involved in adhesion.
NINJ1	3.5	Adhesion molecule. Role in wound healing.
ADAM11	3.3	Promotes cell-cell-matrix interactions. Important in tissue repair and development
MMP25	2.8	Metalloprotease. Promotes inflammatory cell invasion of tissue
SOST	2.7	BMP antagonist
CD47	2.3	Cell adhesion signaling molecule
COL17A1	2.2	Hemidesmosome component

RNA was collected and analyzed by Agilent RNA array. Key upregulated genes associated with inflammation, remodeling, and proliferation and repair are shown.

indicated strong enrichment for cytokine-mediated signaling pathways (GO:0019221, $p = 2 \times 10^{-6}$), positive regulation of cytokine production (GO:0001819, $p = 6 \times 10^{-5}$), and inflammatory responses (GO:0006954, $p = 6 \times 10^{-6}$).

An array of inflammatory signals were up-regulated in human tenocytes isolated the flexor carpi radialis (Table 7.1), showing that autocrine/paracrine IL-1β signaling can lead to the production of a powerful inflammatory cocktail. Concurrently, the increased production of genes associated with remodeling can allow for the infiltration of inflammatory cells and promote enzymatic degradation of damaged matrix following injury. TGF-β3 expression was reduced more than sevenfold after IL-1β stimulation, which would reduce the production and deposition of collagen I and III [58]. Expression of secreted (PLA2G3) and cytosolic (PLA2G4) phospholipases supports the role of COX-2 and

PGE_2 in tendon repair. Expression of matrix laminin and growth factors (such as EREG and FGF-2) rounds out the beginnings of a set of controlled signaling sequences leading from inflammation to resolution to repair.

Conclusion

Further research revealing how tenocytes, *in vitro* and *in vivo*, respond to strain and ligands in health and disease will be revealed as the field focuses on altered pathways discerned from array and metabolomics data. For tendon biology, fundamental signaling pathways and well known pathways in strain responses and inflammation have been elucidated. Future work must reveal pathways that can be manipulated to prevent matrix degradation and even support functional matrix replacement with muscle, bone and nerve sparing strategies. Further studies are also needed to better elucidate the role of tendon-derived stem cells, a potential source for endogenous repair [65], and the role these cells play in tendon development, tendinopathy, rehabilitation as well as mechanotransduction [20, 80, 106].

References

1. Ascenzi MG, Lenox M, Farnum C (2007) Analysis of the orientation of primary cilia in growth plate cartilage: a mathematical method based on multiphoton microscopical images. J Struct Biol 158(3):293–306
2. Al-Sadi O, Schulze-Tanzil G, Kohl B, Lohan A, Lemke M, Ertel W, John T (2012) Tenocytes, pro-inflammatory cytokines and leukocytes: a relationship? Muscles Ligaments Tendons J 1(3):68–76
3. Archambault JM, Jelinsky SA, Lake SP, Hill AA, Glaser DL, Soslowsky LJ (2007) Rat supraspinatus tendon expresses cartilage markers with overuse. J Orthop Res 25(5):617–624
4. Arnoczky S, Lavagnino M, Gardner K, Tian T, Vaupel ZM, Stick JA (2004) Ex vivo static tensile loading inhibits MMP-1 expression in rat tail tendon cells through a cytoskeletally based mechanotransduction mechanism. J Orthop Res 22:328–333
5. Arnoczky SP, Lavagnino M, Egerbacher M, Caballero O, Gardner K, Shender MA (2008) Loss of homeostatic strain alters mechanostat "set point" of tendon cells in vitro. Clin Orthop Relat Res 466:1583–1591. doi:10.1007/s11999-008-0264-x
6. Arnoczky SP, Lavagnino M, Whallon JH, Hoonjan A (2002) In situ cell nucleus deformation in tendons under tensile load; a morphological analysis using confocal laser microscopy. J Orthop Res 20:29–35
7. Arnoczky SP, Tian T, Lavagnino M, Gardner K, Schuler P, Morse P (2002) Activation of stress-activated protein kinases (SAPK) in tendon cell following cyclic strain: the effects of strain frequency, strain magnitude, and cytosolic calcium. J Orthop Res 20:947–952
8. Backman LJ, Andersson G, Fong G, Alfredson H, Scott A, Danielson P (2013) Alpha-2 adrenergic stimulation triggers Achilles tenocyte hypercellularity: comparison between two model systems. Scand J Med Sci Sports 23(6):687–696. doi:10.1111/j.1600-0838.2011.01442.x
9. Baffi MO, Slattery E, Sohn P, Moses HL, Chytil A, Serra R (2004) Conditional deletion of the TGF-β type II receptor in Col2a expressing cells results in defects in the axial skeleton without alterations in chondrocyte differentiation or embryonic development of long bones. Dev Biol 276(1):124–142. doi:10.1016/j.ydbio.2004.08.027
10. Bagchi RA, Czubryt MP (2012) Synergistic roles of scleraxis and smads in the regulation of collagen 1α2 gene expression. Biochim Biophys Acta Mol Cell Res 1823(10):1936–1944. doi:10.1016/j.bbamcr.2012.07.002
11. Banes AJ, Enterline D, Bevin AG, Salisbury RE (1981) Repair of flexor tendon: effects of trauma and devascularization on collagen synthesis. J Trauma 21:505–512
12. Banes AJ, Horesovsky G, Tsuzaki M, Boitano S, Lawrence WT, Brown T, Weinhold P, Kenamond C, Benjamin M, Ralphs JR, McNeilly C, Burt J, Miller L (1999) The connexin 43 gap junction is a mechanosensitive gene in avian flexor tendon cells. In: Caterson B, Archer C, Benjamin M, Ralphs J (eds) The biology of the synovial joint. Harwood Academic Publishers, Amsterdam, pp 279–299
13. Banes AJ, Hu P, Xiao H, Sanderson MJ, Boitano S, Brigman B, Fischer T, Tsuzaki M, Brown TD, Almekinders LC, Lawrence WT (1995) Tendon cells of the epitenon and internal tendon compartment communicate mechanical signals through gap junctions and respond differentially to mechanical load and growth factors. In: Gordon SL, Blair SJ, Fine LJ (eds) Repetitive motion disorders of the upper extremity. American Academy of Orthopaedic Surgeons, Rosemont, pp 231–245
14. Banes AJ, Lee G, Graff R, Otey C, Archambault J, Tsuzaki M, Elfervig M, Qi J (2001) Mechanical forces and signaling in connective tissue cells: cellular mechanisms of detection, transduction, and responses to mechanical deformation. Curr Opin Orthop 12:389–396
15. Banes AJ, Tsuzaki M, Yamamoto J, Fischer T, Brigman B, Brown T, Miller L (1995)

Mechanoreception at the cellular level: the detection, interpretation, and diversity of responses to mechanical signals. Biochem Cell Biol 73(7–8):349–365
16. Banes AJ, Tsuzaki M, Yang X, Faber J, Bottlang M, Pederson D, Brown T (1998) Equibiaxial strain activates AP-1 and CRE transcription factors but not NF-κB or SSRE and up regulates Cx43 mRNA in tendon cells in vitro [abstract]. Transactions of the 44th annual meeting of the Orthopaedic Research Society 23:182
17. Banes AJ, Weinhold P, Yang X, Tsuzaki M, Bynum D, Bottlang M, Brown T (1999) Gap junctions regulate responses of tendon cells ex vivo to mechanical loading. Clin Orthop Relat Res 367: S356–S370
18. Bénazet J, Pignatti E, Nugent A, Unal E, Laurent F, Zeller R (2012) Smad4 is required to induce digit ray primordia and to initiate the aggregation and differentiation of chondrogenic progenitors in mouse limb buds. Development (Cambridge) 139 (22):4250–4260. doi:10.1242/dev.084822
19. Berthet E, Chen C, Butcher K, Schneider RA, Alliston T, Amirtharajah M (2013) Smad3 binds scleraxis and mohawk and regulates tendon matrix organization. J Orthop Res 31(9):1475–1483. doi:10.1002/jor.22382
20. Bi Y, Ehirchiou D, Kilts TM, Inkson CA, Embree MC, Sonoyama W, Li L, Leet AI, Seo BM, Zhang L, Shi S, Young MF (2007) Identification of tendon stem/progenitor cells and the role of the extracellular matrix in their niche. Nat Med 13(10):1219–1227. doi:10.1038/nm1630
21. Bjurholm A, Kreicbergs A, Schultzberg M, Lerner UH (1988) Parathyroid hormone and noradrenaline-induced enhancement of cyclic AMP in a cloned osteogenic sarcoma cell line (UMR 106) is inhibited by neuropeptide Y. Acta Physiol Scand 134:451–452
22. Blitz E, Sharir A, Akiyama H, Zelzer E (2013) Tendon-bone attachment unit is formed modularly by a distinct pool of scx-and Sox9-positive progenitors. Development (Cambridge) 140 (13):2680–2690. doi:10.1242/dev.093906
23. Blitz E, Viukov S, Sharir A, Shwartz Y, Galloway JL, Pryce BA, Johnson RL, Tabin CJ, Schweitzer R, Zelzer E (2009) Bone ridge patterning during musculoskeletal assembly is mediated through SCX regulation of Bmp4 at the tendon-skeleton junction. Dev Cell 17(6):861–873. doi:10.1016/j.devcel.2009.10.010
24. Bodle JC, Rubenstein CD, Phillips ME, Bernacki SH, Qi J, Banes AJ, Loboa EG (2013) Primary cilia: the chemical antenna regulating human adipose-derived stem cell osteogenesis. PLoS One 8(5), e62554
25. Boitano S, Dirksen ER, Sanderson MJ (1992) Intercellular propagation of calcium waves mediated by inositol trisphosphate. Science 258:292–295
26. Breidenbach AP, Aschbacher-Smith L, Lu Y, Dyment NA, Liu CF, Liu H, Wylie C, Rao M, Shearn JT, Rowe DW, Kadler KE, Jiang R, Butler DL (2015) Ablating hedgehog signaling in tenocytes during development impairs biomechanics and matrix organization of the adult murine patellar tendon enthesis. J Orthop Res 33(8):1142–1151. doi:10.1002/jor.22899
27. Bruzzone R, White TW, Paul DL (1996) Connections with connexins: the molecular basis of direct intercellular signaling. Eur J Biochem 238:1–27
28. Charles A (1998) Intercellular calcium waves in glia. Glia 24:39–49
29. Chen S, Yuan W, Mori Y, Levenson A, Trojanowska M, Varga J (1999) Stimulation of type I collagen transcription in human skin fibroblasts by TGF-β: Involvement of smad 3. J Investig Dermatol 112(1):49–57. doi:10.1046/j.1523-1747.1999.00477.x
30. Communi D, Govaerts C, Parmentier M, Boeynaems JM (1997) Cloning of a human purinergic P2Y receptor coupled to phospholipase C and adenylyl cyclase. J Biol Chem 272(51):31969–31973
31. Communi D, Janssens R, Suareq-Huerta N, Robaye B, Boeynaems J (2000) Advances in signaling by extracellular nucleotides: the role and transduction mechanisms of P2Y receptors. Cell Signal 12:351–360
32. Cotrina ML, Lin JH, Alves-Rodrigues A, Liu S, Li J, Azmi-Ghadimi H, Kang J, Naus CC, Nedergaard M (1998) Connexins regulate calcium signaling by controlling ATP release. Proc Natl Acad Sci U S A 95:15735–15740
33. Danielson P, Alfredson H, Forsgren S (2007) Studies on the importance of sympathetic innervation, adrenergic receptors, and a possible local catecholamine production in the development of patellar tendinopathy (tendinosis) in man. Microsc Res Tech. doi:10.1002/jemt
34. Docherty JR (1998) Subtypes of functional alpha1- and alpha2-adrenoceptors. Eur J Pharmacol 361:1–15
35. Donnelly E, Ascenzi MG, Farnum C (2010) Primary cilia are highly oriented with respect to collagen direction and long axis of extensor tendon. J Orthop Res 28(1):77–82. doi:10.1002/jor.20946
36. Dyment NA, Breidenbach AP, Schwartz AG, Russell RP, Aschbacher-Smith L, Liu H, Haqiwara Y, Jiang R, Thomopoulos S, Butler D, Rowe DW (2015) Gdf5 progenitors give rise to fibrocartilage cells that mineralize via hedgehog signaling to form the zonal enthesis. Dev Biol 405 (1):96–107. doi:10.1016/j.ydbio.2015.06.020
37. Dyment NA, Hagiwara Y, Matthews BG, Li Y, Kalajzic I, Rowe DW (2014) Lineage tracing of resident tendon progenitor cells during growth and natural healing. PLoS ONE 9(4), e96113. doi:10.1371/journal.pone.0096113
38. Eden E, Navon R, Steinfeld I, Lipson D, Yakhini Z (2009) GOrilla: a tool for discovery and visualization of enriched GO terms in ranked gene lists. BMC Bioinformatics 10:48

39. Evans WH, De Vuyst E, Leybaert L (2006) The gap junction cellular internet: connexin hemichannels enter the signalling limelight. Biochem J 397:1–14
40. Evans WH, Martin PE (2002) Gap junctions: structure and function. Mol Membr Biol 19:121–136
41. Farnum CE, Williams RM, Donnelly E (2009) Analyzing primary cilia by multiphoton microscopy. Methods Cell Biol 94:117–135
42. Farnum CE, Wilsman NJ (2011) Orientation of primary cilia of articular chondrocytes in three-dimensional space. Anat Rec (Hoboken) 294 (3):533–549
43. Ferry ST, Afshari HM, Lee JA, Dahners LE, Weinhold PS (2012) Effect of prostaglandin E2 injection on the structural properties of the rat patellar tendon. Sports Med Arthrosc Rehabil Ther Technol 4(1):2
44. Francke E, Sood A, Kenamond C, Yang X, Faber J, Boitano S, Bynum D, Sanderson M, Banes AJ (1998) ATP stimulates an increase in intracellular calcium in human tendon cells via purinergic receptors and temporally blocks gap junction signaling [abstract]. Transactions of the 44th annual meeting of the Orthopaedic Research Society 23:91
45. Gardner K, Arnoczky SP, Lavagnino M (2011) Effect of in vitro stress-deprivation and cyclic loading on the length of tendon cell cilia in situ. J Orthop Res 29:582–587
46. Goldberg S, Lampe PD, Nicholson BJ (1999) Selective transfer of endogenous metabolites through gap junctions composed of different connexins. Nat Cell Biol 1:457–459
47. Goodenough DA, Paul DL (2003) Beyond the gap: functions of unpaired connexon channels. Nat Rev Mol Cell Biol 4:1–10
48. Graham RM, Perez DM, Hwa J, Piascik MT (1996) Alpha 1-adrenergic receptor subtypes. Molecular structure, function, and signaling. Circ Res 78:737–749
49. Hart DA, Frank CB, Bray RC (1995) Inflammatory processes in repetitive motion and overuse syndromes: potential role of neurogenic mechanisms in tendons and ligaments. In: Gordon SL, Blair SJ, Fine LJ (eds) Repetitive motion disorders of the upper extremity. American Academy of Orthopaedic Surgeons, Rosemont, pp 247–262
50. Hoey DA, Tormey S, Ramcharan S, O'Brien FJ, Jacobs CR (2012) Primary cilia-mediated mechanotransduction in human mesenchymal stem cells. Stem Cells 30(11):2561–2570. doi:10.1002/stem.1235
51. Ingber DE, Wang N, Stamenovic D (2014) Tensegrity, cellular biophysics, and the mechanics of living systems. Rep Prog Phys 77(4):046603
52. Jelinsky SA, Li L, Ellis D, Archambault J, Li J, St. Andre M, Morris C, Seeherman H (2011) Treatment with rhBMP12 or rhBMP13 increase the rate and the quality of rat Achilles tendon repair. J Orthop Res 29(10):1604–1612
53. John T, Lodka D, Kohl B, Ertel W, Jammrath J, Conrad C, Stoll C, Busch C, Schulze-Tanzil G (2010) Effect of pro-inflammatory and immunoregulatory cytokines on human tenocytes. J Orthop Res 28(8):1071–1077
54. Jones B, Yang X, Koller BH, Banes AJ (2005b) P2Y1/P2Y2-Null tendons exhibit a decreased intracellular calcium response to uniaxial strain and ATP [abstract]. Transactions of the 51st annual meeting of the Orthopaedic Research Society 30:753
55. Jones BF, Wall ME, Carroll RL, Washburn S, Banes AJ (2005) Ligament cells stretch-adapted on a microgrooved substrate increase intercellular communication in response to a mechanical stimulus. J Biomech 38:1653–1664
56. Jorgensen NR, Geist ST, Civitelli R, Steinberg TH (1997) ATP- and gap junction-dependent intercellular calcium signaling in osteoblastic cells. J Cell Biol 139:497–506
57. Jorgensen NR, Henriksen Z, Brot C, Eriksen EF, Sorensen OH, Civitelli R, Steinberg TH (2000) Human osteoblastic cells propagate intercellular calcium signals by two different mechanisms. J Bone Miner Res 15:1024–1032
58. Klein MB, Yalamanchi N, Pham H, Longaker MT, Chang J (2002) Flexor tendon healing in vitro: effects of TGF-β on tendon cell collagen production. J Hand Surg Am 27A:615–620
59. Kumagai H, Sakamoto H, Guggino S, Filburn CR, Sacktor B (1989) Neurotransmitter regulation of cytosolic calcium in osteoblast-like bone cells. Calcif Tissue Int 45:251–254
60. Kuzma-Kuzniarska M, Yapp C, Pearson-Jones TW, Jones AK, Hulley PA (2014) Functional assessment of gap junctions in monolayer and three-dimensional cultures of human tendon cells using fluorescence recovery after photobleaching. J Biomed Opt 19 (1):15001. doi:10.1117/1.JBO.19.1.015001
61. Lavagnino M, Arnoczky SP, Gardner K (2011) In situ deflection of tendon cell-cilia in response to tensile loading: an in vitro study. J Orthop Res 29 (6):925–930. doi:10.1002/jor.21337
62. Lavagnino M, Arnoczky S (2005) In vitro alterations in cytoskeletal tensional homeostasis control gene expression in tendon cells. J Orthop Res 23:1211–1218
63. Lavagnino M, Arnoczky SP, Tian T, Vaupel Z (2003) Effect of amplitude and frequency of cyclic tensile strain on the inhibition of MMP-1 mRNA expression in tendon cells: an in vitro study. Connect Tissue Res 44:181–187
64. Lavagnino M, Wall ME, Little D, Banes AJ, Guilak F, Arnoczky SP (2015) Tendon mechanobiology: current knowledge and future research opportunities. J Orthop Res 33 (6):813–822. doi:10.1002/jor.22871
65. Lee CH, Lee FY, Tarafder S, Kao K, Jun Y, Yang G, Mao JJ (2015) Harnessing endogenous stem/

progenitor cells for tendon regeneration. J Clin Invest 125(7):2690–2701. doi:10.1172/JCI81589
66. Léjard V, Brideau G, Blais F, Salingcarnboriboon R, Wagner G, Roehrl MH, Noda M, Duprez D, Houillier P, Rossert J (2007) Scleraxis and NFATc regulate the expression of the pro-α1(I) collagen gene in tendon fibroblasts. J Biol Chem 282 (24):17665–17675. doi:10.1074/jbc.M610113200
67. Lin TW, Cardenas L, Glaser DL, Soslowsky LJ (2006) Tendon healing in interleukin-4 and interleukin-6 knockout mice. J Biomech 39(1):61–69
68. Liu C, Breidenbach A, Aschbacher-Smith L, Butler D, Wylie C (2013) A role for hedgehog signaling in the differentiation of the insertion site of the patellar tendon in the mouse. PLoS ONE 8(6), e65411. doi:10.1371/journal.pone.0065411
69. Maeda E, Ohashi T (2015) Mechano-regulation of gap junction communications between tendon cells is dependent on the magnitude of tensile strain. Biochem Biophys Res Commun 465(2):281–286. doi:10.1016/j.bbrc.2015.08.021
70. Malone AM, Anderson CT, Tummala P, Kwon RY, Johnston TR, Stearns T, Jacobs CR (2007) Primary cilia mediate mechanosensing in bone cells by a calcium-independent mechanism. Proc Natl Acad Sci U S A 104(33):13325–13330
71. McMurray RJ, Wann AK, Thompson CL, Connelly JT, Knight MM (2013) Surface topography regulates Wnt signaling through control of primary cilia structure in mesenchymal stem cells. Sci Rep 3:3545
72. McNeilly CM, Banes AJ, Benjamin M, Ralphs JR (1996) Tendon cells in vivo form a three dimensional network of cell processes linked by gap junctions. J Anat 189(Pt 3):593–600
73. Molla-Herman A, Ghossoub R, Blisnick T, Meunier A, Serres C, Silbermann F, Emmerson C, Romeo K, Bourdoncle P, Schmitt A, Saunier S, Spassky N, Bastin P, Benmerah A (2010) The ciliary pocket: an endocytic membrane domain at the base of primary and motile cilia. J Cell Sci 123 (Pt 10):1785–1795
74. Muhammad H, Rais Y, Miosge N, Ornan EM (2012) The primary cilium as a dual sense of mechanochemical signals in chondrocytes. Cell Mol Life Sci 69 (13):2101–2107
75. Nguyen AM, Jacobs CR (2013) Emerging role of primary cilia as mechanosensors in osteocytes. Bone 54(2):196–204
76. Ni M, Lui PP, Rui YF, Lee YW, Lee YW, Tan Q, Wong YM, Kong SK, Lau PM, Li G, Chan KM (2012) Tendon-derived stem cells (TDSCs) promote tendon repair in a rat patellar tendon window defect model. J Orthop Res 30(4):613–619. doi:10.1002/jor.21559
77. Nourissat G, Berenbaum F, Duprez D (2015) Tendon injury: from biology to tendon repair. Nat Rev Rheumatol 11:223–233
78. Poole CA, Jensen CG, Snyder JA, Gray CG, Hermanutz VL, Wheatley DN (1997) Confocal analysis of primary cilia structure and colocalization with the Golgi apparatus in chondrocytes and aortic smooth muscle cells. Cell Biol Int 21(8):483–494
79. Poole CA, Flint MH, Beaumont BW (1985) Analysis of the morphology and function of primary cilia in connective tissues: a cellular cybernetic probe? Cell Motil 5(3):175–193
80. Popov C, Burggraf M, Kreja L, Ignatius A, Schieker M, Docheva D (2015) Mechanical stimulation of human tendon stem/progenitor cells results in upregulation of matrix proteins, integrins and MMPs, and activation of p38 and ERK1/2 kinases. BMC Mol Biol 16:6. doi:10.1186/s12867-015-0036-6
81. Praetorius HA (2015) The primary cilium as sensor of fluid flow: new building blocks to model. A review in the theme: cell signaling: proteins, pathways and mechanisms. Am J Physiol Cell Physiol 308(3):C198–C208
82. Pryce BA, Watson SS, Murchison ND, Staverosky JA, Dunker N, Schweitzer R (2009) Recruitment and maintenance of tendon progenitors by TGFβ signaling are essential for tendon formation. Development 136:1351–1361. doi:10.1242/dev.027342
83. Qi J, Chi L, Bynum D, Banes AJ (2011) Gap junctions in IL-1β-mediated cell survival response to strain. J Appl Physiol (1985) 110(5):1425–1431. doi:10.1152/japplphysiol.00477.2010
84. Qi J, Chi L, Faber J, Koller B, Banes AJ (2007) ATP reduces gel compaction in osteoblast-populated collagen gels. J Appl Physiol 102(3):1152–1160
85. Qi J, Chi L, Wang J, Sumanasinghe R, Tsuzaki M, Bynum D, Banes AJ (2009) Primary cilia are modulated by serum, interleukin-1β and strain in human tenocytes [abstract]. Transactions of the 55th annual meeting of the Orthopaedic Research Society 35
86. Saez JC, Berthoud VM, Branes MC, Martinez AD, Beyer EC (2003) Plasma membrane channels formed by connexins: their regulation and functions. Physiol Rev 83:1359–1400
87. Sanderson MJ, Charles AC, Boitano S, Dirksen ER (1994) Mechanisms and function of intercellular calcium signaling. Mol Cell Endocrinol 98:173–187
88. Scott A, Cook JL, Hart DA, Walker DC, Duronio V, Khan KM (2007) Tenocyte responses to mechanical loading in vivo: a role for local insulin-like growth factor 1 signaling in early tendinosis in rats. Arthritis Rheum 56(3):871–881
89. Scott A, Khan KM, Duronio V (2005) IGF-I activates PKB and prevents anoxic apoptosis in Achilles tendon cells. J Orthop Res 23(5):1219–1225
90. Schwartz AG, Long F, Thomopoulos S (2015) Enthesis fibrocartilage cells originate from a population of hedgehog-responsive cells modulated by the loading environment. Development (Cambridge) 142 (1):196–206. doi:10.1242/dev.112714
91. Schweitzer R, Chyung JH, Murtaugh LC, Brent AE, Rosen V, Olson EN, Lassar A, Tabin CJ (2001) Analysis of the tendon cell fate using scleraxis, a

specific marker for tendons and ligaments. Development 128(19):3855–3866
92. Singla V, Reiter JF (2006) The primary cilium as the cell's antenna: signaling at a sensory organelle. Science 313(5787):629–633
93. Spiesz EM, Thorpe CT, Chaudhry S, Riley GP, Birch HL, Clegg PD, Screen HR (2015) Tendon extracellular matrix damage, degradation and inflammation in response to in vitro overload exercise. J Orthop Res 33(6):889–897
94. Sugimoto Y, Takimoto A, Akiyama H, Kist R, Scherer G, Nakamura T, Hiraki Y, Shukunami C (2012) Scx+/Scx9+ progenitors contribute to the establishment of the junction between cartilage and tendon/ligament. Development (Cambridge) 140 (11):2280–2288. doi:10.1242/dev.096354
95. Sun HB, Li Y, Fung DT, Majeska RJ, Schaffler MB, Flatow EL (2008) Coordinate regulation of IL-1beta and MMP-13 in rat tendons following subrupture fatigue damage. Clin Orthop Relat Res 466 (7):1555–1561
96. Tsuzaki M, Bynum D, Almekinders L, Faber J, Banes AJ (2005) Mechanical loading stimulates ecto-ATPase activity in human tendon cells. J Cell Biochem 96:117–125
97. Tsuzaki M, Bynum D, Almekinders L, Yang X, Faber J, Banes AJ (2003) ATP modulates load-inducible IL-1beta, COX 2, and MMP-3 gene expression in human tendon cells. J Cell Biochem 89:556–562
98. Tummala P, Arnsdorf EJ, Jacobs CR (2010) The Role of primary cilia in mesenchymal stem cell differentiation: a pivotal switch in guiding lineage commitment. Cell Mol Bioeng 3(3):207–212
99. Varma DR, Deng XF (2000) Cardiovascular alpha1-adrenoceptor subtypes: functions and signaling. Can J Physiol Pharmacol 78:267–292
100. Waggett AD, Benjamin M, Ralphs JR (2006) Connexin 32 and 43 gap junctions differentially modulate tenocyte response to cyclic mechanical load. Eur J Cell Biol 85:1145–1154
101. Wall ME, Banes AJ (2005) Early responses to mechanical load in tendon: role for calcium signaling, gap junctions and intercellular communication. J Musculoskelet Neuronal Interact 5:70–84
102. Wall ME, Faber JE, Yang X, Tsuzaki M, Banes AJ (2004) Norepinephrine-induced calcium signaling and expression of adrenoceptors in avian tendon cells. Am J Physiol Cell Physiol 287:C912–C918
103. Wall ME, Otey C, Qi J, Banes AJ (2007) Connexin 43 is localized with actin in tenocytes. Cell Motil Cytoskeleton 64(2):121–130
104. Wall ME, Weinhold PS, Siu T, Brown TD, Banes AJ (2007) Comparison of cellular strain with applied substrate strain in vitro. J Biomech 40(1):173–181
105. Whitfield JF (2008) The solitary (primary) cilium – a mechanosensory toggle switch in bone and cartilage cells. Cell Signal 20(6):1019–1024
106. Xu Y, Wang Q, Li Y, Gan Y, Li P, Li S, Zhou Y, Zhou Q (2015) Cyclic tensile strain induces tenogenic differentiation of tendon-derived stem cells in bioreactor culture. Biomed Res Int 2015:790804. doi:10.1155/2015/790804
107. Young NJ, Becker DL, Fleck RA, Goodship AE, Patterson-Kane JC (2009) Maturational alterations in gap junction expression and associated collagen synthesis in response to tendon function. Matrix Biol 28(6):311–323
108. Yu W, Dahl G, Werner R (1994) The connexin43 gene is responsive to oestrogen. Proc Biol Sci 255:125–132
109. Zimmerman KW (1898) Beitrage zur kenntnis einiger Drusen und Epithelien. Arch Mikrosk Anat 52:552–706

Methods of Assessing Human Tendon Metabolism and Tissue Properties in Response to Changes in Mechanical Loading

8

Katja M. Heinemeier, Michael Kjaer, and S. Peter Magnusson

Abstract

In recent years a number of methodological developments have improved the opportunities to study human tendon. Microdialysis enables sampling of interstitial fluid in the peritendon tissue, while sampling of human tendon biopsies allows direct analysis of tendon tissue for gene- and protein expression as well as protein synthesis rate. Further the ^{14}C bomb-pulse method has provided data on long-term tissue turnover in human tendon. Non-invasive techniques allow measurement of tendon metabolism (positron emission tomography (PET)), tendon morphology (magnetic resonance imaging (MRI)), and tendon mechanical properties (ultrasonography combined with force measurement during movement). Finally, 3D cell cultures of human tendon cells provide the opportunity to investigate cell-matrix interactions in response to various interventions.

Keywords

Collagen • Turnover • Metabolism • Protein synthesis • Bomb-pulse • Microdialysis • Magnetic resonance imaging (MRI) • Ultrasonography

K.M. Heinemeier (✉)
Institute of Sports Medicine, Department of Orthopedic Surgery M, Bispebjerg Hospital, Copenhagen, Denmark

Department of Biomedical Sciences, Centre for Healthy Ageing, Faculty of Health and Medical Sciences, University of Copenhagen, Copenhagen, Denmark
e-mail: KH@sund.ku.dk

M. Kjaer
Institute of Sports Medicine, Department of Orthopedic Surgery M, Bispebjerg Hospital, Copenhagen, Denmark

Center for Healthy Aging, Faculty of Health and Medical Sciences, University of Copenhagen, Copenhagen, Denmark
e-mail: michaelkjaer@sund.ku.dk

S.P. Magnusson
Institute of Sports Medicine, Department of Orthopedic Surgery M, Bispebjerg Hospital, Center for Healthy Aging, Faculty of Health and Medical Sciences, University of Copenhagen, Copenhagen, Denmark

Department of Physical Therapy, Musculoskeletal Rehabilitation Research Unit, Bispebjerg Hospital, Copenhagen, Denmark
e-mail: P.Magnusson@sund.ku.dk

P.W. Ackermann and D.A. Hart (eds.), *Metabolic Influences on Risk for Tendon Disorders*, Advances in Experimental Medicine and Biology 920, DOI 10.1007/978-3-319-33943-6_8

Abbreviations

Asp	Aspartate
CSA	Cross-sectional area
FGD	Fluorodeoxyglucose
MMP	Matrix metallo proteinase
MRI	Magnetic resonance imaging
PET	Positron emission tomography
PINP	Pro-collagen I N-terminal peptide
PICP	Pro-collagen I C-terminal peptide
ICTP	Human C-telopeptide of type I collagen
UTC	Ultrasound Tissue Characterization

Introduction

Until recently much of what was known about tendon physiology and homeostasis has largely been based on animal studies. Clearly such models allow for a well controlled and substantial tissue sampling and testing, as well as for marked potential manipulations (e.g. genetic) and perturbations (e.g. mechanical and chemical damage). Several important proteins for the tendon structure have been identified (e.g. decorin, collagen type V or XI) by using specific genetic knock out mice where certain proteins were lacking and where the tendon structure was demonstrated to be markedly altered [1]. Using such models have often the advantage of a reproducible intervention often of marked nature, but in the end will be dependent upon whether or not the animal model reflects the human phenotype and dynamics of adaptation. Only occasional observations of human phenotypes that genetically are impaired in regards to certain proteins (e.g. collagen type V) can demonstrate altered tendon structure and tendon mechanical properties [2], but its not possible to get human models where genetic changes can be induced acutely.

Using almost exclusively animal and in vitro cell models, we have obtained – and can still obtain – several important findings of relevance for tendon metabolism and tissue mechanical properties, but over the later years it has also become clear that animal models of mechanical loading or unloading do not adequately reflect the tissue adaptation (or lack thereof) in human tendon, nor does tendon overloading in animal models in any satisfactory manner reflect the pathological changes observed in clinical tendon overuse (tendinopathy). Thus, we are currently facing a dilemma in tendon research. Do we on one side want to study changes in animal and in vitro study models that allow for controlled changes but do not reflect human phenotypical events, or do we want the limitations that study of human tendon in vivo confronts us with? With regards to the latter, in the last two decades a number of methodological developments have provided the opportunity to examine the response of human tendon cellular and tissue adaptation to mechanical loading (Table 8.1). Herein we will briefly outline the various methods available to examine tendon in humans.

Table 8.1 General overview of some methods that can be used in assessing collagen synthesis and degradation in tendon tissue, primarily human

Synthesis
mRNA collagen
Enzyme activity
Procollagen peptides (PICP, PINP)
(microdialysis)
Incorporation of amino/imino acids
(unstable or stable isotopes) (e.g. ^{13}C-proline)
Bomb-pulse technique (^{14}C in tissue)
Degradation
Protease activity (MMP)
Collagen degradation fragments (ICTP)
Stable isotope disappearance (D_20)

Given the slow turnover of tendon tissue in general, several methods have a limited use to detect small changes in collagen turnover, and some are more directly than others

Estimation of Tendon Tissue Metabolism and Tissue Turnover

Although tendons possess a relatively low metabolic activity (see Chap. 2), physical loading of tendon can demonstrate increased interstitial tissue concentrations of both carbohydrate (glucose and lactate), lipid (free fatty acids) and protein (urea) metabolism, either when determined with the use of semi-permeabel plastic tubes placed either intratendinously (typical in animals) [3] or peri-tendinously (typical in humans) [4]. This technique is called microdialysis and allows for passage of specific molecules based on their size and weight across the plastic membrane, and by sampling the fluid from the plastic tubes, and controlling for the exchange of the substance of interest (recovery) it allows for estimation of interstitial tissue concentration of substances either within or in close proximity of the tendon. Interestingly, this technique can also be used to sample collagen-related proteins, as well as nociceptive or inflammatory substances and can thus be informative of the interplay between molecules in relation to tendon tissue. Increased metabolic activity in human tendon has also been demonstrated in relation to acute exercise by using positron emission tomography (PET) and injection of radiolabeled fluorodeoxyglucose (^{18}FGD), whereby local radioactivity incorporation into the tissue was larger when tissue was exercising [5].

Several methods exist for estimating the turnover of structural proteins (e.g. collagen and other matrix proteins) in biological systems. Many of these are quite indirect in nature, and the most simple one is to determine enzyme concentration or better activity of substances involved in formation of the main target proteins (e.g. P-4-H enzyme activity involved in collagen formation) [6]. Similar to using the enzymes involved in matrix protein formation – also the enzymes involved in proteolytic activity of collagen and other proteins of the tendon – the matrix metallo proteinase (MMP's) activity can be determined in tendon tissue. This has been used not only to demonstrate dynamic changes with exercise, but also in relation to pathological situations in relation to tendon (rupture or tendinopathy) [7, 8].

Next in the line of methods used to determine tissue protein turnover of tendon is the determination of mRNA of specific proteins (e.g. pro-collagen type I) that reflects activation of certain signaling pathways and is often correlated with increased protein synthesis. Clearly the ability of detecting an increase in mRNA depends upon the timing of tissue sampling after a certain perturbation as mRNAs are transient in nature, and further it depends upon the biochemical "need" for an increase, and often sufficient amounts of mRNA is already present for the relevant protein synthesis to take place. To take it a step further from this approach, the determination of protein concentration itself (e.g. collagen content) will provide information of the present situation in the tissue. However it does not reveal any dynamics of proteins in the tissue as not does reflect any turnover activity. Using the before mentioned microdialysis technique, one can also get an indirect estimate of the collagen turnover in relation to tendon. As pro-collagen pro-peptides either at the C- or N-terminal end of the molecule are cleaved off during formation of collagen , microdialysis sampling of these peptides from respective procollagen types (e.g. PINP or PICP from procollagen I) and subsequent determination of the interstitial concentration of these substances will provide an estimate of changes in collagen synthesis [9]. Likewise, the degradation products when collagen is broken down (e.g. ICTP) can also be sampled with microdialysis and used as an estimate for protein degradation.

A way to determine synthesis of collagen (or other matrix proteins) includes administration of labeled amino acids (either with radioactive- or stable isotopes), to animals or humans. The labeled amino acids can be traced in the tissue-proteins, into which they are incorporated, and a high level of incorporation will indicate a high protein synthesis rate and indirectly a high turnover rate. In tendon tissue, early animal studies investigating the incorporation of radioactively marked amino acids showed almost no turnover of tendon collagen in adult rats [10] and similarly, a relatively slow collagen synthesis was seen in rabbit cartilage [11]. Opposing this, more recent studies on humans indicated a surprisingly high collagen turnover in tendon tissue with levels corresponding to the turnover of muscle contractile protein (0.045 % /h)

[12, 13]. However, the measure of short-term incorporation of labeled amino acids suffers from the complication that it registers synthesis of all new collagen even though this newly synthesized collagen may often be rapidly degraded, and thus never incorporated into the tissue [14]. In addition, one of the limitations by using this method in human tendon tissue is the need for tissue sampling, and it has been attempted to use a more tissue-wise non-invasive method and to detect the positron emission tomography (PET) after injection of short term radio-labeled proline. So far when used in animal models the short lived isotope could not detect any increased dynamics of proline in tendon tissue after exercise, but could in combination with direct determination of incorporated proline into collagen demonstrate that approximately 20 % of the proline taken up in the tissue was incorporated into tendon tissue [15]. Whether this methodology can be developed further for human studies is yet to be shown. In regards to collagen breakdown in tendon tissue of humans, a new attempt has been used, by providing oral deuterium ("heavy water", D_2O) over a period of weeks to humans. When stopping the ingestion of this, the decay of tissue content of this substance will estimate the decay and thus breakdown for specific proteins in the tissue [16].

A method of determining protein age is to detect levels of conversion of the amino acid aspartate from the L-isomer (L-Asp) to the D-isomer (D-Asp). This process, called racemization, happens over time and thus, old proteins will have higher levels of D-Asp than newly synthesized proteins. One study that employed this method on horse tendon indicated a turnover of tendon collagen of 0.25 %/year – equivalent to a half life of almost 200 years [17], which is a clear contradiction to the results obtained with amino acid tracers in humans [12, 13]. The racemization measure has also been used in cartilage and these studies support at slow turnover of cartilage collagen, compared to other tissues, with half-lives of 100–400 years [18, 19]. Unfortunately, the rate of racemization depends on several factors, including the amino acid sequence of the protein, the pH and not least the temperature, which has a very high impact on the degree of conversion from L-asp to D-asp (a 4 °C increase doubles the rate of racemization) [17]. Because of this, racemization may be a good measure if identical proteins in identical environments are compared but lacks accuracy if these parameters vary.

The Carbon-14 bomb-pulse method is a unique method that can accurately determine life-long tissue replacement rates. We have recently taken advantage of this method to solve the controversy of tendon tissue turnover rates (described above) and found that at least the core of the Achilles tendon in adult humans is essentially inert [20]. This method takes advantage of the dramatic increase in atmospheric ^{14}C levels peaking in 1963, due to testing of nuclear bombs in the 1960s and the subsequent exponential decline after the testing of nuclear bombs was stopped (Fig. 8.1). This dramatic change of the atmospheric ^{14}C is called the "bomb pulse". Organisms that have lived through this pulse will have incorporated corresponding amounts of ^{14}C in their tissues, since all living organisms incorporate ^{14}C from the atmospheric CO_2 and equilibrium exists between levels of ^{14}C in the atmosphere and in the organism (Fig. 8.1).

Most tissues in living organisms are gradually replaced over weeks or months and will have a content of ^{14}C corresponding to the current level in the atmosphere [21] *(the half life of ^{14}C is 5730 years, and thus radioactive decay of the isotope is unimportant in this context)*. However, if one imagines a tissue or tissue component that is not turned over after the initial formation, this tissue/component will contain a level of ^{14}C corresponding to the amount that was present in the atmosphere when the tissue was formed. In other words the changes over time in ^{14}C caused by the bomb pulse can be utilized for determining the turnover of tissue from organisms borne in the years spanning the bomb-pulse [22]. The method of ^{14}C bomb pulse gives a unique possibility to get a better understanding of the slow turnover tissues such as tendon and cartilage, and has already proved important in bringing tendon research an important step forward [20]. The advantages of the bomb pulse method are that:

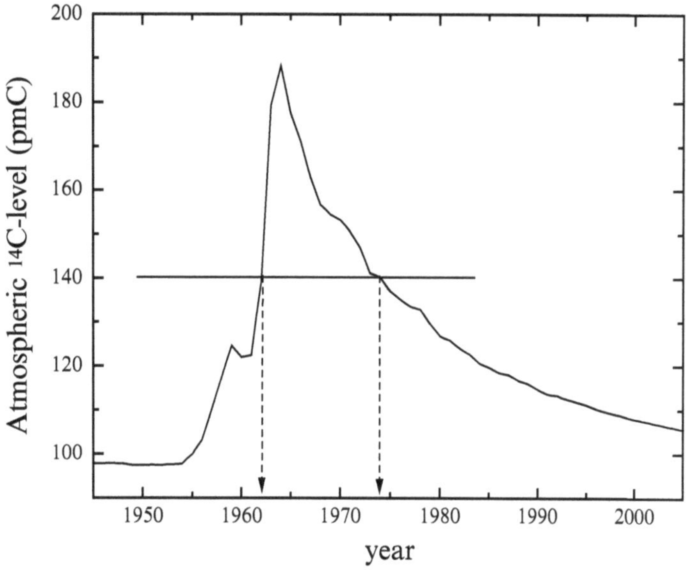

Fig. 8.1 "The bomb pulse", levels of ^{14}C in the atmosphere since 1945 (pmC = percent modern Carbon). The solid line indicates how a certain amount of ^{14}C may be used for dating (two possible dates)

(1) The "labeling" of tissue with ^{14}C starts from the very beginning embryogenesis, and thus, the turnover of tissue over an entire life-span can be estimated (in contrast to the acute measures of incorporation of stable/radioactive isotopes discussed above). (2) This long-term labeling overcomes the problem of acute changes in collagen synthesis – with possible subsequent fast breakdown. (3) The incorporation of ^{14}C is not influenced by parameters such as temperature, pH and amino acid sequence as is the case for the racemization of aspartate (discussed above). Naturally, the use of this method will be limited to the next few decades since ^{14}C levels have now almost returned to "pre-bomb" levels, and only organisms born before ca. 1975 are relevant for analyses of tissue turnover.

The successful use of the ^{14}C bomb-pulse to determine turnover of connective tissue in humans opens up a large range of opportunities for further studies that can substantially increase our basic knowledge of these tissues that play an essential role in human movement.

Cell Models

The use of tendon cells to study in isolated cultures and allowing these to form tendon constructs, either in 2D or 3D has been widely evaluated, and lately succesful formation of 3D tendon constructs from human tendon cells has been used [23–25]. The cell models in relation to tendon have been very informative in regards to testing cell communication and cell activity when stimulated by potential hormonal growth factors [26]. Further, the stimulation of tendon constructs with mechanical loading or the opposite namely de-tensioning has demonstrated what mechanical factors are very important for tendon cell signaling [27, 28]. The drawback in using cell models clearly are that the cells are taken completely out of their regular environment and may respond very different than when studied in vivo (see Chap. 7).

Size of the Whole Human Tendon, *in vivo*

Both strength training and habitual loading of tendons appear to be associated with an increase in tendon size [29–33], suggesting that the responses to elevated loading results in a net increase of tendon tissue. Studies that have reported increased tendon cross-sectional area (CSA) were all measured by magnetic resonance imaging (MRI). Because of its high resolution, and seemingly good contrast between different tissues compared to other available imaging modalities, including ultrasound, it is the preferable assessment tool to detect modest changes in tendon CSA [34]. But until recently it is unknown how well the contrast in the image corresponds with the actual borders of the tendon and thereby whether MRI measurements may under or overestimate tendon CSA. It has been shown that an underestimation of 2.8 % of the CSA can be expected using 3.0 Tesla MRI compared to the direct measurements from a mold, indicating that using this optimized CSA measuring procedure the contrast of MRI is largely able to represent the tissue uniquely. By comparison, measurements on the gray scale images resulted in significant greater underestimation of tendon CSA. The reason for this underestimation is likely that the high intensity signal of the surrounding tissue overlaps the low intensity of the tendon. Using a higher contrast with 3.0 Tesla enables the examiner to better trace the outline of the tendon, and thereby reduce the error of measuring tendon CSA in vivo [34]. In this way the typical error percent of measures between 2 examiners can below 2 %. Thus, with a well defined protocol it is possible to detect those modest changes (5–6 %) that can be expted to accompany short-term resistance training interventions.

Ultrasonography has emerged as an attractive alternative to the more costly and time-consuming MRI in order to determination of tendon CSA. However, the reproducibility is highly questionable [35], and therefore MRI continues to be a far better approach to accurately assess CSA of tendon. Moreover, a recent and novel ultrasonography technique, called Ultrasound Tissue Characterization (UTC) has emerged a popular clinical tool. It semi-quantifies the echo information from the tendon into four echo types that have been suggested to relate to various stages in tendon pathology [36]. However, to date there is no firm data to confirm the validity of this technique.

Mechanical Properties of Whole Human Tendon in vivo

Measurements of mechanical properties of tendon in vitro have been performed for more than 70 years [37]. However, relatively recent advances in imaging techniques such as ultrasonography has enabled the examination of human tendon tissues under physiological conditions *in vivo* [38]. This technique, based on B-mode ultrasonography, enable the estimation of tendon-aponeurosis mechanical properties by combining force and the corresponding tissue deformation. These techniques normally apply load via voluntary muscle contraction and determine the magnitude of the applied load by externally measuring the generated joint moments on the limb in question. Tendon tissue deformation is measured using a non-invasive ultrasound. In addition to providing the proper living physiological environment, testing in vivo also overcomes the gripping issues encountered in vitro, however, although other difficulties are introduced. First, it is of course impossible to perform tests to failure, which limits the mechanical parameters that can be determined. Furthermore, the force on the tendon is back-calculated from external moments and there can be some uncertainty in determining the relevant moment arms across which the tendon acts [39]. In addition, the external moment may not be generated solely through the tendon in question and agonist/antagonist co-activation may have to be accounted for [40]. With respect to deformation, ultrasound produces a 2D image and the measured deformation can therefore be affected by out-of-plane motion. Measurement of deformation requires tracking of anatomical landmarks such as the insertions into muscle and bone, but for long tendons it may not be possible to image both ends of the tendon simultaneously [41]. For this reason, the motion of just the muscle insertion is sometimes measured, while assuming that the bone insertion remains fixed during isometric contractions, however, this assumption may not be valid [42]. A comprehensive outline of technical challenges with the technique are outlined in a recent review [43]. Taken together, the methodological challenges in determining human tendon properties in vivo, the determination of patella tendon characteristics is at this stage far better that for Achilles tendon or other tendon types.

A fairly new tracking method called speckle tracking makes use of ultrasound contrast in the form of bright dots within the tendon. This may be a better way of assessing tendon strains without the need for anatomical landmarks, and in addition, speckle tracking enables local strain measures within the tendon [44, 45].

Mechanical Properties of Human Tendon Fascicles

If tendon tissue can be obtained during surgery, then it is possible to measure the mechancial properties of isolated fascicles in a tensile testing apparatus in which the isolated tissue can be gripped and placed under controlled load and deformation [46–49]. Within such a setup there are a number of factors that may influence the measured properties. One primary concern is the gripping where slippage, damage and stress concentration may occur (i.e. end-effects), leading to overestimates of deformation and underestimates of strength and stiffness [50]. These effects are exacerbated with shorter and thicker test specimens [51, 52]. Different gripping methods have been attempted with varying success, but even the best cannot entirely avoid these issues [53]. To further alleviate end-effects and to separate region-specific strains, local strain gauges (often optical) can be applied to the tendon [54, 55]. This can either be to specifically determine mechanical response at different structural levels [56, 57], to do tests where whole tissues are unavailable (biopsies) [58], to enable multiple experiments on the same tendon [59] or to obtain samples with greater aspect ratio in order to minimize end-effects. It is also possible to go even further down in scale to make measurements on individual collagen fibrils and molecules using different techniques such as atomic force microscopy, optical tweezers or custom made systems [60–62]. Such experiments can be technically demanding and are only just beginning to be applied to clinical questions [63]. Measuring mechanical properties at smaller scales has the potential to reveal underlying mechanistic relationships and filter out some of the structural and compositional complexity at the higher levels of hierarchy [46].

Conclusion

Taken together, several methods do exist to study molecular, biochemical, physiological and biomechanical adaptation of tendon tissue to either improved or decreased amounts of mechanical loading. We have today much better tools to assess tendon metabolism and structural changes, and many of these techniques have provided new insight into tendon adaptive behavior. However, they also have limitations in regards to accuracy, reproducibility and relevance in relation to "real life" tendon behavior in the living human.

References

1. Mienaltowski MJ, Birk DE (2014) Mouse models in tendon and ligament research. Adv Exp Med Biol 802:201–230
2. Nielsen RH, Couppe C, Jensen JK, Olsen MR, Heinemeier KM, Malfait F et al (2014) Low tendon stiffness and abnormal ultrastructure distinguish classic Ehlers-Danlos syndrome from benign joint hypermobility syndrome in patients. FASEB J 28 (11):4668–4676
3. Langberg H, Olesen JL, Bulow J, Kjaer M (2002) Intra- and peri-tendinous microdialysis determination of glucose and lactate in pigs. Acta Physiol Scand 174 (4):377–380
4. Langberg H, Skovgaard D, Karamouzis M, Bulow J, Kjaer M (1999) Metabolism and inflammatory mediators in the peritendinous space measured by microdialysis during intermittent isometric exercise in humans. J Physiol 515(Pt 3):919–927
5. Bojsen-Moller J, Kalliokoski KK, Seppanen M, Kjaer M, Magnusson SP (2006) Low-intensity tensile loading increases intratendinous glucose uptake in the Achilles tendon. J Appl Physiol 101(1):196–201

6. Kovanen V (1989) Effects of ageing and physical training on rat skeletal muscle. An experimental study on the properties of collagen, laminin, and fibre types in muscles serving different functions. Acta Physiol Scand Suppl 577:1–56
7. Riley GP, Curry V, DeGroot J, van El B, Verzijl N, Hazleman BL et al (2002) Matrix metalloproteinase activities and their relationship with collagen remodelling in tendon pathology. Matrix Biol 21 (2):185–195
8. Jones GC, Corps AN, Pennington CJ, Clark IM, Edwards DR, Bradley MM et al (2006) Expression profiling of metalloproteinases and tissue inhibitors of metalloproteinases in normal and degenerate human achilles tendon. Arthritis Rheum 54(3):832–842
9. Langberg H, Skovgaard D, Petersen LJ, Bulow J, Kjaer M (1999) Type I collagen synthesis and degradation in peritendinous tissue after exercise determined by microdialysis in humans. J Physiol 521 (Pt 1):299–306
10. Neuberger A, Perrone JC, Slack HG (1951) The relative metabolic inertia of tendon collagen in the rat. Biochem J 49(2):199–204
11. Repo RU, Mitchell N (1971) Collagen synthesis in mature articular cartilage of the rabbit. J Bone Joint Surg Br 53(3):541–548
12. Miller BF, Olesen JL, Hansen M, Dossing S, Crameri RM, Welling RJ et al (2005) Coordinated collagen and muscle protein synthesis in human patella tendon and quadriceps muscle after exercise. J Physiol 567 (Pt 3):1021–1033
13. Babraj JA, Cuthbertson DJ, Smith K, Langberg H, Miller B, Krogsgaard MR et al (2005) Collagen synthesis in human musculoskeletal tissues and skin. Am J Physiol Endocrinol Metab 289(5):E864–E869
14. McAnulty RJ, Laurent GJ (1987) Collagen synthesis and degradation in vivo. Evidence for rapid rates of collagen turnover with extensive degradation of newly synthesized collagen in tissues of the adult rat. Coll Relat Res 7(2):93–104
15. Skovgaard D, Kjaer A, Heinemeier KM, Brandt-Larsen M, Madsen J, Kjaer M (2011) Use of cis-[18 F]fluoro-proline for assessment of exercise-related collagen synthesis in musculoskeletal connective tissue. PLoS One 6(2), e16678
16. Holm L, O'Rourke B, Ebenstein D, Toth MJ, Bechshoeft R, Holstein-Rathlou NH et al (2013) Determination of steady-state protein breakdown rate in vivo by the disappearance of protein-bound tracer-labeled amino acids: a method applicable in humans. Am J Physiol Endocrinol Metab 304(8): E895–E907
17. Thorpe CT, Streeter I, Pinchbeck GL, Goodship AE, Clegg PD, Birch HL (2010) Aspartic acid racemization and collagen degradation markers reveal an accumulation of damage in tendon collagen that is enhanced with aging. J Biol Chem 285 (21):15674–15681
18. Sivan SS, Wachtel E, Tsitron E, Sakkee N, van der Ham F, DeGroot J et al (2008) Collagen turnover in normal and degenerate human intervertebral discs as determined by the racemization of aspartic acid. J Biol Chem 283(14):8796–8801
19. Maroudas A, Palla G, Gilav E (1992) Racemization of aspartic acid in human articular cartilage. Connect Tissue Res 28(3):161–169
20. Heinemeier KM, Schjerling P, Heinemeier J, Magnusson SP, Kjaer M (2013) Lack of tissue renewal in human adult Achilles tendon is revealed by nuclear bomb (14)C. FASEB J 27(5):2074–2079
21. Goodsite ME, Rom W, Heinemeier J, Lange T, Ooi S, Appleby PG et al (2001) High-resolution AMS C-14 dating of post-bomb peat archives of atmospheric pollutants. Radiocarbon 43(2B):495–515
22. Lynnerup N, Kjeldsen H, Heegaard S, Jacobsen C, Heinemeier J (2008) Radiocarbon dating of the human eye lens crystallines reveal proteins without carbon turnover throughout life. PLoS One 3(1), e1529
23. Bayer ML, Yeung CY, Kadler KE, Qvortrup K, Baar K, Svensson RB et al (2010) The initiation of embryonic-like collagen fibrillogenesis by adult human tendon fibroblasts when cultured under tension. Biomaterials 31(18):4889–4897
24. Kalson NS, Holmes DF, Herchenhan A, Lu Y, Starborg T, Kadler KE (2011) Slow stretching that mimics embryonic growth rate stimulates structural and mechanical development of tendon-like tissue in vitro. Dev Dyn 240(11):2520–2528
25. Paxton JZ, Grover LM, Baar K (2010) Engineering an in vitro model of a functional ligament from bone to bone. Tissue Eng Part A 16(11):3515–3525
26. Banes AJ, Horesovsky G, Larson C, Tsuzaki M, Judex S, Archambault J et al (1999) Mechanical load stimulates expression of novel genes in vivo and in vitro in avian flexor tendon cells. Osteoarthritis Cartilage 7(1):141–153
27. Bayer ML, Schjerling P, Herchenhan A, Zeltz C, Heinemeier KM, Christensen L et al (2014) Release of tensile strain on engineered human tendon tissue disturbs cell adhesions, changes matrix architecture, and induces an inflammatory phenotype. PLoS One 9 (1), e86078
28. Hagerty P, Lee A, Calve S, Lee CA, Vidal M, Baar K (2012) The effect of growth factors on both collagen synthesis and tensile strength of engineered human ligaments. Biomaterials 33:6355–6361
29. Kongsgaard M, Aagaard P, Kjaer M, Magnusson SP (2005) Structural Achilles tendon properties in athletes subjected to different exercise modes and in Achilles tendon rupture patients. J Appl Physiol 99 (5):1965–1971
30. Kongsgaard M, Reitelseder S, Pedersen TG, Holm L, Aagaard P, Kjaer M et al (2007) Region specific patellar tendon hypertrophy in humans following resistance training. Acta Physiol (Oxf) 191(2):111–121

31. Couppe C, Kongsgaard M, Aagaard P, Hansen P, Bojsen-Moller J, Kjaer M et al (2008) Habitual loading results in tendon hypertrophy and increased stiffness of the human patellar tendon. J Appl Physiol 105 (3):805–810
32. Arampatzis A, Karamanidis K, Albracht K (2007) Adaptational responses of the human Achilles tendon by modulation of the applied cyclic strain magnitude. J Exp Biol 210(Pt 15):2743–2753
33. Seynnes OR, Erskine RM, Maganaris CN, Longo S, Simoneau EM, Grosset JF et al (2009) Training-induced changes in structural and mechanical properties of the patellar tendon are related to muscle hypertrophy but not to strength gains. J Appl Physiol 107(2):523–530
34. Couppe C, Svensson RB, Sodring-Elbrond V, Hansen P, Kjaer M, Magnusson SP (2014) Accuracy of MRI technique in measuring tendon cross-sectional area. Clin Physiol Funct Imaging 34(3):237–241
35. Ekizos A, Papatzika F, Charcharis G, Bohm S, Mersmann F, Arampatzis A (2013) Ultrasound does not provide reliable results for the measurement of the patellar tendon cross sectional area. J Electromyogr Kinesiol 23(6):1278–1282
36. van Schie HT, de Vos RJ, de Jonge S, Bakker EM, Heijboer MP, Verhaar JA et al (2010) Ultrasonographic tissue characterisation of human Achilles tendons: quantification of tendon structure through a novel non-invasive approach. Br J Sports Med 44 (16):1153–1159
37. Cronkite AE (1936) The tensile strength of human tendons. Anat Rec 64(2):173–186
38. Fukunaga T, Ito M, Ichinose Y, Kuno S, Kawakami Y, Fukashiro S (1996) Tendinous movement of a human muscle during voluntary contractions determined by real-time ultrasonography. J Appl Physiol 81 (3):1430–1433
39. An KN, Takahashi K, Harrigan TP, Chao EY (1984) Determination of muscle orientations and moment arms. J Biomech Eng 106(3):280–282
40. Aagaard P, Simonsen EB, Andersen JL, Magnusson SP, Bojsen-Moller F, Dyhre-Poulsen P (2000) Antagonist muscle coactivation during isokinetic knee extension. Scand J Med Sci Sports 10(2):58–67
41. Kubo K, Kanehisa H, Miyatani M, Tachi M, Fukunaga T (2003) Effect of low-load resistance training on the tendon properties in middle-aged and elderly women. Acta Physiol Scand 178(1):25–32
42. Hansen P, Bojsen-Moller J, Aagaard P, Kjaer M, Magnusson SP (2006) Mechanical properties of the human patellar tendon, in vivo. Clin Biomech (Bristol, Avon) 21(1):54–58
43. Seynnes OR, Bojsen-Moller J, Albracht K, Arndt A, Cronin NJ, Finni T et al (2015) Ultrasound-based testing of tendon mechanical properties: a critical evaluation. J Appl Physiol 118(2):133–141
44. Korstanje JW, Selles RW, Stam HJ, Hovius SE, Bosch JG (2010) Development and validation of ultrasound speckle tracking to quantify tendon displacement. J Biomech 43(7):1373–1379
45. Arndt A, Bengtsson AS, Peolsson M, Thorstensson A, Movin T (2012) Non-uniform displacement within the Achilles tendon during passive ankle joint motion. Knee Surg Sports Traumatol Arthrosc 20 (9):1868–1874
46. Svensson RB, Hansen P, Hassenkam T, Haraldsson BT, Aagaard P, Kovanen V et al (2012) Mechanical properties of human patellar tendon at the hierarchical levels of tendon and fibril. J Appl Physiol 112 (3):419–426
47. Haraldsson BT, Aagaard P, Qvortrup K, Bojsen-Moller J, Krogsgaard M, Koskinen S et al (2008) Lateral force transmission between human tendon fascicles. Matrix Biol 27(2):86–95
48. Hansen P, Haraldsson BT, Aagaard P, Kovanen V, Avery NC, Qvortrup K et al (2010) Lower strength of the human posterior patellar tendon seems unrelated to mature collagen cross-linking and fibril morphology. J Appl Physiol 108(1):47–52
49. Hansen P, Kovanen V, Holmich P, Krogsgaard M, Hansson P, Dahl M et al (2012) Micromechanical properties and collagen composition of ruptured human achilles tendon. Am J Sports Med
50. Bennett MB, Ker RF, Dimery NJ, Alexander RM (1986) Mechanical properties of various mammalian tendons. J Zool 209:537–548
51. Anssari-Benam A, Legerlotz K, Bader DL, Screen HRC (2012) On the specimen length dependency of tensile mechanical properties in soft tissues: Gripping effects and the characteristic decay length. J Biomech 45(14):2481–2482
52. Atkinson TS, Ewers BJ, Haut RC (1999) The tensile and stress relaxation responses of human patellar tendon varies with specimen cross-sectional area. J Biomech 32(9):907–914
53. Ng BH, Chou SM, Krishna V (2005) The influence of gripping techniques on the tensile properties of tendons. Proc Inst Mech Eng H 219(H5):349–354
54. Butler DL, Grood ES, Noyes FR, Zernicke RF, Brackett K (1984) Effects of structure and strain measurement technique on the material properties of young human tendons and fascia. J Biomech 17 (8):579–596
55. Dressler MR, Butler DL, Wenstrup R, Awad HA, Smith F, Boivin GP (2002) A potential mechanism for age-related declines in patellar tendon biomechanics. J Orthop Res 20(6):1315–1322
56. Yamamoto E, Hayashi K, Yamamoto N (1999) Mechanical properties of collagen fascicles from the rabbit patellar tendon. J Biomech Eng 121 (1):124–131
57. Thorpe CT, Udeze CP, Birch HL, Clegg PD, Screen HRC (2012) Specialization of tendon mechanical properties results from interfascicular differences. J R Soc Interface 9(76):3108–3117
58. Hansen P, Kovanen V, Holmich P, Krogsgaard M, Hansson P, Dahl M et al (2013) Micromechanical properties and collagen composition of ruptured human achilles tendon. Am J Sports Med 41 (2):437–443

59. Legerlotz K, Riley GP, Screen HRC (2013) GAG depletion increases the stress-relaxation response of tendon fascicles, but does not influence recovery. Acta Biomater 9(6):6860–6866
60. van der Rijt JAJ, van der Werf KO, Bennink ML, Dijkstra PJ, Feijen J (2006) Micromechanical testing of individual collagen fibrils. Macromol Biosci 6 (9):697–702
61. Eppell SJ, Smith BN, Kahn H, Ballarini R (2006) Nano measurements with micro-devices: mechanical properties of hydrated collagen fibrils. J R Soc Interface 3(6):117–121
62. Sun YL, Luo ZP, Fertala A, An KN (2002) Direct quantification of the flexibility of type I collagen monomer. Biochem Biophys Res Commun 295 (2):382–386
63. Kemp AD, Harding CC, Cabral WA, Marini JC, Wallace JM (2012) Effects of tissue hydration on nanoscale structural morphology and mechanics of individual Type I collagen fibrils in the Brtl mouse model of Osteogenesis Imperfecta. J Struct Biol 180 (3):428–438

Part II

Tendon Disorders Associated with Altered Metabolism and Metabolic Disorders

Towards an Understanding of the Genetics of Tendinopathy

9

Alison September, Masouda Rahim, and Malcolm Collins

Abstract

To date, more than 18 genomic intervals, which underpin the complex myriad of extracellular matrix interactions of tendons, have been implicated in risk models for tendinopathy. It is these relationships that most likely regulate the tissue's response to loading and unloading, thereby dictating the overall capacity of tendons and influencing injury susceptibility. The evidence suggesting a genetic contribution to the susceptibility of sustaining a tendon injury is growing. However, only a few of the loci have been repeated in independent studies, of which some have included a range of musculoskeletal soft tissues injuries. Case-control study designs can be effective in capturing risk, provided that the cases and controls are equally well-defined and carefully considered. The genome consists of 3.6×10^9 sequences and therefore we realise that we are far from decoding all the genomic signatures. We are indeed fortunate to be living in such exciting times where *high-throughput* technologies are at our disposal. Through collaboration, our chances of harnessing these "*omics*" technologies to further our clinical understanding of tendinopathy will increase.

Introduction

We are living in the era where "*precision medicine*" is becoming a large focus underpinning research. We are faced with the challenge of (i) making a differential diagnosis and prognosis and (ii) the design and prescription of personalised treatment and therapeutic strategies to fit the clinical disease/injury/risk profile. The technologies of the *omics* era has promised to assist us and to this end revolutionised our capacity to interrogate these complex phenotypes. We are experiencing an unparalleled wave of huge data collection at the genomics, proteomics, transcriptomics and epigenomics levels and bioinformaticians are currently challenged with the collective synthesis of this information into both biological and clinical significance.

A. September (✉) • M. Rahim • M. Collins
Division of Exercise Science and Sports Medicine, Department of Human Biology, Faculty of Health Sciences, University of Cape Town, Cape Town, 115, Newlands 7725, South Africa
e-mail: alison.september@uct.ac.za

P.W. Ackermann and D.A. Hart (eds.), *Metabolic Influences on Risk for Tendon Disorders*, Advances in Experimental Medicine and Biology 920, DOI 10.1007/978-3-319-33943-6_9

This high throughput scale of genomics has, as yet, not filtrated into the research of tendinopathies; however, we predict that this scenario will change. Evidence suggesting a genetic component to risk of tendinopathy injuries is mounting and to date more than 30 genetic loci have been described in the literature (Fig. 9.1). This chapter will summarise the genetic loci implicated and describe some of the proposed biological mechanisms underpinning these associations.

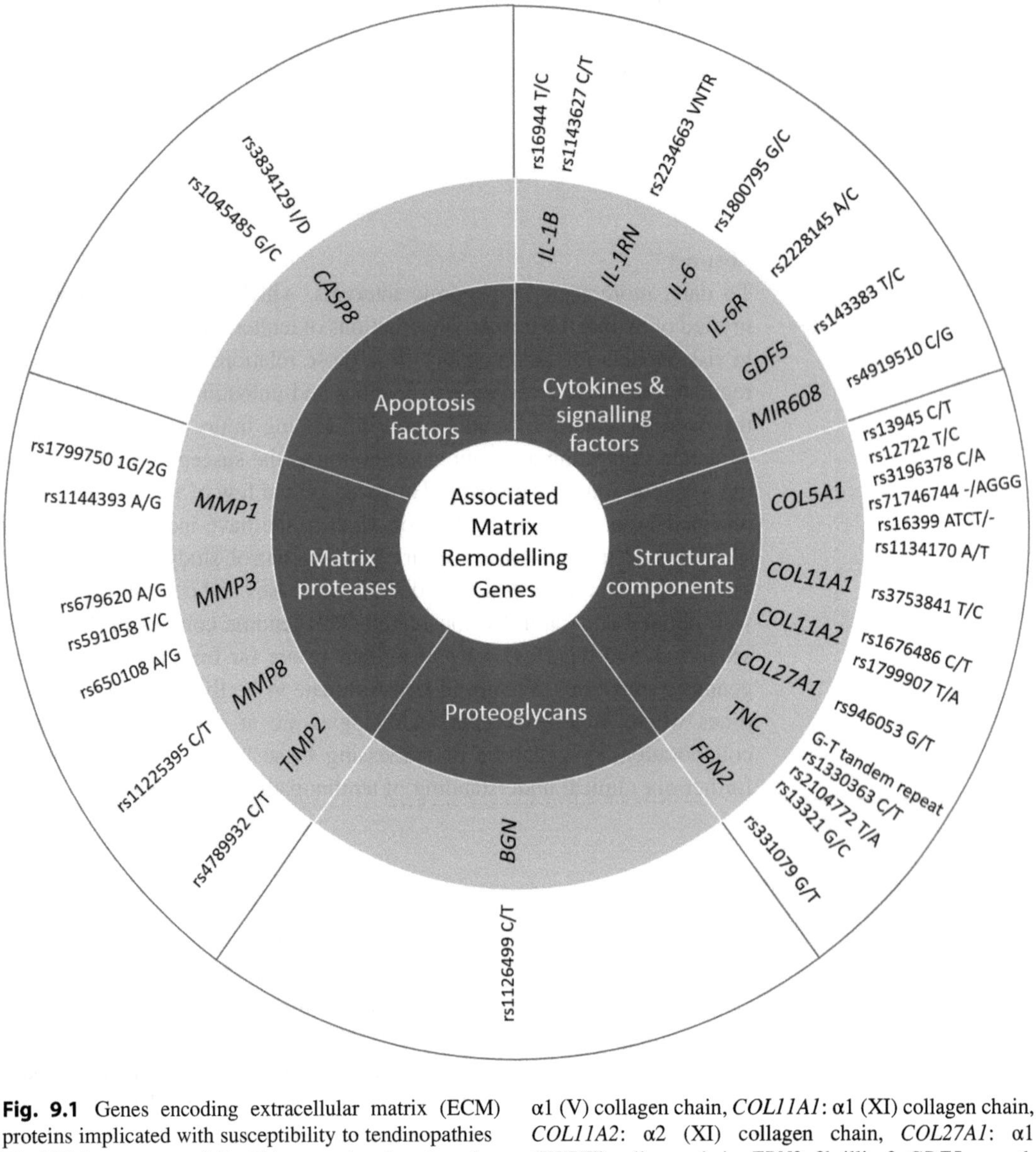

Fig. 9.1 Genes encoding extracellular matrix (ECM) proteins implicated with susceptibility to tendinopathies The ECM components fall within several main categories, as depicted in the inner ring, namely: structural components, proteoglycans, matrix proteases, apoptosis factors and cytokines & signalling factors. The implicated genes, along with the accession numbers for the associated polymorphisms, are shown in the two outermost rings. The proteins encoded by the implicated genes include: [*BGN*: biglycan, *CASP8*: caspase 8, *COL5A1*: α1 (V) collagen chain, *COL11A1*: α1 (XI) collagen chain, *COL11A2*: α2 (XI) collagen chain, *COL27A1*: α1 (XXVII) collagen chain, *FBN2*: fibrillin-2, *GDF5*: growth differentiation factor 5, *IL-1B*: interleukin-1β, *IL-1RN*: interleukin-1 receptor antagonist, *IL-6*: interleukin-6, *IL-6R*: interleukin-6 receptor, *MIR608*: microRNA 608, *MMP1*: matrix metalloproteinase 1, *MMP3*: matrix metalloproteinase 3, *MMP8*: matrix metalloproteinase 8, *TIMP2*: tissue inhibitor of metalloproteinase 2 and *TNC*: tenascin-C glycoprotein]

The implicated genes (Fig. 9.1) encode a variety of proteins ranging from structural components of tendons such as the collagens and glycoproteins to regulators of the extracellular matrix (ECM) and include proteoglycans, matrix metalloproteases, tissue inhibitors of matrix metalloproteases, cytokines, growth factors and caspases to mention a few (see Chap. 1). Although each of these proteins have their own unique function, they also function as a collective with the overarching aim to maintain ECM integrity. It is therefore not surprising that in some instances individual genetic loci have been implicated in a risk profile whilst in other instances a combination of genetic loci, encoding proteins functioning in a common biological pathway, have been implicated in a risk model [1–5].

Structural Components

To date, variants within four collagen encoding genes (*COL5A1*, *COL11A1*, *COL11A2* and *COL27A1*) have been implicated in risk models of Achilles tendinopathy (AT) either independently or as part of a haplotype. A haplotype is the combination of a set of alleles which are inherited together and represent a specific region on a chromosome. Of particular interest is the *COL5A1* 3'-UTR region for which several variants (rs12722 T/C, rs71746744 -/AGGG, rs16399 -/ATCT, rs1134170 A/T), including single nucleotide polymorphisms (SNPs), were found to be implicated with altered risk of chronic Achilles tendinopathy in individuals from South Africa and Australia; independently and as part of a haplotype [1, 6–8]. Interestingly, some of these variants and additional variants, rs13946 T/C and rs14776422/rs5574880 W/M, have also been implicated with altered risk of carpal tunnel syndrome (CTS) [9].

The *COL5A1* gene encodes the α1(V) chain of type V collagen. Type V collagen is a minor fibrillar collagen implicated in regulating the collagen fibril diameter and its lateral growth during fibrillogenesis. It has been suggested that the *COL5A1* 3'-UTR implicated in these genetic association studies, may harbour DNA signatures which may either individually or collectively alter the secondary structure of the mRNA thereby potentially affecting the mRNA stability of this gene [8]. Laguette et al., (2011) identified two major *COL5A1* 3'-UTR functional forms (namely the "C functional form" and the "T functional form") where the T functional form was associated with an overall increase in mRNA stability [8]. Further to support this hypothesis of gene regulation within the *COL5A1*-3'UTR region, Abrahams et al. (2013) identified the *MIR608* rs4919510 CC genotype to be associated with an increased risk of AT [7]. The *MIR608* gene encodes a 25 bp microRNA which binds to the Hsa-miR-608 binding site in the *COL5A1* 3'-UTR. Collins and Posthumus et al. (2011) hypothesised that altered *COL5A1* mRNA stability results in altered type V collagen production which in turn may alter the collagen fibril diameter and density potentially affecting the biomechanical properties of tendon [10].

Functional SNPs within the genes encoding the α1(XI) and α2(XI) chains of type XI collagen (*COL11A1* rs3753841 T/C and rs1676486 C/T and *COL11A2* rs1799907 T/A) have been implicated in a risk-associated allele combination with AT. More specifically, the T-C-T inferred haplotype (rs3753841 T/C; rs1676486 C/T; rs17999079 T/A) was associated with an increased risk of AT [11]. The authors further observed that individuals diagnosed with AT were more likely to have the inferred haplotype T-C-T-(AGGG) for the *COL11A1* rs3753841 T/C – *COL11A1* rs1676486 C/T – *COL11A2* rs1799907 T/A and *COL5A1* rs71746744 -/AGGG variants [11]. The cumulative biological effect of this increased risk haplotype needs to be explored. Interestingly, the encoded type XI collagen shares structural and functional homology with type V collagen and also plays an important role in collagen fibril assembly in developing tendons [12].

In a similar study by Saunders et al. (2013) although no independent associations were noted, the authors too highlighted a combination of alleles for *COL27A1* and its neighbouring gene, *TNC*, to be associated with an increased risk profile for AT [2]. The *COL27A1* variant (rs946053 G/T), and the two proximal *TNC* variants (rs13321 G/C and rs2104772 T/A) as defined by the G-C-A inferred haplotype implicated a critical region on chromosome 9q33 to be associated with an increased risk of AT in both a South African and an Australian group [2]. *COL27A1* encodes the α1(XXVII) chain of type XXVII collagen, an atypical fibrillar collagen predominantly expressed in cartilage. The *TNC* gene encodes for the tenascin-C glycoprotein, a component of tendons implicated in regulating cell-matrix interactions.

Fibrillin-2, encoded by the *FBN2* gene, is a component of connective tissue microfibrils and the rs331079 GG genotype and G allele were noted to be significantly over-represented within the AT group in comparison to the control group [13].

Structurally, tendon composition varies along the length of the tendon to accommodate the functional differences of the tissue. Accordingly, the distribution of ECM components also varies and is linked to biological function. Tenocytes, interspersed between collagen fibres, are able to respond to mechanical and biological stimuli by synthesizing ECM components to enable matrix remodelling in an attempt to adapt and/or heal [14]. Thus, tendon structural integrity is regulated by other ECM components, such as proteoglycans, growth factors and proteases, which collectively work to maintain matrix homeostasis. Several genetic polymorphisms within genes encoding ECM regulators have been studied and are discussed below.

Regulators of the Extracellular Matrix

Proteoglycans

Proteoglycans form part of the ground substance and give it its characteristic gel-like appearance, facilitating resistance to compressive forces (see Chap. 1). They facilitate matrix-matrix interactions, cell-matrix organisation and cell-matrix signalling interactions and hereby contribute to the structural and functional integrity of the tendon [15, 16]. To date, a variant in the biglycan encoding gene, the *BGN* rs1126499 CC genotype was specifically implicated in decreased risk of CTS [17]. The *BGN* gene, on chromosome Xq28, plays a role in collagen fibrillogenesis and the regulation of bone formation [16]. Recent evidence has shown that this ubiquitously expressed small leucine-rich proteoglycan may also function as a signalling molecule [16].

Matrix Proteases

The 11q22 chromosomal locus harbours a cluster of matrix metalloproteinase (MMP) encoding genes which have been implicated in risk models for tendinopathy related conditions; specifically variants within *MMP1*, *MMP3* and *MMP8*. MMPs are zinc-dependent endopeptidases that play key roles in ECM degradation after mechanical loading, facilitating the direct or indirect synthesis of ECM components, cell proliferation and differentiation thereby allowing the maintenance of the ECM [18]. They are able to degrade a range of substances including collagenous and non-collagenous matrix components [19]. Altered MMP expression profiles were previously noted in both tendinopathic and ruptured tendons [20, 21].

Raleigh et al. (2009) noted that the *MMP3* genotypes: rs679620 GG, rs591058 CC and rs650108 AA were significantly over-represented in the control group [3]. Moreover, the alternate alleles, as noted in the inferred A-T-G haplotype, was associated with increased risk of AT. Their analyses further suggested that there is a combined effect between variants *MMP-3* rs679620 and *COL5A1* rs12722 on risk of AT (A-C inferred haplotype associated with decreased risk and the G-T inferred haplotype associated with increased risk) [3]. MMP-3, stromelysin 1, degrades several ECM components such as collagen types (II, III, IV, IX and X), proteoglycans, fibronectin, elastin and laminin [18].

Variants within *MMP1* and *MMP8* have been implicated with posterior tibialis tendon dysfunction (PTT) [22–24]. Both the TT genotype and T allele of the functional *MMP-8* - (collagenase-2) rs11225395 variant were implicated with increased risk [24]. Two functional promoter polymorphisms within *MMP*-1 were similarly associated with increased risk of PTT independently and as a collective haplotype. Specifically the risk alleles included the G allele of rs1144393 A/G and the functional 2G allele of rs1799750 1G/2G, and the A-2G and G-2G risk haplotypes of rs1144393 and rs1799750 [23].

MMP activity is regulated by the tissue inhibitors of metalloproteinases (TIMP) and therefore it is not surprising to note that the promotor polymorphism, *TIMP-2* rs4789932 C/T was implicated in a risk model for Achilles tendon pathology (ATP) in a combined Caucasian cohort [25]. Specifically, the CC genotype was significantly over-represented in the controls and the CT genotype was significantly over-represented in the ATP group [25]. TIMPs contribute to matrix turnover in tendon [18].

Cytokines and Signalling Factors

Tendons are dynamic structures that have the unique ability to withstand forces up to ten times an individual's body weight. Thus, it is reasonable to propose that the ECM of these structures are under great pressure to continuously remodel in response to the various loads. It is therefore critical that the synthesis and degradation of the ECM components needs to be highly regulated. There are a myriad of interactions, which regulate ECM homeostasis at both the cellular (tenocytes) and molecular levels. To date, there is growing evidence to suggest that tenocytes, inflammatory gene expression profiles, cytokines, growth factors and signalling factors individually or collectively contribute to several processes, which include mediating matrix turnover, inflammation, wound healing, angiogenesis, hypoxia and pain perception amongst others [26]. All these processes facilitate effective response to load/forces as experienced by tendon.

The interleukins play key roles in cell signalling pathways, upregulating several critical downstream cascades as well as having key roles in inflammation. September et al. (2011) conducted pathway based case–control genetic association analyses and implicated variants within three interleukin encoding genes together with the *COL5A1* rs12772 polymorphism as part of a risk profile for AT [4]. The inferred allele combination included the genes encoding interleukin-1β (*IL-1B*: rs16944 T/C and rs1143627 C/T), interleukin-1 receptor antagonist (*IL-1RN*: rs2234663 86-bp VNTR) and interleukin-6 (*IL-6*: rs1800795 G/C) [4]. IL-1β together with IL-6, collectively upregulate several downstream biochemical mediators such as COX-2, PGE_2, TGF-β and several MMPs [27–29]. More recently, a SNP within the IL-6 receptor, encoded by the *IL-6R* gene (rs2228145 A/C) was implicated with increased risk of CTS [30]. Further analyses by the authors also implicated a combined risk profile between the *IL-6R* rs2228145 and *COL5A1* rs12722 and *BGN* rs1276499 variants with risk of CTS [17, 30].

A case-control genetic association study by Posthumus et al. (2010) reported the association of a functional polymorphism within the gene encoding growth differentiation factor 5 (*GDF-5* rs143383 C > T) with AT; the TT genotype was associated with risk [5]. Growth differentiation factor 5 (GDF-5), a member of the TGF-β superfamily, plays a role in regulating cell growth and differentiation. [31, 32].

Receiver operating curve analyses (ROC) has further suggested that in considering all the above risk-implicated polymorphisms described within the pro-inflammatory pathway, it is most likely the two functional variants (rs3834129 ins/del and rs1045485 G/C) within the caspase-8 encoding gene (*CASP*-8) together with sex, which are best at discriminating risk for AT [33]. These findings further support the hypothesis that apoptosis is an important biological pathway to be considered when trying to understand the mechanisms underlying AT. It has previously been suggested that excessive apoptosis can potentially weaken the collagenous structure of tendon [34]. The initiator caspase-8, upregulated by IL-6, is responsible for the activation of caspase 3, an effector caspase that triggers the apoptosis cascade.

Saunders et al. (2015) recently considered 20 factors which included polymorphisms within genes (i) encoding structural components (*COL3A1*, *COL5A1*, *COL5A2*, *COL5A3*, *COL27A1* and *TNC*) and pro-inflammatory regulators (*IL-1B*, *IL-6* and *CASP8*) of tendon in an attempt to identify a finite set of informative biomarkers to inform a risk model for AT [35]. They highlighted two risk models from the logistic regression and ROC analyses. Model A included variables: sex, *COL27A1* rs1249744, *COL5A1* rs12722, and *COL5A3* rs1559186 and Model B included the variables: sex, three *COL27A1* variants (rs4143245, rs1249744, rs946053), *COL5A1* rs12722, and both *CASP8* variants (rs1045485, rs3834129). The authors proposed that Model B was more effective than Model A in predicting risk, with a sensitivity of 70 % and specificity of 64 %. It does seem that the more variables included in the risk assessment model, the more effective it becomes in assigning risk. Furthermore, the inclusion of cell signaling markers and structural components in both models highlights the potential intrinsic relationship between ECM signaling pathways and the structural integrity of the collagen network of tendon. These seemingly, dependent functional relationships need to be considered when designing effective clinical management strategies.

Precision Medicine

There is an ever-increasing demand for the expansion of current testing models to increase the sensitivity of differential diagnoses and the prescription of treatment. Precision medicine is one of the current models, which tries to exploit the genetic and non-genetic clinical indicators to improve the quality of medical care through comprehensive diagnostics, improved treatment intervention and clinical management [36]. To date, there are several genetic tests available on the market to assist with injury risk assessment. However, there are limitations to consider, which includes the testing panel may be lagging behind current research in the field and it may not have been validated in (i) different populations or (ii) with the injury clinical profile being tested. These tests are largely premature as researchers are still unravelling the intricacies and complexity of the biological pathways leading to tendinopathy. Moreover, the effects of regulatory non-coding regions, epigenetics and environmental factors on injury susceptibility need to be considered [36, 37]. It is therefore an added responsibility of healthcare professionals to (i) be aware of the various tests, (ii) the validity of the clinical utility, (iii) understand the injury specificity, if defined, (iv) caution their patients on the limitations of such tests and (v) guide them to identify the most informative but also the most clinically relevant tests based on biological functional data. Healthcare professionals are therefore reminded to integrate the complete clinical profile of the individual, which includes their unique medical history, family history, all known environmental exposures and the genetic contribution, to facilitate improved diagnoses, direct treatment intervention and clinical management. Genetic profiling should only be interpreted in the context of an integrated clinical assessment platform.

Concluding Remarks

The extracellular matrix of a tendon is a dynamic reservoir of components, which are able to respond and adapt to load. The effects of loading and unloading on the capacity of tendon, however, remains a mystery [38]. Biologists continue to be intrigued and perplexed by the subsequent dynamics leading to failure of the tendon to heal and/or adapt. Decoding this myriad of potential matrix interactions has therefore become an obsession in the pursuit of unravelling the structural-functional organization of tendon and this relationship to its biomechanical properties.

Genetics has tried to identify some of the key biological variables, which need consideration

(Fig. 9.1). To date, 18 genetic intervals and 32 polymorphisms have been associated with risk of tendon pathologies such as AT, CTS and or PTT. The large majority of these studies have followed a case-control genetic association approach, which is a limitation. Only a small proportion of these associations have been replicated in independent populations having the same or similar phenotype [39]. Associations cannot be viewed independently and have to be part of a phased approach where the functional hypothesis is also explored. Evidence for a functional hypothesis is emerging for the 3'-UTR of the *COL5A1* gene [8]. One of the steps towards this functional end is mining these genetic boundaries more carefully across all populations. Identifying critical regions of biological relevance in the large majority of populations will inform us of the major biological role players in tendinopathy disorders [40, 41].

In future, it is going to be imperative that we start pooling resources and combining expertise to more effectively interrogate tendinopathy at the genomics, cellular, functional, physiological and biomechanical levels to gain a comprehensive understanding of the mechanisms underpinning injury and tendon capacity. The era of *omics* and *high-throughput* technologies is well established, but only through collaboration will we effectively exploit these arrays towards an improved clinical understanding of tendinopathy.

References

1. Mokone GG, Schwellnus MP, Noakes TD, Collins M (2006) The COL5A1 gene and Achilles tendon pathology. Scand J Med Sci Sports 16:19–26
2. Saunders CJ, van der Merwe L, Posthumus M, Cook J, Handley CJ, Collins M et al (2013) Investigation of variants within the COL27A1 and TNC genes and Achilles tendinopathy in two populations. J Orthop Res 31:632–637
3. Raleigh SM, van der Merwe L, Ribbans WJ, Smith RKW, Schwellnus MP, Collins M (2009) Variants within the MMP3 gene are associated with Achilles tendinopathy: possible interaction with the COL5A1 gene. Br J Sports Med 43:514–520
4. September AV, Nell E-M, O'Connell K, Cook J, Handley CJ, van der Merwe L et al (2011) A pathway-based approach investigating the genes encoding interleukin-1, interleukin-6 and the interleukin-1 receptor antagonist provides new insight into the genetic susceptibility of Achilles tendinopathy. Br J Sports Med 45:1040–1047
5. Posthumus M, Collins M, Cook J, Handley CJ, Ribbans WJ, Smith RKW et al (2010) Components of the transforming growth factor-beta family and the pathogenesis of human Achilles tendon pathology – a genetic association study. Rheumatology 49:2090–2097
6. September AV, Cook J, Handley CJ, van der Merwe L, Schwellnus MP, Collins M (2009) Variants within the COL5A1 gene are associated with Achilles tendinopathy in two populations. Br J Sports Med 43:357–365
7. Abrahams Y, Laguette M-J, Prince S, Collins M (2013) Polymorphisms within the COL5A1 3'-UTR that alters mRNA structure and the MIR608 gene are associated with Achilles tendinopathy. Ann Hum Genet 77:204–214
8. Laguette M-J, Abrahams Y, Prince S, Collins M (2011) Sequence variants within the 3'-UTR of the COL5A1 gene alters mRNA stability: implications for musculoskeletal soft tissue injuries. Matrix Biol 30:338–345
9. Burger M, de Wet H, Collins M (2015) The COL5A1 gene is associated with increased risk of carpal tunnel syndrome. Clin Rheumatol 34:767–774
10. Collins M, Posthumus M (2011) Type V collagen genotype and exercise-related phenotype relationships. Exerc Sport Sci Rev 39:191–198
11. Hay M, Patricios J, Collins R, Branfield A, Cook J, Handley CJ et al (2013) Association of type XI collagen genes with chronic Achilles tendinopathy in independent populations from South Africa and Australia. Br J Sports Med 47:569–574
12. Wenstrup RJ, Smith SM, Florer JB, Zhang G, Beason DP, Seegmiller RE et al (2011) Regulation of collagen fibril nucleation and initial fibril assembly involves coordinate interactions with collagens V and XI in developing tendon. J Biol Chem 286:20455–20465
13. Khoury L, Posthumus M, Collins M, van der Merwe W, Handley C, Cook J et al (2014) ELN and FBN2 gene variants as risk factors for two sports-related musculoskeletal injuries. Int J Sports Med 36:333–337
14. Docking S, Samiric T, Scase E, Purdam C, Cook J (2013) Relationship between compressive loading and ECM changes in tendons. Muscles Ligaments Tendons J 3:7–11
15. Yanagishita M (1993) Function of proteoglycans in the extracellular matrix. Acta Pathol Jpn 43:283–293
16. Nastase MV, Young MF, Schaefer L (2012) Biglycan: a multivalent proteoglycan providing structure and signals. J Histochem Cytochem 60:963–975
17. Burger MC, De Wet H, Collins M (2014) The BGN and ACAN genes and carpal tunnel syndrome. Gene 551:160–166

18. Pasternak B, Aspenberg P (2009) Metalloproteinases and their inhibitors-diagnostic and therapeutic opportunities in orthopedics. Acta Orthop 80:693–703
19. Somerville RPT, Oblander SA, Apte SS (2003) Matrix metalloproteinases: old dogs with new tricks. Genome Biol 4:216
20. Jones GC, Corps AN, Pennington CJ, Clark IM, Edwards DR, Bradley MM et al (2006) Expression profiling of metalloproteinases and tissue inhibitors of metalloproteinases in normal and degenerate human Achilles tendon. Arthritis Rheum 54:832–842
21. de Mos M, van El B, DeGroot J, Jahr H, van Schie HTM, van Arkel ER et al (2007) Achilles tendinosis: changes in biochemical composition and collagen turnover rate. Am J Sports Med 35:1549–1556
22. Godoy-Santos A, Cunha MV, Ortiz RT, Fernandes TD, Mattar R, dos Santos MCLG (2013) MMP-1 promoter polymorphism is associated with primary tendinopathy of the posterior tibial tendon. J Orthop Res 31:1103–1107
23. Baroneza JE, Godoy-Santos A, Ferreira Massa B, Boçon de Araujo Munhoz F, Diniz Fernandes T, Leme Godoy Dos Santos MC (2014) MMP-1 promoter genotype and haplotype association with posterior tibial tendinopathy. Gene 547:334–337
24. Godoy-Santos A, Ortiz RT, Junior RM, Fernandes TD, Santos MCLG (2014) MMP-8 polymorphism is genetic marker to tendinopathy primary posterior tibial tendon. Scand J Med Sci Sport 24:220–223
25. El Khoury L, Posthumus M, Collins M, Handley CJ, Cook J, Raleigh SM (2013) Polymorphic variation within the ADAMTS2, ADAMTS14, ADAMTS5, ADAM12 and TIMP2 genes and the risk of Achilles tendon pathology: a genetic association study. J Sci Med Sport 16:493–498
26. Kjaer M (2004) Role of extracellular matrix in adaptation of tendon and skeletal muscle to mechanical loading. Physiol Rev 84:649–698
27. Yang G, Im H-J, Wang JH-C (2005) Repetitive mechanical stretching modulates IL-1β induced COX-2, MMP-1 expression, and PGE2 production in human patellar tendon fibroblasts. Gene 363:166–172
28. Tsuzaki M, Guyton G, Garrett W, Archambault JM, Herzog W, Almekinders L et al (2003) IL-1B induces COX2, MMP-1, −3 and −13, ADAMTS-4, IL-1B and IL-6 in human tendon cells. J Orthop Res 21:256–264
29. Thampatty BP, Li H, Im H-J, Wang JH-C (2007) EP4 receptor regulates collagen type-I, MMP-1, and MMP-3 gene expression in human tendon fibroblasts in response to IL-1β treatment. Gene 386:154–161
30. Burger MC, de Wet H, Collins M (2015) Interleukin and growth factor gene variants and risk of carpal tunnel syndrome. Gene 564:67–72
31. Buxton P, Edwards C, Archer CW, Francis-West P (2001) Growth/differentiation factor-5 (GDF-5) and skeletal development. J Bone Joint Surg Am 83-A (Suppl):S23–S30
32. Francis-West PH, Abdelfattah A, Chen P, Allen C, Parish J, Ladher R et al (1999) Mechanisms of GDF-5 action during skeletal development. Development 126:1305–1315
33. Nell E-M, van der Merwe L, Cook J, Handley CJ, Collins M, September AV (2012) The apoptosis pathway and the genetic predisposition to Achilles tendinopathy. J Orthop Res 30:1719–1724
34. Millar NL, Wei AQ, Molloy TJ, Bonar F, Murrell GAC (2009) Cytokines and apoptosis in supraspinatus tendinopathy. J Bone Joint Surg (Br) 91:417–424
35. Saunders CJ, van der Merwe L, Cook J, Handley CJ, Collins M, September AV (2015) Extracellular matrix proteins interact with cell-signaling pathways in modifying risk of achilles tendinopathy. J Orthop Res 33:898–903
36. September AV, Posthumus M, Collins M (2012) Application of genomics in the prevention, treatment and management of Achilles tendinopathy and anterior cruciate ligament ruptures. Recent Pat DNA Gene Seq 6:216–223
37. Collins M, September AV, Posthumus M (2015) Biological variation in musculoskeletal injuries: current knowledge, future research and practical implications. Br J Sports Med 49:1497–1503
38. Cook JL, Docking SI (2015) Rehabilitation will increase the 'capacity' of your . . .insert musculoskeletal tissue here. . . Defining "tissue capacity": a core concept for clinicians. Br J Sport Med 49 (23):1484–1485
39. Raleigh SM, Collins M (2012) Gene variants that predispose to Achilles tendon injuries: an update on recent advances. Achilles Tendon. InTech, pp 25–40
40. September AV, Mokone GG, Schwellnus MP, Collins M (2006) Genetic risk factors for Achilles tendon injuries. Int Sport J 7:201–215
41. September AV, Schwellnus MP, Collins M, Gibson W (2007) Tendon and ligament injuries: the genetic component. Br J Sports Med 41:241–246

Tendons Involvement in Congenital Metabolic Disorders

10

Michele Abate, Vincenzo Salini, and Isabel Andia

Abstract

Congenital metabolic disorders are consequence of defects involving single genes that code for enzymes. Blocking metabolic pathways, the defect leads to the shortage of essential compounds, and/or to the accumulation of huge quantities of precursors, which interfere with normal functions. Only few of these diseases are characterized by a clinically significant tendon involvement.

Heterozygous Familial Hypercholesterolaemia results from the inheritance of a mutant low-density lipoprotein receptor gene; patients show high cholesterol levels, precocious coronary artery disease, and may develop tendon xanthomata (mainly in Achilles tendon). The detection of xanthomata is important, because it allows an early diagnosis and treatment of the disorder. Cerebrotendinous Xanthomatosis is a rare genetic metabolic disorder of cholesterol and bile acid metabolism, characterized by accumulation of cholestanol in brain and tendons. Tendon abnormalities are similar to those reported in Heterozygous Familial Hypercholesterolaemia. Alkaptonuria is caused by a deficiency of the enzyme homogentisic acid oxidase. Due to the accumulation of the homogentisic acid, tendons and ligaments are characterized by a typical ochre/yellow pigmentation (ochronosis), with ensuing inflammation, calcification and rupture. In Congenital Hypergalactosemia an increased tendon collagen cross-linking by non-enzymatic galactosylation can be observed. Finally, Congenital Hypophosphatasia may be associated to deposition of hydroxyapatite crystals in rotator cuff, elbow, and Achilles tendons.

M. Abate (✉) • V. Salini
Department of Medicine and Science of Aging, University G. d'Annunzio, Via dei Vestini 31, Chieti-Pescara, 66013 Chieti Scalo (CH), Italy
e-mail: m.abate@unich.it

I. Andia
Regenerative Medicine Laboratory, BioCruces Health Research Institute, Cruces University Hospital, 48903 Barakaldo, Spain

P.W. Ackermann and D.A. Hart (eds.), *Metabolic Influences on Risk for Tendon Disorders*, Advances in Experimental Medicine and Biology 920, DOI 10.1007/978-3-319-33943-6_10

Keywords
Alkaptonuria • Cerebrotendinous Xanthomatosis • congenital metabolic disorders • Hypercholesterolaemia • Hypergalactosemia • Hypophosphatasia • Tendon

Introduction

Congenital metabolic disorders (CMDs) represent a wide spectrum of genetic diseases of metabolism. Since the seminal works about alkaptonuria published by a British physician, Archibald Garrod, in the early twentieth century, who suggested the "one gene-one enzyme" hypothesis [13], much has been learned about the histopathological, biochemical and clinical findings of these diseases. Briefly, CMDs are produced by alterations of single genes that code for enzymes, with the impairment of the conversion of some substrates into products, and their consequent accumulation, in addition to deficit of the ultimate product of metabolism. As result, the large increase of these substances is toxic, interferes with normal function, and reduces the ability to synthesize essential compounds [12].

CMDs include a plethora of diseases, related to defects of carbohydrate, aminoacid, organic acids, porphyrin, purine/pyrimidine, steroids metabolism etc., and their cumulative incidence is estimated to be 40/100,000 live births, overall representing more than 15 % of single gene disorders in the population [36].

Because of the wide range of systems involved, the clinical presentation is heterogeneous, with nervous, cardiovascular, pulmonar, gastro-enteral, uro-genital, immunological, endocrinological, and skeletal disorders [12]. Usually, CMDs can be suspected on the basis of clinical presentation and are easily detectable by newborn screening and specific tests, which favor an earlier treatment and a better outcome. However, in some cases a definitive diagnosis cannot be made, resulting in the need to rely on the patient's clinical course.

A detailed report on CMDs is beyond the scope of the present chapter, which deals exclusively with only few diseases characterized by tendon involvement, important in the diagnostic and clinical perspective.

Familial Hypercholesterolemia

Heterozygous Familial Hypercholesterolaemia (HeFH) is a monogenic disorder that affects about 1/500 people, usually resulting from the inheritance of a mutation on the gene codifying the receptor for low density lipoprotein (LDL), which normally removes LDL from the circulation [30]. Untreated, HeFH is associated with a high mortality and morbidity from coronary heart disease, but when intensive treatment is precociously undertaken, life expectancy can be substantially improved.

Most patients with HeFH develop tendon xanthomata (mainly in Achilles tendon), which become increasingly common from the third decade onwards [9]. Histologically, cholesterol deposition is seen both extracellularly and inside histiocytic and other foam cells, which show numerous intracytoplasmic lipid vacuoles, lysosomes, and myelin-figures. An inflammatory cell infiltrate, and a fibrous reaction may be associated [2, 8, 40].

The early detection of xanthomata is thus exceptionally important. Unfortunately, the clinical diagnosis in several cases is not easy, because the nodules are too small to be detected or otherwise because pain is ascribed to an unspecific tendon damage. At this regard, Beeharry et al. [4] have shown that episodes of Achilles tendon pain, lasting more than 3 days, are very common in patients with HeFH, yet in absence of apparent xantomatosis. Therefore, the authors suggest that measurement of serum cholesterol in young patients presenting with

Achilles tendon pain is mandatory, because could lead to an early diagnosis of HeFH.

Sonography is very useful for detecting tendon abnormalities. The technique may visualize xanthomata located deep within the tendon, which cannot be detected by palpation. Tsouli et al., in a large case-control study, found moderate and high Achilles tendon damage in 30/80 and in 8/80 patients, respectively [34, 35]. The thickness of the tendon was increased in patients with HeFH compared to controls, proportionally to the echostructural abnormalities. Only patients with minor sonographic changes showed significant reduction in Achilles tendon thickness after statin treatment, whereas in patients with moderate and high Achilles tendon damage no significant reduction was observed [20, 34]. Exacerbations of Achilles tendon pain can occur when statin treatment is started, and it is associated with the rapid lowering of cholesterol. The condition would seem to be akin to the exacerbations of gout, which occur when allopurinol treatment is begun and the serum uric acid level decreases rapidly. The mobilisation of cholesterol, like that of uric acid crystals, presumably provokes an inflammatory cell reaction [17, 34]. Nodular excision of the xanthomatas, followed by tendon reconstruction, is sometimes necessary [1].

Homozygous FH is very uncommon, occurring in 1 on a million births. The symptoms are similar to those of HeFH, but more severe [3, 11].

Cerebrotendinous Xanthomatosis

Cerebrotendinous Xanthomatosis (CTX) is a rare autosomal recessive disorder involving mutations on the CYP27A1 gene, and subsequent alterations in cholesterol and bile acid metabolism that results in systemic and neurologic abnormalities [27]. The genes causing CTX were mapped to the STSL locus on human chromosome 2p21, and mutations in either of the two genes that comprise this locus, ABCG5 or ABCG8, cause this disease. So far, more than 50 different mutations have been identified on the CYP27A1 gene, which encodes the mitochondrial enzyme 27-esterol hydrolase, involved in the oxidation of cholesterol intermediates as part of the bile acid synthesis pathway. This enzyme is crucial in metabolic homeostasis because it catalyzes many reactions in the synthesis of cholesterol, steroids and other lipids. The prevalence has been roughly estimated in 3-5/100,000 in the Caucasian population [21]. No more than 300 cases have been described worldwide [28].

The disease is characterized by normal to moderately elevated cholesterol levels, accumulation of the metabolite cholestanol in brain and tendons, whereas chenodeoxycholic acid, an important bile acid, is virtually absent. CTX diagnosis is based on clinical symptoms. Cognitive impairment, dysphagia, dysarthria, dystonia, spasticity, muscle weakness and ataxia are the more common neurologic signs. Tendon abnormalities are similar to those reported in HeFH ([5, 23], Pudhiavan [28, 32, 38, 39]). At early stages, during the first or second decade, diarrhea and bilateral cataracts are diagnosed. Later in the third decade, tendon xanthomas and progressive neurologic dysfunction arise [33].

Alkaptonuria

Alkaptonuria is a rare disease with a prevalence of 1/100.000–250.000 [24]. Mutations on homogentisate 1,2-dioxygenase (HGD) gene which encodes the enzyme HGD impair the catabolism of tyrosine and phenylalanine. Transmission is autosomal recessive with variable expression. The HGD has been mapped to chromosome 3q21–q23. More than 115 mutations impairing the HGD enzyme translations have been described [29]. Consequently, a deficiency of the HGD enzyme favors the accumulation of homogentisic acid (a metabolic product of the aromatic amino acids phenylalanine and tyrosine) into the fibrillary collagens. After irreversible binding to collagen fibers, homogentisic acid becomes polymerized to form a dark pigment, conferring a characteristic ochre/yellow appearance to connective tissues (ochronosis). As consequence, the structural integrity of collagen is reduced due to inhibition of collagen cross-linking, thus increasing the likelihood of spontaneous rupture [6, 22]. Histopathology shows extensive degenerative changes with swollen, rigid, and fragmented collagen fibers, jagged edges and ochronotic deposition [18]. The disease is clinically characterized by homogentisinic aciduria, bluish-black discoloration of connective tissues, vertebral and large joints arthropathy, and tendon abnormalities. Less common manifestations are cardiovascular disorders, renal, urethral and prostate calculi, and scleral and ear involvement. Alkaptonuria may be recognized during childhood, but the musculoskeletal involvement is usually present only after the age of 30. Early arthropatic changes include diffuse cartilage pigmentation and chondrocyte necrosis, while in advanced stages degenerative changes in synovial and intervertebral joints are noted. Longstanding disease may result into severe kyphosis, obliteration of intervertebral spaces and marginal osteophytes [25, 37].

Tendons and ligaments are heavily pigmented due to their collagen content, resulting in inflammation, calcification and rupture. Usually, the tendons with synovial sheath are not involved. Enthesopathy at the insertion site is the more frequent finding. At ultrasound evaluation, tendons appear thick and hypoechoic, with coarse hyperechoic mass at bone insertion, which may represent detached cartilage fragments, or pigmented tissue debris aggregations [10].

Galactosemia

This rare autosomal recessive disorder is caused by the absence of the enzyme responsible for adequate galactose metabolism. Galactose accumulation can cause serious damage to the eyes, liver, kidneys and brain. Increased tendon collagen cross-linking by non-enzymatic galactosylation has been observed in cases of Congenital Hypergalactosemia, even if clinical tendinopathies have never been reported [7, 26, 31].

Hypophosphatasia

Hypophosphatasia (HPP) is an inborn error of metabolism, characterized by hypercalcemia and skeletal disease. The condition is attributed to low serum alkaline phosphatase (ALP) activity, caused by functional mutation(s) within the gene codifying the tissue nonspecific (liver/bone/kidney) isoenzyme of ALP (TNSALP) located on chromosome 1. Both, autosomal dominant and recessive defects, involving more than 260 TNSALP mutations, have been described, and this explain why the spectrum and severity of clinical features can be very large [19].

The diagnosis can be performed in the prenatal stadium by ultrasonography, which shows small thoracic circumference with pulmonary hypoplasia, severe micromelia, generalized demineralization and osteochondral spurs [15]. In pediatric patients, a common clinical manifestation is the premature exfoliation of deciduous teeth, attributable to deficiency of mineralized cementum covering the tooth roots [16]. In middle aged patients, poorly healing metatarsal stress fractures and proximal femoral pseudofractures have been observed [14].

Lacking of the enzyme TNSALP leads to the extracellular accumulation of inorganic pyrophosphate (PPi), which in turn causes, in affected adults, calcium pyrophosphate dihydrate deposition, PPi arthropathy, or pseudogout, or seemingly paradoxical deposition of hydroxyapatite crystals in ligaments or around joints (calcific periarthritis), as reported by Guañabens et al. [14] in rotator cuff, elbow and wrist extensor/flexors, hips and Achilles tendons in three middle-aged sisters.

Conclusions

Tendons involvement in congenital metabolism disorders occurs in a very limited number of subjects. Actually, the prevalence of these diseases is very low in the general population, and only few of them show tendon damage among the wide range of their clinical manifestations. The common pathogenetic mechanism is the block of an enzymatic pathway with accumulation of huge amounts of precursors. Tendon abnormalities are of clinical relevance in HeFH, where unexplained Achilles tendon pain in young subjects may allow an early diagnosis. Tendinopathy can be observed also in Alkaptonuria, however as a relatively late manifestation associated to systemic pathologies. In Congenital Hypergalactosemia and HPP tendon involvement is a minor complication compared to more important systemic disorders.

References

1. Ahn JH, Chun TJ, Lee S (2011) Nodular excision for painful localized Achilles tendon xanthomas in type II hyperlipoproteinemia: a case report. J Foot Ankle Surg 50:603–606
2. Artieda M, Cenarro A, Junquera C et al (2005) Tendon xanthomas in familial hypercholesterolemia are associated with a differential inflammatory response of macrophages to oxidized LDL. FEBS Lett 579:4503–4512
3. Asai A, Kohli R (2015) Familial homozygous hypercholesterolemia: when to turn to transplant? Pediatr Transplant 19:577–579
4. Beeharry D, Coupe B, Benbow EW et al (2006) Familial hypercholesterolaemia commonly presents with Achilles tenosynovitis. Ann Rheum Dis 65:312–315
5. Bhojwani RA, Khot R (2011) Cerebrotendinous xanthomatosis: a rare genetic disorder. BMJ Case Rep, pii: bcr0820114582
6. Chua SY, Chang HC (2006) Bilateral spontaneous rupture of the quadriceps tendon as an initial presentation of alkaptonuria – a case report. Knee 13:408–410
7. Coss KP, Treacy EP, Cotter EJ et al (2014) Systemic gene dysregulation in classical Galactosaemia: is there a central mechanism? Mol Genet Metab 113:177–187
8. Cuchel M, Bruckert E, Ginsberg HN, European Atherosclerosis Society Consensus Panel on Familial Hypercholesterolaemia et al (2014) Homozygous familial hypercholesterolaemia: new insights and guidance for clinicians to improve detection and clinical management. A position paper from the Consensus Panel on Familial Hypercholesterolaemia of the European Atherosclerosis Society. Eur Heart J 35:2146–2157
9. Dagistan E, Canan A, Kizildag B, Barut AY (2013) Multiple tendon xanthomas in patient with heterozygous familial hypercholesterolaemia: sonographic and MRI findings. BMJ Case Rep, pii: bcr2013200755
10. Damian LO, Felea I, Boloşiu C, Botar-Jid C, Fodor D, Rednic S (2013) A case of alkaptonuria – ultrasonographic findings. Med Ultrason 15:321–325
11. Dawoud NM, Bakry OA, Seleit I (2015) Homozygous familial hypercholesterolemia associated with symmetric subcutaneous lipomatosis. Indian J Dermatol 60:420
12. El-Hattab AW (2015) Inborn errors of metabolism. Clin Perinatol 42:413–439
13. Garrod AE (1902) About Alkaptonuria. Med Chir Trans 85:69–78
14. Guañabens N, Mumm S, Möller I et al (2014) Calcific periarthritis as the only clinical manifestation of hypophosphatasia in middle-aged sisters. J Bone Miner Res 29:929–934
15. Guguloth A, Aswani Y, Anandpara KM (2015) Prenatal diagnosis of hypophosphatasia congenita using ultrasonography. Ultrasonography. doi: 10.14366/usg.15008
16. Haliloglu B, Guran T, Atay Z et al (2013) Infantile loss of teeth: odontohypophosphatasia or childhood hypophosphatasia. Eur J Pediatr 172:851–853
17. Harada-Shiba M, Arai H, Oikawa S et al (2012) Guidelines for the management of familial hypercholesterolemia. J Atheroscler Thromb 19:1043–1060
18. Helliwell TR, Gallagher JA, Ranganath L (2008) Alkaptonuria – a review of surgical and autopsy pathology. Histopathology 53:503–512
19. Hollis A, Arundel P, High A, Balmer R (2013) Current concepts in hypophosphatasia: case report and literature review. Int J Paediatr Dent 23:153–159
20. Jarauta E, Junyent M, Gilabert R et al (2009) Sonographic evaluation of Achilles tendons and carotid atherosclerosis in familial hypercholesterolemia. Atherosclerosis 204:345–347
21. Lorincz MT, Thomas D, Fink JK (2005) Cerebrotendinous xanthomatosis: possible higher prevalence than previously recognized. Arch Neurol 62(9):1459–1463
22. Manoj Kumar RV, Rajasekaran S (2003) Spontaneous tendon ruptures in alkaptonuria. J Bone Joint Surg (Br) 85:883–886
23. Mirzanli C, Esenyel CZ, Ozturk K, Baris A, Imren Y (2013) Cerebrotendinous xanthomatosis presenting with bilateral achilles tendon xanthomata: a case report. J Am Podiatr Med Assoc 103:152–155
24. Mistry JB, Bukhari M, Taylor AM (2013) Alkaptonuria. Rare Dis 1, e27475
25. Perry MB, Suwannarat P, Furst GP, Gahl WA, Gerber LH (2006) Musculoskeletal findings and disability in alkaptonuria. J Rheumatol 33:2280–2285
26. Porta F, Pagliardini S, Pagliardini V, Ponzone A, Spada M (2015) Newborn screening for galactosemia: a 30-year single center experience. World J Pediatr 11:160–164
27. Preiss Y, Santos JL, Smalley SV, Maiz A (2014) Cerebrotendinous xanthomatosis: physiopathology, clinical manifestations and genetics. Rev Med Chil 142:616–622
28. Pudhiavan A, Agrawal A, Chaudhari S, Shukla A (2013) Cerebrotendinous xanthomatosis – the spectrum of imaging findings. J Radiol Case Rep 7:1–9

29. Ranganath LR, Cox TF (2011) Natural history of alkaptonuria revisited: analyses based on scoring systems. J Inherit Metab Dis 34:1141–1151
30. Reiner Ž (2015) Management of patients with familial hypercholesterolaemia. Nat Rev Cardiol 16
31. Richard S, Tamas C, Sell DR, Monnier VM (1991) Tissue-specific effects of aldose reductase inhibition on fluorescence and cross-linking of extracellular matrix in chronic galactosemia. Relationship to pentosidine cross-links. Diabetes 40:1049–1056
32. Rubio-Agusti I, Kojovic M, Edwards MJ et al (2012) Atypical parkinsonism and cerebrotendinous xanthomatosis: report of a family with corticobasal syndrome and a literature review. Mov Disord 27:1769–1774
33. Smalley SV, Preiss Y, Suazo J, Vega JA, Angellotti I, Lagos CF, Rivera E, Kleinsteuber K, Campion J, Martínez JA, Maiz A, Santos JL (2015) Novel splice-affecting variants in CYP27A1 gene in two Chilean patients with Cerebrotendinous Xanthomatosis. Genet Mol Biol 38(1):30–36
34. Tsouli SG, Xydis V, Argyropoulou MI, Tselepis AD, Elisaf M, Kiortsis DN (2009) Regression of Achilles tendon thickness after statin treatment in patients with familial hypercholesterolemia: an ultrasonographic study. Atherosclerosis 205:151–155
35. Tsouli SG, Kiortsis DN, Argyropoulou MI, Mikhailidis DP, Elisaf MS (2005) Pathogenesis, detection and treatment of Achilles tendon xanthomas. Eur J Clin Invest 35:236–244
36. Verma IC, Puri RD (2015) Global burden of genetic disease and the role of genetic screening. Semin Fetal Neonatal Med, pii: S1744-165X(15)00078-5
37. Verma SB (2005) Early detection of alkaptonuria. Indian J Dermatol Venereol Leprol 71:189–91
38. Wallon D, Guyant-Maréchal L, Laquerrière A et al (2010) Clinical imaging and neuropathological correlations in an unusual case of cerebrotendinous xanthomatosis. Clin Neuropathol 29:361–364
39. Yoshinaga T, Sekijima Y, Koyama S et al (2014) Clinical and radiological findings of a cerebrotendinous xanthomatosis patient with a novel p.A335V mutation in the CYP27A1 gene. Intern Med 53:2725–2729
40. Yuan G, Wang J, Hegele RA (2006) Heterozygous familial hypercholesterolemia: an underrecognized cause of early cardiovascular disease. CMAJ 174:1124–1129

Hyperuricemia in Tendons

11

Isabel Andia and Michele Abate

Abstract

Hyperuricemia, particularly gout, and the immune inflammatory response are highly integrated. Both, long standing hyperuricemia and monosodium urate (MSU) crystal deposition can challenge tendon homeostasis because of their potential to cause inflammation to the host. Knowledge is emerging from clinical imaging research depicting where MSU crystals deposit, including patellar tendon, triceps and quadriceps tendons. Remarkably, subclinical tendon inflammation and damage are also present in asymptomatic hyperuricemia. Monosodium urate crystals act as danger activating molecular patterns (DAMPs), activating the inflammasome and inducing the secretion of IL-1beta, a key mediator of the inflammatory response. The crucial role of IL-1beta in driving the inflammatory events during gout attacks is supported by the clinical efficacy of IL-1beta blockade. Some data implicating IL-1beta as an initiator of tendinopathy exist, but the link between hyperuricemia and the development of tendinopathy remains to be validated. Further knowledge about the interactions of uric acid with both innate immune and tendon cells, and their consequences may help to determine if there is a subclass of hyperuricemic-tendinopathy.

Abbreviations

ADAMTS1	A Disintegrin-Like And Metalloprotease (Reprolysin Type) With Thrombospondin Type 1 Motif, 1
ADAMTS4	A Disintegrin-Like And Metalloprotease (Reprolysin Type) With Thrombospondin Type 1 Motif, 4

I. Andia (✉)
Regenerative Medicine Laboratory, BioCruces Health Research Institute, Cruces University Hospital, 48903 Barakaldo, Spain
e-mail: iandia2010@hotmail.com

M. Abate
Department of Medicine and Science of Aging, University G. d'Annunzio, Via dei Vestini 31, Chieti-Pescara, 66013 Chieti Scalo (CH), Italy

P.W. Ackermann and D.A. Hart (eds.), *Metabolic Influences on Risk for Tendon Disorders*, Advances in Experimental Medicine and Biology 920, DOI 10.1007/978-3-319-33943-6_11

ADAMTS5	A Disintegrin-Like And Metalloprotease (Reprolysin Type) With Thrombospondin Type 1 Motif, 5
CRP	C reactive protein
DAMP	Danger Associated Molecular patterns
DECT	dual energy computerized tomography
DNA	deoxynucleic acid
IL-1beta, IL-1alpha, IL-6, IL-8, IL-18	Interleukins
MCP-1/ CCL2	macrophage chemotactic protein
MMP-2, MMP-3, MMP-13	metalloproteinases
MSU	monosodium urate
NLRP3	NOD-like receptor protein 3
NSAIDs	non-steroidal anti-inflammatory drugs
PGE2	prostaglandin E2
RNA	ribonucleic acid
TLR2, TLR4	Toll like receptors
TNF-alpha	tumor necrosis factor alpha
VEGF	vascular endothelial growth factor

Introduction

People with hyperuricemia related tendinopathy represent the intersection between two prevalent entities: millions with hyperuricemia and the millions of people who suffer tendinopathy. The rising frequency of both, hyperuricemia and tendinopathy, is associated to a combination of factors including increased longevity, and shifts in diet and lifestyle.

Hyperuricemia is the gateway for gout, an old disease that concerns almost all civilizations over the ages. The increasing prevalence of hyperuricemia, along with the doubling of the rate of gout in the past decades [56], and its association with important comorbidities such as cardiovascular disease, metabolic syndrome or diabetes among others is creating a heavy burden by driving up medical expenditures.

Gout describes an inflammatory arthritis resulting from monosodium urate (MSU) crystal deposition in specific anatomical sites including joints, and tendons. Currently, medical English has retained old Greek and Latin names such as podagra (from *podos*, foot and *agreos* attack), to describe acute gout attacks, more specifically MSU crystal deposition in the first metatarso-phalangeal joint. Tophus, (from the Latin *tofus* meaning porous stone), depicts nodular masses of MSU crystals, most commonly occurring at the base of the great toe and fingers.

Hyperuricemia is defined as a serum urate level higher than 6.8 mg/dL (0.40 μmol/L). Remarkably, patients who overproduce uric acid represent fewer than 20 % of those with gout [11]. As described below, not only gout and associated inflammation, but hyperuricemia can be an intrinsic element that directs immune activities, favoring the development and progression of tendinopathy. Indeed, hyperuricemia can induce cellular and/or metabolic distress [1], thereafter tendon extracellular matrix degeneration as well as sub-clinical inflammation.

In this chapter, we first summarize the main characteristics of hyperuricemia/gout and tendinopathy. Then, we discuss clinical and biological information that can mean an association between both pathologies, and can help us in substantiating a more precise medicine by explaining different subclasses of tendinopathies driven by new diagnostic.

Hyperuricemia and Tendinopathy?

The hypothesis that hyperuricemia can have a role in the development of tendinopathy is based on recent advances of crystal biology and the pathophysiology of tendinopathy.

Hyperuricemia/Gout

Uric acid, (a heterocyclic purine derivative 7,9-dihydro-1H-purine-2, 6, 8 (3H)-trione), is the degradation product of purine metabolism after xanthine oxidase oxidizes purines (Fig. 11.1). In mammals, other than humans and primates, uricase (urate oxidase) forms soluble allantoin, thereby lowering uric acid levels below 2 mg/dL. Instead, human subjects lack uricase, thus reference range of serum urate is higher, i.e. 3.4–7.2 mg/dL for men and 2.4–6.1 mg/dL for women.

Two important physiological mechanisms decide the levels of uric acid in body fluids. On one hand, purine metabolism in the liver, i.e. 2/3 of uric acid is formed naturally and purines in the diet produce 1/3. Thus, excessive intake of food rich in purines can induce overproduction of uric acid, and acute hyperuricemia. On the other hand, renal function is involved in the systemic adjustments of urate levels, i.e. glomerular filtration, tubular reabsorption, secretion and post-secretory reabsorption. Alterations in these mechanisms can cause chronic hyperuricemia.

The fact that native levels of serum urate are very close to saturation point (7.0 mg/dL) suggests that this molecule has important roles in maintaining tissue homeostasis. Indeed, serum urate is a reducing agent that accounts for almost half of the antioxidant potential of blood, influences redox potential and protects against oxidative damage. Uric acid prevents the toxicity by reactive oxygen and nitrogen species with negative influence on critical cell functions. Recent data show that uric acid is protective against Alzheimer, Parkinson, amyotrophic lateral sclerosis, and multiple sclerosis [27, 49]. Instead, the crystalline form that develops after chronic (longstanding) hyperuricemia is well-known because of its inflammatory properties.

Hyperuricemia is the most important risk factor for inflammatory gout, but peculiarly many people with hyperuricemia do not follow crystal deposition and gout attacks. To add complexity, sometimes serum urate concentration is within the normal range during acute gout attacks [36]. In one cohort study with a follow up of 5 years, gout developed in only 22 % of patients with urate levels above 9 mg/dL (535 umol/L) [11]. Figure 11.2a depicts the clinical stages from hyperuricemia to chronic gout.

Vulnerability to gout is also attributed to genetic influences. The kidneys reabsorb about 90 % of the daily load of filtered urate, and specific anion transporters mediate the process. Actually, genome wide studies have identified polymorphisms, associated with serum urate levels, in various loci codifying glucose and urate protein transporters in the kidney, including

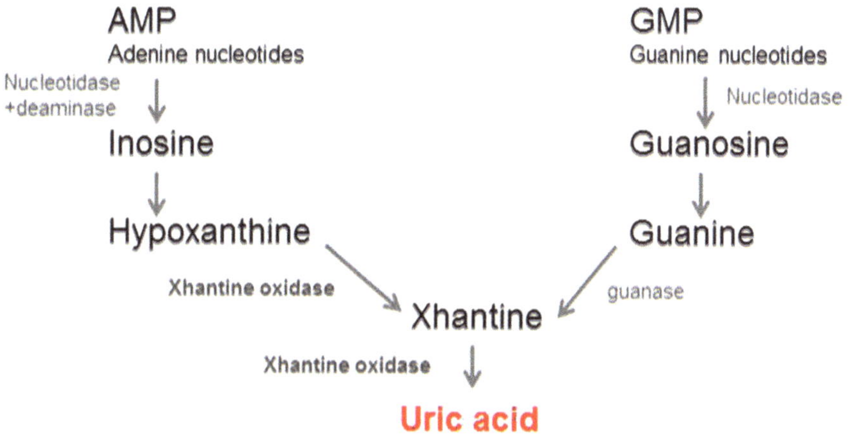

Fig. 11.1 Purine degradation and formation of uric acid
Uric acid is a normal intracellular constituent, generated as part of the normal turnover of nucleic acids following purine degradation in the cell cytosol. Purines, adenine and guanine, are essential components of DNA and RNA

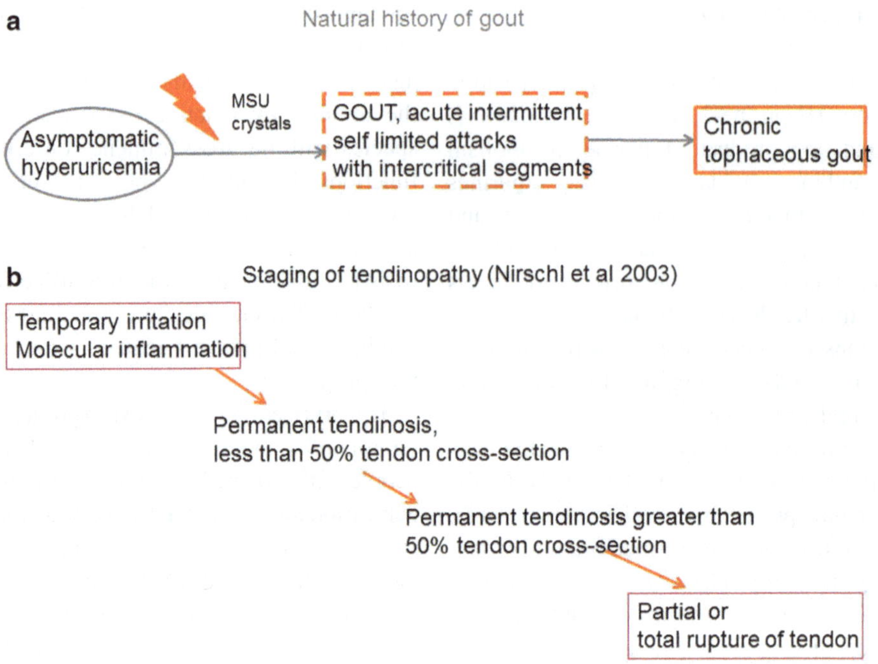

Fig. 11.2 Clinical outcome of hyperuricemia/gout (a), and tendinopathy (b)
(**a**) Asymptomatic stages, without or with urate deposition, are followed by acute intermittent painful flares that occur most often in lower extremities. Inter-critical periods become shorter as disease progresses. People with advanced chronic gout can have tophi, usually in distal areas of the body. (**b**) Tendinopathy, initially characterized by temporary irritation and molecular inflammation, is considered a degenerative disease with changes in extracellular matrix composition and loss of tissue architecture. Ensuing detrimental mechanical properties can result in partial tears or total rupture [42]

GLUT9 (SLC2A9), and ABCG2 [35, 52, 58]. However, data derived from rheumatologic studies in identical twins show a concordance rate of 53 % in monozygotic and 24 % in dizygotic twin pairs, indicating that the genetic susceptibility requires other concomitant factors [25]. Despite increasing knowledge, the factors controlling formation and deposition of MSU crystals are not fully understood.

Inflammatory Cell Reactions to MSU Crystals

Broadly speaking two processes occur after MSU deposition, first cell detection of the crystals, and secondly activation of the inflammatory response driven by chemokines and cytokines (please also see Chap. 20).

Both immune cells and local cells can detect MSU crystals (Fig. 11.3). As described later, patients with hyperuricemia have elevated levels of circulating MCP-1, a chemokine that mobilizes monocyte/macrophages [21]. Monocyte/macrophages can phagocytose crystals that interact with the inflammasome, of which NLRP3 (NOD-like receptor protein-3) is the best characterized. This interaction triggers activation of caspase 1 in the cytosol, followed by cleavage of pro-IL-1beta into mature IL-1beta. Secreted IL-1beta binds to the IL-1R, present in the membrane of different cell phenotypes. In doing so, IL-1beta activates an inflammatory cascade by promoting the expression and secretion of additional inflammatory molecules, mainly cytokines and chemokines. Other reported inflammatory pathways involve the engagement of TLR2 and TLR4 receptors [31, 32, 53, 54].

The crucial role of IL-1beta in initiating the inflammatory cascade during gout attacks is supported by the efficacy of IL-1beta blockade

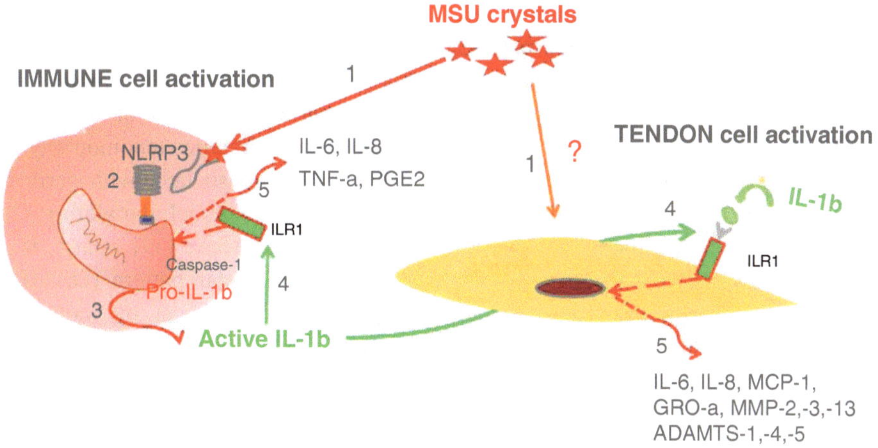

Fig. 11.3 Cartoon showing a tentative model that links hyperuricemia with tendinopathy, mainly through activation an inflammatory cascade driven by IL-1beta on immune and tendon cells
MSU crystal detected and phagocyted by macrophages (1), interact with the inflammasome (2) inducing the activation of caspase 1. Ensuing cleavage of pro-IL-1b by caspase 1 (3), generation and secretion of active IL-1beta switch on the inflammatory cascade. IL-1beta binds to IL-1R in macrophages (4) and tendon cells (4) and set in motion gene transcription and production of inflammatory molecules

therapy in patients with gout initiating urate-lowering therapy [40]. In fact, several anti-IL-1beta therapies have been tested to prevent gout flares after the sudden urate decline (at the beginning of urate lowering treatment) which induces the release of MSU crystals from tophus.

Supporting the innate immune response to gout, MSU crystals induced cultured monocytes to express the chemokine IL-8, a potent neutrophil chemoattractant [21]. Actually, neutrophil infiltration is a feature of crystal-induced inflammation.

MSU deposition affects not only innate immune cells but also tendon cells. Indeed, the presence of MSU crystals induced a reduction of tendon cell viability and reduced the expression of ECM proteins, specifically type 1 collagen. Besides, MSU crystals up-regulate MMP-2, −3, −13, ADAMTS1, ADAMTS4 and ADAMTS5 [14] (Chhana 2014). Despite up-regulation of ECM detrimental enzymes, tendon rupture in gout is less frequent than cartilage damage and bone erosions. The latter are frequent in gout, as well as decreased viability and function of osteoblast [13]. Of note, altered bone functions may have implication at the enthesis.

Both, MSU and hyperuricemia, are inflammatory adjuvants triggering an adaptive immune response (i.e. antigen-specific response mediated by T, and B cells). Remarkably, it is believed that MSU induced sterile inflammation is different from uric acid mediated adjuvancity, [34] but our understanding of the immunological mechanisms underlying adjuvanticity is still incomplete.

Tendinopathy

The etiology of tendinopathy is multifactorial often triggered by vulnerability factors. Since the myth of the Greek hero Achilles of the Trojan War, telling how vulnerable a tendon is, much has been learned, and current research has produced several biological hypothesis based on histopathological, biochemical and clinical findings that show cell apoptosis, angiofibroblastic features or abnormal biochemical adaptations. These findings suggest that a failed healing response underlies the condition [4].

Here we focus on two mechanisms underlying the loss of tendon homeostasis, i.e. inflammation and cell death, because they are particularly relevant to provide a tentative relationship between hyperuricemia and tendinopathy.

Inflammation

Classical signs of active inflammation, such as histological evidence of leukocyte infiltrates, are scarce in degenerative tendinopathy [3]. For example, the presence of mast cells shows in the patellar tendon of patients with pain and swelling [53] and, in calcaneal overuse tendinopathy [46]. Instead, in acute reactive tendinitis, such as De Quervain disease, inflammatory components, including neutrophil elastase, cyclooxygenase, as well as macrophage infiltration affecting collagen structure is evident [26].

The presence of an inflammatory molecular milieu in degenerative tendinopathy is more obvious, as local tendon cells can synthesize inflammatory molecules in response to tissue stress. This consideration differs from the classical view of inflammation, embodied by the presence of inflammatory cells. Essentially, the biochemical adaptation of prolonged repetitive mechanical loading produces cytokines such as IL-1beta, IL-6, IL-8 and TNF-alpha, prostaglandins such as PGE2, and neuropeptides such as substance P [17, 18, 29, 38]. Some harmful effects of inflammatory cytokines include up-regulated VEGF production along with enhanced production of metalloproteinases, such as MMP-1 MMP-3 and MMP-13 [57] that cause matrix destruction.

In accordance with the concept of intrinsic tendon healing deficits and the weak response to injury, during acute early healing Achilles, tendons did not display detectable levels of pro-inflammatory molecules. Merely, pleiotropic cytokines including IL-6 and IL-8 were elevated [2].

Cell Death and Local Hyperuricemia

Another histopathological feature that emerges in tendinopathy is cell death, suggesting that the problem is the loss of homeostasis, and failed healing response that creates a vicious circle between cell death and progressive matrix disruption leading to chronic degeneration. At least one study has shown many apoptotic cells in ruptured supraspinatus tendon [59], and other studies showed excessive apoptosis in patellar tendinopathic specimens in athletes [30], and noninsertional Achilles tendinopathy [43].

Cell death is associated with sterile inflammation induced by intracellular alarmins released to the extracellular space [39]. Alarmins are danger activating molecular pattern (DAMP), because they can alert the host by mobilizing innate immune cells, and ensuing inflammatory response through Toll-like receptors (TLR). Uric acid is one of these alarmins. Cells produce even more uric acid when they die and DNA and RNA are metabolized. Actually, large-scale cell death induces robust MSU precipitation caused by intracellular urate released to the extracellular space, creating a supersaturated solution in the high sodium extracellular environment [55].

In what circumstances monosodium urate precipitate, nucleate and form inflammatory crystals have been the focus of recent experimental research. Indeed, the molecular composition of the milieu may influence both the amount and size of crystals, and their inflammatory properties. Besides, crystallization depends on hydration, and extracellular pH, and often occurs in areas of compromised vascularity where the ability to regulate temperature is hampered. Animal studies have proposed that immunoglobulins may be part of a positive feedback loop promoting uric acid crystallization and immunogenicity [23].

Activated tendon cells drive immune cell infiltration [7], the duration and intensity of inflammation; they also control the switch from acute to chronic inflammation. In this context, we have explored whether tendon cells can sense hyperuricemia in their biological milieu, and whether hyperuricemic PRP can incite tendon cells to switch to an inflammatory phenotype. Actually, tenocytes express the main receptors involved in sterile inflammation, TLR2 and TLR4 [16], but it is not clear in what circumstances these receptors are functional. Because serum amyloid protein primed synovial fibroblasts to produce active IL-1beta and IL-1alpha when exposed to high uric acid and

MSU crystals, we hypothesized that hyperuricemic PRP with native levels of amyloid protein could trigger a molecular inflammatory response by tendon cells. But, we found that hyperuricemia is a minor stressor for tendon cells as it mitigates the modest inflammatory effect induced by PRP reducing the expression and synthesis of IL-6 and IL8 [5, 6].

Instead, the presence of MSU crystals induce the reduction of tendon cell viability and reduce expression of ECM proteins, specifically type 1 collagen and in parallel up-regulates catabolic ECM proteins [14]. Interestingly, MSU crystals were identified next to and invading the tendon, and at the enthesis [14].

Clinical Observations

The hypothesis that uric acid may play a role in the development, and progression of tendinopathy is predicated on the presence of low-grade inflammation in hyperuricemic patients, and recent findings on tendon imaging.

Low-Grade Inflammation

One potential explanation for how longstanding hyperuricemia can modify tendon homeostasis is the presence of low-grade systemic inflammation. For instance, elevated levels of MCP-1/CCL2, a chemokine involved in leukocyte trafficking, are displayed in patients with acute gout and hyperuricemia [21]. Moreover, in these patients MCP-1/CCL2 concentrations in serum correlates with the increased number of circulating CD14+ monocytes, and the adhesion molecule CD11b. In asymptomatic subjects, there is an association between serum levels of uric acid with inflammatory markers including CRP, IL-6, IL-18 and TNF-a [51].

These findings are consistent with human studies showing enhanced levels of circulating $CD14^+$ monocytes, not only in patients with gout, but also in asymptomatic hyperuricemia compared to normouricemic patients.

Besides, neutrophils from gout patients are primed for enhanced MSU crystal induced superoxide production. Moreover, this neutrophil function persists in asymptomatic hyperuricemic, and the inflammatory environment likely contributes to higher IL-8 production and neutrophil survival in the absence of direct crystal stimulation [33].

Tendon Imaging

Advances of imaging science permits taking a closer look at soft tissues [12]. A recent ultrasound study has identified not only cartilage, but also tendons as tissues with high frequency deposition of MSU crystals. In particular, the patellar tendon, triceps and quadriceps tendons are often affected [19, 41].

Besides, tophi causing flexor tenosynovitis along with dactylitis is often seen in patients with gout [8, 10]. Gouty tophi causing ruptures of tendons and ligaments are more unusual. Even so, anecdotic tophaceous depositions have shown in the anterior cruciate ligament [37], and distal quadriceps tendon [9]. Moreover, peculiar acute podagra happened in the Achilles tendon of a young long distance runner with normouricemia [22].

Interestingly, subclinical tendon inflammation and subclinical structural damage have been reported in people with asymptomatic hyperuricemia [45, 47]. For example, enthesopathy in the patellar tendon was found in 12 % of hyperuricemic patients and 2 % of normouricemic. Accordingly, Achilles tendon enthesopathy was present in 15 % of hyperuricemic contrasting with 1.9 % of normouricemic subjects. Corroborating these data, Achilles tendon ruptures, characterized as acute trauma of chronically degenerated tendons, occur more often in asymptomatic patients with hyperuricemia than in asymptomatic normouricaemic subjects [45].

Imaging continues to yield information relevant to tendinopathy, and in the past years ultrasound is routinely used by sports physician as a tool for the diagnosis of tendon pathology. When treating tendinopathic symptoms, we shall keep in mind the odds of gouty tophus within patellar tendon [15, 20, 50]. Experience in visualizing the ultrasonographic features of MSU crystals or gouty tophus may accelerate the diagnosis and treatment modality depending on clinical stage.

In the same way, DECT (dual energy computerized tomography) provides correct views in patients with tophaceous gout and has shown MSU deposition affecting the Achilles tendon and peroneal tendons [15, 24, 28, 48].

Conclusions

Hyperuricemia is an increasingly important metabolic condition. However, its clinical repercussions in tendons are often underestimated, and not clearly understood. While little evidence is available to implicate hyperuricemia in the pathogenesis of tendinopathy, it is more obvious that crystal deposition in and around tendons during gout attacks can trigger cell death, as a consequence of the loss of homeostatic collagen tension owing to microscopic collagen breakdown. Indeed, tendon cells lacking appropriate ECM attachment are rapidly eliminated.

But, not only mechanical interference of crystals with the extracellular matrix, also inflammatory reactions of monocytes/macrophages and tendon cells, ensuing from crystal deposition and IL-1beta induced inflammation, can favor the progression of tendinopathy.

Nevertheless, while the biological mechanisms underlying MSU activation of inflammation are understood, there is little information about hyperuricemia-mediated adjuvancity in tendinopathy. Knowledge about the interactions of urate with both innate immune and local cells, may help the research community to determine if there is a subclass of hyperuricemic-tendinopathy, and set the grounds for clarifying the biology and mechanism behind hyperuricemia linked tendinopathy.

There is, however, much to investigate because most concepts exposed here are still speculative, and future research has to focus on how hyperuricemia-mediated adjuvancity works in tendon inflammation, cell death and extracellular matrix deterioration.

A two-stage approach, firstly urate-lowering therapy designed to dissolve MSU crystals and, secondly keeping uric acid below saturation point for long-life has been recommended [44]. Imaging can be used to evaluate outcomes until MSU crystals are dissolved. Rheumatologists often prescribe prophylactic treatments such as colchicine to minimize inflammatory flares during the initial stages of urate lowering therapy. Investigational prophylactic treatments based on anti-IL-1b blockade are promising in patients who have contraindications for colchicine and NSAIDs [40] and may be beneficial for tendons.

Contrasting with current trends of precision medicine, tendinopathy management is unspecific and merely palliative. Patient heterogeneity hinders advances in novel treatments and clinical trial design claims subpopulations are more clearly understood. Exploring potential connections between tendinopathy and hyperuricemia, and determining whether or not there is a subtype of tendinopathy induced by hyperuricemia may help to tackle part of this important problem. Should the growing evidence that the high urate level is a risk factor for tendinopathy become accepted would have a major impact on the diagnostic and treatment of tendinopathy.

References

1. Abate M, Schiavone C, Salini V, Andia I (2013) Occurrence of tendon pathologies in metabolic disorders. Rheumatology (Oxford) 52(4):599–608
2. Ackermann PW, Domeij-Arverud E, Leclerc P, Amoudrouz P, Nader GA (2013) Anti-inflammatory cytokine profile in early human tendon repair. Knee Surg Sports Traumatol Arthrosc 21(8):1801–1806
3. Alfredson H (2005) The chronic painful Achilles and patellar tendon: research on basic biology and treatment. Scand J Med Sci Sports 15(4):252–259
4. Andia I, Sanchez M, Maffulli N (2010) Tendon healing and platelet-rich plasma therapies. Expert Opin Biol Ther 10(10):1415–1426
5. Andia I, Rubio-Azpeitia E, Maffulli N (2014) Hyperuricemic PRP in tendon cells. Biomed Res Int 2014:926481

6. Andia I, Rubio-Azpeitia E (2014) Angiogenic and innate immune responses triggered by PRP in tendon cells are not modified by hyperuricemia. Muscles Ligaments Tendons J 4(3):292–297
7. Andia I, Rubio-Azpeitia E, Maffulli N (2015) Platelet-rich plasma modulates the secretion of inflammatory/angiogenic proteins by inflamed tenocytes. Clin Orthop Relat Res 473(5):1624–1634
8. Aslam N, Lo S, McNab I (2004) Gouty flexor tenosynovitis mimicking infection: a case report emphasising the value of ultrasound in diagnosis. Acta Orthop Belg 70(4):368–370
9. Bond JR, Sim FH, Sundaram M (2004) Radiologic case study. Gouty tophus involving the distal quadriceps tendon. Orthopedics 27(1):18, 90–92
10. Bullocks JM, Downey CR, Gibler DP, Netscher DT (2009) Crystal deposition disease masquerading as proliferative tenosynovitis and its associated sequelae. Ann Plast Surg 62(2):128–133
11. Campion EW, Glynn RJ, DeLabry LO (1987) Asymptomatic hyperuricemia. Risks and consequences in the normative aging study. Am J Med 82(3):421–426
12. Chen CK, Chung CB, Yeh L, Pan HB, Yang CF, Lai PH, Liang HL, Resnick D (2000) Carpal tunnel syndrome caused by tophaceous gout: CT and MR imaging features in 20 patients. AJR Am J Roentgenol 175 (3):655–659
13. Chhana A, Callon KE, Pool B, Naot D, Watson M, Gamble GD, McQueen FM, Cornish J, Dalbeth N (2011) Monosodium urate monohydrate crystals inhibit osteoblast viability and function: implications for development of bone erosion in gout. Ann Rheum Dis 70(9):1684–1691
14. Chhana A, Callon KE, Dray M, Pool B, Naot D, Gamble GD, Coleman B, McCarthy G, McQueen FM, Cornish J, Dalbeth N (2014) Interactions between tenocytes and monosodium urate monohydrate crystals: implications for tendon involvement in gout. Ann Rheum Dis 73(9):1737–1741
15. Dalbeth N, Kalluru R, Aati O, Horne A, Doyle AJ, McQueen FM (2013) Tendon involvement in the feet of patients with gout: a dual-energy CT study. Ann Rheum Dis 72(9):1545–1548
16. de Mos M, Joosten LA, Oppers-Walgreen B, van Schie JT, Jahr H, van Osch GJ, Verhaar JA (2009) Tendon degeneration is not mediated by regulation of Toll-like receptors 2 and 4 in human tenocytes. J Orthop Res 27(8):1043–1047
17. Fredberg U, Stengaard-Pedersen K (2008) Chronic tendinopathy tissue pathology, pain mechanisms, and etiology with a special focus on inflammation. Scand J Med Sci Sports 18(1):3–15
18. Fedorczyk JM, Barr AE, Rani S, Gao HG, Amin M, Amin S, Litvin J, Barbe MF (2010) Exposure-dependent increases in IL-1beta, substance P, CTGF, and tendinosis in flexor digitorum tendons with upper extremity repetitive strain injury. J Orthop Res 28 (3):298–307
19. Forbess LJ, Fields TR (2012) The broad spectrum of urate crystal deposition: unusual presentations of gouty tophi. Semin Arthritis Rheum 42(2):146–154
20. Gililland JM, Webber NP, Jones KB, Randall RL, Aoki SK (2011) Intratendinous tophaceous gout imitating patellar tendonitis in an athletic man. Orthopedics 34(3):223
21. Grainger R, McLaughlin RJ, Harrison AA, Harper JL (2013) Hyperuricaemia elevates circulating CCL2 levels and primes monocyte trafficking in subjects with inter-critical gout. Rheumatology (Oxford) 52 (6):1018–1021
22. Gunawardena H, Churn P, Blake DR (2005) Running for gout research. Rheumatology (Oxford) 44 (8):1073–1074
23. Kanevets U, Sharma K, Dresser K (2009) Shi Y A role of IgM antibodies in monosodium urate crystal formation and associated adjuvanticity. J Immunol 182 (4):1912–1918
24. Kimura-Hayama E, Criales-Vera S, Nicolaou S, Betanzos JL, Rivera Y, Alberú J, Rull-Gabayet M, Hernández-Molina G (2014) A pilot study on dual-energy computed tomography for detection of urate deposits in renal transplant patients with asymptomatic hyperuricemia. J Clin Rheumatol 20(6):306–309
25. Krishnan E, Lessov-Schlaggar CN, Krasnow RE, Swan GE (2012) Nature versus nurture in gout: a twin study. Am J Med 125(5):499–504
26. Kuo YL, Hsu CC, Kuo LC, Wu PT, Shao CJ, Wu KC, Wu TT, Jou IM (2015) Inflammation is present in de quervain disease–correlation study between biochemical and histopathological evaluation. Ann Plast Surg 74(Suppl 2):S146–S151
27. Kutzing MK, Firestein BL (2008) Altered uric acid levels and disease states. J Pharmacol Exp Ther 324 (1):1–7
28. Lagoutaris ED, Adams HB, DiDomenico LA, Rothenberg RJ (2005) Longitudinal tears of both peroneal tendons associated with tophaceous gouty infiltration. A case report. J Foot Ankle Surg 44 (3):222–224, Muscles Ligaments Tendons J. 2014 Nov 17;4(3):292–297
29. Legerlotz K, Jones ER, Screen HR, Riley GP (2012) Increased expression of IL-6 family members in tendon pathology. Rheumatology (Oxford) 51 (7):1161–1165
30. Lian Ø, Scott A, Engebretsen L, Bahr R, Duronio V, Khan K (2007) Excessive apoptosis in patellar tendinopathy in athletes. Am J Sports Med 35 (4):605–611
31. Liu-Bryan R, Pritzker K, Firestein GS, Terkeltaub R (2005) TLR2 signaling in chondrocytes drives calcium pyrophosphate dihydrate and monosodium urate crystal-induced nitric oxide generation. J Immunol 174(8):5016–5023
32. Liu-Bryan R, Scott P, Sydlaske A, Rose DM, Terkeltaub R (2005) Innate immunity conferred by Toll-like receptors 2 and 4 and myeloid differentiation factor 88 expression is pivotal to monosodium urate monohydrate crystal-induced inflammation. Arthritis Rheum 52(9):2936–2946
33. Martin WJ, Grainger R, Harrison A, Harper JL (2010) Differences in MSU-induced superoxide responses by neutrophils from gout subjects compared to healthy

controls and a role for environmental inflammatory cytokines and hyperuricemia in neutrophil function and survival. J Rheumatol 37(6):1228–1235
34. Martinon F, Pétrilli V, Mayor A, Tardivel A, Tschopp J (2006) Gout-associated uric acid crystals activate the NALP3 inflammasome. Nature 440 (7081):237–241
35. Matsuo H, Yamamoto K, Nakaoka H, Nakayama A, Sakiyama M, Chiba T, Takahashi A, Nakamura T, Nakashima H, Takada Y, Danjoh I, Shimizu S, Abe J, Kawamura Y, Terashige S, Ogata H, Tatsukawa S, Yin G, Okada R, Morita E, Naito M, Tokumasu A, Onoue H, Iwaya K, Ito T, Takada T, Inoue K, Kato Y, Nakamura Y, Sakurai Y, Suzuki H, Kanai Y, Hosoya T, Hamajima N, Inoue I, Kubo M, Ichida K, Ooyama H, Shimizu T, Shinomiya N (2015) Genome-wide association study of clinically defined gout identifies multiple risk loci and its association with clinical subtypes. Ann Rheum Dis 75 (4):652–659. pii: annrheumdis-2014-206191
36. McCarty DJ (1994) Gout without hyperuricemia. JAMA 271(4):302–303
37. Melloni P, Valls R, Yuguero M, Sáez A (2004) An unusual case of tophaceous gout involving the anterior cruciate ligament. Arthroscopy 20(9):e117–e121
38. Millar NL, Wei AQ, Molloy TJ, Bonar F, Murrell GA (2009) Cytokines and apoptosis in supraspinatus tendinopathy. J Bone Joint Surg (Br) 91(3):417–424
39. Millar NL, Murrell GA, McInnes IB (2013) Alarmins in tendinopathy: unravelling new mechanisms in a common disease. Rheumatology (Oxford) 52 (5):769–779
40. Mitha E, Schumacher HR, Fouche L, Luo SF, Weinstein SP, Yancopoulos GD, Wang J, King-Davis S, Evans RR (2013) Rilonacept for gout flare prevention during initiation of uric acid-lowering therapy: results from the PRESURGE-2 international, phase 3, randomized, placebo-controlled trial. Rheumatology (Oxford) 52(7):1285–1292
41. Naredo E, Uson J, Jiménez-Palop M, Martínez A, Vicente E, Brito E, Rodríguez A, Cornejo FJ, Castañeda S, Martínez MJ, Sanz J, Möller I, Batlle-Gualda E, Garrido J, Pascual E (2014) Ultrasound-detected musculoskeletal urate crystal deposition: which joints and what findings should be assessed for diagnosing gout? Ann Rheum Dis 73 (8):1522–1528
42. Nirschl RP, Ashman ES (2003) Elbow tendinopathy: tennis elbow. Clin Sports Med 22(4):813–836
43. Pearce CJ, Ismail M, Calder JD (2009) Is apoptosis the cause of noninsertional achilles tendinopathy? Am J Sports Med 37(12):2440–2444
44. Perez-Ruiz F, Dalbeth N, Bardin T (2015) A review of uric acid, crystal deposition disease, and gout. Adv Ther 32(1):31–41
45. Pineda C, Amezcua-Guerra LM, Solano C, Rodriguez-Henríquez P, Hernández-Díaz C, Vargas A, Hofmann F, Gutiérrez M (2011) Joint and tendon subclinical involvement suggestive of gouty arthritis in asymptomatic hyperuricemia: an ultrasound controlled study. Arthritis Res Ther 13(1):R4
46. Pingel J, Wienecke J, Kongsgaard M, Behzad H, Abraham T, Langberg H, Scott A (2013) Increased mast cell numbers in a calcaneal tendon overuse model. Scand J Med Sci Sports 23(6):e353–e360
47. Puig JG, de Miguel E, Castillo MC, Rocha AL, Martínez MA, Torres RJ (2008) Asymptomatic hyperuricemia: impact of ultrasonography. Nucleosides Nucleotides Nucleic Acids 27(6):592–595
48. Radice F, Monckeberg JE, Carcuro G (2011) Longitudinal tears of peroneus longus and brevis tendons: a gouty infiltration. J Foot Ankle Surg 50(6):751–753
49. Rock KL, Kataoka H, Lai JJ (2013) Uric acid as a danger signal in gout and its comorbidities. Nat Rev Rheumatol 9(1):13–23
50. Rodas G, Pedret C, Català J, Soler R, Orozco L, Cusi M (2013) Intratendinous gouty tophus mimics patellar tendonitis in an athlete. J Clin Ultrasound 41 (3):178–182
51. Ruggiero C, Cherubini A, Ble A, Bos AJ, Maggio M, Dixit VD, Lauretani F, Bandinelli S, Senin U, Ferrucci L (2006) Uric acid and inflammatory markers. Eur Heart J 27(10):1174–1181
52. Scharpf RB, Mireles L, Yang Q, Köttgen A, Ruczinski I, Susztak K, Halper-Stromberg E, Tin A, Cristiano S, Chakravarti A, Boerwinkle E, Fox CS, Coresh J, Linda Kao WH (2014) Copy number polymorphisms near SLC2A9 are associated with serum uric acid concentrations. BMC Genet 15:81
53. Scott A, Lian Ø, Bahr R, Hart DA, Duronio V, Khan KM (2008) Increased mast cell numbers in human patellar tendinosis: correlation with symptom duration and vascular hyperplasia. Br J Sports Med 42 (9):753–757
54. Scott P, Ma H, Viriyakosol S, Terkeltaub R, Liu-Bryan R (2006) Engagement of CD14 mediates the inflammatory potential of monosodium urate crystals. J Immunol 177(9):6370–6378
55. Shi Y, Evans JE, Rock KL (2003) Molecular identification of a danger signal that alerts the immune system to dying cells. Nature 425(6957):516–521
56. Shields GE, Beard SM (2015) A systematic review of the economic and humanistic burden of gout. Pharmacoeconomics 33(10):1029–1047
57. Thampatty BP, Li H, Im HJ, Wang JH (2007) EP4 receptor regulates collagen type-I, MMP-1, and MMP-3 gene expression in human tendon fibroblasts in response to IL-1 beta treatment. Gene 386(1–2):154–161
58. Wen CC, Yee SW, Liang X, Hoffmann TJ, Kvale MN, Banda Y, Jorgenson E, Schaefer C, Risch N, Giacomini KM (2015) Genome-wide association study identifies ABCG2 (BCRP) as an allopurinol transporter and a determinant of drug response. Clin Pharmacol Ther 97(5):518–525
59. Yuan J, Murrell GA, Wei AQ, Wang MX (2002) Apoptosis in rotator cuff tendonopathy. J Orthop Res 20(6):1372–1379

Influence of Thyroid Hormones on Tendon Homeostasis

12

Francesco Oliva, Eleonora Piccirilli, Anna C. Berardi, Umberto Tarantino, and Nicola Maffulli

Abstract

Tendinopathies have a multifactorial etiology driven by extrinsic and intrinsic factors. Recent studies have elucidated the importance of thyroid hormones in the alteration of tendons homeostasis and in the failure of tendon healing after injury. The effects of thyroid hormones are mediated by receptors (TR)-α and –β that seem to be ubiquitous. In particular, T3 and T4 play an antiapoptotic role on tenocytes, causing an increase in vital tenocytes isolated from tendons *in vitro* and a reduction of apoptotic ones; they are also able to influence extra cellular matrix proteins secretion *in vitro* from tenocytes, enhancing collagen production. From a clinical point of view, disorders of thyroid function have been investigated only for rotator cuff calcific tendinopathy and tears. In this complex scenario, further research is needed to clarify the role of thyroid hormones on the onset of tendinopathies.

General Aspects and Epidemiology

The natural history of tendinopathies is difficult to define. Different factors, such as genetic background, biomechanics, intrinsic degeneration within the tendon itself and co-morbidities, are involved in the pathogenesis of tendon diseases. The etiology of tendon tears remains multifactorial and attempts have been made to unify intrinsic and extrinsic theories. Recent evidence strongly suggests that most of rotator cuff tears are caused by primary failed healing response [1], but it is not clear whether circulating hormones can act on tendons and which is the link between hormonal and metabolic diseases and the development of tendinopathy [2]. In this complex framework,

F. Oliva (✉) • E. Piccirilli • U. Tarantino
Department of Orthopaedics and Traumatology, University of Rome "Tor Vergata", Rome, Italy
e-mail: olivafrancesco@hotmail.com

A.C. Berardi
UOC Immunohematology and Transfusion Medicine Laboratories, Laboratory of Stem Cells, Spirito Santo Hospital, Pescara, Italy

N. Maffulli
Centre for Sports and Exercise Medicine, Queen Mary University of London Barts and The London School of Medicine and Dentistry, Mile End Hospital, London, UK

Head of Department of Physical and Rehabilitation Medicine, University of Salerno, Salerno, Italy

P.W. Ackermann and D.A. Hart (eds.), *Metabolic Influences on Risk for Tendon Disorders*, Advances in Experimental Medicine and Biology 920, DOI 10.1007/978-3-319-33943-6_12

thyroid hormones (Ths) imbalance can deeply affect the structural setting and the homeostasis processes of tendons, but the precise mechanism is still unknown [3]. Thyroid hormones play an important role in the regulation of adult metabolism influencing genomic actions, such transcriptional processes and phosphorylation and non genomic actions such as thermogenesis, cellular growth and mitochondrial pathways [4]. In both cases, The start a transduction cascade with a strong impact on cellular metabolism by activation of multiple mechanisms [5]. These hormones are essential both in early and adult life and they strongly influence changes in oxygen consumption, protein, carbohydrate, lipid and vitamin metabolism [6]. In particular, T4 (thyroxine) is important for both collagen synthesis and extra cellular matrix (ECM) metabolism. Hypothiroidism causes accumulation of glycosaminoglycans (GAGs) in the ECM, which may predispose to tendon injury [7]. Tendinopathy can be the presenting complaint in hypothyroidism, and symptomatic relief can be obtained by appropriate management of the primary thyroid deficiency.

An epidemiological study recently showed the possible role of thyroid hormones on modifying and increasing the rate of non traumatic rotator cuff tear in male and female [8]. A higher frequency (63 %) of disorders of thyroid function was detected in females: according to these data, gender can be considered as an increased risk factor for rotator cuff diseases, enhanced by the higher incidence (56 % of the total female group between 60 and 80 year) of thyroid pathologies among the females. According to this study, even though aging or genetic predisposing factors or traumas are the main actors in the tendons tears, an important role is played by metabolic substrate. Thyroid disorders strongly influence the failed healing response in some subjects, causing the final symptomatic tears [9].

Basic Science

The effects of Ths are mainly mediated by T3, which regulates gene expression by binding the TH receptors (TR)-α and –β that seem to be ubiquitous [10]. The importance of Ths on the genesis of the tendinopathies has been confirmed by *in vitro* studies that have detected the presence of Ths receptors (TRα/β isoforms) on tenocytes and confirmed the essential role of T3 and T4 in regulating cell proliferation. Tenocytes were isolated from human tendon tissues obtained from healthy subjects and cultured for 72 h with or without Ths. As expected, both T3 and T4 induced cell growth and the higher increase was obtained by 72 h of hormone treatment, being 19 % for T3 and 10 % for T4. The addition of the Ths in the culture medium led to stimulation of cell growth with a reduction of the doubling time. These findings stress the physiological action of THs in the homeostasis of tendons: T3 and T4 play an antiapoptotic action on tenocytes, causing an increase in vital tenocytes isolated from tendons *in vitro* and a reduction of apoptotic ones in a dose- and time-dependent manner [11] (Fig. 12.1).

Ths are also able to influence tenocyte secretion of ECM proteins *in vitro*. Tendons are fibrous connective tissues composed of cells surrounded by a complex extra cellular matrix rich of collagens, proteoglycans, glycoprotein and water [12]. When primary tenocyte-like cells were cultivated in the presence of T3 or T4 individually or in combination with ascorbic acid (AA), Ths together with AA significantly increase the expression of collagen I, the major fibrillar collagen of the tendon, in ECM. In addition, among the proteins, Cartilage Oligomeric Matrix Protein (COMP) was enhanced. This is an abundant glycoprotoein, first identified in cartilage, particularly present in tendon exposed to compressive load. COMP belongs to the thrombospondin gene family with the ability to bind to type I, II and IX collagen molecules as well as fibronectin (see Chap. 2). COMP modulates the organization of collagen fibrils. In contrast, Ths do not affect the expression of collagen III, which is normally less abundant in tendon and increases only during the early phases of remodeling or in tendinopathies. Therefore, Ths play a role on the extra cellular matrix of tendons, enhancing *in vitro* the production of several proteins [13].

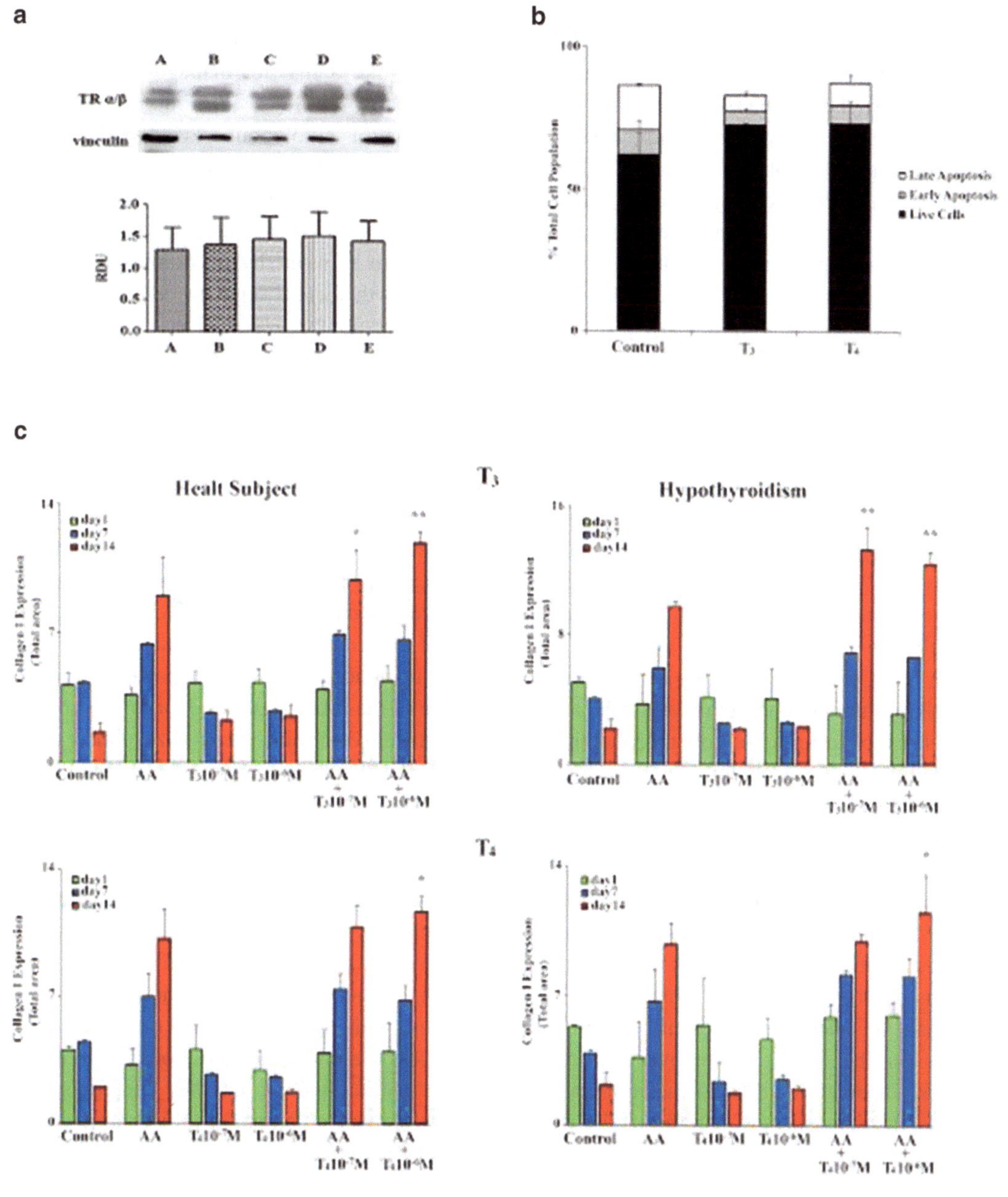

Fig. 12.1 (a) Western blot analysis of TRα/β isoforms. *A* indicates patients with healthy rotator cuff tendons; *B–C* indicate patients with rotator cuff tears without thyroid disease, *D–E* represent patients with rotator cuff tears and thyroid disease. The polyclonal antibodies against TRs α/β recognize two specific bands at 47 and 55 kDa, respectively. Densitometric absorbance values from three separate experiments were averaged (± SD), after they had been normalized to Vinculin for equal loading. Data relative to each protein are presented in the histogram of the Western Blot as Relative Densitometric Units (*y axis*). (**b**) Determination of apoptosis by Annexin V assay. At 48 h after culture, the tenocytes in serum deprived medium, double staining with Annexin V/PI, apoptotic cell population were evaluated by flow cytometry. Results from one representative experiment of three independent experiments with similar finding are shown. Annexin V-PI− cells were considered as vital, Annexin V + PI− cells were considered as early apoptotic, Annexin V + PI+ cells as late apoptotic. The results are expressed as percentage of total cells. The percentage of control cells (untreated) and treated (T3 or T4) are presented in the histogram. All the data are presented as mean ± SD, and are the results from five individual experiments. (**c**) Collagen I expression of primary tenoyte-like cells *in vitro* culture isolated from five healthy subjects and three hypothyroidism. Collagen I Intensity (Total Area was quantified by anti-collagen I) it was measured by Nikon software. F) Data are expressed as mean ± SD for four independent experiments for samples run in triplicate (n = 4, *P <0.05, **P <0.01)

Tendons are affected by thyroid disorders because of the heterogeneous composition of their matrix: in fact, the T3-mediated up-regulation of collagen is reflected in a change of the cross-linking pattern, with a resulting improvement of collagen quality. Thyroid diseases compromise collagen quality affecting transduction mechanism involved in collagen maturation: in particular, in hypothyroidism genes coding for collagen I and collagen Va1 are poorly expressed, with a marked disruption of the architecture of the ECM [14]. This is in contrast with the endocrine theory according to which hyperthyroidism may be accompanied by an increased catabolism of proteins and collagen, probably because other intrinsic factors should be considered, such as other metabolic pathways and the global hormonal status.

Some studies have demonstrated that T3 negatively influence normal human skin fibroblasts by inhibiting the accumulation of glycosaminoglycan (GAGs) and the synthesis of fibronectin [15]. Skin fibroblasts and tenocytes have similar characteristics in terms of cell morphology and extracellular matrix components, and this may explain why hypothyroidism leads to accumulation of GAGs in the ECM, with a high risk of tendon calcification and nerve compression by mixedema as in the carpal tunnel syndrome [16].

The pattern of worsening hypoxia-induced apoptosis supports a continuous failure of the tendons in the rotator cuff; this is why it is important to determine the actual size of these effects on the functional recovery in tendon injuries. Disorders of thyroid metabolism are a most interesting aspect of calcific tendinopathy. One of the most reliable theories on the genesis of calcific tendinopathies concerns the hypothyroidism related reduction of oxygenation in the rotator cuff tendons with a consequent metaplasia that leads to calcium deposition [17].

Clinical Aspects

It is important to underline the similitude between skin fibroblast and tenocytes. The latter are specialized fibroblasts that regulate the homeostasis of ECM in mesenchymal tissues. Recent studies used laboratory-amplified tenocyte-like cells to manage lateral elbow tendinopathy [18], while ultrasound-guided injections of autologous skin-derived tendon-like cells have been used to treat patellar tendinopathy [19]. Proliferation of fibroblasts is related to thyroid status. The complex interactions between thyroid hormones and fibroblast growth factors regulate cells proliferation and differentiation [20]. Ths deficiency also leads to a decrease of tendon fibroblasts activity, with down-regulation of the mechanism involved in ECM/cytoskeletal remodeling. The ECM structure becomes weaker, and tenocytes reduce their migration under mechanical stress stimuli.

Hypothyroidism leads to hypoxia and apoptosis: these two conditions are greatest in mild impingement and in partial, small, medium and large rotator cuff tears with macroscopic deterioration of the tendons [21]. The pattern of worsening hypoxia-induced apoptosis supports a continuous failure of the rotator cuff and, for this reason, it is important to determine the real size of these effects on the functional recovery in tendon injuries.

In addition, disorders of thyroid function are most interesting aspects of calcific tendinopathy [22]. Tendon healing includes many sequential processes, and disturbances at different stages of healing may lead to different combinations of histopathological changes, shifting the normal healing processes to an abnormal pathway. Patients with associated endocrine and thyroid disorders present earlier onset of symptoms, and longer natural history, and they undergo surgery more frequently compared to a control population. Hypothyroidism induces accumulation of glycosaminoglycans (GAGs) in the extracellular matrix, which may, in turn, predispose to tendon calcification [3, 23]. One of the most reliable theories on the genesis of calcific tendinopathies concerns the reduction of oxygenation in the rotator-cuff tendons with consequent metaplasia that leads to calcium deposition. The association between hypothyroidism and tendinous calcium deposits was first analyzed in myxedematous patients who showed synovitis of the wrists, metacarpal joints, and flexor tendon sheaths thickening. Synovial fluid analysis demonstrated the presence of intra- and extracellular crystals, and the presence of chondrocalcinosis was distinguished from acute attacks of pseudogout, which was not observed in most patients [24]. In addition, a significantly increased expression of tissue transglutaminase (tTG) 2 and its substrate, osteopontin, was detected in the calcific areas: this shows that a variation in expression of different genes could be peculiar in calcific tendinopathy [25]. The etiopathogenesis of the mechanism whereby thyroid disease could contribute to enhance the influx of calcium deposits into tendons and joints is still unknown.

Conclusions

The relationship between thyroid hormones and tendons disorders is clinically relevant. The presence of high levels of thyroid receptors isoforms, their protective action during tenocytes apoptosis and their ability to enhance tenocyte proliferation *in vitro* in healthy tendons enhance the idea of a physiological action of THs in the homeostasis of tendons, but does not allow to clarify the role of THs in the pathogenesis of the rotator cuff tears. There is increasing recognition of the prevalence of autoimmune thyroid diseases in patients with connective tissue disorder, highlighting a common mechanism for the pathogenesis of these conditions [26]. THs may have a substantial role in failing the healing response during tendinopathies [27] Much research remains to be performed to clarify the exact role of the Ths in tendon tissue and their implications in tendon tears, tendinopathies and tendon healing after injury. Common tendinopathies affecting different sites could be associated to disorders of thyroid function, but the relationship has not been clearly defined. If this association is further confirmed, assessment and treatment of patients with tendon pathologies may have to be revisited and integrated. The natural history of tendon tears is to progress with time. These may well produce tendon retraction and fatty degeneration, which make it more difficult to successfully repair a tear. Tendons, including rotator cuff tears, seem able to heal after surgical repair, but many factors influence their healing: age, sex and co-morbidities, including metabolic and endocrine disorders. If orthopaedic surgeons plan surgical repair of ruptured tendons, it is necessary to address the underlying conditions, because co-morbidities increase the probability of failure of surgical repair [28]. The knowledge of the mechanisms that underlie the onset of tendon pathology provides a comprehensive and multidisciplinary approach to the patient and leads to a better outcome of the therapeutic strategy.

References

1. Giai VA, De Cupis M, Spoliti M, Oliva F (2013) Clinical and biological aspects of rotator cuff tears. Muscles Ligaments Tendons J 3(2):70–9
2. Oliva F, Piccirilli E, Berardi AC, Frizziero A, Tarantino U, Maffulli N (2015) Hormones and tendinopathies: the current evidence. Br Med Bull 117(1):39–58
3. Oliva F, Giai Via A, Maffulli N (2012) Physiopathology of intratendinous calcific deposition. BMC Med 10:95
4. Duncan Bassett JH, Harvey CB, Williams GR (2003) Mechanisms of thyroid hormone receptor-specific nuclear and extra nuclear actions. Mol Cell Endocrinol 213(1):1–11
5. Oetting A, Yen PM (2007) New insights into thyroid hormone action. Best Pract Res Clin Endocrinol Metab 21(2):193–208
6. Brent GA (2000) Tissue-specific actions of thyroid hormone: insights from animal models. Rev Endocr Metab Disord 1(1–2):27–33
7. Berardi AC, Oliva F (2014) How thyroid hormones modify tendons, Metabolic diseases and tendinopathies book, Fondazione Ibsa, IV forum 21 June 2014, Lugano.
8. Oliva F, Osti L, Padulo J, Maffulli N (2014) Epidemiology of the rotator cuff tears: a new incidence related to thyroid disease, muscles. Ligaments Tendons J 4 (3):309–314
9. Harvie P, Pollard TC, Carr AJ (2007) Calcific tendinitis: natural history and association with endocrine disorders. J Shoulder Elbow Surg 16(2):169–173
10. Flamant F, Samarut J (2003) Thyroid hormone receptor: lessons from knock-out and knock-in mice. Trends Endoc Metab 14:85–90
11. Oliva F, Berardi AC, Misiti S, Falzacappa CV, Iaconeand A, Maffulli N (2013) Thyroid hormones enhance growth and counteract apoptosis in human tenocytes isolated from rotator cuff tendons. Cell Death Dis 4, e705

12. Tresoldi I, Oliva F, Benvenuto M, Fantini M, Masuelli L, Bei R, Modesti A (2013) Tendon's ultrastructure. Muscles Ligaments Tendons J 3(1):2–6
13. Berardi AC, Oliva F, Berardocco M, la Rovere M, Accorsi P, Maffulli N (2014) Thyroid hormones increase collagen I and cartilage oligomeric matrix protein (COMP) expression *in vitro* human tenocytes, muscles. Ligaments Tendons J 4(3):285–291
14. Varga F, Rumpler M, Zoehrer R, Turecek C, Spitzer S, Thaler R, Paschalis EP, Klaushofer K (2010) T3 affects expression of collagen I and collagen cross-linking in bone cell cultures. Biochem Biophys Res Commun 402(2–3):180–185
15. Murata Y, Refetoff S, Horwitz AL, Smith TJ (1983) Hormonal regulation of glycosaminoglycan accumulation in fibroblasts from patients with resistance to thyroid hormone. J Clin Endocrinol Metab 57 (6):1233–9
16. Purnell DC, Daly DD, Lipscomb PR (1961) Carpal-tunnel syndrome associated with myxedema. Arch Intern Med 108:751–756
17. Oliva F, Piccirilli E, Berardi AC, Frizziero A, Tarantino U, Maffulli N (2016) Hormones and tendinopathies: the current evidence. Br Med Bull 117(1):39–58. doi:10.1093/bmb/pii:ldv054
18. Connell D, Datir A, Alyas F, Curtis M (2009) Treatment of lateral epicondylitis using skin-derived tenocyte-like cells. Br J Sports Med 43:293–298
19. Clarke AW, Alyas F, Morris T, Robertson CJ, Bell J, Connell DA (2011) Skin-derived tenocyte-like cells for the treatment of patellar tendinopathy. Am J Sports Med 39(3):614–23, Epub 2010 Dec 7
20. Williams AJ, O'Shea PJ, Williams GR (2007) Complex interactions between thyroid hormone and fibroblast growth factor signalling. Curr Opin Endocrinol Diabetes Obes 14(5):410–5
21. Mackley JR, Ando J, Herzyk P, Winder SJ (2006) Phenotypic responses to mechanical stress in fibroblasts from tendon, cornea and skin. Biochem J 396(2):307–16
22. Benson RT, McDonnell SM, Knowles HJ, Rees JL, Carr AJ, Hulley PA (2010) Tendinopathy and tears of the rotator cuff are associated with hypoxia and apoptosis. J Bone Joint Surg (Br) 92(3):448–453
23. Oliva F, Via AG, Maffulli N (2011) Calcific tendinopathy of the rotator cuff tendons. Sports Med Arthrosc 19(3):237–43
24. Anwar S, Gibofsky A (2010) Musculoskeletal manifestations of thyroid disease. Rheum Dis Clin North Am 36(4):637–46
25. Oliva F, Barisani D, Grasso A, Maffulli N (2011) Gene expression analysis in calcific tendinopathy of the rotator cuff. Eur Cell Mater 21:548–57
26. Knopp WD, Bohm ME, McCoy JC (1997) Hypothyroidism presenting as tendinitis. Phys Sportsmed 25:47–55
27. Oliva F, Berardi AC, Misiti S, Maffulli N (2013) Thyroid hormones and tendon: current views and future perspectives concise review. Muscles Ligaments Tendons J 3(3):2013
28. Linee Guida IS, Mu LT (2014) Rotture della cuffia dei rotator, Percorsi Editoriali Carocci Publisher, Francesco Oliva Editor, Roma, Italy, pp 1–239. ISBN 978-88-430-7683-3

Sex Hormones and Tendon

13

Mette Hansen and Michael Kjaer

Abstract

The risk of overuse and traumatic tendon and ligament injuries differ between women and men. Part of this gender difference in injury risk is probably explained by sex hormonal differences which are specifically distinct during the sexual maturation in the teenage years and during young adulthood. The effects of the separate sex hormones are not fully elucidated. However, in women, the presence of estrogen in contrast to very low estrogen levels may be beneficial during regular loading of the tissue or during recovering after an injury, as estrogen can enhance tendon collagen synthesis rate. Yet, in active young female athletes, physiological high concentration of estrogen may enhance the risk of injuries due to reduced fibrillar crosslinking and enhanced joint laxity. In men, testosterone can enhance tendon stiffness due to an enhanced tendon collagen turnover and collagen content, but testosterone has also been linked to a reduced responsiveness to relaxin. The present chapter will focus on sex difference in tendon injury risk, tendon morphology and tendon collagen turnover, but also on the specific effects of estrogen and androgens.

Keywords

Sex hormones • Sex • Gender • Injury risk • ACL rupture • Estrogen • Estradiol • Testosterone • Progesterone • Relaxin • Insulin-like growth factor I (IGF-I) • Collagen • Fascicles • Knee laxity • Joint laxity • Tendinopathy • Cross-links • Biomechanical properties

M. Hansen (✉)
Department for Public Health, Section for Sport Science, Aarhus University, Dalgas Avenue 4, 8000 Aarhus, Denmark
e-mail: mhan@ph.au.dk

M. Kjaer
Institute of Sports Medicine, Department of Orthopedic Surgery M, Bispebjerg Hospital, Copenhagen, Denmark

Center for Healthy Aging, Faculty of Health and Medical Sciences, University of Copenhagen, Copenhagen, Denmark

P.W. Ackermann and D.A. Hart (eds.), *Metabolic Influences on Risk for Tendon Disorders*, Advances in Experimental Medicine and Biology 920, DOI 10.1007/978-3-319-33943-6_13

Introduction

Sex differences in tendon and ligament injury risk are reported. The relative risk of tearing the anterior cruciate ligament (ACL) is 2–6 times greater in young female compared to male athletes [1–3] even after socioeconomic, health and lifestyle background variables and the level of sports participation have been taken into account [4, 5]. Especially, when circulating estrogen is peaking around the time of ovulation during the menstrual cycle the risk of an ACL-rupture seems to be enhanced [6, 7]. In contrast, the risk of sustaining an Achilles rupture [8] or developing tendon pathology [9] seems to be lower in premenopausal women compared to that in men, whereas this sex discrepancy in risk disappears after menopause [10] where the level of estrogen is comparable between the sexes [11]. Furthermore, the transition to the postmenopausal state characterized by low estrogen levels is associated with a dramatic increase in the prevalence of asymptomatic rotator cuff tears [12]. These observations have led to a search for the underlying mechanism(s) involved in the discrepancy in the risk of injuries between men and women and between young and elderly women. A complex interaction involving several risk factors is probably in play (e.g. ligamentous laxity and size, limb alignment, notch dimensions and decreased neuromuscular control of knee, motion skill level and muscular strength) [13]. However, since estrogen receptors have been localized in tendons and ligaments [14, 15] it has been suggested, that at least part of this gender difference in risk might be explained by sex hormonal differences influencing tendon and ligament structure and biomechanical properties.

Sex Differences in Tendon Structure and Biomechanical Properties

Tendon stiffness has been reported to be lower in young women than men during voluntary isometric contractions using ultrasonography [16, 17]. Similarly, joint laxity has repeatedly been reported to be greater in *post-pubertal* women compared to age-matched men [18–20], whereas no differences seem to exits between pre-pubertal boys and girls [20] and between postmenopausal men and women [21]. Taken together, these observations indicate that sex differences in tendon and ligament biomechanical properties are significant in the period of life where the sex hormonal profile is markedly different between women and men. In line with this, the sex disparity in ACL injury risk is peaking in the teenage years where sex hormonal differences are the greatest [5, 22].

Part of the sex difference in tendon and ligament biomechanical properties is explained by a general greater tendon size in men compared to women. Yet, also after adjustment for body weight and control of activity level Achilles and patellar tendon cross-sectional area (CSA) is greater in young male runners compared to equally trained young female runners [17] (Fig. 13.1). Furthermore, tendon CSA is greater in trained male runners compared to that in untrained men, whereas no difference in tendon CSA is reported between untrained and trained female runners [17]. These cross-sectional data indicate that the ability to adapt to regular training in regards to patellar and Achilles tendon size might be reduced in women in contrast to in men [23]. However, it should be noted that low energy availability is a common phenomenon in endurance running, especially among young female athletes [24, 25]. Therefore, sex difference in tendon adaptation to training may be confounded by hormonal disturbances related to low energy availability such as low estrogen and insulin-like growth factor I (IGF-I) levels as these hormones are coupled to a stimulating effect on tendon collagen synthesis [26, 27]. Actually, other data supports that female handball players seem to be able to adapt in tendon size after long-termed regular strenuous loading of the tissue [28]. Thus, long-term intervention studies are needed to clarify if the is a sex difference in tendon adaptation to training.

Tendon structural quality also seems to differ between young men and women. Mechanical testing of single human patellar tendon collagen fascicles from young men and women have shown that during loading the ultimate stress before rupture is greater in fascicles from males compared to those from females [23]. The sex differences in fascicle biomechanical properties may be related to structural differences. In young

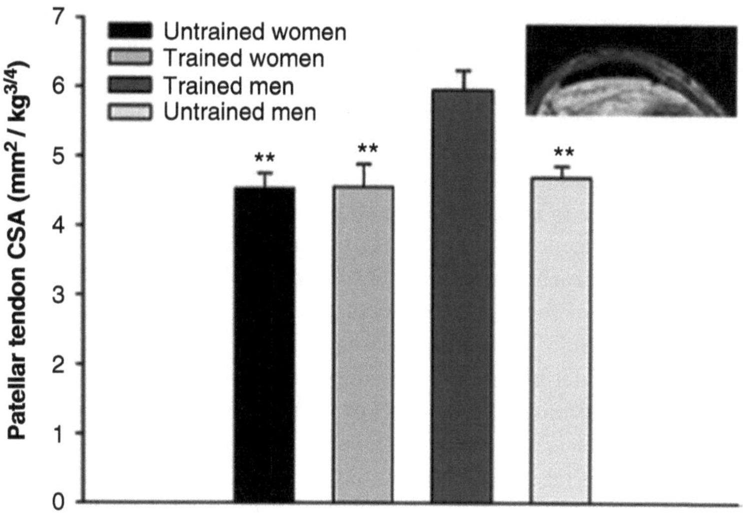

Fig. 13.1 The magnetic resonance imaging (MRI) determined patellar tendon cross-sectional area (CSA) for trained and untrained men and women normalized to body mass. Trained men had a greater CSA than untrained men ($P < 0.01$); however, note that trained women had a similar CSA compared with untrained women. An MRI of the patellar tendon [21]. (Reprinted from [21]. Copyright © 2007 John Wiley and Sons. Used with permission.)

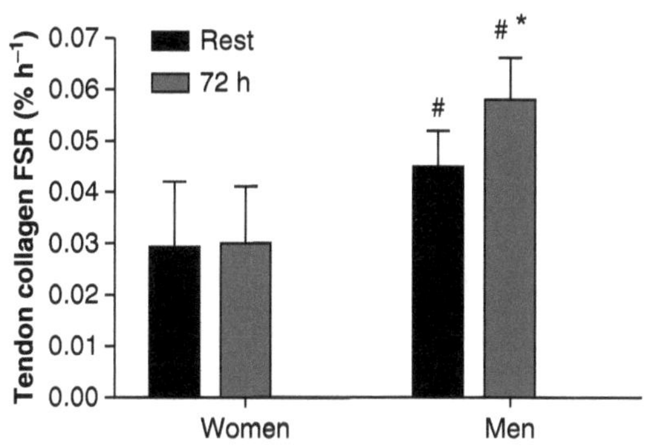

Fig. 13.2 Comparison of patellar tendon collagen fractional synthesis rates (FSR) at rest and 72 h after exercise in women and men. *Significantly different from rest value, $P < 0.05$. #Significantly different from women, $P < 0.05$ [24]. (Reprinted [24]. Copyright © 2007 The American Physiological Society. Used with permission.)

women, expression of Type III collagen in the patellar tendon is higher than in men [29], which may induce a greater elastic flexibility. In addition, women patellar tendon dry mass and collagen content per tendon weight has been reported to be lower compared to in men [30].

The tendon collagen content is determined by the balance between collagen protein synthesis rate and collagen protein breakdown rate (see Chap. 8). The tendon collagen turnover is slow and specifically recent data suggest that the tendon core of collagen is the same from the late teenage years until death [31], whereas the outer layer seems to responsive to anabolic stimuli such as training [32–34] and probably also hormonal stimuli [26, 27, 35, 36]. In young healthy women, a lower tendon collagen synthesis rate has been reported compared to age-matched men both at rest and in response to exercise at the same relative intensity [37]. Patellar tendon collagen synthesis rate was still enhanced over resting values 72 h after exercise in men whereas the synthesis rate was not different from resting values in women (Fig. 13.2). Unfortunately, the balance between synthesis and breakdown was not determined. Only limited validated methods exist in relating to determining tendon collagen breakdown rate *in vivo* in humans. However, these data similarly to cross-sectional including untrained and trained female runners [17] indicate that the response to mechanical loading may be reduced in women [23].

Influence of Estrogen on Tendon Structural and Biomechanical Properties

Introduction; Estrogen and Estrogen Signaling

The effect of sex hormones may differ between animals and human since the sex hormonal and menstrual profile varies considerably between species, which questions the transferability of results from animal studies to humans. Based on this, the following is primarily based on human data.

In premenopausal women 17-β estradiol is the dominating type of estrogen. The estradiol level in women is in general several folds higher than in men until menopause [38, 39], but the concentrations of sex hormones fluctuate during the menstrual cycle. During the menses period in the early follicular phase (FP) the concentration of estradiol and progesterone are low, whereas the concentration is peaking just before ovulation (~day 14 of the regular cycle) and remains at a high level during the following luteal phase (LP) until next menses.

Estrogens primarily bind to the estrogen receptors (α & β) in the nucleus, but there are also bindings sites in the plasma membrane [40, 41]. Estrogen receptor distribution and the relative isoform (α- and β estrogen receptors) predominance varies within different tissue types and the expression is influenced by the estrogen level [42–45]. Activation of the two types of estrogen receptors seems to induce both similar and opposing effects [46, 47]. - Estrogen-binding to the receptor induce receptor activation, but also other stimuli can independently of estrogen-binding influence signaling pathways linked to estrogen receptor activation [48, 49].

Influence of Estrogen on Tendon and Ligaments

To get closer to an elucidation of the effect of estradiol on tendon and ligament it is relevant to compare results from women tested in the distinct menstrual phases or women with low compared to high endogenous concentrations of estradiol independent of menstrual phases.

In a systematic review from 2006, six of nine studies observed no significant effect of menstrual cycle on knee laxity [50]. Nevertheless, a meta-analysis of the data show that knee laxity was significantly greater around ovulation, when estrogen is peaking, medium in the LP and lowest in the FP [50]. The findings are supported by later data [51–53], but not all [18, 54]. The cyclic variation in knee laxity seems to be repeated in each menstrual cycle, but the magnitude and pattern of cyclic changes varying greatly among women [55]. An inverse relationship between circulating estradiol concentration and tendon stiffness has been observed in female handball players independent of cycle phase [28]. Similarly, cyclic changes in Genu recurvatum and general joint laxity have been reported, which underlines a systemic effect [55]. The greater knee laxity associated with high levels of estrogen seems influence landing biomechanics [56] and thereby the risk of injury. In support, exposure to high level of estrogen reduce maximal load to failure in rabbit anterior cruciate ligaments [57].

Until recently, a mechanistic explanation for the rapid changes in tendon and ligament laxity during the cycle was missing, since it is very unlikely that it is caused by any significant change in collagen content within days. Though, new *in vitro* findings have shown that short-term (24 h or 48 h) exposure to physiological concentration of estrogen decreases mechanical function of engineered ligaments by inhibiting the activity (61–77 %) of the crosslinking enzyme lysyl oxidase [58]. If the enzyme activity is similarly reduced *in vivo* when exposed to high level of estrogen it may explain the observations of greater knee laxity and a risk of sustaining an ACL injury around ovulation [58].

Inter-individual difference in the average level of estrogen between women may also influence the tendon and ligament structure and biomechanical properties over time. A positive correlation between circulating estrogen and human tendon collagen synthesis has been reported [27] (Fig. 13.3). Furthermore, in elderly hysterectomized women who were long-term users of estrogen replacement therapy (ERT), tendon collagen synthesis was higher and patellar tendon relative stiffness was lower compared to age matched postmenopausal women with very low levels of estrogen [27]. The latter observation may be explained by a general high tendon

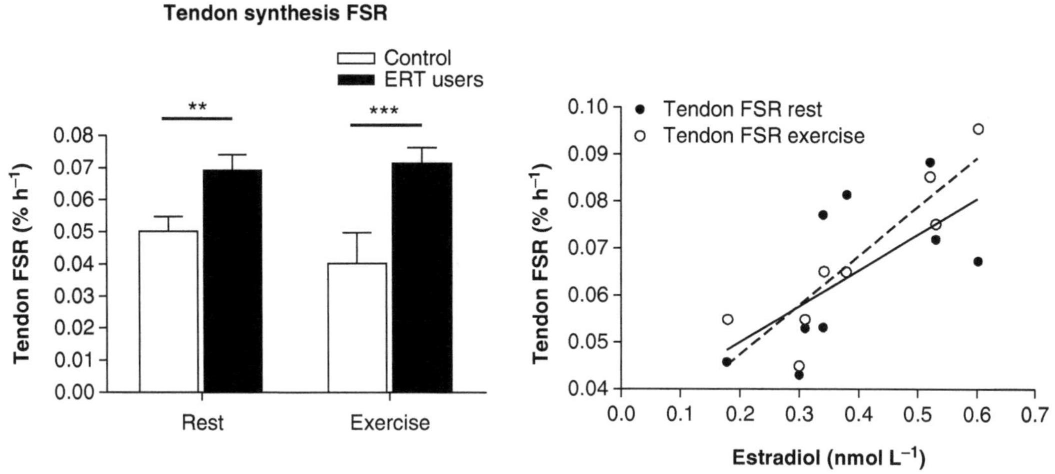

Fig. 13.3 *Top left:* patellar tendon collagen fractional synthesis rates (FSR) at rest and 24 h after exercise in postmenopausal women who used ERT and postmenopausal women who did not use ERT (control). $^{**}P < 0.01$ and $^{**}P < 0.001$, unpaired *t*-test, control vs. ERT users. *Top right*: relationship between tendon FSR and serum (s)-estradiol in ERT users at rest ($r^2 = 0.41$, $P = 0.06$) and postexercise ($r^2 = 0.80$, $P < 0.001$)

collagen turnover at high levels of estrogen since the collagen content was not changed. An enhanced tendon collagen turnover may indirectly reduce the possibilities for intra- and intermolecular cross-linking by enhancing the tendon collagen turnover on the top of a probably direct inhibiting effect on crosslinking enzymes [58]. In line with a lower tendon collagen synthesis in postmenopausal women not using ERT, estrogen deficiency in rats has been associated with down-regulation of fibroblast cell proliferation and density in the Achilles tendon [59].

It should be noted that estrogen may have divergent effect on the biomechanical properties of collagen rich tissues placed in anatomical distinct positions due to a disparity in loading profile and differences in relative distribution and numbers of estrogen receptors (α and β) [46, 60]. Knowledge in this field is very limited, but as an example, high versus low circulating estrogen in postmenopausal women has been found to be associated with reduced relative stiffness of the patellar tendon [27], which is rich in type I collagen as most tendon and ligaments (~60 %) [61]. Similarly, the rise in estradiol around ovulation [50] and during pregnancy [62] in premenopausal women is associated which enhanced knee joint laxity. In contrast, in arcus tendinous fasciae pelvis, which is rich in type III collagen (82 % of the total amount of collagen) the tensile strength of the tissue and the susceptivility to anterior vaginal wall prolapse are low in postmenopausel women charachterized by low estrogen level [63]. Similarly, hip joint flexibility was observed to be lower at low estrogen levels. The latter observation may be related to the content of type I collagen and the type I to (Type III + IV) ratio was observed to be lower in postmenopausal women compared to postmenopausal women on hormone replacement therapy (HRT) [63]. Similarly, low estrogen status may be detrimental for tendon and ligament healing [59]. In oophorectomized rats maximal stress upon the Achilles tendon, cell proliferation and density was reduced compared to sham-operated rats [59].

To sum up, effects of estrogen on tendon and ligaments are not fully elucidated and seem to depend on the age and the hormonal status of the women, the specific type of tendon and ligament in focus, and the degree of mechanical loading of the tissue. In women, presence of estrogen in contrast to low or lack of estrogen may be beneficial when tendon and ligaments are exposure to regular tissue loading [64] or recovering after an injury as shown in animal studies [59, 65], since estrogen seem to have a stimulating effect on type I collagen synthesis in tendon [27]. Conversely, in active young female athletes physiological high concentration of estrogen may enhance the risk of injuries since a high level of estrogen seems to reduce the responsiveness to mechanical loading in collagen synthesis [49, 66] and is coupled to enhanced joint laxity [6]. The latter may be coupled to a greater risk of tendon and ligament injuries [5].

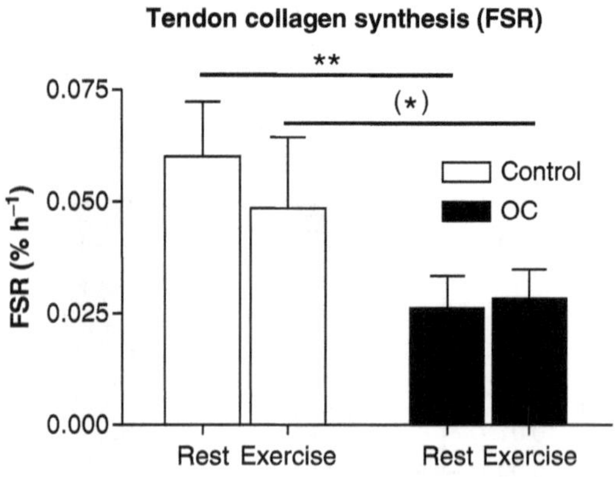

Fig. 13.4 Patellar tendon collagen fractional synthesis rates (FSR) at rest and 24 h after exercise in controls and oral contraceptive (OC) users. Values are means ± SE. Controls, women who had never used oral contraceptives who were tested in the follicular phase; OC, OC users. $**P < 0.01$, (*) $P = 0.13$, unpaired t-test, control vs. OC

Influence of Oral Contraceptives on Tendon and Ligaments

Worldwide more than 100 million premenopausal women are using oral contraceptives (OC) [67]. Nevertheless, knowledge regarding the effect of OC on tendon and ligaments is sparse. Yet, use of OC in premenopausal women seems to influence tendon and ligaments differently than use of oral administration of ERT in elderly women [27, 36]. High endogenous estradiol is associated with high tendon collagen synthesis [27], whereas tendon collagen synthesis is lower in young OC users compared to control [36]. This may be related to OC has a marked inhibiting effect on the IGF-I level [26, 36]. Furthermore, in contrast to endogenous estradiol or ERT, cross-sectional data suggest that OC use reduces knee joint laxity in young athletes at ambient temperature [53, 68, 69], but not in all studies [28, 52, 70] and not after warming up the leg to 38° [53]. If tendon and ligament laxity is reduced in OC users compared to non-OC users, as in men, it might explain why OC users and men experience greater muscle damage and more delayed muscle soreness after heavy exercise than non-OC users [69, 71, 72], since a flexible tendon reduces the tensile loading on the connected muscle-tendon junctions and contractile muscle filaments during muscle contractions [73].

Use of OC inhibits endogenous secretion of estradiol and thereby the fluctuations in estrogen during the menstrual cycle [74]. In conjunction with this, the fluctuation in knee joint laxity during the menstrual cycle disappears [53], which diminished the enhanced risk of ACL injury in the ovulation phase [75]. Likewise, prospective data show no periodicity in non-contact ACL injury and ankle sprains in OC users in contrast to non-OC users. However, in the latter trial no overall significant difference in injury risk between groups was observed [76]. Actually, use of OC may have negative impact on tendon adaptation to regular training and thereby the risk of injury in athletes based on a lower tendon collagen synthesis rate (Fig. 13.4) and a reduced responsiveness to mechanical loading [35, 36].

Data on the effect of OC on tendon and ligament injury risk is limited. In a 12 month prospective study including Swedish female soccer players a lower rate of traumatic knee and ankle injuries was reported in the group of OC users (13 of 33 women) compared to in non-OC users (29 of 51 women) [77]. Furthermore, in a study including 4497 operatively treated ACL cases and 8858 age-matched controls with no ACL injury, OC use was linked to a lower relative ACL injury risk [78]. Nevertheless, in the latter study they did not control for sport participation which may differ between OC and non OC users. Furthermore, two studies have reported that OC does not seem to exert any protective effect on

ACL injury risk in skiers [7, 79]. Also, a higher risk of symptomatic Achilles tendinopathy has been reported in OC users [80]. Therefore, additional clinical studies are needed to clarify the biological and causal association between OC use and injury risk.

Influence of Androgen and Relaxin on Tendon Structural and Biomechanical Properties

Knowledge in relation to the effect of androgens on human tendon and ligament is limited and the exact role remains elusive. Androgen receptors have been localized in human ACL tissue [81], which seems to be a androgen-responsive tissue.

In women, testosterone, the most abundant androgen, fluctuates – although at low concentrations - during the menstrual cycle and is enhanced around the ovulation [81]. The level of testosterone is positively correlated with ACL stiffness, whereas estrogen and the estrogen to testosterone ratio are negatively correlated to ACL stiffness [81]. Since knee laxity is enhanced around ovulation [50] this suggests that estrogen seems to have the dominating influence on tendon and ligament biomechanical properties in women. Nevertheless, testosterone may have a potential counteracting role. In support of the later, animal data indicates that testosterone changes estrogen receptor α and β expression in upper site directions [82, 83] and reduces the stimulating effect of estrogen on mammary epithelial proliferation [83, 84].

In men, the level of testosterone is 7–8 times higher in men than women [85] and this is coupled to a higher tendon synthesis [37], a lower knee laxity [18], whereas tendon stiffness is higher [17]. It should be noted that the circulating concentration of estrogen is positively correlated to high tendon collagen synthesis rates in women [27]. Therefore, it can be hypothesized that testosterone has a more pronounced role on tendon and ligaments in men than estrogen in women, but this statement needs to be confirmed. An anabolic effect on testosterone on tendon collagen content and thereby tissue stiffness is supported by animal findings showing that testosterone increases collagen content in prostate [84, 86] and hip joint capsule [87], and reduces passive range of motion of the patellar tendon and ligament [88]. Furthermore, *in vitro* findings have shown that testosterone administration in high doses to human cultured tenocytes increased tenocytes number and changed the phenotype after short-term exposure [89].

Animal data suggest that testosterone may indirectly reduce tendon and ligament laxity by downregulating the expression of relaxin receptors [88] and thereby the effects of relaxin. Relaxin is linked to activation of collagenases, increased tendon laxity, enhanced risk of ACL injury, but also to improved ligament healing [90–92]. Interestingly, in contrast to testosterone, recent animal data suggest that progesterone and high doses of estrogen administration are coupled to enhanced knee laxity and enhanced expression of relaxin receptors [93]. Therefore, on the top of a likely direct effect of estrogen and androgens on tendon the sex hormones may indirectly influence tissue stiffness through changing the responsiveness to relaxin.

Testosterone may also have a positive influence on tendon and ligament adaptation to training, but the human data are primarily based on cross-sectional data comparing men and women [23]. Tendon synthesis rate in response to exercise is greater in men than women [23]. In addition, in elderly men a greater increase in tendon stiffness after 12 weeks of alpine skiing training has been observed compared to in postmenopausal women [94]. The latter should be seen in light of circulating estrogen levels is comparable between elderly men and women, whereas circulating testosterone is still enhanced in the men [11].

Similarly to estrogen, effect of testosterone on tendon and ligament injury risk is probably dependent of the physiological concentration and balance between unbeneficial effects of low, but also of high testosterone concentrations induced by administration of androgens.

Case reports indicate that high levels of testosterone or synthetic derivatives of testosterone may enhance the risk of tendon and ligament injuries [95–98]. Furthermore, analysis of human patellar tendon morphology and mechanical properties indicates that adaptation to strength training is influenced by long-term use of androgenic anabolic steroids [99]. This is supported by animal data which suggests that anabolic androgenic steroids reverse the beneficial effect of exercise on Achilles tendon adaptation, thus resulting in inferior maximal stress values [100]. The higher risk of tendon rupture in anabolic steroid users may be directly caused by a direct effect on tendon or may also be indirectly related to the enhanced muscle hypertrophy and gain in muscle strength which is not balanced by a similar degree of adaptation in the connected tendon [101, 102].

Conclusion

Sex differences in tendon biomechanical properties, tendon morphology and tendon collagen turnover suggests that sex hormones play an explanatory role in relation to injury risk. However, it is difficult to sort out the isolated effect of the sex hormones in cross-sectional studies comparing women and men or users and non-users of sex hormonal administration, since the groups may differ in many other ways e.g. training status. Nevertheless, both estrogen and testosterone in balanced physiological concentration seems to be important for tendon health and physical function, whereas very low or high concentrations of endogenous or exogenous administrated sex hormones may led to an enhanced risk of injuries and inadequate adaptation to mechanical loading.

The research up to now has primarily focused on the ACL, Achilles, and the patellar tendon. The future may elucidate whether sex hormones influence tendons and ligaments differently depending on their localization and function.

References

1. Arendt E, Dick R (1995) Knee injury patterns among men and women in collegiate basketball and soccer. Ncaa data and review of literature. Am J Sports Med 23:694–701
2. Prodromos CC, Han Y, Rogowski J et al (2007) A meta-analysis of the incidence of anterior cruciate ligament tears as a function of gender, sport, and a knee injury-reduction regimen. Arthroscopy 23:1320–1325, e6
3. Powell JW, Barber-Foss KD (2000) Sex-related injury patterns among selected high school sports. Am J Sports Med 28:385–91
4. Parkkari J, Pasanen K, Mattila VM et al (2008) The risk for a cruciate ligament injury of the knee in adolescents and young adults: a population-based cohort study of 46 500 people with a 9 year follow-up. Br J Sports Med 42:422–6
5. Renstrom P, Ljungqvist A, Arendt E et al (2008) Non-contact acl injuries in female athletes: an international olympic committee current concepts statement. Br J Sports Med 42:394–412
6. Hewett TE, Zazulak BT, Myer GD (2007) Effects of the menstrual cycle on anterior cruciate ligament injury risk: a systematic review. Am J Sports Med 35:659–68
7. Lefevre N, Bohu Y, Klouche S et al (2013) Anterior cruciate ligament tear during the menstrual cycle in female recreational skiers. Orthop Traumatol Surg Res 99:571–5
8. Huttunen TT, Kannus P, Rolf C et al (2014) Acute achilles tendon ruptures: incidence of injury and surgery in Sweden between, 2001 and 2012. Am J Sports Med 42:2419–23
9. Cook JL, Khan KM, Kiss ZS et al (2000) Patellar tendinopathy in junior basketball players: a controlled clinical and ultrasonographic study of 268 patellar tendons in players aged 14–18 years. Scand J Med Sci Sports 10:216–220
10. Maffulli N, Waterston SW, Squair J et al (1999) Changing incidence of Achilles tendon rupture in Scotland: a 15-year study. Clin J Sport Med 9:157–60
11. Smith GI, Atherton P, Villareal DT et al (2008) Differences in muscle protein synthesis and anabolic signaling in the postabsorptive state and in response to food in 65–80 year old men and women. PLoS ONE 3, e1875
12. Abate M, Schiavone C, Di Carlo L et al (2014) Prevalence of and risk factors for asymptomatic rotator cuff tears in postmenopausal women. Menopause 21:275–80
13. Hewett TE, Myer GD, Ford KR (2006) Anterior cruciate ligament injuries in female athletes: part

1, mechanisms and risk factors. Am J Sports Med 34:299–311
14. Liu SH, Shaikh a R, Panossian V et al (1996) Primary immunolocalization of estrogen and progesterone target cells in the human anterior cruciate ligament. J Orthop Res 14:526–533
15. Hart DA, Archambault JM, Kydd A et al (1998) Gender and neurogenic variables in tendon biology and repetitive motion disorders. Clin Orthop Relat Res 44–56
16. Onambele GN, Burgess K, Pearson SJ (2007) Gender-specific in vivo measurement of the structural and mechanical properties of the human patellar tendon. J Orthop Res 25:1635–1642
17. Westh E, Kongsgaard M, Bojsen-Moller J et al (2008) Effect of habitual exercise on the structural and mechanical properties of human tendon, in vivo, in men and women. Scand J Med Sci Sports 18:23–30
18. Pollard CD, Braun B, Hamill J (2006) Influence of gender, estrogen and exercise on anterior knee laxity. Clin Biomech (Bristol, Avon) 21:1060–1066
19. Deep K (2014) Collateral ligament laxity in knees: what is normal? Clin Orthop Relat Res 472:3426–31
20. Quatman CE, Ford KR, Myer GD et al (2008) The effects of gender and pubertal status on generalized joint laxity in young athletes. J Sci Med Sport 11:257–63
21. Burgess KE, Pearson SJ, Breen L et al (2009) Tendon structural and mechanical properties do not differ between genders in a healthy community-dwelling elderly population. J Orthop Res 27:820–825
22. Shea KG, Pfeiffer R, Wang JH et al (2004) Anterior cruciate ligament injury in pediatric and adolescent soccer players: an analysis of insurance data. J Pediatr Orthop 24:623–8
23. Magnusson SP, Hansen M, Langberg H et al (2007) The adaptability of tendon to loading differs in men and women. Int J Exp Pathol 88:237–240
24. Melin A, Tornberg AB, Skouby S et al (2015) Energy availability and the female athlete triad in elite endurance athletes. Scand J Med Sci Sports 25:610–22
25. Mountjoy M, Sundgot-Borgen J, Burke L et al (2014) The ioc consensus statement: beyond the female athlete triad–relative energy deficiency in sport (red-s). Br J Sports Med 48:491–7
26. Hansen M, Boesen A, Holm L et al (2012) Local administration of insulin-like growth factor-i (igf-i) stimulates tendon collagen synthesis in humans. Scand J Med Sci Sports 23(5):614–619, Epub ahead of print
27. Hansen M, Kongsgaard M, Holm L et al (2009) Effect of estrogen on tendon collagen synthesis, tendon structural characteristics, and biomechanical properties in postmenopausal women. J Appl Physiol 106:1385–1393
28. Hansen M, Couppe C, Hansen CS et al (2013) Impact of oral contraceptive use and menstrual phases on patellar tendon morphology, biochemical composition, and biomechanical properties in female athletes. J Appl Physiol (1985) 114:998–1008
29. Sullivan BE, Carroll CC, Jemiolo B et al (2009) Effect of acute resistance exercise and sex on human patellar tendon structural and regulatory mrna expression. J Appl Physiol 106:468–475
30. Lemoine JK, Lee JD, Trappe TA (2009) Impact of sex and chronic resistance training on human patellar tendon dry mass, collagen content, and collagen cross-linking. Am J Physiol Regul Integr Comp Physiol 296:R119–R124
31. Heinemeier KM, Schjerling P, Heinemeier J et al (2013) Lack of tissue renewal in human adult achilles tendon is revealed by nuclear bomb (14)c. FASEB J 27:2074–9
32. Miller BF, Olesen JL, Hansen M et al (2005) Coordinated collagen and muscle protein synthesis in human patella tendon and quadriceps muscle after exercise. J Physiol 567:1021–1033
33. Kongsgaard M, Reitelseder S, Pedersen TG et al (2007) Region specific patellar tendon hypertrophy in humans following resistance training. Acta Physiol (Oxf) 191:111–121
34. Couppe C, Kongsgaard M, Aagaard P et al (2008) Habitual loading results in tendon hypertrophy and increased stiffness of the human patellar tendon. J Appl Physiol 105:805–810
35. Hansen M, Koskinen S, Petersen SG et al (2008) Ethinyl estradiol administration in women suppresses synthesis of collagen in tendon in response to exercise. J Physiol 586:3005–3016
36. Hansen M, Miller BF, Holm L et al (2009) Effect of administration of oral contraceptives in vivo on collagen synthesis in tendon and muscle connective tissue in young women. J Appl Physiol 106:1435–1443
37. Miller BF, Hansen M, Olesen JL et al (2006) Tendon collagen synthesis at rest and after exercise in women. J Appl Physiol 102:541–546
38. Stricker R, Eberhart R, Chevailler MC et al (2006) Establishment of detailed reference values for luteinizing hormone, follicle stimulating hormone, estradiol, and progesterone during different phases of the menstrual cycle on the abbott architect analyzer. Clin Chem Lab Med 44:883–7
39. Dighe AS, Moy JM, Hayes FJ et al (2005) High-resolution reference ranges for estradiol, luteinizing hormone, and follicle-stimulating hormone in men and women using the axsym assay system. Clin Biochem 38:175–9
40. Hewitt SC, Deroo BJ, Korach KS (2005) Signal transduction. A new mediator for an old hormone? Science 307:1572–1573
41. Revankar CM, Cimino DF, Sklar LA et al (2005) A transmembrane intracellular estrogen receptor mediates rapid cell signaling. Science 307:1625–1630

42. Hoyland JA, Baris C, Wood L et al (1999) Effect of ovarian steroid deficiency on oestrogen receptor alpha expression in bone. J Pathol 188:294–303
43. Lim SK, Won YJ, Lee HC et al (1999) A PCR analysis of ERalpha and ERbeta mRNA abundance in rats and the effect of ovariectomy. J Bone Miner Res 14:1189–1196
44. Press MF, Nousek-Goebl N, King WJ et al (1984) Immunohistochemical assessment of estrogen receptor distribution in the human endometrium throughout the menstrual cycle. Lab Invest 51:495–503
45. Sciore P, Frank CB, Hart DA (1998) Identification of sex hormone receptors in human and rabbit ligaments of the knee by reverse transcription-polymerase chain reaction: evidence that receptors are present in tissue from both male and female subjects. J Orthop Res 16:604–610
46. Matthews J, Gustafsson JA (2003) Estrogen signaling: a subtle balance between ER alpha and ER beta. Mol Interv 3:281–292
47. Saxon LK, Turner CH (2005) Estrogen receptor beta: the antimechanostat? Bone 36:185–192
48. Lee KC, Lanyon LE (2004) Mechanical loading influences bone mass through estrogen receptor alpha. Exerc Sport Sci Rev 32:64–68
49. Tobias JH (2003) At the crossroads of skeletal responses to estrogen and exercise. Trends Endocrinol Metab 14:441–443
50. Zazulak BT, Paterno M, Myer GD et al (2006) The effects of the menstrual cycle on anterior knee laxity: a systematic review. Sports Med 36:847–862
51. Park SK, Stefanyshyn DJ, Loitz-Ramage B et al (2009) Changing hormone levels during the menstrual cycle affect knee laxity and stiffness in healthy female subjects. Am J Sports Med 37:588–98
52. Hicks-Little CA, Thatcher JR, Hauth JM et al (2007) Menstrual cycle stage and oral contraceptive effects on anterior tibial displacement in collegiate female athletes. J Sports Med Phys Fitness 47:255–60
53. Lee H, Petrofsky JS, Daher N et al (2014) Differences in anterior cruciate ligament elasticity and force for knee flexion in women: oral contraceptive users versus non-oral contraceptive users. Eur J Appl Physiol 114:285–94
54. Eiling E, Bryant AL, Petersen W et al (2007) Effects of menstrual-cycle hormone fluctuations on musculotendinous stiffness and knee joint laxity. Knee Surg Sports Traumatol Arthrosc 15:126–132
55. Shultz SJ, Levine BJ, Nguyen AD et al (2010) A comparison of cyclic variations in anterior knee laxity, genu recurvatum, and general joint laxity across the menstrual cycle. J Orthopaedic Res 28:1411–7
56. Shultz SJ, Schmitz RJ, Kong Y et al (2012) Cyclic variations in multiplanar knee laxity influence landing biomechanics. Med Sci Sports Exerc 44:900–9
57. Slauterbeck J, Clevenger C, Lundberg W et al (1999) Estrogen level alters the failure load of the rabbit anterior cruciate ligament. J Orthop Res 17:405–408
58. Lee CA, Lee-Barthel A, Marquino L et al (2015) Estrogen inhibits lysyl oxidase and decreases mechanical function in engineered ligaments. J Appl Physiol (1985) 118:1250–7
59. Circi E, Akpinar S, Balcik C et al (2009) Biomechanical and histological comparison of the influence of oestrogen deficient state on tendon healing potential in rats. Int Orthop 33(5):1461–1466
60. Huisman ES, Andersson G, Scott A et al (2014) Regional molecular and cellular differences in the female rabbit achilles tendon complex: potential implications for understanding responses to loading. J Anat 224:538–47
61. Kjaer M (2004) Role of extracellular matrix in adaptation of tendon and skeletal muscle to mechanical loading. Physiol Rev 84:649–698
62. Charlton WP, Coslett-Charlton LM, Ciccotti MG (2001) Correlation of estradiol in pregnancy and anterior cruciate ligament laxity. Clin Orthop Relat Res 165–170
63. Moalli PA, Talarico LC, Sung VW et al (2004) Impact of menopause on collagen subtypes in the arcus tendineous fasciae pelvis. Am J Obstet Gynecol 190:620–627
64. Cook JL, Bass SL, Black JE (2007) Hormone therapy is associated with smaller achilles tendon diameter in active post-menopausal women. Scand J Med Sci Sports 17:128–132
65. Torricelli P, Veronesi F, Pagani S et. al (2013) In vitro tenocyte metabolism in aging and oestrogen deficiency. Age (Dordr) 35:2125–36
66. Lee CY, Liu X, Smith CL et al (2004) The combined regulation of estrogen and cyclic tension on fibroblast biosynthesis derived from anterior cruciate ligament. Matrix Biol 23:323–329
67. Christin-Maitre S (2013) History of oral contraceptive drugs and their use worldwide. Best Pract Res Clin Endocrinol Metab 27:3–12
68. Martineau PA, Al-Jassir F, Lenczner E et al (2004) Effect of the oral contraceptive pill on ligamentous laxity. Clin J Sport Med 14:281–286
69. Lee H, Petrofsky JS, Yim J (2015) Do oral contraceptives alter knee ligament damage with heavy exercise? Tohoku J Exp Med 237:51–6
70. Pokorny MJ, Smith TD, Calus SA et al (2000) Self-reported oral contraceptive use and peripheral joint laxity. J Orthop Sports Phys Ther 30:683–692
71. Minahan C, Joyce S, Bulmer AC et al (2015) The influence of estradiol on muscle damage and leg strength after intense eccentric exercise. Eur J Appl Physiol 115:1493–500
72. Savage KJ, Clarkson PM (2002) Oral contraceptive use and exercise-induced muscle damage and recovery. Contraception 66:67–71

73. Hicks KM, Onambele-Pearson GL, Winwood K et al (2013) Gender differences in fascicular lengthening during eccentric contractions: the role of the patella tendon stiffness. Acta Physiol (Oxf) 209:235–44
74. Jung-Hoffmann C, Fitzner M, Kuhl H (1991) Oral contraceptives containing 20 or 30 micrograms ethinylestradiol and 150 micrograms desogestrel: pharmacokinetics and pharmacodynamic parameters. Horm Res 36:238–246
75. Wojtys EM, Huston LJ, Boynton MD et al (2002) The effect of the menstrual cycle on anterior cruciate ligament injuries in women as determined by hormone levels. Am J Sports Med 30:182–188
76. Agel J, Bershadsky B, Arendt EA (2006) Hormonal therapy: ACL and ankle injury. Med Sci Sports Exerc 38:7–12
77. Moller-Nielsen J, Hammar M (1989) Women's soccer injuries in relation to the menstrual cycle and oral contraceptive use. Med Sci Sports Exerc 21:126–129
78. Rahr-Wagner L, Thillemann TM, Mehnert F et al (2014) Is the use of oral contraceptives associated with operatively treated anterior cruciate ligament injury? A case-control study from the Danish Knee Ligament Reconstruction Registry. Am J Sports Med 42:2897–905
79. Ruedl G, Ploner P, Linortner I et al (2009) Are oral contraceptive use and menstrual cycle phase related to anterior cruciate ligament injury risk in female recreational skiers? Knee Surg Sports Traumatol Arthrosc 17:1065–9
80. Holmes GB, Lin J (2006) Etiologic factors associated with symptomatic achilles tendinopathy. Foot Ankle Int 27:952–959
81. Lovering RM, Romani WA (2005) Effect of testosterone on the female anterior cruciate ligament. Am J Physiol Regul Integr Comp Physiol 289:R15–R22
82. Asano K, Maruyama S, Usui T et al (2003) Regulation of estrogen receptor alpha and beta expression by testosterone in the rat prostate gland. Endocr J 50:281–7
83. Dimitrakakis C, Zhou J, Wang J et al (2003) A physiologic role for testosterone in limiting estrogenic stimulation of the breast. Menopause 10:292–8
84. Srinivasan N, Aruldhas MM, Govindarajulu P (1986) Sex steroid-induced changes in collagen of the prostate and seminal vesicle of rats. J Androl 7:55–8
85. Torjesen PA, Sandnes L (2004) Serum testosterone in women as measured by an automated immunoassay and a ria. Clin Chem 50:678, author reply 678-9
86. Zhou J, Ng S, Adesanya-Famuiya O et al (2000) Testosterone inhibits estrogen-induced mammary epithelial proliferation and suppresses estrogen receptor expression. FASEB J 14:1725–30
87. Hama H, Yamamuro T, Takeda T (1976) Experimental studies on connective tissue of the capsular ligament. Influences of aging and sex hormones. Acta Orthop Scand 47:473–479
88. Dehghan F, Muniandy S, Yusof A et al (2014) Testosterone reduces knee passive range of motion and expression of relaxin receptor isoforms via 5alpha-dihydrotestosterone and androgen receptor binding. Int J Mol Sci 15:4619–34
89. Denaro V, Ruzzini L, Longo UG et al (2010) Effect of dihydrotestosterone on cultured human tenocytes from intact supraspinatus tendon. Knee Surg Sports Traumatol Arthrosc 18:971–6
90. Dehghan F, Haerian BS, Muniandy S et al (2014) The effect of relaxin on the musculoskeletal system. Scand J Med Sci Sports 24:e220–9
91. Dragoo JL, Castillo TN, Braun HJ et al (2011) Prospective correlation between serum relaxin concentration and anterior cruciate ligament tears among elite collegiate female athletes. Am J Sports Med 39:2175–80
92. Pearson SJ, Burgess KE, Onambele GL (2011) Serum relaxin levels affect the in vivo properties of some but not all tendons in normally menstruating young women. Exp Physiol 96:681–688
93. Dehghan F, Yusof A, Muniandy S et al (2015) Estrogen receptor (ER)-alpha, beta and progesterone receptor (PR) mediates changes in relaxin receptor (RXFP1 and RXFP2) expression and passive range of motion of rats' knee. Environ Toxicol Pharmacol 40:785–791
94. Seynnes OR, Koesters A, Gimpl M et al (2011) Effect of alpine skiing training on tendon mechanical properties in older men and women. Scand J Med Sci Sports 21(Suppl 1):39–46
95. Freeman BJ, Rooker GD (1995) Spontaneous rupture of the anterior cruciate ligament after anabolic steroids. Br J Sports Med 29:274–5
96. Hill JA, Suker JR, Sachs K et al (1983) The athletic polydrug abuse phenomenon. A case report. Am J Sports Med 11:269–71
97. Kramhoft M, Solgaard S (1986) Spontaneous rupture of the extensor pollicis longus tendon after anabolic steroids. J Hand Surg (Br) 11:87
98. Kanayama G, DeLuca J, Meehan WP 3rd et al (2015) Ruptured tendons in anabolic-androgenic steroid users: a cross-sectional cohort study. Am J Sports Med 43:2638–44
99. Seynnes OR, Kamandulis S, Kairaitis R et al (2013) Effect of androgenic-anabolic steroids and heavy strength training on patellar tendon morphological and mechanical properties. J Appl Physiol (1985) 115:84–9
100. Tsitsilonis S, Chatzistergos PE, Mitousoudis AS et al (2014) Anabolic androgenic steroids reverse the beneficial effect of exercise on tendon biomechanics: an experimental study. Foot Ankle Surg 20:94–9
101. Bhasin S, Storer TW, Berman N et al (1996) The effects of supraphysiologic doses of testosterone on muscle size and strength in normal men. N Engl J Med 335:1–7
102. Bhasin S, Woodhouse L, Storer TW (2001) Proof of the effect of testosterone on skeletal muscle. J Endocrinol 170:27–38

14 Tendon Homeostasis in Hypercholesterolemia

Louis J. Soslowsky and George W. Fryhofer

Abstract

Hypercholesterolemia is a serious health problem that is associated not only with heart disease, but also tendon pathology. In high cholesterol environments (e.g. familial hyperlipidemia), lipids accumulate within the tendon extracellular matrix and form deposits called xanthomas. Lipid-related changes are known to affect several tendon mechanical properties, including stiffness and modulus, in uninjured and injured tendons, alike. Mechanisms to explain these cholesterol-related changes are multiple, including alterations in tenocyte gene and protein expression, matrix turnover, tissue vascularity, and cytokine production. Clinically, rotator cuff tear and Achilles tendon rupture are clearly associated with metabolic derangements, and elevated total cholesterol is often among the specific metabolic parameters implicated. Treatment of hypercholesterolemia using statin medications has also been shown to affect tendon properties, resulting in normalization of tendon thickness and improved tendon healing. Despite current work, the pathophysiology of lipid-related tendon pathology remains incompletely understood, and additional hypothesis-generating studies, including those incorporating whole-genome and whole-transcriptome technologies, will help to point the field in new directions.

Keywords

Tendons • Hypercholesterolemia • Statins • Xanthoma • Tendinopathy • Hyperlipidemia

List of Acronyms & Abbreviations

A-20	low-dose atorvastatin
A-80	high-dose atorvastatin
ApoE	apolipoprotein E
ATV	atorvastatin

L.J. Soslowsky (✉) • G.W. Fryhofer
McKay Orthopaedic Laboratory, University of Pennsylvania, Philadelphia, PA, USA
e-mail: soslowsk@upenn.edu; fryhofer@upenn.edu

P.W. Ackermann and D.A. Hart (eds.), *Metabolic Influences on Risk for Tendon Disorders*, Advances in Experimental Medicine and Biology 920, DOI 10.1007/978-3-319-33943-6_14

BMI	body mass index
BMP-2	bone morphogenetic protein 2
CEL	celecoxib
COX2	cyclooxygenase 2
CSA	cross-sectional area
CXCL3	chemokine (C-X-C motif) ligand 3
FH	familial hyperlipidemia
GTPase	hydrolase enzyme that binds and hydrolyzes guanosine triphosphate (GTP)
HC	high cholesterol
HDL	high-density lipoprotein
IL-6,-8	interleukin-6,-8
LDL	low-density lipoprotein
MMP	matrix metalloproteinase
PGE2	prostaglandin E2
Rap1a	Ras-related protein Rap-1A
RC	rotator cuff
S-20	low-dose simvastatin
S-80	high-dose simvastatin
TNF-alpha	tumor necrosis factor alpha
TX-	patients without tendon xanthomas
TX+	patients with tendon xanthomas

Introduction

Clinical and Scientific Motivation for Study of Lipid-Related Tendon Pathology

Hypercholesterolemia – defined by the American Heart Association as a total serum cholesterol concentration of 240 mg/dL or higher – is a serious health problem in the United States and around the world. As of 2012, 12.9 % of the US population aged 20 and older had high total cholesterol [1].

Hypercholesterolemia is associated with heart disease [2, 3] and is why cholesterol-lowering drugs are some of the most widely prescribed medications in our healthcare system. Hypercholesterolemia is also associated with tendon pathology, and, in patients genetically predisposed to hypercholesterolemia, there are increased rates of tendon rupture [2–4] (see Chap. 10).

Given the dual associations of high cholesterol with both heart disease and tendon pathology, it is perhaps not surprising that changes in tendon structure have also been associated with ischemic heart disease [5]. Importantly, patients who suffer from lipid-related tendon pathology may not actually be aware of their high cholesterol state and accompanying risk for heart disease. For example, in one case series of 41 patients who underwent surgical treatment for Achilles tendon rupture, serum total cholesterol was elevated in 83 %, but only 19 % of patients were aware of their condition [6].

Thus, the motivation to understand the link between tendinopathy and hypercholesterolemia in humans is twofold. First, the continued accumulation of clinical data demonstrating an irrefutable link between hypercholesterolemia and tendon pathology may provide additional clinical motivation for patients with tendon-related complaints (who are also at risk for heart disease) to seek treatment for their high cholesterol. Second, the study of lipid-related tendon dysfunction provides a clinically-relevant model for clinicians and researchers to better understand the underlying pathophysiology of lipid-related tendinopathy as well as tendinopathy *in general*, neither of which is fully understood.

How Lipids Accumulate Within Tendons; Formation of Tendon Xanthoma

High serum cholesterol (also referred to as "lipid") levels allow for the accumulation of *oxidized* - low-density lipoproteins (LDL), which are cholesterol carrier proteins that include modified phospholipids and cholesterols, isoprostanes, oxidized arachidonoyl residues, lysolipids, and lysophosphatidic acid [7]. Many of these oxidized lipids possess platelet stimulatory properties, which in the context of heart disease contribute to arterial thrombus formation after rupture of atherosclerotic plaques. In addition to platelet activation, LDL also produces a more general local inflammatory response and contributes to cell death through activation of the intrinsic apoptotic signaling pathway within mitochondria [8].

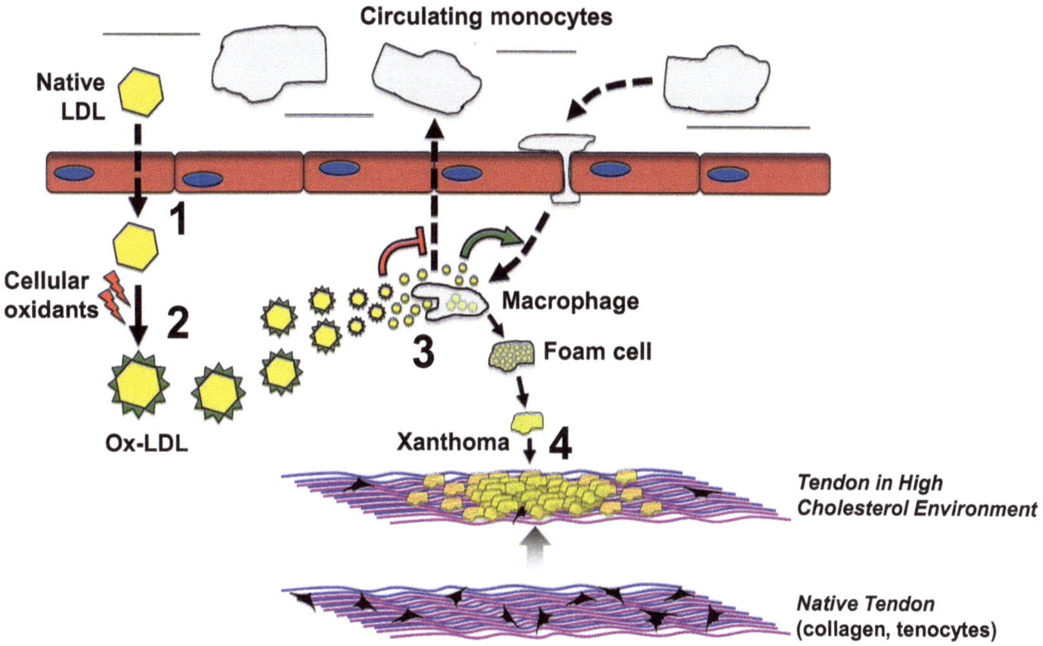

Fig. 14.1 *Diagram illustrating how accumulation of oxidized LDL in a high cholesterol environment results in tendon pathology.* (**1**) Entry of native LDL from circulation into interstitial compartment. (**2**) Cellular oxidants produced by macrophages and other inflammatory cells react with native LDL to form oxidized LDL (Ox-LDL). Ox-LDL recruits monocytes from the circulation (*green arrow*). Ox-LDL also impairs macrophage motility and blocks the macrophage's return to the circulation (*red arrow*). (**3**) Macrophages engulf Ox-LDL via scavenger receptors, resulting in foam cell formation. (**4**) Foam cells accumulate within the tendon substance, forming a tendon xanthoma

The clinical manifestation of lipid accumulation in human tendons is a tendon xanthoma and represents the major tendon pathology observed in patients with familial dyslipidemias. Xanthomas are collections of lipid-laden macrophages around tendon. By dry weight, xanthomas are composed of 33 % lipids and 24 % collagen [9]. The lipid component is made up of 55 % free cholesterol, 28 % cholesterolesters, and 13 % phospholipids [10]. *Unesterified* cholesterol accumulates mostly within the extracellular space, whereas *esterified* cholesterol accumulates in both extra- and intracellular spaces in the form of intracytoplasmic lipid vacuoles, lysosomes, and myelin figures [8, 11, 12].

Lipids found within xanthomas are derived from circulating plasma rather than synthesized locally [13]. It has been proposed that, after leaving the circulation, initially unmodified LDL becomes trapped in the collagen and glycosaminoglycans of the tendon matrix, at which point the LDL is oxidized by local factors produced by macrophages. Macrophages then take up the now oxidized LDL, which results in the accumulation of lipid-laden macrophages, known as foam cells (Fig. 14.1) [14–18].

Hypercholesterolemia and Tendon Pathology are Clinically Linked

Familial Dyslipidemias and Their Association With Tendon Pathology

Lipid-related tendon pathology is most prevalent among patients affected by familial dyslipidemias. The major familial hyperlipidemias can be classified into 5–6 types (Table 14.1) (see Chap. 10) [19, 20].

Types II and III are associated with xanthoma formation (Fig. 14.2). Type II is an autosomal dominant disorder and is associated with tendinopathy and xanthomas that occur *within* the tendon substance. Xanthomas are more

Table 14.1 Fredrickson/World Health Organization classification of hyperlipidemias

Type	Elevated lipoprotein(s)	Lipids elevated	Tendon pathology?
I	Chylomicrons	Triglycerides	–
IIa	LDL	Cholesterol	+, *Within* tendon
IIb	LDL, VLDL	Cholesterol > triglycerides	*Rare*
III	IDL (VLDL remnants), chylomicrons	Cholesterol, triglycerides	±, *Over* tendon
IV	VLDL	Triglycerides	–
V	VLDL, chylomicrons	Triglycerides > cholesterol	–

Abbreviations: LDL low-density lipoprotein, *VLDL* very low-density lipoprotein, *IDL* intermediate-density lipoprotein

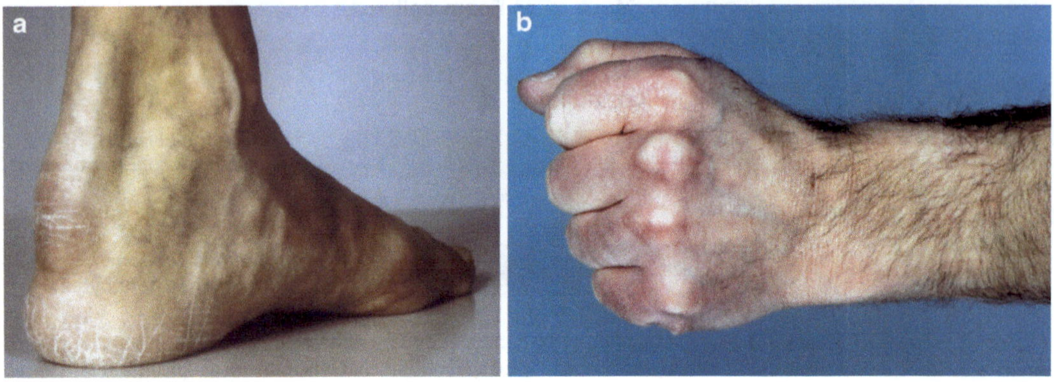

Fig. 14.2 *Tendon xanthomas in familial hyperlipidemia.* (**a**), Achilles tendon xanthoma (type IIa hyperlipidemia). (**b**), Tendon xanthomata on dorsum of hand (type IIa hyperlipidemia) (Reproduced from Durrington [19], Copyright © 2003 Elsevier Ltd., with permission from Elsevier)

common in *homozygous* type II hyperlipidemia and typically develop in childhood. In contrast, xanthomas present later and are less common in patients with *heterozygous* type II hyperlipidemia [20]. Type III familial hyperlipidemia is an autosomal recessive disorder and is associated with xanthomas that occur *superficial to* tendons and involve extensor surfaces as well as the plantar crease. Unlike xanthomas in type II disease, xanthomas in type III disease are typically asymptomatic.

In addition to the 5 types of familial hyperlipidemia already described, cerebrotendinous xanthomatosis and sitoserolemia are two additional rare autosomal disorders, which, respectively, result in accumulation of either dihydrocholesterol or plant sterols. These two disorders are associated with tendon xanthomas, but not tendonitis [21]. Finally, hyperlipidemia secondary to thyroid disorders or diabetes has also been associated with xanthoma formation [22].

Table 14.2 Sonographic grading of tendon xanthoma

Grade	Description
1	Normal, or minor sonographic changes
2	Diffuse heterogenous echo pattern
3	Focal hypoechoic lesions

Clinically, xanthomas may or may not be symptomatic and can be detected by MRI or ultrasound earlier than by clinical exam alone (Fig. 14.2). On ultrasound, Achilles abnormalities may be graded 1–3 (Table 14.2) [8, 14]. Tendon involvement can be unilateral or bilateral, and patients who are symptomatic may report up to 12 attacks of pain per year [4]. In one study of 73 patients with heterozygous type II hyperlipidemia, 18 % of patients reported Achilles pain, and in 11 % of patients there was evidence of tendonitis [14]. A subsequent cross-sectional study reported that 46.6 % of patients with familial heterozygous type II hyperlipidemia had one or more episodes

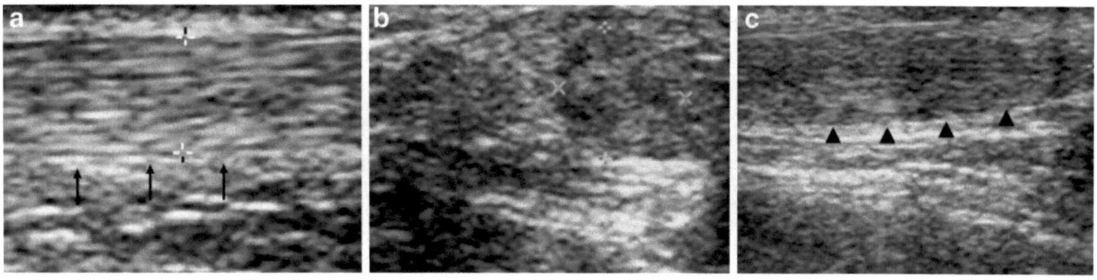

Fig. 14.3 *Sonographic appearance of the Achilles tendon.* (**a**), 22-year-old normal male subject – sagittal sonogram of the Achilles tendon (→) shows a fibrillar pattern characterized by the presence of multiple parallel linear echoes. (**b**), 21-year-old patient with familial hypercholesterolaemia (FH) – axial sonogram of the Achilles tendon demonstrates a focal hypoechoic lesion (*white markers*) suggestive of Achilles tendon xanthomas (ATX). (**c**), 25- year-old male patient with FH – sagittal sonogram of the Achilles tendon (*black arrowheads*) demonstrates a diffuse heterogeneous echo pattern (Reproduced from Tsouli et al. [14], Copyright © 2015 by John Wiley & Sons, Inc, with permission from Wiley)

of Achilles pain lasting longer than 3 days, as compared to 6.9 % of unaffected controls [23].

Hypercholesterolemia is Also Associated With Tendon Pathology in Patients Without Familial Hyperlipidemia

A number of clinical observational studies have also been performed to identify an association between tendon injury and hypercholesterolemia in patients who do *not* have familial hyperlipidemia.

Hypercholesterolemia has been implicated as a risk factor for rotator cuff injury, but the evidence is mixed. For example, in one case–control study, patients with rotator cuff (RC) tendon rupture were age-matched against patients seen for non-tendon-related shoulder complaints [24]. Patients with RC tear had significantly elevated serum triglyceride, LDL, and total cholesterol; there was a trend toward decreased serum high-density lipoprotein (HDL). However, in another study, there was *no* association between tendinopathy and hypercholesterolemia [25]. And in a population of postmenopausal women, RC tear was associated with increased body mass index (BMI), increased fasting glucose, and decreased HDL – but not total cholesterol or triglycerides [26].

Hypercholesterolemia has also been identified as a risk factor for Achilles tendon injury in patients without familial hyperlipidemia (Fig. 14.3). In one case–control study, the association between tendon injury and hypercholesterolemia was significant. Among patients with Achilles tendon rupture, concentrations of total cholesterol and LDL cholesterol were higher; triglyceride and very low-density lipoprotein cholesterol were higher; and HDL cholesterol was lower compared to controls [27]. In a separate study of Achilles tendinopathy, ultrasound detected entheseal alteration in 100 % of symptomatic patients compared to 13.3 % of asymptomatic patients [28]. Symptomatic patients had significantly higher BMI, higher blood glucose, and increased risk for osteoarthritis; in symptomatic patients, there were also trends toward increased total cholesterol and decreased HDL.

Taken together, these data suggest that at least some degree of *metabolic derangement* is associated with tendinopathy. However, the degree to which elevated total cholesterol *specifically* contributes to tendinopathy may be less predictable among varying patient populations and differing injury patterns.

In the remaining sections of this chapter, we will discuss:

1. How properties of *uninjured* tendon are altered by hypercholesterolemia
2. How properties of *injured* tendon, including healing response, are altered by hypercholesterolemia
3. How tendon properties are altered by pharmacologic treatment of hypercholesterolemia

4. Summary and future directions for the study of tendon homeostasis in hypercholesterolemia

Uninjured Tendon Properties In Hypercholesterolemia

Pathophysiology of Cholesterol-Related Changes in Uninjured Tendon

Changes in Tendon Microenvironment

Structural alterations of the tendon microenvironment may represent one mechanism by which high cholesterol environments alter tendon homeostasis. For example, in chick embryonic fibroblasts, exposure to hypercholesterolemic serum was associated with decreased synthesis of non-collagenous proteins and decreased incorporation of extracellular matrix components [29]. Changes in the tendon microenvironment in response to hypercholesterolemia have also been studied in a rabbit model of heritable hyperlipidemia [30]. In this model, lipid-related tendon xanthomas were identified in the plantar side of the plantaris tendon, the flexor retinaculum of the carpus, and around the digital flexor tendons of each joint level. Xanthomatous components were localized to the superficial paratenon of tendons wrapping around bony or fibrous pulleys, and there was increased vascularity throughout tendon tissues. Together, these data suggest that elevated cholesterol levels may alter the tendon microenvironment via local *changes in protein synthesis* and *extracellular matrix composition/turnover*. Furthermore, areas of *increased mechanical stress* and *increased vascularity* within tendon tissue may predispose to lipid-related tendon pathology.

Changes in Gene Expression

In addition to local changes in the tendon microenvironment, *differential gene expression* may represent another mechanism by which high cholesterol environments alter tendon homeostasis. For example, macrophages derived from patients *with* tendon xanthomas (TX+) are more likely than macrophages from patients *without* tendon xanthoma (TX−) to form foam cells with differential gene expression [31]. TX+ patients also have higher plasma levels of inflammatory cytokines (TNF-alpha, IL-8, IL-6), and TX+ macrophages produce higher concentrations of IL-8 in response to oxidized LDL when compared to TX- macrophages [32]. CXCL3 gene expression is also higher in macrophages from TX+ patients and is positively correlated with thickness of the Achilles tendon in the specific patient from which the macrophage is derived. These data suggest a role for *increased inflammatory gene expression* and *increased cytokine production* to explain why some patients with familial hyperlipidemia experience lipid-related tendon pathology, while other patients do not.

Clinical observational studies have also attempted to uncover a genetic basis for xanthoma formation among patients with familial hyperlipidemia. In patients with heterozygous familial hyperlipidemia, the presence of tendon xanthoma is associated with age, male gender, LDL cholesterol levels, and triglyceride level. The presence of xanthoma is also associated with a 3.2-fold higher risk of cardiovascular disease, and the odds of having tendon xanthoma increase with the number of cardiovascular risk alleles in pathways for reverse cholesterol transport and LDL oxidation [33–38]. These data suggest that xanthoma formation and cardiovascular disease not only are associated with each other, but also share common pathophysiologic mechanisms via genes governing *reverse cholesterol transport* and *LDL oxidation.*

Biomechanics of Uninjured Tendon are Altered in a High Cholesterol Environment

Animal models have been used to quantify the effect of hypercholesterolemia on biomechanical properties of uninjured tendon.

In a hypercholesterolemic rat rotator cuff model, uninjured supraspinatus tendon *stiffness* and *modulus* were increased in a high cholesterol

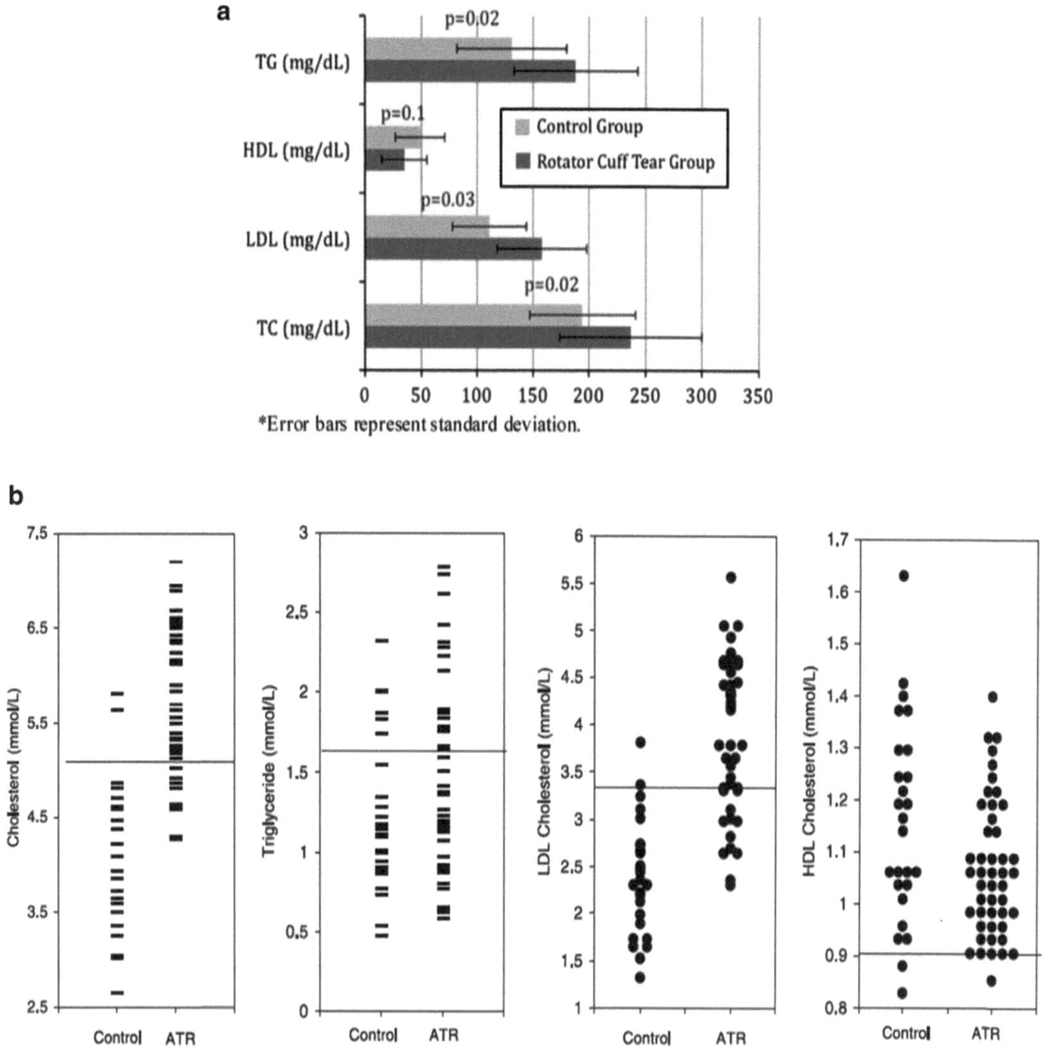

Fig. 14.4 *High cholesterol is associated with Achilles tendon rupture and rotator cuff tear.* (**a**), Patients with rotator cuff tears had a higher incidence of hypercholesterolemia compared to patients with normal rotator cuff tendons (Reproduced with permission from Abboud and Kim [24], © The Association of Bone and Joint Surgeons® 2009, with permission from Springer Science + Business Media). (**b**), Patients with Achilles tendon rupture had *higher* serum cholesterol, *higher* serum triglyceride, *higher* serum LDL, and *lower* serum HDL compared to control patients without Achilles tendon pathology (Reproduced from Ozgurtas et al. [27], © 2003 by Elsevier Science B.V., with permission from Elsevier)

(HC) environment [39]. In contrast with these results, uninjured tendon from a porcine biceps tendon model exhibited decreased *stiffness* and *modulus* in a HC environment; the differences between these two models may be explained by differences in loading environment experienced by supraspinatus and biceps tendon, respectively [40]. The effect of hypercholesterolemia on uninjured tendon has also been studied across multiple species using supraspinatus tendons from Apolipoprotein E (ApoE) knockout mice, rats (HC diet), and monkeys (HC diet) [41]. Tendon *stiffness* was significantly increased in HC mice and rats, and there was a trend toward increased stiffness in HC monkeys. Tendon *modulus* was significantly increased in HC mice and monkeys, and there was a trend toward increased modulus in HC rats (Fig. 14.4) (Table 14.3).

Table 14.3 Changes in mechanical properties of uninjured tendon in high cholesterol environment [41]

	Stiffness	Modulus
Mouse	↑	↑
Rat	↑	↑ (trend)
Monkey	↑ (trend)	↑

Taken together, these data suggest that biomechanical properties in uninjured tendon, including tendon *stiffness* and *modulus*, are altered in a high cholesterol environment.

Injured Tendon Properties & Healing Response In Hypercholesterolemia

Biomechanical Properties of Healing Tendon are Altered in a High Cholesterol Environment

Animal models have also been used to determine the effect of hypercholesterolemia on biomechanical properties of injured and healing tendon.

In a mouse patellar tendon injury model, control mice were compared to mice deficient for ApoE, which produces a hypercholesterolemic state [42]. In tendons from mice that were young (14 weeks) at time of injury, there was no difference in tendon cross-sectional area (CSA) between ApoE mice and controls. Tendons from young ApoE mice exhibited improved healing strength and improved percent relaxation compared to controls, with a higher normalized (injured/sham) maximum stress and a lower normalized percent relaxation that was closer to normal baseline values. In tendons from young ApoE mice, there was no difference in normalized modulus. Tendon properties of older ApoE-knockout mice have also been studied. In *uninjured* tendons from 40-week old ApoE mice, there was decreased modulus and a trend toward increased CSA compared to controls [43]. In *injured* tendons from older mice injured at 40 weeks, ApoE tendons exhibited decreased normalized maximum stress compared to controls; there was no difference in normalized area or modulus.

A hypercholesterolemic rat rotator cuff model has also been used to study the effect of hypercholesterolemia on tendon healing [44]. In rats fed a HC diet, normalized tendon stiffness was decreased at 4 weeks post-injury compared to rats fed a normal diet. There was no difference in modulus or angular deviation. There were also no histological differences in collagen organization, cellularity, or cell shape. Interestingly, in rats fed a HC diet, CSA was significantly closer to normal by 8 weeks after injury compared to rats fed a normal diet; one potential (but untested) explanation for this finding is the possibility of a reduced inflammatory fibrotic response in HC animals.

Taken together, these data suggest that exposure to a high cholesterol environment serves to negatively impact tendon healing, as demonstrated by *decreased normalized maximum tendon stress* in ApoE mice and *decreased normalized stiffness* in rats fed a HC diet. Moreover, the reduction in tendon healing observed in the 40-week (but not 14-week) age group of mice may be explained by a process that allows for accumulation over time, such as the intratendinuous cholesterol deposition or relative tissue ischemia due to vascular compromise that are seen clinically in older patients with hypercholesterolemia. This suggests that *long-term exposure to high cholesterol* may be necessary for tendons to develop appreciable deficits in a high cholesterol environment.

Tendon Properties in Response to Pharmacologic Treatment of Hypercholesterolemia

Statin Therapy is Associated with Tendon-Related Side Effects in Patients with Hypercholesterolemia

The pharmacological treatment of hypercholesterolemia – using statins, bile acid sequestrants, ezetimibe, and niacin – has been associated with regression of tendon xanthomas [45–47].

One large case–control study classified Achilles tendon morphology and measured tendon thickness in 80 patients with heterozygous familial hyperlipidemia (FH) before and after 12 months of statin therapy [48]. Ultrasonic evaluation of Achilles tendon in the FH population identified grade 1 (52.5 %), grade 2 (37.5 %), and grade 3 (10 %) echostructures. In an age- and sex-matched population without familial hyperlipidemia, 100 % of patients were found to have grade I echostructures. After initial imaging, statin therapy was initiated in the patients with FH; after 12 months of therapy, there was a significant reduction in Achilles tendon thickness in response to statin treatment only in those patients who had grade 1 abnormal echostructures on initial imaging. Tendon thickness remained unchanged in patients whose initial imaging showed grade 2 and grade 3 echostructures.

Despite allowing for normalization of tendon thickness in at least a subset of patients, the use of lipid-lowering medication to treat hyperlipidemia may also result in at least transient changes that are detrimental to tendon structure and function. Of all lipid-lowering medications, statins are the most commonly prescribed. Tendon injury accounts for 2.1 % of reported statin-related side effects. 59 % of these cases occur within the first year of therapy, and median time from statin initiation to tendon injury is 8–-10 months [49]. Among patients who report tendon-related side effects, approximately two-thirds of patients experience tendinopathy alone, while the remaining one-third experience frank tendon rupture. The Achilles tendon is most commonly affected (52 % of cases) [50]; other areas of involvement include rotator cuff tendon and biceps tendon. In one retrospective study of 100 patients with biceps tendon rupture, there was a nearly twofold increased risk of biceps tendon rupture among statin-treated patients compared to patients not taking statins [51]. And yet, not all data is so conclusive in identifying a link between statin therapy and tendon rupture. In a multivariate logistic regression model of tendon rupture that controlled for diabetes, renal disease, rheumatologic disease, and steroid use, there was no significant association between statin use and tendon rupture [52]. However, on subgroup analysis, statin exposure was associated with increased odds of tendon rupture in women but not in men.

In addition to statins, other lipid-lowering medications, including niacin and bile acid sequestrants, have also been implicated in producing tendon-related side effects. There are case reports of patients with documented familial hyperlipidemia and Achilles tendon xanthomas who experience new-onset Achilles tendon pain following treatment intensification with niacin and bile acid sequestrants [53]. It has been suggested that these medications may disrupt tissue architecture and stability through the removal of lipids from an existing xanthoma, and that such a phenomenon might explain the tendon pain experienced by some patients.

Finally, in addition to the demonstrated effects of pharmacotherapy on tendon structure and function [46, 47], lipid apheresis therapy in patients with severe familial hyperlipidemia has also been shown to significantly decrease tendon thickness by 18.8 % over a 3-year period [54].

Pathophysiology of Statin-Related Side Effects in Tendon

Multiple laboratory-based models have been used to identify potential pathophysiologic mechanisms by which statins alter tendon homeostasis.

In human primary tenocytes, exposure to statin at therapeutic concentrations is not associated with any changes in cell viability or morphology [55]. At supratherapeutic concentrations, however, short-term exposure results in *reduced cell migration*; prolonged exposure results in *increased cell rounding*, *decreased mRNA for matrix proteins*, *increased BMP-2 expression*, *impaired gap junction communication*, and *inhibition of prenylation of Rap1a GTPase*. The clinical relevance of observing cell changes in response to treatment with

Fig. 14.5 *Mechanical testing of uninjured tendon isolated from high cholesterol environment.* Supraspinatus tendons with stain lines applied from mouse (**1**), rat), (**2**) and monkey (**3**) (Reproduced from Beason et al. [41], © 2013 Journal of Shoulder and Elbow Surgery Board of Trustees, with permission from Elsevier)

statin at supratherapeutic levels – but not at therapeutic levels – is difficult to interpret.

Animal models have also been used to study the effects of statin treatment on tendon homeostasis (Fig. 14.5). A rabbit model of Achilles tendon injury and repair demonstrates that statin therapy may alter the tendon healing response by *decreasing vascularity* and *increasing production of irregular collagen structures*. In this model, rabbits received a 6-week course of statin therapy post-injury and were compared against control rabbits that were also injured but did not receive statin treatment [56]. The statin-treated group trended toward decreased revascularization and increased collagenization compared to controls. Collagen constructions were observed to be significantly more irregular in tendon from the statin-treated group.

In a rat Achilles tendon model, the effects of different formulations of high-dose and low-dose chronic statin therapy have also been studied [57]. Rats were divided into 5 groups: low- (S-20) and high-dose (S-80) simvastatin groups; low- (A-20) and high-dose (A-80) atorvastatin groups, and untreated control. In the S-20 group, non-collagenous protein levels were decreased versus control. In A-80, there was increased pro-MMP-2 activity and latent MMP-9 activity versus control. In S-20, there was increased active MMP-2, latent MMP-9 activity, and hydroxyproline versus control. In A-20, there was decreased collagen type I and increased glycosaminoglycans versus control. These data suggest that statins cause *imbalance of ECM components* and that this imbalance may predispose to tendon microdamage. Finally, in contrast to previous studies, data from a rat rotator cuff injury & repair model suggests that statin therapy may, in fact, enhance tendon healing [58]. Rats were divided into four groups according to medication received during a 3-week post-operative course. These groups were atorvastatin (ATV), celecoxib (CEL; COX2 inhibitor), ATV + CEL, or saline alone. Healed tendons in the ATV group demonstrated higher max load and stiffness as compared to the saline and CEL groups. Celecoxib alone did not affect tendon healing. Tenocytes treated with ATV demonstrated increased proliferation, migration, and adhesion; these effects were blocked by co-treatment with a PGE2 receptor 4 antagonist. CEL treatment also diminished the effect of statin treatment on the isolated tenocytes; however, treatment with PGE2 stimulated tenocyte proliferation even in the presence of CEL. These data suggest that statins may enhance tendon healing during the acute inflammatory phase – as demonstrated by *increased tendon max load and stiffness* and *improved tenocyte function* – and COX2 inhibitors may diminish these statin-mediated effects. Additionally, *activation of the COX2/PGE2 pathway* may serve as a useful target for pharmacotherapy designed to enhance early tendon healing.

Summary & Future Directions

Hypercholesterolemia is an important clinical problem and is associated with significant tendon pathology. Lipid-related tendon pathology is most prevalent among patients with familial dyslipidemias; however, associations between hypercholesterolemia and tendinopathy have also been identified in patients without familial hypercholesterolemia. Rotator cuff tear and Achilles tendon rupture are clearly associated with metabolic derangements, and elevated total cholesterol is sometimes (but not always) among the specific metabolic parameters implicated in patients with tendon pathology.

In *uninjured* tendon, high cholesterol environments alter specific biomechanical properties, including tendon stiffness and modulus. These lipid-related changes in tendon properties have been demonstrated in mice, rats, monkeys, and pigs and hold true across multiple tendon types, including supraspinatus tendon and biceps tendon. A few overarching pathophysiological mechanisms have been identified to help explain these cholesterol-related changes in uninjured tendon biomechanics (Fig. 14.6). These mechanisms include: protein synthesis; local extracellular matrix composition and turnover; mechanical stress; vascularity; inflammatory gene expression; cytokine production; expression of genes related to LDL oxidation; and expression of genes related to reverse cholesterol transport.

In *injured and healing* tendon, the detrimental effects of hypercholesterolemia have also been demonstrated. In healing tendon, high cholesterol environments have been associated with decreased maximum tendon stress and decreased tendon stiffness, and these changes occur after long-term (rather than short-term) exposure to high cholesterol.

The effects of lipid-lowering pharmacotherapy on tendon were also discussed. Statins help to normalize tendon thickness in some patients affected by tendon xanthomas; however, statin therapy has also been implicated as the cause of new-onset tendon pain and as a risk factor for tendon rupture. Statin therapy was shown to improve tendon healing in a rat model with increased tendon max load and increased tendon stiffness. Mechanisms by which statin therapy may alter tendon homeostasis have also been identified. These mechanisms include: changes in cell migration; cell rounding; mRNA expression for matrix proteins; BMP-2 expression; gap junction communication; and signaling within the COX2/PGE2 pathway.

Future studies in the field of hypercholesterolemia and tendon homeostasis should continue to explore the pathophysiological mechanisms and pathways that have already been identified. Additionally, in order to continue to characterize new pathways by which alterations in biomechanical properties arise in the setting of hypercholesterolemia, the field may benefit from additional hypothesis-generating studies. For example, studies that measure gene expression across a wide array of genes, including those not already implicated in lipid-related tendon pathology, will help to point the field in new directions.

The field would also benefit from incorporation of more clinically translatable study elements. For example, in the context of statin therapy and its effect on tendon homeostasis, one group studied not only the effect of atorvastatin on *ex vivo* tendon and *in vitro* tenocytes, but also the degree to which a clinically relevant COX2 inhibitor modulates these statin-mediated effects [58]. In this manner, the study accrues additional clinical relevance, which allows the study conclusions to be more readily translated to a clinical setting.

Finally, clinical data indicate there are increased rates of tendon pathology in patients who have elevation of other metabolic

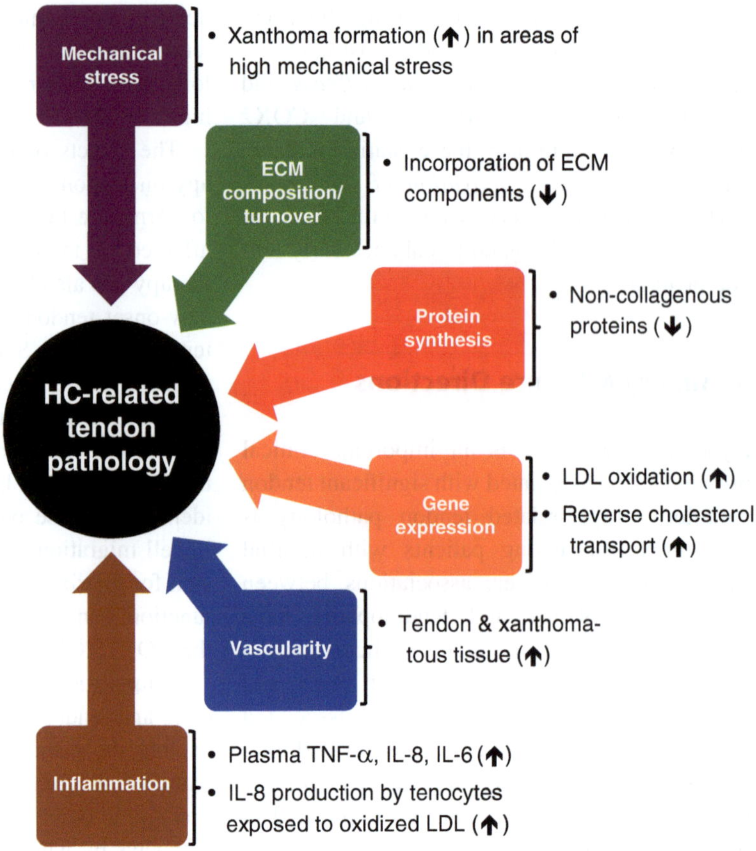

Fig. 14.6 Overview of pathophysiological mechanisms by which hypercholesterolemia alters tendon homeostasis

parameters in addition to elevated cholesterol [28]; among these additional parameters is elevated blood glucose (hyperglycemia). Hyperglycemia is one of the hallmarks of diabetes mellitus. Given the increasing prevalence of diagnosed diabetes in the United States [59], the study of tendon homeostasis in a combined high lipid *and* high glucose environment would generate knowledge that is clinically useful for this significant subset of patients with hypercholesterolemia and hyperglycemia. Animal models of diabetes do exist and could be studied alongside the animal models of hypercholesterolemia already discussed in this chapter (see Chaps. 16, 17, 18 and 19). Along these lines, future research could, for example, compare tendon biomechanics among "normal" animals, hyperglycemic animals, hypercholesterolemic animals, and animals that are both hypercholesterolemic and hyperglycemic. Such a study would allow for more specific delineation of the differential – or, perhaps, similar – roles of hypercholesterolemia and hyperglycemia in tendon homeostasis and may better reflect the more complex metabolic environment manifested in a human patient population.

Glossary

Apolipoprotein E (ApoE) a class of lipoprotein important for normal catabolism of triglyceride-rich lipoprotein components; some genetic variants are associated with high levels of circulating cholesterol.

Celecoxib an anti-inflammatory drug that blocks transformation of arachidonic acid to prostaglandin precursors via reversible inhibition of the cyclooxygenase-2 (COX-2) enzyme.

Familial hyperlipidemia a heritable medical condition characterized by abnormally high levels of lipid or lipoprotein in the blood; often caused by mutation of a lipoprotein receptor or lipoprotein lipase, which prevents normal clearance of lipid and lipoprotein from the blood.

Elastic modulus a material property that, like stiffness, describes a substance's ability to resist deformation; however, unlike stiffness, modulus also takes into account cross-sectional area and is defined as the slope of the stress-strain curve in the elastic deformation region for a given substance. Modulus is measured in units of pascals (Pa).

Glycosaminoglycans negatively charged, linear carbohydrate chains composed of repeating disaccharides; glycosaminoglycans are an important component of tendon extracellular matrix that serve to occupy space and resist compressive forces.

Hypercholesterolemia medical condition defined as a total serum cholesterol concentration $\geq$ 240 mg/dL.

Lipoproteins carrier proteins that bind cholesterol and allow for transport throughout the body

MMP-2, MMP-9 matrix metalloproteinases involved in normal breakdown of the extracellular matrix, including type IV collagen (a main structural component of basement membranes).

Statin a class of cholesterol-lowering drugs that function through competitive inhibition of the enzyme HMG-CoA reductase, which effectively blocks cholesterol synthesis in the liver and leads to a reduction in circulating cholesterol levels.

Stiffness the degree to which a substance resists deformation. Stiffness is measured in units of Newtons per meter (N/m).

Xanthoma on histological examination, a collection of lipid-laden macrophages (foam cells) in the tendon extracellular matrix; xanthomas may also be appreciated on gross examination as subcutaneous nodules tracking along a tendon.

References

1. Carroll MD, Kit BK, Lacher DA (2013) Total and high-density lipoprotein cholesterol in adults: National Health and Nutrition Examination Survey, 2011–2012: US Department of Health and Human Services, Centers for Disease Control and Prevention, National Center for Health Statistics
2. Langer T, Strober W, Levy RI (1972) The metabolism of low density lipoprotein in familial type II hyperlipoproteinemia. J Clin Invest 51(6):1528–36
3. Mensink RP, Katan MB (1990) Effect of dietary trans fatty acids on high-density and low-density lipoprotein cholesterol levels in healthy subjects. N Engl J Med 323(7):439–45
4. Glueck CJ, Levy RI, Frederickson DS (1968) Acute tendinitis and arthritis. A presenting symptom of familial type II hyperlipoproteinemia. Jama 206(13):2895–7
5. Mabuchi H, Tatami R, Haba T et al (1978) Achilles tendon thickness and ischemic heart disease in familial hypercholesterolemia. Metab Clin Exp 27 (11):1672–9
6. Mathiak G, Wening JV, Mathiak M, Neville LF, Jungbluth K (1999) Serum cholesterol is elevated in patients with Achilles tendon ruptures. Arch Orthop Trauma Surg 119(5–6):280–4
7. Siess W (2006) Platelet interaction with bioactive lipids formed by mild oxidation of low-density lipoprotein. Pathophysiol Haemost Thromb 35 (3–4):292–304
8. Abate M, Schiavone C, Salini V, Andia I (2013) Occurrence of tendon pathologies in metabolic disorders. Rheumatology 52(4):599–608
9. Tall AR, Small DM, Lees RS (1978) Interaction of collagen with the lipids of tendon xanthomata. J Clin Invest 62(4):836–46
10. Vermeer BJ, Mateysen AA, van Gent CM, van Sabben RM, Emeis JJ (1982) The lipid composition and localization of free and esterified cholesterol in different types of xanthomas. J Invest Dermatol 78(4):305–8
11. von Bahr S, Movin T, Papadogiannakis N et al (2002) Mechanism of accumulation of cholesterol and cholestanol in tendons and the role of sterol 27-hydroxylase (CYP27A1). Arterioscler Thromb Vasc Biol 22(7):1129–35
12. Kruth HS (1985) Lipid deposition in human tendon xanthoma. Am J Pathol 121(2):311–5
13. Bhattacharyya AK, Connor WE, Mausolf FA, Flatt AD (1976) Turnover of xanthoma cholesterol in hyperlipoproteinemia patients. J Lab Clin Med 87(3):503–18
14. Tsouli SG, Kiortsis DN, Argyropoulou MI, Mikhailidis DP, Elisaf MS (2005) Pathogenesis, detection and treatment of Achilles tendon xanthomas. Eur J Clin Investig 35(4):236–44
15. Mori M, Itabe H, Higashi Y et al (2001) Foam cell formation containing lipid droplets enriched with free cholesterol by hyperlipidemic serum. J Lipid Res 42 (11):1771–81

16. Sugiyama N, Marcovina S, Gown AM, Seftel H, Joffe B, Chait A (1992) Immunohistochemical distribution of lipoprotein epitopes in xanthomata from patients with familial hypercholesterolemia. Am J Pathol 141(1):99–106
17. Ginsberg HN, Goldsmith SJ, Vallabhajosula S (1990) Noninvasive imaging of 99mtechnetium-labeled low density lipoprotein uptake by tendon xanthomas in hypercholesterolemic patients. Arteriosclerosis 10 (2):256–62
18. Rapp JH, Connor WE, Lin DS, Inahara T, Porter JM (1983) Lipids of human atherosclerotic plaques and xanthomas: clues to the mechanism of plaque progression. J Lipid Res 24(10):1329–35
19. Durrington P (2003) Dyslipidaemia. Lancet 362 (9385):717–31
20. Kedar E, Gardner GC (2013) Lipid-associated rheumatologic syndromes. Rheum Dis Clin N Am 39 (2):481–93
21. Handel ML, Simons L (2000) Rheumatic manifestations of hyperlipidaemia. Bailliere's Best Pract Res Clin Rheumatol 14(3):595–8
22. Parker F (1985) Xanthomas and hyperlipidemias. J Am Acad Dermatol 13(1):1–30
23. Beeharry D, Coupe B, Benbow EW et al (2006) Familial hypercholesterolaemia commonly presents with Achilles tenosynovitis. Ann Rheum Dis 65 (3):312–5
24. Abboud JA, Kim JS (2010) The effect of hypercholesterolemia on rotator cuff disease. Clin Orthop Relat Res 468(6):1493–7
25. Longo UG, Franceschi F, Spiezia F, Forriol F, Maffulli N, Denaro V (2010) Triglycerides and total serum cholesterol in rotator cuff tears: do they matter? Br J Sports Med 44(13):948–51
26. Abate M, Schiavone C, Di Carlo L, Salini V (2014) Prevalence of and risk factors for asymptomatic rotator cuff tears in postmenopausal women. Menopause 21(3):275–80
27. Ozgurtas T, Yildiz C, Serdar M, Atesalp S, Kutluay T (2003) Is high concentration of serum lipids a risk factor for Achilles tendon rupture? Clin Chim Acta 331(1–2):25–8
28. Abate M, Di Carlo L, Salini V, Schiavone C (2014) Metabolic syndrome associated to non-inflammatory Achilles enthesopathy. Clin Rheumatol 33(10):1517–22
29. Ronnemaa T, Juva K, Kulonen E (1975) Effect of hyperlipidemic rat serum on the synthesis of collagen by chick embryo fibroblasts. Atherosclerosis 21 (3):315–24
30. Nakano A, Kinoshita M, Okuda R, Yasuda T, Abe M, Shiomi M (2006) Pathogenesis of tendinous xanthoma: histopathological study of the extremities of Watanabe heritable hyperlipidemic rabbits. J Orthop Sci 11(1):75–80
31. Artieda M, Cenarro A, Junquera C et al (2005) Tendon xanthomas in familial hypercholesterolemia are associated with a differential inflammatory response of macrophages to oxidized LDL. FEBS Lett 579 (20):4503–12
32. Martin-Fuentes P, Civeira F, Solanas-Barca M, Garcia-Otin AL, Jarauta E, Cenarro A (2009) Overexpression of the CXCL3 gene in response to oxidized low-density lipoprotein is associated with the presence of tendon xanthomas in familial hypercholesterolemia. Biochem Cell Biol 87(3):493–8
33. Oosterveer DM, Versmissen J, Yazdanpanah M, Hamza TH, Sijbrands EJ (2009) Differences in characteristics and risk of cardiovascular disease in familial hypercholesterolemia patients with and without tendon xanthomas: a systematic review and meta-analysis. Atherosclerosis 207(2):311–7
34. Ohashi R, Mu H, Wang X, Yao Q, Chen C (2005) Reverse cholesterol transport and cholesterol efflux in atherosclerosis. QJM 98(12):845–56
35. Lai CQ, Parnell LD, Ordovas JM (2005) The APOA1/C3/A4/A5 gene cluster, lipid metabolism and cardiovascular disease risk. Curr Opin Lipidol 16(2):153–66
36. Kiss RS, Kavaslar N, Okuhira K et al (2007) Genetic etiology of isolated low HDL syndrome: incidence and heterogeneity of efflux defects. Arterioscler Thromb Vasc Biol 27(5):1139–45
37. Garasto S, Rose G, Derango F et al (2003) The study of APOA1, APOC3 and APOA4 variability in healthy ageing people reveals another paradox in the oldest old subjects. Ann Hum Genet 67(Pt 1):54–62
38. Frikke-Schmidt R, Nordestgaard BG, Jensen GB, Steffensen R, Tybjaerg-Hansen A (2008) Genetic variation in ABCA1 predicts ischemic heart disease in the general population. Arterioscler Thromb Vasc Biol 28(1):180–6
39. Beason D, Hsu J, Edelstein L, Lee C, Tucker J, Abboud J (2011) Effect of diet-induced hypercholesterolemia on rotator cuff tendon mechanics in a rat model. Trans Orthop Res Soc 36:223
40. Beason D, Kuntz A, Hamamdzic D, Wilensky R, Mohler E, Abboud J (2009) High cholesterol adversely affects biceps tendon mechanical properties in a porcine model. Trans Orthop Res Soc 34:184
41. Beason DP, Hsu JE, Marshall SM et al (2013) Hypercholesterolemia increases supraspinatus tendon stiffness and elastic modulus across multiple species. J Shoulder Elbow Surg 22(5):681–6
42. Beason DP, Abboud JA, Bassora R, Soslowsky LJ (2008) Tendon healing in a mouse injury model: the role of hypercholesterolemia. Poster No 814. Orthopaedic Research Society Annual Meeting, San Francisco, CA
43. Beason D, Abboud J, Bassora A, Kuntz A, Soslowsky L (2009) Hypercholesterolemia is detrimental to tendon properties and healing in a mouse injury model. Trans Orthop Res Soc 34:1418
44. Beason DP, Tucker JJ, Lee CS, Edelstein L, Abboud JA, Soslowsky LJ (2014) Rat rotator cuff tendon-to-bone healing properties are adversely affected by hypercholesterolemia. J Shoulder Elbow Surg 23(6):867–72

45. Lind S, Olsson AG, Eriksson M, Rudling M, Eggertsen G, Angelin B (2004) Autosomal recessive hypercholesterolaemia: normalization of plasma LDL cholesterol by ezetimibe in combination with statin treatment. J Intern Med 256(5):406–12
46. Illingworth DR, Cope R, Bacon SP (1990) Regression of tendon xanthomas in patients with familial hypercholesterolemia treated with lovastatin. South Med J 83(9):1053–7
47. Kane JP, Malloy MJ, Tun P et al (1981) Normalization of low-density-lipoprotein levels in heterozygous familial hypercholesterolemia with a combined drug regimen. N Engl J Med 304(5):251–8
48. Tsouli SG, Xydis V, Argyropoulou MI, Tselepis AD, Elisaf M, Kiortsis DN (2009) Regression of Achilles tendon thickness after statin treatment in patients with familial hypercholesterolemia: an ultrasonographic study. Atherosclerosis 205(1):151–5
49. Marie I, Delafenetre H, Massy N, Thuillez C, Noblet C (2008) Network of the French Pharmacovigilance C. Tendinous disorders attributed to statins: a study on ninety-six spontaneous reports in the period 1990–2005 and review of the literature. Arthritis Rheum 59(3):367–72
50. Kirchgesner T, Larbi A, Omoumi P et al (2014) Drug-induced tendinopathy: from physiology to clinical applications. Joint Bone Spine 81(6):485–92
51. Savvidou C, Moreno R (2012) Spontaneous distal biceps tendon ruptures: are they related to statin administration? Hand Surg 17(2):167–71
52. Beri A, Dwamena FC, Dwamena BA (2009) Association between statin therapy and tendon rupture: a case–control study. J Cardiovasc Pharmacol 53 (5):401–4
53. Lakey WC, Greyshock N, Guyton JR (2013) Adverse reactions of Achilles tendon xanthomas in three hypercholesterolemic patients after treatment intensification with niacin and bile acid sequestrants. J Clin Lipidol 7(2):178–81
54. Scheel AK, Schettler V, Koziolek M et al (2004) Impact of chronic LDL-apheresis treatment on Achilles tendon affection in patients with severe familial hypercholesterolemia: a clinical and ultrasonographic 3-year follow-up study. Atherosclerosis 174(1):133–9
55. Kuzma-Kuzniarska M, Cornell HR, Moneke MC, Carr AJ, Hulley PA (2015) Lovastatin-mediated changes in human tendon cells. J Cell Physiol 230 (10):2543–51
56. Esenkaya I, Sakarya B, Unay K, Elmali N, Aydin NE (2010) The influence of atorvastatin on tendon healing: an experimental study on rabbits. Orthopedics 33(6):398
57. de Oliveira LP, Vieira CP, Da Re GF, de Almeida MS, Pimentel ER (2013) Statins induce biochemical changes in the Achilles tendon after chronic treatment. Toxicology 311(3):162–8
58. Dolkart O, Liron T, Chechik O et al (2014) Statins enhance rotator cuff healing by stimulating the COX2/PGE2/EP4 pathway: an in vivo and in vitro study. Am J Sports Med 42(12):2869–76
59. Control CfD (2012) Prevention. Increasing prevalence of diagnosed diabetes–United States and Puerto Rico, 1995–2010. MMWR Morb Mortal Wkly Rep 61 (45):918

How Obesity Affects Tendons?

15

Michele Abate, Vincenzo Salini, and Isabel Andia

Abstract

Several epidemiological and clinical observations have definitely demonstrated that obesity has harmful effects on tendons. The pathogenesis of tendon damage is multi-factorial. In addition to overload, attributable to the increased body weight, which significantly affects load-bearing tendons, systemic factors play a relevant role. Several bioactive peptides (chemerin, leptin, adiponectin and others) are released by adipocytes, and influence tendon structure by means of negative activities on mesenchymal cells. The ensuing systemic state of chronic, sub-clinic, low-grade inflammation can damage tendon structure. Metabolic disorders (diabetes, impaired glucose tolerance, and dislipidemia), frequently associated with visceral adiposity, are concurrent pathogenetic factors. Indeed, high glucose levels increase the formation of Advanced Glycation End-products, which in turn form stable covalent cross-links within collagen fibers, modifying their structure and functionality.

Sport activities, so useful for preventing important cardiovascular complications, may be detrimental for tendons if they are submitted to intense acute or chronic overload. Therefore, two caution rules are mandatory: first, to engage in personalized soft training program, and secondly to follow regular check-up for tendon pathology.

Keywords

Obesity • Metabolic disorders • Tendon • Therapeutic exercise

M. Abate (✉) • V. Salini
Department of Medicine and Science of Aging, University G. d'Annunzio, Via dei Vestini 31, Chieti-Pescara, 66013 Chieti Scalo (CH), Italy
e-mail: m.abate@unich.it

I. Andia
Regenerative Medicine Laboratory, BioCruces Health Research Institute, Cruces University Hospital, 48903 Barakaldo, Spain

Introduction

Tendon disorders are mostly seen among people practicing sports, but can be observed also in the sedentary population. Genetic vulnerability, overuse, aging and metabolic disorders have

P.W. Ackermann and D.A. Hart (eds.), *Metabolic Influences on Risk for Tendon Disorders*, Advances in Experimental Medicine and Biology 920, DOI 10.1007/978-3-319-33943-6_15

been identified as risk factors that predispose to tendinopathies [6]. Among them, obesity is important because it represents a common and potentially modifiable condition [20]. Indeed, obesity is one of the major public health problems in western countries and its prevalence has increased dramatically in recent decades. When considered cumulatively, overweight (BMI >25) and obesity (BMI >30) affect up to 60 % of the population [19, 20, 50].

In this chapter, we summarize recent findings useful for understanding how obesity can modify tendons. First, we report observations coming from epidemiology, clinics, histopathology, biochemistry, and biomechanical studies, for both load-bearing and non-load-bearing tendons. Then, we discuss the pathogenesis of tendon damage in obese patients, combining data drawn from experimental and clinical investigations. Finally, we provide suggestions useful to overweight and obese subjects who practice sport activities.

Epidemiology

Several studies suggest that individuals with clinically overt or asymptomatic tendinopathies, or failure to respond to conservative and surgical treatments, have significantly higher adiposity levels than controls without tendon damage [23, 25, 40, 46, 60].

Load-Bearing-Tendons

Plantar fasciitis is very common in overweight and obese subjects, due to the increased stress on foot structures [21, 31]. Riddle et al. [45], in a study enrolling 50 subjects with plantar fasciitis and 100 controls, observed that participants with a BMI >30 were 5,6 times (CI: 1,9-16,6) more likely to be affected when compared with subjects with BMI <25. Also in asymptomatic people a significant relationship between plantar fascia thickness (which is considered the expression of tendon degeneration) and BMI values ($r = 0{,}749$, $p < 0{,}001$) was observed at the ultrasound evaluation [3]. In the large cohort of patients, studied by Galli et al. [27] by Magnetic Resonance Imaging (MRI), the prevalence of tendon damage in the hindfoot-ankle district was approximately twice among overweight compared with non-overweight patients ($p < 0{,}001$). Similarly, patients with Achilles pathology exhibited significantly higher BMI than controls, even after accounting for age. In particular, overweight and obese were 2,6 to 6,6 times respectively more likely than patients with normal BMI to be affected by Achilles tendinopathy ($p < 0{,}001$) [47]. Further studies have shown that in men waist circumference is a better predictor of tendon pathology than BMI [22, 30, 33]. Indeed, asymptomatic Achilles tendon pathology was found associated with central fat distribution in men (i.e. an increased waist/hip ratio, WHR), whereas it was associated with peripheral fat distribution in women. Considering surgery, overweight patients with recalcitrant Achilles tendinopathy experience more prolonged recovery times, more post-intervention complications, and a greater risk of further procedure than normal weight subjects [35] (Fig. 15.1).

Excess of body weight is also a significant risk factor for both acute and chronic patellar tendon pathologies [26, 36, 54]. Although no statistically valid information on the possible association between obesity and quadriceps/patellar tendon rupture is available, some authors report cases of spontaneous rupture of patellar/quadriceps tendon in morbidly obese patients [32, 42]. The association is stronger when chronic patellar tendon damage is considered. Indeed, individuals with WHR > 83 cm had a 74 % chance of patellar tendon pathology, compared with 15 % of those with < 83 cm waistline. These observations are supported by a recent study where the relationship between obesity and prevalence of MRI-defined patellar tendinopathy was investigated. In a group of asymptomatic patients, (age 50–79) obesity measures were collected (weight, BMI, self-reported weight at the age of 18–21 years, heaviest lifetime weight, fat-free mass and fat mass), and a MRI of the dominant knee was performed. The results showed that MRI features of patellar tendinopathy were common in the general population (28 %), and that current

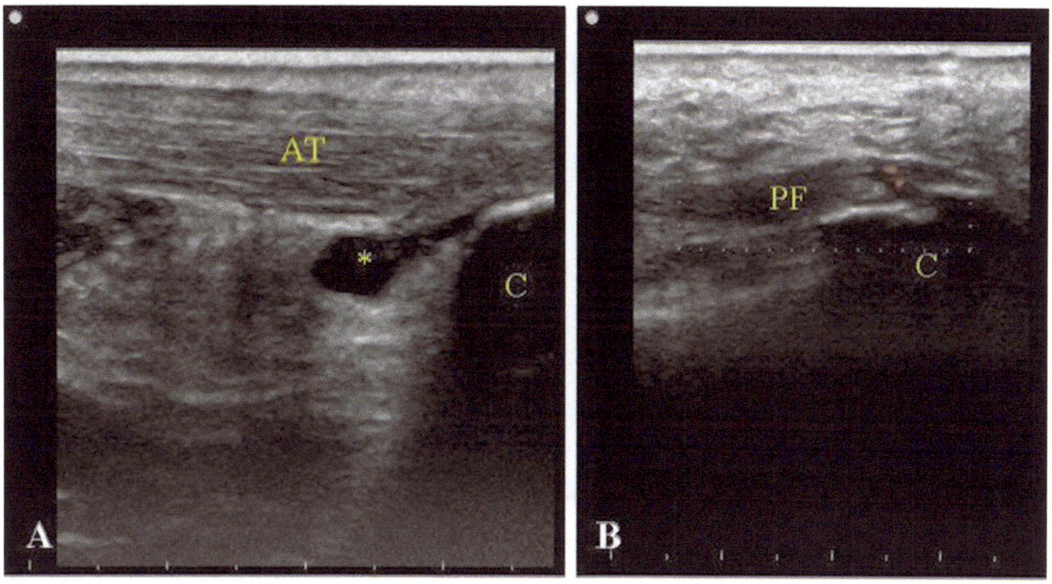

Fig. 15.1 Ultrasound appearance of ankle degenerative tendon disorders in obese subjects.
(**a**) Thickening of the Achilles Tendon (*AT*), hypoechoic echotexture, especially in the ventral portion, are expression of degeneration. An effusion into the retrocalcaneal bursa (*) is also present (longitudinal scan). *C* = Calcaneal bone; (**b**) In this longitudinal scan Plantar fascia (PF) shows degenerative features and intra- tendinous vascularization. The calcaneal bone (**c**) is irregular

weight (p = 0,002), BMI (p = 0,002), heaviest lifetime weight (p = 0,007) and weight at age of 18–21 years (p = 0,05) were all positively associated with the prevalence of the tendinopathy [18] (Fig. 15.2).

Non-Load Bearing Tendons

Rotator cuff disease is a multifactorial condition. Some studies suggest a link between shoulder pain and metabolic disorders [20]. The association with diabetes is clear, whereas for obesity we do not dispose of the same amount of evidence. There are few studies in literature and none of them is a level I study. A cross-sectional investigation, performed on a large cohort of patients, showed that having a BMI >25 and abdominal obesity was a significant risk factor for rotator cuff disease (p <0,01) [44, 52]. Moreover, obesity influenced the tear size (determined intra-operatively), as reported in a case-control study performed on 381 consecutive patients who underwent arthroscopic rotator cuff repair [29]. In these patients, the occurrence and severity of rotator cuff tear were positively associated with a higher BMI (p = 0,001) and percentage of body fat (%BF). Interestingly, BMI and %BF significantly increased from patients with a small to those with a massive tear (BMI: 27,8 *versus* 29,9; %BF: 37,6 *versus* 39,4, p = 0,031).

The impact of obesity on the outcomes after surgery is debated. Warrander et al. [56] report that obesity lengthen the operative times of arthroscopic rotator cuff repair and of hospital stay, and is associated to worse functional outcomes, mainly in subjects with a BMI >35 [59]. In contrast, in Namdari's study [39], BMI was not significantly related to outcomes, after controlling for confounding variables.

Finally, some evidence exists about the association of obesity with elbow disorders. Descatha et al. [16] and Shiri et al. [48] demonstrated an increased prevalence of epicondylitis in patients with a BMI >30 or a waist circumference > 100 cm. According to Titchener et al. [51] only morbid obesity (BMI >40) was associated with this condition (Fig. 15.3).

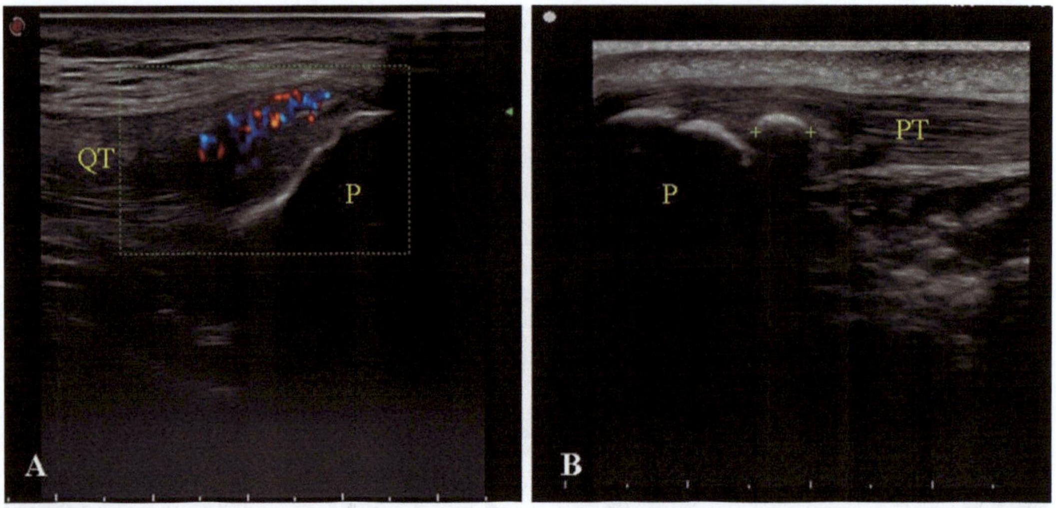

Fig. 15.2 Ultrasound appearance of knee degenerative tendon disorders in obese subjects.
In panel (**a**) the quadriceps tendon (*QT*) is thickened, hypoechoic and dishomogeneous; intra-tendinous vascularization is detected at color-doppler examination (longitudinal scan); In panel (**b**) degenerative features of the patellar tendon (*PT*) are present; a calcification (calipers) into the proximal portion can be also seen (longitudinal scan). P = Patella

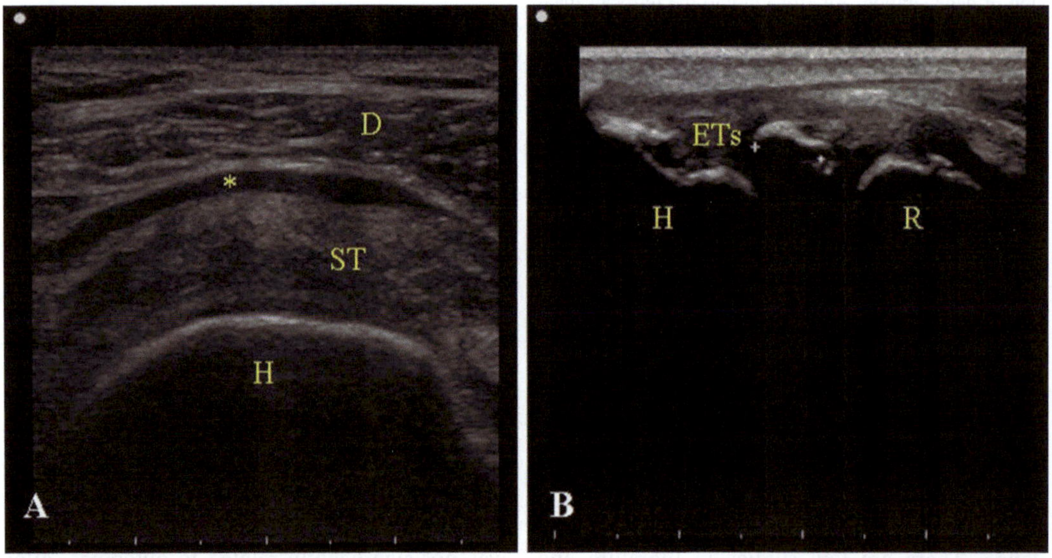

Fig. 15.3 Ultrasound appearance of shoulder and elbow degenerative tendon disorders in obese subjects.
(**a**) In this trasverse scan the supraspinatus tendon (*ST*) appears thickened and dishomogeneous. An effusion into the subacromial bursa can be also observed (*). H = Humeral head; D = Deltoid muscle; (**b**) This longitudinal scan shows degenerative features of extensor tendons (ETs) and the presence of intra-tendinous calcification (calipers). Humeral (H) and radial (R) bones are also irregular

In conclusion, the majority of studies suggest that overweight and obesity are significantly associated to symptomatic tendinopathies or to abnormal tendon structural features, which in turn are important predictors of a future clinically evident pathology. The association, in general, is more evident for lower-limb than for upper-limb tendons. However, most of published studies are observational, and therefore do not allow a precise identification of a cause-effect relationship between adiposity and each type of tendinopathy. Therefore, further research is necessary to better clarify the role of adiposity "*in se*", excluding the pathogenetic relevance of associated factors, such as aging, the level of overuse and metabolic disorders.

Histopathology, Biochemical and Biomechanical Features

Histopathology

Structural modifications of tendons in obesity have been observed in animal studies. Biancalana et al. [10], in a research performed on the deep digital flexor tendon (DFT) of lean and genetically obese Zucker rats, put in evidence different morphological features. In normal weight rats, free to move around the cages, the collagen fibrils showed a bimodal pattern characterized by both large and small fibrils. In obese rats, the collagen fibrils showed an unimodal distribution, due to the relative prevalence of large on small fibers, the expression of an impaired remodeling process. Because thin fibers confer greater elasticity to tendons, the authors supposed that their relative paucity in obese animals could be responsible for increased stiffness and micro-ruptures as a consequence of excessive loads. Moreover, disorganized and tangled collagen fibrils were observed in the tension region of tendons from obese animals [9].

Boivin et al. [11, 12] studied the structure of Achilles tendon in mice submitted to a high fat diet. No significant morphological abnormalities were found, and only in a small percentage of animals a mild degeneration and neutrophil infiltration was observed near the Achilles tendon insertion on the heel. According to Biancalana's observations [9], the ultrastructural analysis by transmission electron microscopy revealed disorganized and tangled collagen fibrils in the tension region of the tendon.

In an experimental model of acute DFT damage, performed in mice submitted to a high fat diet, the histological examination showed at 14–28 days consistently small cellular and fibrous tissue at the injury site, with a lesser degree of collagen remodeling and fiber alignment [14]. Besides, tenocytes adopted an abnormal, proliferative behaviour, reducing type I collagen production and increasing metalloproteinases expression. So, tendon healing was significantly impaired.

Some important limits of these experimental studies must be acknowledged. First, the experimental model of obese Zucker rats cannot be compared to that of animals submitted to a high fat diet. Indeed, Zucker rats, used in Biancalana's experiments [9], represent a model of genetic obesity, related to an autosomal recessive gene encoding defective leptin receptor, and suffer from both hyperinsulinemia and hyperlipidemia. Therefore the changes observed in the distribution pattern of collagen fibrils could be genetic in origin, and not due to obesity "*in se*".

Second, the animal experiments explore only short-term structural changes, whereas in humans the damage due to adiposity is a chronic condition. In this regard, we lack morphologic studies in obese, but we can get indirect information from MRI or ultrasound tendon evaluation. Actually, the reported abnormalities (increased thickness, disorganized echotexture, hypoechoic areas), observed in subjects with increased BMI, are expression of morphologic alterations, i.e. increased deposition and packing of collagen fibrils, which appear twisted, overlapping and disorganized, and/or signs of hyaline, mucoid or lipoid degeneration [3, 17].

Biochemical and Biomechanical Alterations

The abnormal deposition and disruption of collagen fibrils previously reported are in agreement with biochemical findings, which put in evidence a reduced concentration of glycosaminoglycans (chondroitin and dermatan sulfate) in the tendons' extracellular matrix (ECM) from obese animals. Moreover, isolated lipid droplets have been detected in ECM, expression of the initial phase of tendolipomatosis [9]. On the contrary, the level of hydroxyproline, which provides higher tensile strength associated with lower strain, is increased, probably as a consequence of the higher mechanical requirements [10].

Morphologic and biochemical abnormalities are associated to biomechanical dysfunctions [57, 58]. In experimental conditions, the excess of fat intake, causing enlargement of fibril diameter and shortening of the modulus of the tendon, leads to a stiffer tendon, less able to withstand the loads [12]. After tendon damage, by puncturing with a beveled needle in obese mice, the tendons biomechanical properties show a significant reduction of the normalized maximum force, normalized work to maximum force and normalized stiffness, in comparison with normal weight animals submitted to the same experimental procedure [14].

Similar results have been observed in humans. The increased Achilles tendon thickness, and therefore the larger cross-sectional tendon area, despite the decreased material property (modulus), partially offsets the greater body mass [57]. Therefore, the average stress (force per unit area) experienced by the Achilles tendon is similar in normal and overweight animals. The increased fibril number and diameter, besides the increased collagen cross-links, are both involved in increased tendon stiffness. Interestingly, after strenuous exercise, the tendon thickness decreases, due to the loss of interstitial water, associated with load-induced alignment of collagen fibers [57]. However, the transverse strain in the tendon, calculated as the natural log of the ratio of post to pre-exercise tendon thickness, in obese subjects is inferior to that observed in the normal weight counterpart, suggesting that obesity is associated with structural tendon changes that impair interstitial fluid movement in response to tensile load, and are responsible of a greater transverse stiffness.

Pathogenetic Mechanisms

At present, two different mechanisms have been identified: increased stress on load-bearing tendons, imposed by the bigger body weight; and biochemical disorders associated to systemic dysmetabolic factors (Fig. 15.4).

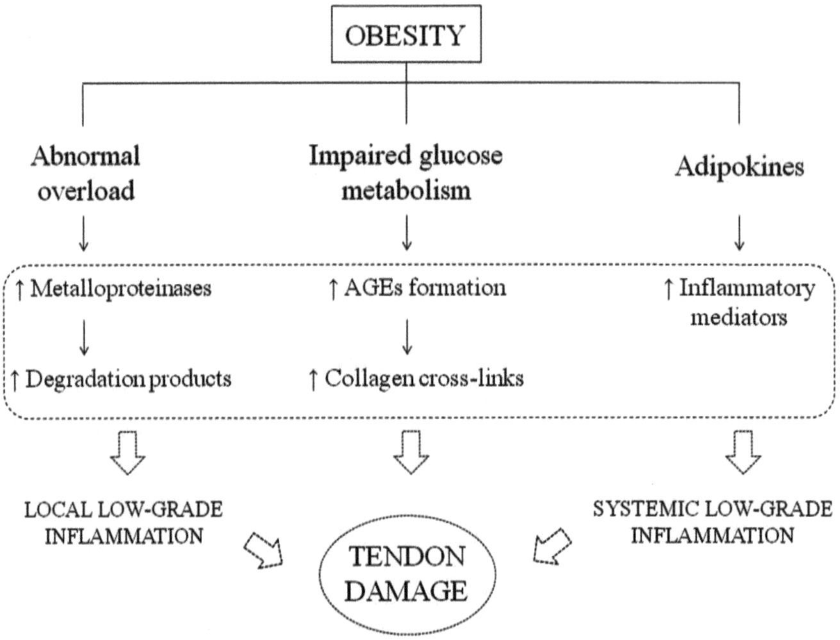

Fig. 15.4 Pathogenetic mechanisms of tendon damage

Overload

It is well known that overweight and obesity lead to alterations of the musculoskeletal system. Obese individuals have higher plantar pressure, especially under the longitudinal arch and on the metatarsal heads both when standing and walking. Moreover, the distribution of forces at the knee during weight bearing is modified and this leads to varus malalignment, and to the development of knee osteoarthritis. The gait pattern is modified, and balance is impaired [40].

The increased body weight has also significant impact on tendons. It is well known that mechanical loading is essential to maintain tendon homeostasis, but, when the loading magnitude is abnormal, as can happen in obese subjects, the tendon response reverses from beneficial towards degenerative. Briefly, there is an exceeding production of proteoglycans and inflammatory molecules, with increased metalloproteinase expression, which leads to the formation of degradation products and water retention. This failed healing process favors a smoldering fibrogenesis, with matrix turnover without normal maturation [2, 8]. This information comes mainly from studies performed on Achilles tendon in runners. Indeed, during running, the tendon is highly solicited and the load can be as high as eight times body weight, so that modest increases in weight are amplified within the tendon [4, 40]. Broadly speaking, these concepts can be applied to all tendons, which are selectively stressed in various athletic activities: the shoulder in swimming and basket, the elbow in tennis and golf, and the knee and ankle in all sports characterized by running and jumping.

Systemic Hypothesis

Adipose tissue can be considered as a major endocrine and signaling organ [2, 43]. Several bioactive peptides and hormones are released by adipocytes, among them a full range of proteins (chemerin, lipocalin 2, serum amyloid A3, leptin and adiponectin), which can influence tendon structures by means of various activities on mesenchymal cells. In particular, inflammatory mediators (cytokines, prostanoids, and metalloproteinases) can be modulated [24], and the subsequent systemic state of chronic, sub-clinic, low-grade inflammation (increased serum levels of PGE2, TNF-α, and LTB4) can damage tendon structure. The cumulative harmful effect of these substances may explain tendon damage both in load-bearing and non load-bearing tendons of obese subjects.

Tendon damage can be amplified by a cluster of metabolic disorders frequently associated with obesity [1]. As shown by epidemiologic studies, obese subjects with visceral fat deposition are more predisposed to tendinopathies. Interestingly, this relationship has been also observed in individuals with an increased fat mass and higher WHR, despite having a normal BMI. These subjects, classified as metabolically obese but normal weight, account from 5 to 45 % of the general population [13]. Therefore, it may be hypothesized that tendon pathology is strongly linked to metabolic syndrome, characterized by insulin resistance, diabetes, or impaired glucose tolerance, hypercholesterolemia, hypertriglyceridemia, and low levels of HDL cholesterol. In the presence of increased glucose levels, the formation of Advanced Glycation End-products (AGEs) is markedly accelerated (see Chap. 14, 16, 17, 18, and 19). [5, 15, 28]. A key characteristic of reactive AGEs is the formation of stable covalent cross-links within collagen fibers, which alter their structure and functionality. Moreover, AGEs react with a variety of AGE-binding receptors on the cell surface, which in turn activates several critical molecular pathways and triggers a number of effects. These include pro-oxidant events via generation of reactive oxygen species and led to a sustained up-regulation of pro-inflammatory mediators [38].

In addition, the systemic influence of excessive blood lipid must be considered. Indeed, dyslipidemia is a well-known cause of tendon pathology [24, 41, 53, 55]. At this purpose, several authors observed that subjects with symptomatic Achilles tendinopathy had higher triglyceride levels, lower HDL cholesterol, a higher triglyceride/HDL-cholesterol ratio and elevated apolipoprotein B concentration, compared with controls matched for gender, age and also BMI. Moreover, dyslipidemia is a negative prognostic factor after surgery (Achilles and rotator cuff tendons) [7, 34, 37].

The Role of Sport Activities

Development of obesity and physical inactivity are closely associated. Obesity rates are higher in sedentary and moderately active persons than in active subjects, as shown by cross-sectional and longitudinal studies [23]. Therefore, sport activities are strongly recommended to slow down, and even stop, the progression of weight gain in obese individuals, or to bring an obese individual into the normal weight range. Physical activity reduces insulin resistance, improves glucose and lipid metabolism, minimize the higher risk of hypercholesterolemia and diabetes, and delay the onset of cardiovascular diseases, as well of hip and knee osteoarthritis [40]. Moreover, in aged people, it counteracts the muscle mass and strength decline, reduces the risk of falls, and enhances the cognitive functions and the quality of life [49]. Therefore, besides an adequate dietary regimen and pharmacological treatments, when necessary, physical exercise is an important therapeutic measure that should be prescribed to all subjects who are obese or overweight.

In this conceptual framework, a number of subjects of all ages practice sport activities. These activities are often performed in 2 or 3 weekly sessions, and not infrequently at high intensity levels to maximize the energy expenditure and to get a significant weight loss. Physicians usually alert their patients about the possible risk of cardiovascular overload, but less attention is paid to the risk of tendon damage. Actually, this is an important issue, because what is beneficial for improving metabolism and preventing important systemic diseases may be detrimental for tendons, which are submitted to acute or chronic overload. Indeed, the risk of developing a symptomatic tendinopathy, and even of undergoing a tendon rupture, is substantially increased in obese subjects, whose tendons are frequently weakened by a sub-clinical damage and by a metabolically disturbed milieu [4]. Therefore, minimizing the impact of tendon pathologies when prescribing exercise and sport activities as medicine for weight reduction in sedentary individuals becomes mandatory. In real life, several persons practice running, for several reasons: it is inexpensive, may be practiced by oneself every time he/she wants, and may be performed by people lacking of any skill in specific sports activities. However, running is deleterious for load-bearing tendons, and Achilles tendon in particular, which is highly solicited. Otherwise, tennis, soccer, basketball, and other contact sports can expose tendons to sudden intense, even short-lasting, stress, favoring microruptures of collagen fibers.

Given these considerations, and after having emphasized again that the practice of sport activities is beneficial, some caution is necessary. First, soft programs of physical activity should be individually prescribed to get positive metabolic outcomes, avoiding the risk of excessive and dangerous overload. Second, the frequency and intensity of sport performance should be increased gradually, in accordance with the progression of weight loss, avoiding agonistic activity and contrast sports, which are more likely to expose to acute injury. Third, some sports, such as swimming and cycling, which have a minor impact on tendons, should be preferred. However, it is evident that the sport choice is strictly linked to the pleasure and gratification deriving from the athletic gesture, and therefore highly subjective. In such perspective, every athletic activity may be accepted, providing the strict observance of caution rules. Finally, frequent controls should be performed in subjects at higher risk, mainly for Achilles tendon pathology, using the ultrasound technique, which is a valid, simple, quick and low cost procedure.

Conclusions

Epidemiological observations and case-control studies show an evident association between adiposity and tendinopathies, both symptomatic and asymptomatic. However, these studies do not allow identifying the pathogenetic role of obesity "*in se*", because the coexistence of several confounding factors: among them, aging, genetic components, overweight duration, and the amount of overload to which tendons have been individually submitted. More importantly, it is very difficult to disentangle the role of metabolic disorders obesity-related, such as impaired glucose tolerance and hyperlipidemia. Actually, more than by the increased BMI, a major role seems to be played by visceral fat deposition, which is associated with the metabolic syndrome and a chronic state of low-grade inflammation. Experimental and clinical research has highlighted several pathogenic mechanisms, which can be responsible of tendon damage. In particular, it has been shown that the biochemical milieu, around tenocytes and other tendon cells and structures, largely influences the tendon response to mechanical loading, reversing it from beneficial to degenerative.

In this framework, some practical issues must be considered. Overweight and obesity, and sedentary lifestyles, are very common in western countries, and account largely for cardiovascular and osteo-articular morbidity. Therefore, the practice of physical activity and sports is mandatory. Indeed, improving glucose and lipid metabolism delay the onset of cardiovascular diseases, as well of hip and knee osteoarthritis. On the other hand, exercise, which is so useful for preventing these important diseases, may be detrimental for tendons, when submitted to intense acute or chronic overload. Among different sports, running and other strenuous activities (tennis, soccer, basket), which expose to sudden and intense overloads, can be particularly dangerous. Swimming and cycling, which have a minor impact on tendons, should be preferred. In general, some caution rules are mandatory: first, to follow a soft individual training program, second to increase gradually the frequency and intensity of sport performance, third to be strictly monitored, mainly for Achilles tendon pathology.

References

1. Abate M, Salini V, Schiavone C (2015) Achilles tendinopathy in elderly subjects with type II diabetes: the role of sport activities. Aging Clin Exp Res 10
2. Abate M, Schiavone C, Salini V, Andia I (2013) Occurrence of tendon pathologies in metabolic disorders. Rheumatology (Oxford) 52(4):599–608
3. Abate M, Schiavone C, Di Carlo L, Salini V (2012) Achilles tendon and plantar fascia in recently diagnosed type II diabetes: role of body mass index. Clin Rheumatol 31(7):1109–1113
4. Abate M, Oliva F, Schiavone C, Salini V (2012) Achilles tendinopathy in amateur runners: role of adiposity (Tendinopathies and obesity). Muscles Ligaments Tendons J 2(1):44–48
5. Abate M, Schiavone C, Pelotti P, Salini V (2011) Limited joint mobility (LJM) in elderly subjects with type II diabetes mellitus. Arch Gerontol Geriatr 53(2):135–140
6. Abate M, Silbernagel KG, Siljeholm C et al (2009) Pathogenesis of tendinopathies: inflammation or degeneration? Arthritis Res Ther 11(3):235
7. Abboud JA, Kim JS (2010) The effect of hypercholesterolemia on rotator cuff disease. Clin Orthop Relat Res 468(6):1493–1497
8. Battery L, Maffulli N (2011) Inflammation in overuse tendon injuries. Sports Med Arthrosc 19(3):213–217
9. Biancalana A, Velloso LA, Taboga SR, Gomes L (2012) Implications of obesity for tendon structure, ultrastructure and biochemistry: a study on Zucker rats. Micron 43(2–3):463–469
10. Biancalana A, Veloso LA, Gomes L (2010) Obesity affects collagen fibril diameter and mechanical properties of tendons in Zucker rats. Connect Tissue Res 51(3):171–178
11. Boivin GP, Elenes EY, Schultze AK, Chodavarapu H, Hunter SA, Elased KM (2014) Biomechanical properties and histology of db/db diabetic mouse Achilles tendon. Muscles Ligaments Tendons J 4(3):280–284
12. Boivin GP, Platt KM, Corbett J et al (2013) The effects of high-fat diet, branched-chainamino acids and exercise on female C57BL/6 mouse Achilles tendon biomechanical properties. Bone Joint Res 2 (9):186–192
13. Conus F, Rabasa-Lhoret R, Péronnet F (2007) Characteristics of metabolically obese normal-weight (MONW) subjects. Appl Physiol Nutr Metab 32 (1):4–12
14. David MA, Jones KH, Inzana JA, Zuscik MJ, Awad HA, Mooney RA (2014) Tendon repair is compromised in a high fat diet-induced mouse model of obesity and type 2 diabetes. PLoS One 9 (3), e91234
15. Del Buono A, Battery L, Denaro V, Maccauro G, Maffulli N (2011) Tendinopathy and inflammation: some truths. Int J Immunopathol Pharmacol 24:45–50

16. Descatha A, Dale AM, Jaegers L, Herquelot E, Evanoff B (2013) Self-reported physical exposure association with medial and lateral epicondylitis incidence in a large longitudinal study. Occup Environ Med 70(9):670–673
17. de Jonge S, Rozenberg R, Vieyra B et al (2015) Achilles tendons in people with type 2 diabetes show mildly compromised structure: an ultrasound tissue characterisation study. Br J Sports Med 49 (15):959–999. doi:10.1136/bjsports-2014-093696
18. Fairley J, Toppi J, Cicuttini FM et al (2014) Association between obesity and magnetic resonance imaging defined patellar tendinopathy in community-based adults: a cross-sectional study. BMC Musculoskelet Disord 15:266
19. Flegal KM, Carroll MD, Ogden CL, Johnson CL (2002) Prevalence and trends in obesity among US adults, 1999–2000. JAMA 288(14):1723–1727
20. Franceschi F, Papalia R, Paciotti M et al (2014) Obesity as a risk factor for tendinopathy: a systematic review. Int J Endocrinol 2014:670262
21. Frey C, Zamora J (2007) The effects of obesity on orthopaedic foot and ankle pathology. Foot Ankle Int 28(9):996–999
22. Gaida JE, Alfredson H, Kiss ZS, Bass SL, Cook JL (2010) Asymptomatic Achilles tendon pathology is associated with a central fat distribution in men and a peripheral fat distribution in women: a cross sectional study of 298 individuals. BMC Musculoskelet Disord 11:41
23. Gaida JE, Ashe MC, Bass SL, Cook JL (2009) Is adiposity an under-recognized risk factor for tendinopathy? A systematic review. Arthritis Rheum 61(6):840–849
24. Gaida JE, Alfredson L, Kiss ZS, Wilson AM, Alfredson H, Cook JL (2009) Dyslipidemia in Achilles tendinopathy is characteristic of insulin resistance. Med Sci Sports Exerc 41(6):1194–1197
25. Gaida JE, Cook JL, Bass SL (2008) Adiposity and tendinopathy. Disabil Rehabil 30(20–22):1555–1562
26. Gaida JE, Cook JL, Bass SL, Austen S, Kiss ZS (2004) Are unilateral and bilateral patellar tendinopathy distinguished by differences in anthropometry, body composition, or muscle strength in elite female basketball players? Br J Sports Med 38(5):581–585
27. Galli MM, Protzman NM, Mandelker EM, Malhotra A, Schwartz E, Brigido SA (2014) Comparing tendinous and ligamentous ankle pathology in atraumatic overweight and nonoverweight patients: a comprehensive MRI review. Foot Ankle Spec 7 (6):449–456
28. Gautieri A, Redaelli A, Buehler MJ, Vesentini S (2014) Age -and diabetes- related nonenzymatic crosslinks in collagen fibrils: candidate amino acids involved in Advanced Glycation End-products. Matrix Biol 34:89–95
29. Gumina S, Candela V, Passaretti D et al (2014) The association between body fat and rotator cuff tear: the influence on rotator cuff tear sizes. J Shoulder Elbow Surg 23(11):1669–1674
30. Holmes GB, Lin J (2006) Etiologic factors associated with symptomatic Achilles tendinopathy. Foot Ankle Int 27(11):952–959
31. Irving DB, Cook JL, Young MA, Menz HB (2007) Obesity and pronated foot type may increase the risk of chronic plantar heel pain: a matched case-control study. BMC Musculoskelet Disord 8:41.
32. Kelly BM, Rao N, Louis SS, Kostes BT, Smith RM (2001) Bilateral, simultaneous, spontaneous rupture of quadriceps tendons without trauma in an obese patient: a case report. Arch Phys Med Rehabil 82 (3):415–418
33. Klein EE, Weil L Jr, Weil LS Sr, Fleischer AE (2013) Body mass index and Achilles tendonitis: a 10-year retrospective analysis. Foot Ankle Spec 6(4):276–282
34. Longo UG, Franceschi F, Spiezia F, Forriol F, Maffulli N, Denaro V (2010) Triglycerides and total serum cholesterol in rotator cuff tears: do they matter? Br J Sports Med 44(13):948–951
35. Maffulli N, Testa V, Capasso G et al (2006) Surgery for chronic Achilles tendinopathy yields worse results in nonathletic patients. Clin J Sport Med 16 (2):123–128
36. Malliaras P, Cook JL, Kent PM (2007) Anthropometric risk factors for patellar tendon injury among volleyball players. Br J Sports Med 41(4):259–263
37. Mathiak G, Wening JV, Mathiak M, Neville LF, Jungbluth K (1999) Serum cholesterol is elevated in patients with Achilles tendon ruptures. Arch Orthop Trauma Surg 119(5–6):280–284
38. Matsuura F, Hirano K, Koseki M et al (2005) Familial massive tendon xanthomatosis with decreased high-density lipoprotein-mediated cholesterol efflux. Metabolism 54:1095–1101
39. Namdari S, Baldwin K, Glaser D, Green A (2010) Does obesity affect early outcome of rotator cuff repair? J Shoulder Elbow Surg 19(8):1250–1255
40. Nantel J, Mathieu ME, Prince F (2011) Physical activity and obesity: biomechanical and physiological key concepts. J Obes 2011:650230
41. Ozgurtas T, Yildiz C, Serdar M, Atesalp S, Kutluay T (2003) Is high concentration of serum lipids a risk factor for Achilles tendon rupture? Clin Chim Acta 331(1–2):25–28
42. Panasiuk M, Groblewski M (2011) Spontaneous patellar tendon rupture as a result of morbid obesity. Chir Narzadow Ruchu Ortop Pol 76(6):353–354
43. Rechardt M, Viikari-Juntura E, Shiri R (2014) Adipokines as predictors of recovery from upper extremity soft tissue disorders. Rheumatology (Oxford) 53(12):2238–2242
44. Rechardt M, Shiri R, Karppinen J, Jula A, Heliövaara M, Viikari-Juntura E (2010) Lifestyle and metabolic factors in relation to shoulder pain and rotator cuff tendinitis: a population-based study. BMC Musculoskelet Disord 11:165

45. Riddle DL, Pulisic M, Pidcoe P, Johnson RE (2003) Risk factors for Plantar fasciitis: a matched case-control study. J Bone Joint Surg Am 85-A(5):872–877
46. Scott A, Zwerver J, Grewal N et al (2015) Lipids, adiposity and tendinopathy: is there a mechanistic link? Critical review. Br J Sports Med 49(15):984–988
47. Scott RT, Hyer CF, Granata A (2013) The correlation of Achilles tendinopathy and body mass index. Foot Ankle Spec 6(4):283–285
48. Shiri R, Viikari-Juntura E, Varonen H, Heliövaara M (2006) Prevalence and determinants of lateral and medial epicondylitis: a population study. Am J Epidemiol 164(11):1065–1074
49. Svantesson U, Jones J, Wolbert K, Alricsson M (2015) Impact of physical activity on the self-perceived quality of life in non-frail older adults. J Clin Med Res 7(8):585–593
50. Thalmann S, Meier CA (2007) Local adipose tissue depots as cardiovascular risk factors. Cardiovasc Res 75(4):690–701
51. Titchener AG, White JJ, Hinchliffe SR, Tambe AA, Hubbard RB, Clark DI (2014) Comorbidities in rotator cuff disease: a case-control study. J Shoulder Elbow Surg 23(9):1282–1288
52. Titchener AG, Fakis A, Tambe AA, Smith C, Hubbard RB, Clark DI (2013) Risk factors in lateral epicondylitis (tennis elbow): a case-control study. J Hand Surg Eur Vol 38(2):159–164
53. Tsouli SG, Kiortsis DN, Argyropoulou MI, Mikhailidis DP, Elisaf MS (2005) Pathogenesis, detection and treatment of Achilles tendon xanthomas. Eur J Clin Invest 35(4):236–244
54. van der Worp H, van Ark M, Roerink S, Pepping GJ, van den Akker-Scheek I, Zwerver J (2011) Risk factors for patellar tendinopathy: a systematic review of the literature. Br J Sports Med 45(5):446–452
55. Viikari-Juntura E, Shiri R, Solovieva S et al (2008) Risk factors of atherosclerosis and shoulder pain–is there an association? A systematic review. Eur J Pain 12(4):412–426
56. Warrender WJ, Brown OL, Abboud JA (2011) Outcomes of arthroscopic rotator cuff repairs in obese patients. J Shoulder Elbow Surg 20(6):961–967
57. Wearing SC, Hooper SL, Grigg NL, Nolan G, Smeathers JE (2013) Overweight and obesity alters the cumulative transverse strain in the Achilles tendon immediately following exercise. J Body Mov Ther 17(3):316–321
58. Wearing SC, Hennig EM, Byrne NM, Steele JR, Hills AP (2006) Musculoskeletal disorders associated with obesity: a biomechanical perspective. Obes Rev 7(3):239–250
59. Wendelboe AM, Hegmann KT, Gren LH, Alder SC, White GL Jr, Lyon JL (2004) Associations between body-mass index and surgery for rotator cuff tendinitis. J Bone Joint Surg Am 86-A(4):743–747
60. Wise BL, Peloquin C, Choi H, Lane NE, Zhang Y (2012) Impact of age, sex, obesity, and steroid use on quinolone-associated tendon disorders. Am J Med 125 (12):1228.e23–1229.e28

Does Diabetes Mellitus Affect Tendon Healing?

16

Aisha Siddiqah Ahmed

Abstract

Diabetes mellitus (DM) is a metabolic disorder resulting from defective insulin production and characterized by chronic hyperglycemia. DM affects around 170 million people worldwide and its incidence is increasing globally. DM can cause a wide range of musculoskeletal disorders such as painful tendinopathies, tendon contracture, tendon rupture, and rotator cuff tear.

In patients with diabetes neuropathy, diminished peripheral blood flow and decreased local angiogenesis are reported which probably are results of abnormalities in the production of collagen production, inflammatory mediators, angiogenic and growth factors and also contribute to lack of healing in damaged tissue. Abnormal or delayed wound healing is one of the main complications of both type-I and type-II DM.

Keywords

Diabetes mellitus • Tendon • Wound healing • Cytokines • Growth factors

Tendon Structure

Tendon consist of cells mainly fibroblasts (tenocytes) and tissue specialized extracellular matrix (ECM). Throughout the whole lifespan tenocytes actively synthesize ECM components in the tendon as well as produce enzymes such as matrix metalloproteinases (MMPs) that can degrade the matrix [1]. Tendons have the capacity to heal and recover from injuries in a process controlled by the tenocytes and their surrounding ECM. However, healed tendons never regain the same mechanical properties as before the injury. ECM mostly consists of collagens, proteoglycans, polysaccharides, glycoproteins and elastin fibers. Specific molecules that constitute ECM vary depending on the cell types embedded within that matrix and the mechanical demands of the tissue (see Chap. 2).

A.S. Ahmed (✉)
Department of Clinical Neuroscience, Karolinska Institutet, 17177 Stockholm, Sweden
e-mail: Aisha.Ahmed@ki.se

P.W. Ackermann and D.A. Hart (eds.), *Metabolic Influences on Risk for Tendon Disorders*, Advances in Experimental Medicine and Biology 920, DOI 10.1007/978-3-319-33943-6_16

Healing Mechanisms in Tendon

Wound healing is highly complex process with interaction of many cell types, inflammatory and neuronal mediators, growth factors and enzymes which is also associated with tendon healing. Generally, there are three main stages of tendon healing; inflammation, repair or proliferation, and remodeling which can further be divided into consolidation and maturation phases. These three stages can also overlap with each other. In the first inflammatory stage, cells such as neutrophils and macrophages are recruited to the site within the first 24 h of injury and phagocytosis of the necrotic material occurs. After the release of vasoactive, angiogenic and chemotactic factors, proliferation of tenocytes occurs and tenocytes start to synthesize collagen III [2, 3]. The inflammatory stage usually lasts for few days, followed by the repair or proliferation stage. During proliferation which usually last for 4–6 weeks, the tenocytes synthesize large amounts of collagen and proteoglycans at the site of injury and their interconnection is increased [4].

After that the remodeling stage begins, consisting of consolidation which lasts from about 6–10 weeks after the injury. The tissue becomes more fibrous with the result of increased production of collagen I and the fibrils become aligned in the direction of mechanical stress [3]. The final maturation stage starts after around 10 weeks, and there is an increase in crosslinking of the collagen fibrils, which causes the tissue to become stiffer. Gradually, over a time period of about 1 year, the tissue will turn from fibrous to scar-like [5] as briefly illustrated in Fig. 16.1.

Effect of DM on Tendon Healing

Diabetes is often associated with neuropathy and decreased angiogenesis resulting in impaired connective tissue healing and reduced biomechanical properties. Clinical and experimental studies strongly indicate that reduced collagen synthesis, abnormal cytokine production, compromised angiogenic and growth factor production occurs during all three inflammatory, proliferative and in remodeling stages and interfere healing process of damaged tissues in diabetes [6, 7]. These studies indicate less fibroblast proliferation and lymphocyte infiltration in healing tendons associated with tendon weakness. Diabetes adversely affects properties of native ECM proteins and delay tendon healing after injury by affecting variety of factors discussed below.

Collagens

The essential structural and functional building block of musculoskeletal ECM is collagen. There are almost 30 collagen subtypes belonging to protein superfamily. Tendon matrix mainly

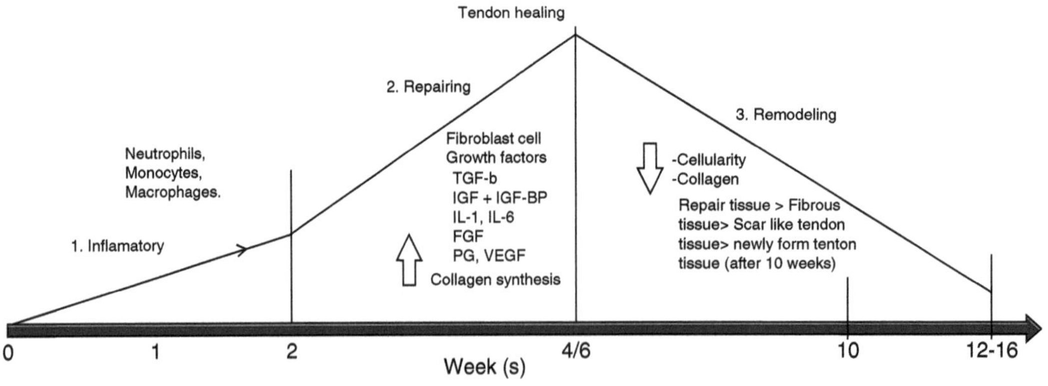

Fig. 16.1 Time frame for tendon healing and remodeling. Illustrated by Alim A

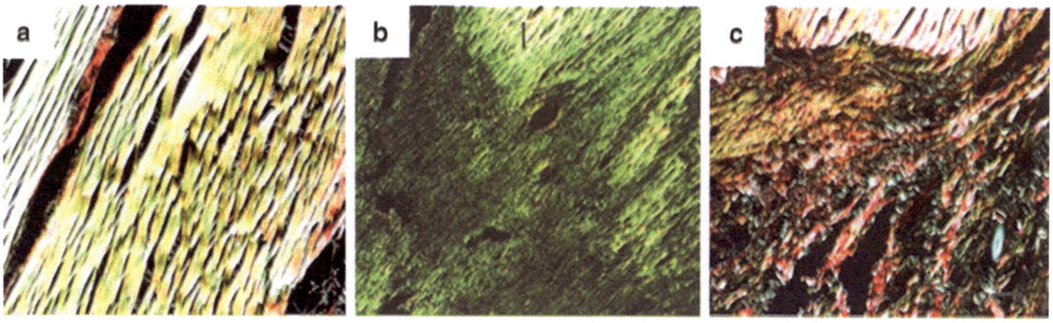

Fig. 16.2 Photomicrographs of the (**a**) normal intact (**b**) normal ruptured and (**c**) diabetic ruptured Achilles tendon of rats stained with *Sirius red* during proliferating phase of healing. Normal intact tendon mainly composed of collagen I fibers of *pale green* colour. Normal ruptured tendons exhibited short thin *pale green* fibrils arranged in a longitudinal direction at the ruptured area denoting collagen III-like structures. In ruptured diabetic tendons very few fibres of *pale green* colour were seen, most fibres were of *yellowish red* colour and arranged irregularly, denoting ruptured collagen I structures

consists of collagen type I-III, V and XI along with other collagens, proteoglycans and glycoproteins with collagen types I as the most abundant one. Numerous preclinical studies indicate that elevated glucose levels impair collagen production and also intensify the presence of advanced glycation end products (AGE), resulting in abnormal ultrastructure of collagen fibrils (see Chap. 18) [8, 9]. During proliferating phase of healing, collagen III is the major components of the callus. In diabetes rat model of tendon healing decreased collagen III production is reported in healing tendon indicating that synthesis of the main component of the callus in the proliferative healing phase is altered [9] (Fig. 16.2). Decreased collagen III and I expression both at gene and protein levels during proliferating phase of healing in rat model of type-II diabetes are reported which associated with decreased tendon strength [9] (Fig. 16.2).

Collagen synthesis at the tendon injury site is dependent on an adequate blood circulation and neuronal supply [10, 11]. Diabetes patients often exhibit neuropathy and also decreased levels of neuronal mediators such as calcitonin gene related peptide (CGRP) and substance P (SP) which might be associated with defective tissue healing [12]. SP and CGRP released from peripheral nerves enhance angiogenesis [13] and also stimulate fibroblast proliferation, angiogenesis and collagen organization in animal model of Achilles tendon healing (see Chap. 5) [14].

Cytokines

Cytokines are small proteins (typically 5–30 kDa) produced by specialized cells in response to variety of stimuli. In tendon, cytokines are produced mainly by the resident fibroblast; the tenocytes and are known to produce several endogenic cytokines and growth factors which work in an autocrine, paracrine, or endocrine manner [15, 16]. Cytokines play essential role in cell proliferation during embryonic development and in later life, in cellular renewal and wound healing, the development of cellular and humoral immunity, and in inflammatory responses. In many disease conditions disruption of these processes are associated with altered regulation of cytokine production and action [16]. Inflammatory cytokines are reported to enhance angiogenesis which is a critical process of wound healing [17]. Endogenous expression of interleukin-1beta (IL-1β), tumor necrosis factor-alpha (TNF-α), interleukin-6 (IL-6), interleukin-10 (IL-10) along with growth factor especially VEGF are reported in healing tendon [18]. IL-1β and TNF-α are thought to play crucial role in initial inflammatory healing phase [19]. Elevated levels of IL-1β and TNF-α are reported in rat model of healing Achilles tendon [20]. Similarly, up-regulated gene expression of hypoxia-inducible factors-1alpha (HIF-1α) is reported during normal healing tendon

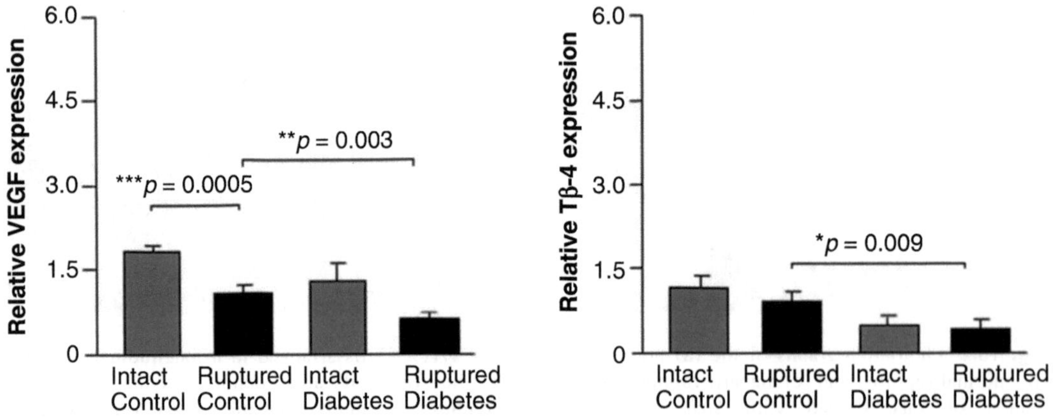

Fig. 16.3 Relative gene expression of VEGF and Tβ-4 in the intact and ruptured Achilles tendons of healthy controls and diabetic rats during proliferating phase of healing

which was not observed in diabetic tendons [19]. Suppressed HIF-1α expression has also been reported in hyperglycaemic conditions [21]. HIF-1α has been demonstrated to be critical for improving wound healing in diabetic mice partly by improving angiogenesis, epidermal regeneration, and recruitment of endothelial precursors [21].

Growth Factors

Growth factors such as insulin-like growth factor 1 (IGF-I), vascular endothelial growth factor (VEGF) , transforming growth factor beta (TGF-β), nerve growth factor (NGF) and brain derived nuclear factor (BDNF) have been shown to play crucial role during all stages of tendon healing from initial inflammatory to late regenerative stage.

IGF-1 has been reported to be upregulated and active during tendon healing by enhancing collagen synthesis in both animals as well as in humans. Local IGF-I administration enhanced tendon collagen synthesis both within and around the human tendon tissue [22, 23]. In diabetes reduced IGF-1 levels have been demonstrated both in clinical and experimental studies. Patients with type-I DM are reported to have low IGF-1 levels [24]. Significant reduction of IGF-1 in serum and bone tissues of animal model of type-II DM has been demonstrated which correlated with skeletal changes like endosteal erosions and osteopenia [25].

VEGF is known to promote angiogenesis and to induce endothelial cell proliferation and migration during the early phase of wound healing. VEGF mRNA are reported to be expressed at the site of tendon injuries along with collagen I mRNA [26, 27]. In diabetes, down-regulated VEGF expression has been reported during proliferating stage in healing tendon in animal model of type-II diabetes which was associated with down-regulated thymosin beta-4 (Tβ-4) expression especially around newly formed blood vessels [19] (Fig. 16.3). Tβ-4 is an essential regulatory factor in angiogenesis [28, 29]. Exogenously applied Tβ-4 is reported to accelerate wound healing with decreased scarring [29]. Decreased VEGF and Tβ-4 expression probably suggest impaired angiogenesis in injured diabetes tendons.

TGF-β. All three isoforms of TGF-β (TGF-β1, TGF-β2, and TGF-β3) are known to play vital role in wound healing and scar formation [30]. Upregulated TGF-β1 has been reported during proliferating stage in healing tendon in animal model of type-II diabetes [19] (Fig. 16.4). Bone morphogenetic proteins (BMPs) are a subgroup of TGF-β superfamily that can induce bone and cartilage formation as well as tissue differentiation, and BMP-12 specifically has been shown to

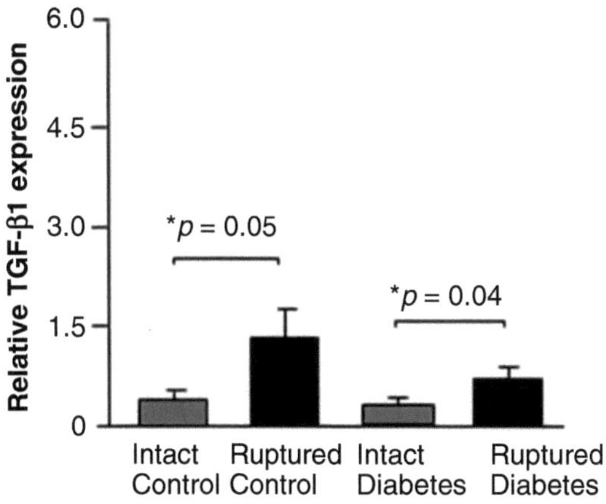

Fig. 16.4 Relative gene expression of TGF-β1 in the intact and ruptured Achilles tendons of healthy controls and diabetic rats during proliferating phase of healing

influence formation and differentiation of tendon tissue and to promote fibrogenesis.

NGF strengthens connective tissue healing by promoting re-innervation, blood flow, and angiogenesis during normal conditions. In diabetes, NGF expression has been down-regulated along with receptor TrkA. NGF is an essential regulator of BDNF expression during tissue healing in diabetes. Down-regulation in DBNF receptor TrkB expression has also been observed (*Ahmed et al. unpublished data*).

Matrix Metalloproteinases (MMPs)

The ability to break down and synthesize new collagen molecules in a tightly controlled and organized fashion in ECM tissue is critical in healing mechanism during injury. Certain MMPs including MMP-1, MMP-2, MMP-8, MMP-13, and MMP-14 have collagenase activity and are capable of degrading collagen fibrils [31]. MMPs play an important role in the degradation and remodeling of the ECM during the healing process after tendon injury. Tissue in the ruptured area undergoes marked rearrangement at the molecular level which is regulated by MMPs [32]. Lower MMP-3 mRNA levels as well as changes in MMP-13 mRNA and protein levels are reported in healing tendons in diabetes animal model which was associated with changes in tendon mechanical strength [9]. MMPs have been suggested to regulate the expression of VEGF and angiogenic factors [33] (see Chap. 17).

Summary

The regenerative capability of tendons is compromised in diabetes with corresponding expressional changes in collagens, matrix metalloproteinase, various inflammatory and growth mediators and their receptors.

References

1. Riley GP, Curry V, DeGroot J, van El B, Verzijl N, Hazleman BL, Bank RA (2002) Matrix metalloproteinase activities and their relationship with collagen remodelling in tendon pathology. Matrix biol 21(2):185–195
2. Sharma P, Maffulli N (2006) Biology of tendon injury: healing, modeling and remodeling. J Musculoskele Neuronal Interact 6(2):181–190
3. Sharma P, Maffulli N (2005) Tendon injury and tendinopathy: healing and repair. J Bone Joint Surg Am 87(1):187–202
4. Raspanti M, Congiu T, Guizzardi S (2002) Structural aspects of the extracellular matrix of the tendon: an atomic force and scanning electron microscopy study. Arch Histol Cytol 65(1):37–43
5. Wang JH (2006) Mechanobiology of tendon. J Biomech 39(9):1563–1582
6. Blakytny R, Jude E (2006) The molecular biology of chronic wounds and delayed healing in diabetes. Diabet Med 23(6):594–608
7. Egemen O, Ozkaya O, Ozturk MB, Sen E, Akan M, Sakiz D, Aygit C (2012) The biomechanical and histological effects of diabetes on tendon healing: experimental study in rats. J Hand Microsurg 4(2):60–64
8. Rosenbloom AL, Silverstein JH (1996) Connective tissue and joint disease in diabetes mellitus. Endocrinol Metab Clin N Am 25(2):473–483
9. Ahmed AS, Schizas N, Li J, Ahmed M, Ostenson CG, Salo P, Hewitt C, Hart DA, Ackermann PW (2012) Type 2 diabetes impairs tendon repair after injury in a rat model. J Appl Physiol (1985) 113(11):1784–1791
10. Dahl J, Li J, Bring DK, Renstrom P, Ackermann PW (2007) Intermittent pneumatic compression enhances neurovascular ingrowth and tissue proliferation during connective tissue healing: a study in the rat. J Orthop Res 25(9):1185–1192

11. Bring DK, Kreicbergs A, Renstrom PA, Ackermann PW (2007) Physical activity modulates nerve plasticity and stimulates repair after Achilles tendon rupture. J Orthopaedic Res 25(2):164–172
12. Pradhan L, Nabzdyk C, Andersen ND, LoGerfo FW, Veves A (2009) Inflammation and neuropeptides: the connection in diabetic wound healing. Expert Rev Mol Med 11, e2
13. Seegers HC, Hood VC, Kidd BL, Cruwys SC, Walsh DA (2003) Enhancement of angiogenesis by endogenous substance P release and neurokinin-1 receptors during neurogenic inflammation. J Pharmacol Exp Ther 306(1):8–12
14. Burssens P, Steyaert A, Forsyth R, van Ovost EJ, Depaepe Y, Verdonk R (2005) Exogenously administered substance P and neutral endopeptidase inhibitors stimulate fibroblast proliferation, angiogenesis and collagen organization during Achilles tendon healing. Foot Ankle Int 26(10):832–839
15. Tsuzaki M, Guyton G, Garrett W, Archambault JM, Herzog W, Almekinders L, Bynum D, Yang X, Banes AJ (2003) IL-1 beta induces COX2, MMP-1, -3 and -13, ADAMTS-4, IL-1 beta and IL-6 in human tendon cells. J Orthopaedic Res 21(2):256–264
16. John T, Lodka D, Kohl B, Ertel W, Jammrath J, Conrad C, Stoll C, Busch C, Schulze-Tanzil G (2010) Effect of pro-inflammatory and immunoregulatory cytokines on human tenocytes. J Orthop Res 28 (8):1071–1077
17. Belperio JA, Keane MP, Arenberg DA, Addison CL, Ehlert JE, Burdick MD, Strieter RM (2000) CXC chemokines in angiogenesis. J Leukoc Biol 68(1):1–8
18. Schulze-Tanzil G, Al-Sadi O, Wiegand E, Ertel W, Busch C, Kohl B, Pufe T (2011) The role of pro-inflammatory and immunoregulatory cytokines in tendon healing and rupture: new insights. Scand J Med Sci Sports 21(3):337–351
19. Ahmed AS, Li J, Schizas N, Ahmed M, Ostenson CG, Salo P, Hewitt C, Hart DA, Ackermann PW (2014) Expressional changes in growth and inflammatory mediators during Achilles tendon repair in diabetic rats: new insights into a possible basis for compromised healing. Cell Tissue Res 357(1):109–117
20. Eliasson P, Andersson T, Aspenberg P (2009) Rat Achilles tendon healing: mechanical loading and gene expression. J Appl Physiol (1985) 107(2):399–407
21. Botusan IR, Sunkari VG, Savu O, Catrina AI, Grunler J, Lindberg S, Pereira T, Yla-Herttuala S, Poellinger L, Brismar K et al (2008) Stabilization of HIF-1alpha is critical to improve wound healing in diabetic mice. Proc Natl Acad Sci U S A 105 (49):19426–19431
22. McCarthy TL, Centrella M, Canalis E (1989) Regulatory effects of insulin-like growth factors I and II on bone collagen synthesis in rat calvarial cultures. Endocrinology 124(1):301–309
23. Hansen M, Boesen A, Holm L, Flyvbjerg A, Langberg H, Kjaer M (2012) Local administration of insulin-like growth factor-I (IGF-I) stimulates tendon collagen synthesis in humans. Scand J Med Sci Sports
24. Janssen JA, Jacobs ML, Derkx FH, Weber RF, van der Lely AJ, Lamberts SW (1997) Free and total insulin-like growth factor I (IGF-I), IGF-binding protein-1 (IGFBP-1), and IGFBP-3 and their relationships to the presence of diabetic retinopathy and glomerular hyperfiltration in insulin-dependent diabetes mellitus. J Clin Endocrinol Metab 82(9):2809–2815
25. Ahmad T, Ugarph-Morawski A, Lewitt MS, Li J, Saaf M, Brismar K (2008) Diabetic osteopathy and the IGF system in the Goto-Kakizaki rat. Growth Horm IGF Res 18(5):404–411
26. Boyer MI, Watson JT, Lou J, Manske PR, Gelberman RH, Cai SR (2001) Quantitative variation in vascular endothelial growth factor mRNA expression during early flexor tendon healing: an investigation in a canine model. J Orthop Res 19(5):869–872
27. Bidder M, Towler DA, Gelberman RH, Boyer MI (2000) Expression of mRNA for vascular endothelial growth factor at the repair site of healing canine flexor tendon. J Orthop Res 18(2):247–252
28. Gupta S, Kumar S, Sopko N, Qin Y, Wei C, Kim IK (2012) Thymosin beta4 and cardiac protection: implication in inflammation and fibrosis. Ann N Y Acad Sci 1269:84–91
29. Ehrlich HP, Hazard SW 3rd (2010) Thymosin beta4 enhances repair by organizing connective tissue and preventing the appearance of myofibroblasts. Ann N Y Acad Sci 1194:118–124
30. Moulin V, Tam BY, Castilloux G, Auger FA, O'Connor-McCourt MD, Philip A, Germain L (2001) Fetal and adult human skin fibroblasts display intrinsic differences in contractile capacity. J Cell Physiol 188(2):211–222
31. Veidal SS, Karsdal MA, Vassiliadis E, Nawrocki A, Larsen MR, Nguyen QH, Hagglund P, Luo Y, Zheng Q, Vainer B et al (2011) MMP mediated degradation of type VI collagen is highly associated with liver fibrosis--identification and validation of a novel biochemical marker assay. PLoS One 6(9), e24753
32. Karousou E, Ronga M, Vigetti D, Passi A, Maffulli N (2008) Collagens, proteoglycans, MMP-2, MMP-9 and TIMPs in human achilles tendon rupture. Clin Orthop Relat Res 466(7):1577–1582
33. Deryugina EI, Soroceanu L, Strongin AY (2002) Up-regulation of vascular endothelial growth factor by membrane-type 1 matrix metalloproteinase stimulates human glioma xenograft growth and angiogenesis. Cancer Res 62(2):580–588

Metalloproteinase Changes in Diabetes 17

Bento João Abreu and Wouber Hérickson de Brito Vieira

Abstract

Matrix metalloproteinases (MMPs) constitute a group of over 20 - structurally-related proteins which include a Zn^{++} ion binding site that is essential for their proteolytic activities. These enzymes play important role in extracellular matrix turnover in order to maintain a proper balance in its synthesis and degradation. MMPs are associated to several physiological and pathophysiological processes, including diabetes mellitus (DM). The mechanisms of DM and its complications is subject of intense research and evidence suggests that MMPs are implicated with the development and progression of diabetic microvascular complications such as nephropathy, cardiomyopathy, retinopathy and peripheral neuropathy. Recent data has associated DM to changes in the tendon structure, including abnormalities in fiber structure and organization, increased tendon thickness, volume and disorganization obtained by image and a tendency of impairing biomechanical properties. Although not fully elucidated, it is believed that DM-induced MMP dysregulation may contribute to structural and biomechanical alterations and impaired process of tendon healing.

Keywords

MMPs • Diabetic alterations • Extracellular matrix • Tendon

B.J. Abreu (✉)
Department of Morphology, Biosciences Center, Federal University of Rio Grande do Norte, Natal, Brazil
e-mail: abreubj@gmail.com

W.H. de Brito Vieira
Department of Physiotherapy, Health Sciences Center, Federal University of Rio Grande do Norte, Natal, Brazil
e-mail: hericksonfisio@yahoo.com.br

List of Abbreviations

AGE	advanced glycation end products
ADAMTS	Disintegrin and Metalloproteinase domain with Thrombospondin-1 motifs
CTGF	connective tissue growth factor
DM	diabetes mellitus

P.W. Ackermann and D.A. Hart (eds.), *Metabolic Influences on Risk for Tendon Disorders*, Advances in Experimental Medicine and Biology 920, DOI 10.1007/978-3-319-33943-6_17

ECM	extracellular matrix
IL	interleukin cytokine family
iNOS	inducible nitric oxide synthase
MMP(s)	matrix metalloproteinase(s)
PGE2	prostaglandin-E2
RAGE(s)	receptor(s) for advanced glycation end products
ROS	reactive oxygen species
TGF-β1	transforming growth factor beta 1
TIMP(s)	tissue inhibitor(s) of metalloproteinase
TNF-α	tumor necrosis factor alpha
STZ	Streptozotocin.

MMPs Structure, Function and Regulation

Matrix metalloproteinases (MMPs) constitute a group of over 20 structurally-related proteins which include a Zn^{++} ion binding site that is essential for their proteolytic activities. These enzymes play important role in extracellular matrix (ECM) turnover in order to maintain a proper balance in its synthesis and degradation [1]. Thus, MMPs substrates are mainly collagens, but also many other ECM proteins, including fibronectin, laminin, tenascin, aggrecan and osteonectin [2].

MMPs can be classified in four groups accordingly to their domain organization, sequence similarities and their substrate specificity. The group of (i) gelatinases comprises MMP-2 and MMP-9 and is characterized by an additional domain, called collagen binding domain, which represents the preferential binding domain for fibrillar collagen I [3]. MMP-2 is generally expressed in endothelial cells, fibroblasts, keratinocytes and chondrocytes; while MMP-9 is present in alveolar macrophages, trophoblasts, osteoclasts and polymorphonuclear leukocytes [4, 5]. The group of (ii) matrilysins is composed of MMP-7 and MMP-26 which are able to process collagen IV but not collagen I [6], and are involved in degradation of ECM constituents in the uterus at postpartum and implantation of embryos [7]. The (iii) typical MMPs group comprises three subgroups, namely stromelysins (MMP-3,-10 and-11), which participate of ECM turnover but are not able to cleave fibrillar collagen I; Collagenases (MMP-1,-8 and -13), which are responsible for degrading several native collagen (type I-V, XI) and other MMPs such as MMP-12, MMP-19, MMP-20 and MMP-27 [4, 8]. The last group is referred as (iv) furin-activated MMPs and consists of membrane type MMPs (MT-1,-2,-3,-4,-5-MMPs), transmembrane MMPs type II and secreted MMPs [4]. With a different structural arrangement from that of MMPs, the disintegrin and metalloproteinase domain with thrombospondin-1 motifs (ADAMTS) is another family of metalloproteinases and consists of 19 members which specifically process proteoglycans, in particular aggrecan, and pro-collagen [9, 10].

The MMPs are synthesized as inactive zymogens with a pro-peptide domain that must be withdrawn by several proteases, such as plasmin, thrombin, chimase, and MT-MMPs for protease activation. Activated MMPs can be either cell-secreted or membrane-bound exhibiting catalytic activity at the cell surface, intracellularly and extracellularly [5]. Due to its high proteolytic potential, MMPs are subjected to tight control and restrictive regulatory mechanisms that maintain homeostasis at the intracellular and extracellular level. One of the regulation mechanisms of the MMPs activity is by their endogenous inhibitors, tissue inhibitors of MMPs (TIMPs), which comprise a family of four members (TIMP-1 to 4) [8]. TIMPs are expressed in a tissue specific pattern and can operate through direct MMPs inhibition or by controlling their activation [11].

MMPs and TIMPs are associated to several physiological processes such as embryogenesis, wound healing, proliferation, cell motility, remodeling, angiogenesis and key reproductive events [4, 8]. On the other hand, recent evidence has shown that pathologically increased MMP activity or imbalance between MMP and TIMP expression can exacerbate a variety of diseases [12]. Indeed, MMPs are correlated to the pathogenesis of abnormalities of the growth plate and wound healing, heart failure, arthritic synovial joints, cancer, periodontal disease, ischemic brain injury [13] and even tendinopathies [14]. This chapter aims to relate MMPs changes in the diabetes mellitus (DM) pathogenesis with special focus on the tendon tissue.

MMPs and Pathogenesis of Diabetic Complications

DM is a group of metabolic disorders considered one of the major health problems worldwide [15, 16]. Chronic hyperglycemia resulting from defects in the secretion and/or insulin action is the main feature of the disease and may lead to vascular damage in vessels, nerves and many internal organs including the eyes, kidneys and heart.

Currently, the development mechanism of DM and its complications is subject of intense research. It has been observed the involvement of the MMPs in the development and progression of various complications of DM [17] and proteases are speculated to have importance for diabetes pathogenesis [18]. Previous studies have demonstrated higher concentrations of MMPs in the serum from type 1 DM [19], type 2 DM patients [20] and in a cell culture model for type 2 DM [21]. By using cultures of endothelial cells, Death and colleagues [22] demonstrated that hyperglycemia enhanced activity and expression of MMP-1,-2, and macrophage-derived MMP-9; while MMP-3 and TIMP-1 had expression and protein levels decreased and unaffected, respectively. These results indicate augmented ECM degradation in hyperglycemic conditions and suggest that MMPs can be useful markers for risk and/or progression of diabetic complications such as diabetic nephropathy and microangiopathy [23].

High levels of blood glucose are also associated to dysfunction of several intracellular signal transduction cascades including generation of reactive oxygen species (ROS), modulation of protein kinases and accumulation of advanced glycation end products (AGEs) (see Chap. 18) [24]. The receptor for AGE (RAGE) represents the main binding site for AGEs on resident and/or inflammatory cells and this interaction is capable to stimulate the production, expression and activity of MMPs in specific cell lines [25]. RAGEs are mainly located on the surface of macrophages, where they bind to the ECM proteins.

It is recognized that AGEs are able to affect the production and structure of ECM proteins, mainly due to formation of collagen cross-links. Aside from altered mechanical properties of collagen, AGEs can modify the interaction of collagen with other molecules such as proteoglycans (PGs), MMPs and cell integrins [26]. This collagen modification by AGEs have been correlated to inhibited wound repair and exacerbated inflammation in a rotator cuff tendon-bone [27] and medial collateral ligament [28] healing models. If the mechanical effects of AGEs in the collagenous tissue is not well understood, it is known that AGE cross-links are involved in reduced remodeling capacity due to reduced sensitivity to collagenases [29, 30], possibly via alterations in collagen solubility and augmented collagen resistance to protease breakdown [26, 31].

In some tissues, matrix-glycation products and AGEs formation can attract and stimulate cells to release cytokines (e.g., IL-1 and IL-6), MMPs and growth factors (e.g., TNF-α) during wound repair, which in turn accelerates the inflammatory response [32]. It should be noted that certain growth factors (e.g., TGF-β and CTGF) seem to have a role in the pathogenesis of long-term diabetic complications since they have increased concentrations and are associated to ECM accumulation in diabetic nephropathy [17] and tenocyte death and scar formation in tendons [33]. In addition, elevated levels of inducible nitric oxide synthase (iNOS) and prostaglandin E2 (PGE2), disturbance of the insulin secretion/action and its key signalling proteins, genetic polymorphism of MMPs and participation of oxidative stress have all been attributed to play a role in the elevated MMP expression during hyperglycemia in diabetic patients and models [25, 34].

MMPs and Diabetic Tendon Disorders: Current Findings

A growing body of evidence has associated DM to changes in the tendon structure, including abnormalities in fiber structure and organization [35], increased tendon thickness, volume and disorganization obtained by image findings [36, 37], and a tendency of impairing biomechanical properties [38, 39]. Interestingly, these tendon alterations may represent features of the ECM, which is in a constant state of dynamic equilibrium between synthesis and degradation [40].

Besides the relevance of MMPs in the ECM turnover and their role in the tendon pathophysiology [41], data linking these proteases to the development and progression of diabetic tendon disorders is still scarce. In a recent review, Shi and colleagues [39] highlighted the participation of the MMPs in the tendon alterations based in two studies. The first work, from Ahmed and co-workers [42], aimed to investigate the mechanical properties and expressional changes of genes involved in the tendon healing process in a diabetic rodent model. The authors found strong expression of MMP-13 and MMP-3 in tenocytes both in diabetic and healthy tendons; whereas MMP-13 expression was upregulated and MMP-3 expression decreased in the tendon healing model. Moreover, Tsai and others [43] found upregulation of MMP-9 and MMP-13 and increased enzymatic activity of MMP-9 in a tendon cells culture treated with high glucose concentration. Overall, these results indicate that hyperglycemia might induce collagen degradation by increasing MMPs activity and lead to a low-quality tendon structure [39].

MMPs are also implicated in the development of diabetic fibrosis, a pathological finding characterized by ECM accumulation with or without changes in the ECM composition. Cardiac fibrosis, for example, is a major feature of diabetic cardiomyopathy and results in cardiac dysfunction [44]. Fibrosis is often a result of an imbalance between MMPs and TIMPs activity after tissue insults such as hyperglycemia, dyslipidemia and hypertension [17], but increased production of collagen and other ECM components may be also involved. To date, it is not clear if DM results in fibrosis in uninjuried or healing tendons. Previously, a streptozotocin (STZ)-induced diabetes study in Achilles tendons of rats demonstrated areas of higher density of collagen I and collagen fibers disorganization when compared to the controls [45]. It was suggested that increased vascularization associated with cell proliferation and possible migration caused hypercellularity and accounted for over-deposition of collagen in the diabetic group [45]. In fact, recruitment of inflammatory cells have been connected to dysregulation of tenocyte homeostasis followed by secretion of ECM proteins, which results in an increased turnover and remodeling of the tendon ECM [41, 46]. Moreover, MMPs are able to increase release of TGF-β1 (resident in the tendon's matrix) which in turn may lead to an increased fibroblasts proliferation and collagen I deposition [47]. In this manner, a relationship between MMP/TIMP system and pro-fibrotic growth factors might contribute to fibrosis development due to ECM turnover dysregulation [17].

As previously mentioned, DM is known to produce a poor wound healing due to complications in the connective tissue metabolism, and higher expression of gelatinases in diabetes might be due to the prolonged inflammatory period [48]. In a recent work with STZ-induced diabetes in rats, Mohsenifar et al [49] showed impaired mechanical properties and altered inflammatory response in the diabetic group after Achilles tenotomy. The authors also found lower collagen content in the diabetic group at 15 days after surgery. Taken together, these results corroborate with other studies [42, 50] that point out for a delayed healing process in tendons of diabetic animals. It is noteworthy to mention that different MMPs operate in tendons, participating not only of the ECM degradation but also promoting micro-trauma healing and maintaining normal tendon function [41]. In this manner, besides augmented collagen turnover, other mechanism associated to the pathogenesis of tendinopathies may involve a "failed healing response" via decreased MMP-3 activity, for example [41]. Thus, although not fully elucidated, it is reasonable to highlight the participation of regulators of the ECM turnover (such as MMPs and others) in the tendon's delayed healing response.

Conclusion

MMPs play a role in the mechanisms of several diabetic complications. Although little studied, evidence suggests that diabetes-induced MMP dysregulation may contribute to structural and biomechanical alterations and impair the healing process in tendons.

References

1. Klein T, Bischoff R (2011) Physiology and pathophysiology of matrix metalloproteases. Amino Acids 41(2):271–290
2. Stamenkovic I (2003) Extracellular matrix remodelling: the role of matrix metalloproteinases. J Pathol 200:448–464
3. Gioia M, Monaco S, Fasciglione GF et al (2007) Characterization of the mechanisms by which gelatinase A, neutrophil collagenase, and membrane-type metalloproteinase MMP-14 recognize collagen I and enzymatically process the two alpha-chains. J Mol Biol 368(4):1101–1113
4. Piperi C, Papavassiliou AG (2012) Molecular mechanisms regulating matrix metalloproteinases. Curr Top Med Chem 12(10):1095–1112
5. Fanjul-Fernandez M, Folgueras AR, Cabrera S et al (2010) Matrix metalloproteinases: evolution, gene regulation and functional analysis in mouse models. Biochim Biophys Acta 1803(1):3–19
6. Wilson CL, Matrisian LM (1996) Matrilysin: an epithelial matrix metalloproteinase with potentially novel functions. Int J Biochem Cell Biol 28 (2):123–136
7. Qiu W, Bai SX, Zhao MR et al (2005) Spatiotemporal expression of matrix metalloproteinase- 26 in human placental trophoblasts and fetal red cells during normal placentation. Biol Reprod 72(4):954–959
8. Amălinei C, Căruntu ID, Bălan RA (2007) Biology of metalloproteinases. Rom J Morphol Embryol 48 (4):323–334
9. Tortorella M, Pratta M, Liu RQ et al (2000) The thrombospondin motif of aggrecanase-1 (ADAMTS-4) is critical for aggrecan substrate recognition and cleavage. J Biol Chem 275(33):25791–25797
10. Colige A, Ruggiero F, Vandenberghe I et al (2005) Domains and maturation processes that regulate the activity of ADAMTS-2, a metalloproteinase cleaving the aminopropeptide of fibrillar procollagens types I-III and V. J Biol Chem 280(41):34397–34408
11. Brew K, Nagase H (2010) The tissue inhibitors of metalloproteinases (TIMPs): an ancient family with structural and functional diversity. Biochim Biophys Acta 1803(1):55–71
12. Tsioufis C, Bafakis I, Kasiakogias A et al (2012) The role of matrix metalloproteinases in diabetes mellitus. Curr Top Med Chem 12(10):1159–1165
13. Malemud CJ (2006) Matrix metalloproteinases (MMPs) in health and disease: an overview. Front Biosci 11:1696–1701
14. Del Buono A, Oliva F, Osti L et al (2013) Metalloproteases and tendinopathy. Muscles Ligaments Tendons J 3(1):51–57
15. Wild S, Roglic G, Green A et al (2004) Global prevalence of diabetes: estimates for the year 2000 and projections for 2030. Diabetes Care 27(5):1047–1053
16. Meetoo D, McGovern P, Safadi R (2007) An epidemiological overview of diabetes across the world. Br J Nurs 16(16):1002–1007
17. Ban CR, Twigg SM (2008) Fibrosis in diabetes complications: pathogenic mechanisms and circulating and urinary markers. Vasc Health Risk Manag 4(3):575–596
18. The Diabetes Control and Complications Trial Research Group (1993) The effect of intensive treatment of diabetes on the development and progression of long-term complications in insulin-dependent diabetes mellitus. N Engl J Med 329:977–986
19. Gharagozlian S, Svennevig K, Bangstad HJ et al (2009) Matrix metalloproteinases in subjects with type 1 diabetes. BMC Clin Pathol 9:7
20. Tayebjee MH, Lip GY, MacFadyen RJ (2005) What role do extracellular matrix changes contribute to the cardiovascular disease burden of diabetes mellitus? Diabet Med 22(12):1628–1635
21. Uemura S, Matsushita H, Li W et al (2001) Diabetes mellitus enhances vascular matrix metalloproteinase activity: role of oxidative stress. Circ Res 88 (12):1291–1298
22. Death AK, Fisher EJ, McGrath KC et al (2003) High glucose alters matrix metalloproteinase expression in two key vascular cells: potential impact on atherosclerosis in diabetes. Atherosclerosis 168(2):263–269
23. Derosa G, Avanzini MA, Geroldi D et al (2005) Matrix metalloproteinase 2 may be a marker of microangiopathy in children and adolescents with type 1 diabetes mellitus. Diabetes Res Clin Pract 70 (2):119–125
24. King GL, Wakasaki H (1999) Theoretical mechanisms by which hyperglycemia and insulin resistance could cause cardiovascular diseases in diabetes. Diabetes Care 22:C31–C37
25. Ryan ME, Ramamurthy NS, Sorsa T et al (1999) MMP-mediated events in diabetes. Ann N Y Acad Sci 878:311–334
26. Snedeker JG, Gautieri A (2014) The role of collagen crosslinks in ageing and diabetes – the good, the bad, and the ugly. Muscles Ligaments Tendons J 4 (3):303–308

27. Bedi A, Fox AJ, Harris PE et al (2010) Diabetes mellitus impairs tendon-bone healing after rotator cuff repair. J Shoulder Elbow Surg 19(7):978–988
28. Frank C, McDonald D, Wilson J (1995) Rabbit medial collateral ligament scar weakness is associated with decreased collagen pyridinoline crosslink density. J Orthop Res 13(2):157–165
29. Reddy GK, Stehno-Bittel L, Enwemeka CS (2002) Glycation-induced matrix stability in the rabbit Achilles tendon. Arch Biochem Biophys 399(2):174–180
30. Reddy GK (2004) Cross-linking in collagen by non-enzymatic glycation increases the matrix stiffness in rabbit Achilles tendon. Exp Diabesity Res 5 (2):143–153
31. Monnier VM, Mustata GT, Biemel KL et al (2005) Cross-linking of the extracellular matrix by the maillard reaction in aging and diabetes: an update on "a puzzle nearing resolution". Ann N Y Acad Sci 1043:533–544
32. Nesto RW, Rutter MK (2002) Impact of the atherosclerotic process in patients with diabetes. Acta Diabetol 2:S22–28
33. Sharir A, Zelzer E (2011) Tendon homeostasis: the right pull. Curr Biol 21:025
34. Kadoglou NP, Daskalopoulou SS, Perrea D et al (2005) Matrix metalloproteinases and diabetic vascular complications. Angiology 56(2):173–189
35. Grant WP, Sullivan R, Sonenshine DE et al (1997) Electron microscopic investigation of the effects of diabetes mellitus on the Achilles tendon. J Foot Ankle Surg 36(4):272–278
36. Papanas N, Courcoutsakis N, Papatheodorou K et al (2009) Achilles tendon volume in type 2 diabetic patients with or without peripheral neuropathy: MRI study. Exp Clin Endocrinol Diabetes 117(10):645–648
37. Abate M, Schiavone C, Pelotti P et al (2010) Limited joint mobility in diabetes and ageing: recent advances in pathogenesis and therapy. Int J Immunopathol Pharmacol 23(4):997–1003
38. de Oliveira RR, Lemos A, de Castro Silveira PV (2011) Alterations of tendons in patients with diabetes mellitus: a systematic review. Diabet Med 28(8):886–895
39. Shi L, Rui YF, Li G et al (2015) Alterations of tendons in diabetes mellitus: what are the current findings? Int Orthop 39(8):1465–1473
40. Jones GC, Corps AN, Pennington CJ et al (2006) Expression profiling of metalloproteinases and tissue inhibitors of metalloproteinases in normal and degenerate human Achilles tendon. Arthritis Rheum 54 (3):832–842
41. Sbardella D, Tundo GR, Fasciglione GF et al (2014) Role of metalloproteinases in tendon pathophysiology. Mini Rev Med Chem 14(12):978–987
42. Ahmed AS, Schizas N, Li J et al (2012) Type 2 diabetes impairs tendon repair after injury in a rat model. J Appl Physiol (1985) 113(11):1784–1791
43. Tsai WC, Liang FC, Cheng JW et al (2013) High glucose concentration up-regulates the expression of matrix metalloproteinase-9 and −13 in tendon cells. BMC Musculoskelet Disord 14:255
44. Asbun J, Villarreal FJ (2006) The pathogenesis of myocardial fibrosis in the setting of diabetic cardiomyopathy. J Am Coll Cardiol 47(4):693–700
45. de Oliveira RR, Martins CS, Rocha YR et al (2013) Experimental diabetes induces structural, inflammatory and vascular changes of Achilles tendons. PLoS One 8(10), e74942
46. Riley GP, Harrall RL, Constant CR, Chard MD et al (1994) Tendon degeneration and chronic shoulder pain: changes in the collagen composition of the human rotator cuff tendons in rotator cuff tendinitis. Ann Rheum Dis 53(6):359–366
47. Mendias CL, Gumucio JP, Davis ME et al (2012) Transforming growth factor-beta induces skeletal muscle atrophy and fibrosis through the induction of atrogin-1 and scleraxis. Muscle Nerve 45(1):55–59
48. Arul V, Kartha R, Jayakumar R (2007) A therapeutic approach for diabetic wound healing using biotinylated GHK incorporated collagen matrices. Life Sci 80(4):275–284
49. Mohsenifar Z, Feridoni MJ, Bayat M et al (2014) Histological and biomechanical analysis of the effects of streptozotocin-induced type one diabetes mellitus on healing of tenotomised Achilles tendons in rats. Foot Ankle Surg 20(3):186–191
50. Egemen O, Ozkaya O, Ozturk MB et al (2012) The biomechanical and histological effects of diabetes on tendon healing: experimental study in rats. J Hand Microsurg 4(2):60–64

How High Glucose Levels Affect Tendon Homeostasis

18

Jess G. Snedeker

Abstract

Among the many factors playing a role in tendon disease, unregulated biochemical reactions between glucose and the collagen extracellular matrix are coming increasingly into focus. We have shown that formation of advanced glycation end-products that cross-link the collagen extracellular matrix can drastically affect cellular level mechanical properties of the matrix, and in turn affect cell-level biomechanical stimuli during physiological loading of the tissue. We suggest that these may adversely affect tendon cell response to matrix damage, as well as the quality of the consequent repair. If such mechanical feedback loops are altered, the ability of tendon cells to maintain tissue in a functional, healthy state may be compromised. Although key foundational elements of biochemical, biomechanical, and biological understanding are now in place, the full extent of how these aspects interact, including the precise mechanisms by which advanced glycation end-products pathologically disrupt connective tissue homeostasis and damage repair, are only beginning to be adequately appreciated.

Keywords

Advanced glycation endproduct • Biomechanics • Mechanobiology • Diabetes • Aging

Connective tissue aging and disease involve complexly interwoven phenomena. Of these, the spontaneous and relatively uncontrolled biochemical interaction of glucose with proteins of the extracellular matrix is increasingly realized to be a factor of importance. The so-called "Maillard reaction", also known as "glycation" or "protein browning" has been linked to the clinically observed mechanical stiffening of connective tissue [5, 22, 36, 46]. This process is a

J.G. Snedeker (✉)
Balgrist University Hospital, Department of Orthopedics, University of Zurich, Zurich, Switzerland

Institute for Biomechanics, ETH Zurich, Zurich, Switzerland
e-mail: snedeker@ethz.ch

P.W. Ackermann and D.A. Hart (eds.), *Metabolic Influences on Risk for Tendon Disorders*, Advances in Experimental Medicine and Biology 920, DOI 10.1007/978-3-319-33943-6_18

progressive and apparently irreversible feature of aging [34]. Extracellular matrix glycation is greatly accelerated in diabetes mellitus, a disorder characterized by hyperglycemia that results from insulin deficiency or insulin resistance. This disease currently affects over 8 % of the adult population worldwide, with rapidly increasing prevalence that borders on a global pandemic [55]. The associated healthcare crisis stems from the fact that diabetic patients are prone to long-term and costly complications that drastically reduce life quality, such as cardiovascular disease, neuropathy, renal failure, retinopathy, cataract development, and poor wound healing. The following chapter will briefly summarize what is known about the potential role of unregulated biochemical reactions between glucose and the collagen extracellular matrix. We focus particularly on glycation driven biophysical changes of tendon collagens and their sequela, keeping in mind that many features of glucose-related tendon disease are shared with other connective tissue of the body.

Tendon is a highly organized fibrous connective tissue dominated by (type-I) collagen that functions to efficiently deliver the muscle forces that move and stabilize our bony skeleton. Like certain tissues of cardiovascular system, the mechanical demands on tendon can be extreme. Tendon must withstand occasionally enormous mechanical forces, and yet transfers these forces within tightly constrained anatomical spaces. This delicate balancing act requires that the extracellular matrix be finely tuned in its basic material properties – namely its elasticity, viscoelasticity, and resistance to mechanical damage. These tissue-level functions fundamentally rely on cell-mediated self-assembly of collagen molecules into functional building blocks that are known as collagen "fibrils". The physical properties of individual collagen fibrils is mostly dependent on collagen molecular packing and inter-molecular. cross-links [15]. As will be introduced later, uncontrolled biochemical reactions between the extracellular matrix and glucose act on collagen tissue function at this sub-cellular size scale, with mechanobiological effects that carry upward to the cell and tissue.

Moving our focus from the molecular scale to the cellular scale, assemblies of bundled collagen fibrils emerge into coherent collagen structures known as "fibers". From the cellular perspective, these fibers can be considered as the functional tendon unit and it is within single fibers and between neighboring fiber structures that glycation-mediated cell-matrix interactions are affected. It is thus at the cell-fiber interface that collagen glycation can be expected to directly provoke disruption of normal tendon tissue homeostasis [32]. Ultimately collagen fibers and their associated cells are grouped within meso-scale tissue structures known as fascicles that are anatomically arranged and recruited to provide mechanical support of muscle-driven joint stabilization and/or joint motion. In turn, it is the loading within each mechanically recruited tendon fascicle that determines the cell-fiber interactions that drive tendon homeostasis and repair. While there is little consensus regarding what constitutes a homeostatic range of cellular level mechanical stimulus, seminal *in vivo* studies on murine Achilles tendon have demonstrated that in vivo cell-level matrix strains are remarkably conserved within any given tendon fascicle – typically on the order of 2 % maximal stretch within any single tendon fascicle – over the course of an imposed full range-of-motion joint excursion [47, 49].

How collagen molecule packing and cross-linking drive the physical properties of the collagen matrix at the cellular scale has been and remains an active area of research; for a focused review see: [48]. It is now understood that the collagen fibril is a helically arranged supramolecular structure that can range in diameter from a few to several hundred nanometers, with lengths that may run on the order of centimeters [13]. The mechanical competence of individual type-I collagen fibrils depends on the enzyme lysyl oxidase that regulates a robust formation of stable inter-molecular collagen cross-links during maturation [7]. The absence of these enzymatically mediated chemical bonds drastically diminishes collagen fibril strength and whole tissue function [27, 33, 39]. Lysyl oxidase specifically acts on lysine or hydroxylysine in the

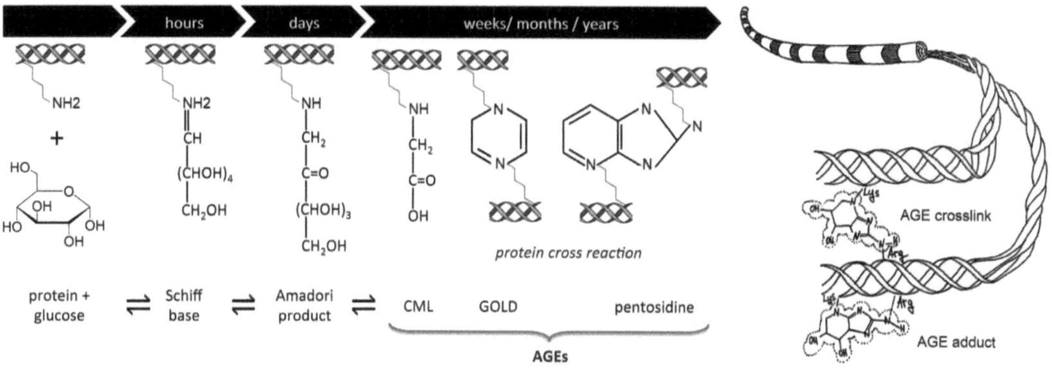

Fig. 18.1 A schematic illustration of the sequence of reactions behind advanced glycation end-product formation (e.g. shown here for AGE adduct CML, and the AGE crosslinks GOLD and pentosidine) (Adapted from: [25, 48])

telopeptide region of the collagen molecule, resulting in divalent immature cross-links between opposing amino-acids in the triple-helical region of the molecule [38]. These immature cross-links later spontaneously convert into trivalent cross-links that more permanently secure collagen molecular interconnectivity, fibril stability and whole tendon biomechanical properties [7, 16].

The essential functional role of enzymatic cross-linking in collagen fibril stability and whole tissue integrity has been convincingly demonstrated by accumulated experimental evidence [27, 29, 33, 37, 39] and has been mechanistically supported by theoretical studies as well [52]. At the core of mechanical cross-link function is the prevention molecular slippage, which in the absence of cross-links manifests as non-reversible fibrillar damage [54]. Given that lysyl oxidase mediated collagen cross-links are so essential to the proper development of fibril structure and mechanical integrity, these are perhaps the best-characterized collagen cross-linkers.

Despite the fact that enzyme driven cross-linking plateaus after tissue maturity, connective tissue stiffness has been shown to further increase with age and diabetes [9, 28, 30, 42, 45, 51]. This non-enzymatically mediated tissue stiffening has been attributed in large part to oxidative reactions between glucose and collagen, and the related formation of so-called advanced glycation end-products – widely known as AGEs [4]. Due to the fact that AGE formation and accumulation are stochastic processes, it is more likely to occur in long-lived proteins [6]. The low biological turnover of collagen thus makes it susceptible to interaction with metabolites such as glucose, and hyperglycemia related to diabetes is suspected to strongly predispose tissues of these patients to AGE accumulation (see Chap. 3) [2, 43].

Very generally, the glycation reaction initiates with the formation of a reversible chemical bond, known as a Schiff base, between a carbohydrate, typically glucose, and a protein amino group (e.g., a collagen lysine side-chain) (Fig. 18.1). The unstable Schiff base may eventually reverse or can become a stable intermediate structure commonly referred to as an Amadori product. Afterwards, a complex series of reactions that take place over the course of months or years lead to various metabolic byproducts of glycolysis including the products glyoxal, methylglyoxal and 3-deoxyglucosone, all of which can interact with extracellular proteins to form AGEs [1]. Some AGEs can bridge between the free amino groups of neighboring proteins to form inter-molecular cross-links, while others known as 'adducts' affect only a single molecule [26]. Among the different AGEs, the most abundant known AGE cross-link in collagen tissues is glucosepane, a lysine-arginine cross-link [35, 44].

The potentially causative relationship between AGE cross-linking, altered tissue properties, and

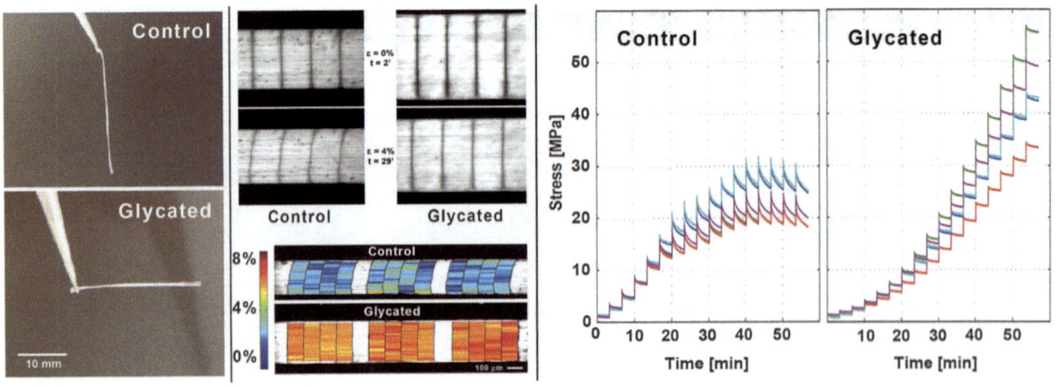

Fig. 18.2 Experimental glycation model using ribose treated rat tail tendon fascicles drastically affects tendon behavior. *Left panels:* tissue level changes are apparent, consistent with tendons seen in very old or diabetic patients. These may be anecdotally reported as "stiffened tissues". *Middle panels:* the mechanical changes are actually not related to stiffening of the tissue, but rather a loss of sliding movements between collagen fibers [18, 32]. *Right panels:* This can be quantified biomechanically at the tissue level as severely altered stress relaxation response

subsequent tissue pathology has long been posited, and in fact seems plausible on the basis of the well documented correlation between AGE markers (pentosidine; auto-fluorescence) and increasing tissue stiffness [2, 8, 14, 23]. However, such *in vivo* correlation is clouded by confounding factors apart from AGE content – as any factor that regulates collagen synthesis, turnover, or enzymatic cross-linking can affect tissue mechanical properties. These confounding effects may explain the often divergent conclusions drawn from experimental studies based on analysis of human tissues [14] and those of animal tissue samples [12, 20, 50, 53]. It is perhaps important to note that while acute diabetic rodent models often show phenotypic changes in tendon mechanical properties, that these do not accord with timescales logically required for functionally relevant accumulation of AGE collagen crosslinks [46].

Our own experimental efforts using tissue explants under carefully controlled conditions of accelerated AGE formation have provided support to previous less direct evidence that collagen fibril stiffness remains unaffected by effects of AGEs, discounting the thesis that increased fibrillar stiffness underlies the tissue level changes reported in diabetic tendon [18]. We have documented, however, that drastic changes in tissue viscoelasticity occur with intensive tissue glycation – particularly the effective elimination of lateral sliding between adjacent collagen fibers within a tendon fascicle [32]. Thus what is often perceived as "tissue stiffening" rather likely reflects loss of tissue gliding mechanisms – the clinically observable effect of which may more precisely be described as "shear stiffening" or "increase bending stiffness" (please also see Chap. 19) (Fig. 18.2).

Despite the recognized importance of AGEs in the development of age-related and diabetes-related conditions, there are still several important open questions around their role in the onset and progression of connective tissue disease. These can be broadly divided into two largely overlapping aspects: biomechanical and biological. The biomechanical consequences mostly relate to tissue fragility and susceptibility to injury [17, 19]. While tissue damage behavior is certainly important, secondary cellular events may be more crucial to the development of tissue pathology as these cellular cues tightly regulate tissue homeostasis, as well as the detection of matrix damage and the downstream orchestration of matrix repair. As such, these aspects are essential to our proper understanding of tissue pathology, and may hold the key to identifying novel therapeutic targets.

Most potentially important cellular consequences of collagen glycation relate to modified collagen-protein and collagen-receptor interactions [6]. Here the formation of AGEs (adducts or cross-links) on specific amino acids involved in intermolecular recognition and binding could lead to dramatically modified interactions between collagen and other molecules. These consequences may involve tendon proteoglycans and bound growth factors, matrix binding cell receptors such as integrins, or the activity of cell secreted enzymes – with all these factors potentially contributing to inhibited wound repair and an aberrant inflammatory response [10, 21].

Altered kinetics of the matrix degrading enzyme family matrix metalloproteinases (MMPs) come under particular scrutiny, as AGE cross-links have been demonstrated *in vitro* to reduce sensitivity to collagenase (see Chap. 17) [23, 40, 41]. Interestingly, recent evidence suggests that this inhibition may be strongly dependent on a mechanically regulated exposure of enzymatic cleavage sites [11]. These experiments are further supported by atomistic molecular modeling studies that have shown that the collagen amino acids most prone to form glucosepane cross-links (on the basis of their position and relative distance) are located close to MMP cleavage sites [24]. This same modeling study suggests that collagen epitopes involved in integrin binding are likely to be affected, as well as heparin and keratan sulphate binding. In addition to all of these cell-matrix interactions, protein glycation is known to provoke cellular production of reactive oxygen species, and further to activate potentially disruptive inflammatory signaling cascades via AGE signaling receptors (broadly classified as RAGEs; for an extensive overview see [56]).

A key aspect to consider with respect to changes in tendon matrix mechanics and tendon matrix biology is where they intersect: transduction of external mechanical signals into intracellular signaling pathways. Given the primarily mechanical function of tendon tissue, it is not surprising that mechanical forces and matrix deformations are highly important regulators of tendon biology. Although mechanical forces and resulting matrix deformations lie at the center of tendon homeostasis, the mechanisms by which extracellular mechanics regulate tissue maintenance and repair are only beginning to be understood. To understand the paradigm by which homeostasis and repair mechanisms are adversely affected by AGEs involves elucidating what is likely to be a one-two punch; First to define the range of causative biophysical changes in the matrix and second to probe how these changes alter biochemical and mechanical regulation of the cells within their matrix.

The most relevant cell stimuli presented by the matrix can be appropriately described in terms of "fiber stretch" and "fiber sliding", with secondary cell distortion effects related to shearing of the tendon cell nuclei [3, 31]. In bench top in vitro studies on tendon explants, we have shown that formation of AGEs fundamentally alters the manner in which tendon collagen structures react to loading at the fiber level, in particular with significantly reduced collagen fiber sliding [32]. Mechanical testing of rat tail tendon with induced AGE cross-links showed nearly complete removal of the lateral fiber–fiber movements inside the fascicle that dominate normal tissue response to mechanical tension, while AGE laden tendons demonstrate a pronounced shift to fiber stretching relative to fiber sliding. While it remains to be investigated, such changes in cell-level stimuli can plausibly be expected to adversely affect the ability of tendon cells to detect local changes in mechanical tissue loads, such as those related to mechanical matrix damage. If such mechanical feedback loops are altered, the ability of tendon cells to maintain tissue in a functional, healthy state may also be impaired.

Thus uncontrolled reactions between glucose and the extracellular matrix, including advanced glycation end-product cross-linking of collagens, have potential to wreak biological and biomechanical havoc on connective tissues of the body. Pioneering work by many has illuminated the nature of these reactions and products [8, 44]. Still the full extent of functional consequences of these reactions, including the mechanisms by which they may pathologically disrupt connective tissue homeostasis and damage repair, are only beginning to be adequately appreciated.

References

1. Ahmed N (2005) Advanced glycation endproducts – role in pathology of diabetic complications. Diabetes Res Clin Pract 67:3–21
2. Andreassen TT, Seyer-Hansen K, Bailey AJ (1981) Thermal stability, mechanical properties and reducible cross-links of rat tail tendon in experimental diabetes. Biochim Biophys Acta 677:313–317
3. Arnoczky SP, Lavagnino M, Whallon JH, Hoonjan A (2002) In situ cell nucleus deformation in tendons under tensile load; a morphological analysis using confocal laser microscopy. J Orthop Res 20:29–35
4. Avery NC, Bailey AJ (2005) Enzymic and non-enzymic cross-linking mechanisms in relation to turnover of collagen: relevance to aging and exercise. Scand J Med Sci Sports 15:231–240
5. Avery NC, Bailey AJ (2006) The effects of the Maillard reaction on the physical properties and cell interactions of collagen. Pathol Biol 54:387–395
6. Avery NC, Bailey AJ (2006) The effects of the Maillard reaction on the physical properties and cell interactions of collagen. Pathol Biol (Paris) 54:387–395
7. Bailey AJ (2001) Molecular mechanisms of ageing in connective tissues. Mech Ageing Dev 122:735–755
8. Bailey AJ, Paul RG, Knott L (1998) Mechanisms of maturation and ageing of collagen. Mech Ageing Dev 106:1–56
9. Bank RA, Bayliss MT, Lafeber FP, Maroudas A, Tekoppele JM (1998) Ageing and zonal variation in post-translational modification of collagen in normal human articular cartilage. The age-related increase in non-enzymatic glycation affects biomechanical properties of cartilage. Biochem J 330:345–351
10. Bedi A, Fox AJS, Harris PE, Deng XH, Ying LA, Warren RF, Rodeo SA (2010) Diabetes mellitus impairs tendon-bone healing after rotator cuff repair. J Shoulder Elb Surg 19:978–988
11. Bourne JW, Lippell JM, Torzilli PA (2014) Glycation cross-linking induced mechanical-enzymatic cleavage of microscale tendon fibers. Matrix Biol 34:179–184
12. Connizzo BK, Bhatt PR, Liechty KW, Soslowsky LJ (2014) Diabetes alters mechanical properties and collagen fiber re-alignment in multiple mouse tendons. Ann Biomed Eng 42:1880–1888
13. Craig AS, Birtles MJ, Conway JF, Parry DA (1989) An estimate of the mean length of collagen fibrils in rat tail-tendon as a function of age. Connect Tissue Res 19:51–62
14. de Jonge S, Rozenberg R, Vieyra B, Stam HJ, Aanstoot HJ, Weinans H, van Schie HT, Praet SF (2015) Achilles tendons in people with type 2 diabetes show mildly compromised structure: an ultrasound tissue characterisation study. Br J Sports Med 49:995–999
15. Depalle B, Qin Z, Shefelbine SJ, Buehler MJ (2015) Influence of cross-link structure, density and mechanical properties in the mesoscale deformation mechanisms of collagen fibrils. J Mech Behav Biomed Mater 52:1–13. doi:10.1016/j.jmbbm.2014.07.008
16. Eyre DR, Weis MA, Wu J-J (2008) Advances in collagen cross-link analysis. Methods 45:65–74
17. Fessel G, Gerber C, Snedeker JG (2012) Potential of collagen cross-linking therapies to mediate tendon mechanical properties. J Shoulder Elbow Surg 21:209–217
18. Fessel G, Li Y, Diederich V, Guizar-Sicairos M, Schneider P, Sell DR, Monnier VM, Snedeker JG (2014) Advanced glycation end-products reduce collagen molecular sliding to affect collagen fibril damage mechanisms but not stiffness. PLoS One 9: e110948
19. Fessel G, Wernli J, Li Y, Gerber C, Snedeker JG (2012) Exogenous collagen cross-linking recovers tendon functional integrity in an experimental model of partial tear. J Orthop Res 30:973–981
20. Fox AJ, Bedi A, Deng XH, Ying L, Harris PE, Warren RF, Rodeo SA (2011) Diabetes mellitus alters the mechanical properties of the native tendon in an experimental rat model. J Orthop Res 29:880–885
21. Frank C, Mcdonald D, Wilson J, Eyre D, Shrive N (1995) Rabbit medial collateral ligament scar weakness is associated with decreased collagen pyridinoline crosslink density. J Orthop Res 13:157–165
22. Fu MX, Wells-Knecht KJ, Blackledge JA, Lyons TJ, Thorpe SR, Baynes JW (1994) Glycation, glycoxidation, and cross-linking of collagen by glucose. Kinetics, mechanisms, and inhibition of late stages of the Maillard reaction. Diabetes 43:676–683
23. Galeski A, Kastelic J, Baer E, Kohn RR (1977) Mechanical and structural changes in rat tail tendon induced by alloxan diabetes and aging. J Biomech 10:775–782
24. Gautieri A, Redaelli A, Buehler MJ, Vesentini S (2013) Age- and diabetes-related nonenzymatic crosslinks in collagen fibrils: candidate amino acids involved in advanced glycation end-products. Matrix Biol
25. Gautieri A, Redaelli A, Buehler MJ, Vesentini S (2014) Age- and diabetes-related nonenzymatic crosslinks in collagen fibrils: candidate amino acids involved in advanced glycation end-products. Matrix Biol 34:89–95
26. Grandhee SK, Monnier VM (1991) Mechanism of formation of the Maillard protein cross-link pentosidine – glucose, fructose, and ascorbate as pentosidine precursors. J Biol Chem 266:11649–11653

27. Haut RC (1985) The effect of a lathyritic diet on the sensitivity of tendon to strain rate. J Biomech Eng 107:166–174
28. Haut RC, Lancaster RL, Decamp CE (1992) Mechanical properties of the canine patellar tendon: some correlations with age and the content of collagen. J Biomech 25:163–173
29. Herchenhan A, Uhlenbrock F, Eliasson P, Weis M, Eyre D, Kadler KE, Magnusson SP, Kjaer M (2015) Lysyl oxidase activity is required for ordered collagen fibrillogenesis by tendon cells. J Biol Chem 290:16440–16450
30. Lai-Fook SJ, Hyatt RE (2000) Effects of age on elastic moduli of human lungs. J Appl Physiol 89:163–168
31. Lavagnino M, Wall ME, Little D, Banes AJ, Guilak F, Arnoczky SP (2015) Tendon mechanobiology: current knowledge and future research opportunities. J Orthop Res 33:813–822
32. Li Y, Fessel G, Georgiadis M, Snedeker JG (2013) Advanced glycation end-products diminish tendon collagen fiber sliding. Matrix Biol 32:169–177
33. Marturano JE, Xylas JF, Sridharan GV, Georgakoudi I, Kuo CK (2014) Lysyl oxidase-mediated collagen crosslinks may be assessed as markers of functional properties of tendon tissue formation. Acta Biomater 10:1370–1379
34. Monnier VM, Mustata GT, Biemel KL, Reihl O, Lederer MO, Dai ZY, Sell DR (2005) Cross-linking of the extracellular matrix by the Maillard reaction in aging and diabetes – an update on "a puzzle nearing resolution". Maillard React Chem Interface Nutr Aging Dis 1043:533–544
35. Monnier VM, Mustata GT, Biemel KL, Reihl O, Lederer MO, Zhenyu DAI, Sell DR (2005) Cross-linking of the extracellular matrix by the maillard reaction in aging and diabetes: an update on "a puzzle nearing resolution". Ann N Y Acad Sci 1043:533–544
36. Monnier VM, Sell DR, Abdul-Karim FW, Emancipator SN (1988) Collagen browning and cross-linking are increased in chronic experimental hyperglycemia. Relevance to diabetes and aging. Diabetes 37:867–872
37. Mosler E, Folkhard W, Knorzer E, Nemetschek-Gansler H, Nemetschek T, Koch MH (1985) Stress-induced molecular rearrangement in tendon collagen. J Mol Biol 182:589–596
38. Orgel JP, Wess TJ, Miller A (2000) The in situ conformation and axial location of the intermolecular cross-linked non-helical telopeptides of type I collagen. Structure 8:137–142
39. Puxkandl R, Zizak I, Paris O, Keckes J, Tesch W, Bernstorff S, Purslow P, Fratzl P (2002) Viscoelastic properties of collagen: synchrotron radiation investigations and structural model. Philos Trans R Soc Lond Ser B Biol Sci 357:191–197
40. Reddy GK (2004) Cross-linking in collagen by non-enzymatic glycation increases the matrix stiffness in rabbit achilles tendon. Exp Diabesity Res 5:143–153
41. Reddy GK, Stehno-Bittel L, Enwemeka CS (2002) Glycation-induced matrix stability in the rabbit achilles tendon. Arch Biochem Biophys 399:174–180
42. Saito M, Marumo K (2010) Collagen cross-links as a determinant of bone quality: a possible explanation for bone fragility in aging, osteoporosis, and diabetes mellitus. Osteoporos Int 21:195–214
43. Schnider SL, Kohn RR (1982) Effects of age and diabetes-mellitus on the solubility of collagen from human-skin, tracheal cartilage and dura mater. Exp Gerontol 17:185–194
44. Sell DR, Biemel KM, Reihl O, Lederer MO, Strauch CM, Monnier VM (2005) Glucosepane is a major protein cross-link of the senescent human extracellular matrix. Relationship with diabetes. J Biol Chem 280:12310–12315
45. Sell DR, Lapolla A, Odetti P, Fogarty J, Monnier VM (1992) Pentosidine formation in skin correlates with severity of complications in individuals with long-standing IDDM. Diabetes 41:1286–1292
46. Sell DR, Monnier VM (2012) Molecular basis of arterial stiffening: role of glycation – a mini-review. Gerontology 58:227–237
47. Snedeker JG, Ben Arav A, Zilberman Y, Pelled G, Gazit D (2009) Functional fibered confocal microscopy: a promising tool for assessing tendon regeneration. Tissue Eng Part C Methods 15:485–491
48. Snedeker JG, Gautieri A (2014) The role of collagen crosslinks in ageing and diabetes – the good, the bad, and the ugly. Muscles Ligaments Tendons J 4:303–308
49. Snedeker JG, Pelled G, Zilberman Y, Ben Arav A, Huber E, Muller R, Gazit D (2009) An analytical model for elucidating tendon tissue structure and biomechanical function from in vivo cellular confocal microscopy images. Cells Tissues Organs 190:111–119
50. Thomas SJ, Sarver JJ, Yannascoli SM, Tucker JJ, Kelly JDT, Ahima RS, Barbe MF, Soslowsky LJ (2014) Effect of isolated hyperglycemia on native mechanical and biologic shoulder joint properties in a rat model. J Orthop Res 32:1464–1470
51. Torp S, Arridge R, Armeniades C, Baer E, Friedman B (1974) Structure-property relationships in tendon as a function of age and effects of age and mechanical deformation on the ultrastructure of tendon. Colston Papers 26:197–249
52. Uzel SGM, Buehler MJ (2011) Molecular structure, mechanical behavior and failure mechanism of the C-terminal cross-link domain in type I collagen. J Mech Behav Biomed Mater 4:153–161
53. Volper BD, Huynh RT, Arthur KA, Noone J, Gordon BD, Zacherle EW, Munoz E, Sorensen MA, Svensson RB, Broderick TL, Magnusson SP, Howden R, Hale TM, Carroll CC (2015) The influence of acute and chronic streptozotocin-induced diabetes on rat tendon extracellular matrix and mechanical properties. Am J Physiol Regul Integr Comp Physiol, ajpregu.00189.2015

54. Willett TL, Labow RS, Aldous IG, Avery NC, Lee JM (2010) Changes in collagen with aging maintain molecular stability after overload: evidence from an in vitro tendon model. J Biomech Eng 132:031002
55. World Health Organization (2012) Global status report on noncommunicable diseases 2014. World Health Organization, Geneva
56. Yan SF, Ramasamy R, Schmidt AM (2008) Mechanisms of disease: advanced glycation end-products and their receptor in inflammation and diabetes complications. Nat Rev Endocrinol 4:285–293

Rehabilitation of Tendon Problems in Patients with Diabetes Mellitus

19

Jonathan Rees, Jamie E. Gaida, Karin Grävare Silbernagel, Johannes Zwerver, Joseph S. Anthony, and Alex Scott

Abstract

Exercise is crucial in the management of diabetes mellitus and its associated complications. However, individuals with diabetes have a heightened risk of musculoskeletal problems, including tendon pathologies. Diabetes has a significant impact on the function of tendons due to the accumulation of advanced glycation end-products in the load-bearing collagen. In addition, tendon vascularity and healing may be reduced due to diabetes-induced changes in the peripheral vascular system, and impaired synthesis of collagen and glycosaminoglycan. The current chapter presents an evidence-based discussion of considerations for the rehabilitation of tendon problems in people with diabetes. The following conditions are discussed in detail – calcific tendinopathy, tenosynovitis, tendon rupture, and non-calcifying tendinopathy. Common diabetes-related findings are presented, along with their potential impact on tendinopathy management and suggested modifications to standard tendinopathy treatment

J. Rees
Department of Rheumatology, Cambridge University Hospitals, Addenbrooke's Hospital, Cambridge, UK

Academic Department of Sport and Exercise Medicine, Queen Mary College, London, UK

Department of Sport and Exercise Medicine, Fortius Clinic, London, UK
e-mail: j.rees@doctors.org.uk

J.E. Gaida
University of Canberra Research Institute for Sport and Exercise (UCRISE), Discipline of Physiotherapy, University of Canberra, Canberra, ACT, Australia
e-mail: Jamie.Gaida@canberra.edu.au

K.G. Silbernagel
Department of Physical Therapy, University of Delaware, Newark, DE, USA
e-mail: kgs@udel.ed

J. Zwerver
Center for Sports Medicine, UMC Groningen, PO Box 30.001, 9700 RB Groningen, The Netherlands
e-mail: j.zwerver@sport.umcg.nl

J.S. Anthony
Department of Physical Therapy, University of British Columbia, Vancouver, BC, Canada
e-mail: joseph.anthony@ubc.ca

A. Scott (✉)
Department of Physical Therapy, Centre for Hip Health and Mobility, University of British Columbia, Vancouver, BC, Canada
e-mail: ascott@interchange.ubc.ca

P.W. Ackermann and D.A. Hart (eds.), *Metabolic Influences on Risk for Tendon Disorders*, Advances in Experimental Medicine and Biology 920, DOI 10.1007/978-3-319-33943-6_19

protocols. A holistic approach should be used to optimize musculotendinous function, including a comprehensive exercise prescription addressing strength, flexibility, and aerobic fitness.

Keywords
Diabetes • Tendon • Tendinopathy • Rupture • Calcification • Rehabilitation

List of Acronyms and Abbreviations

BMI	body mass index
CI	confidence interval
CT	calcifying tendinopathy
DM	diabetes mellitus
IR	insulin resistance
OR	odds ratio
T1DM	type 1 diabetes mellitus
T2DM	type 2 diabetes mellitus

Introduction

Exercise is crucial in the management of diabetes mellitus (DM) and its associated complications. It can assist in improving glycaemic control [1], and delivers additional benefits including a reduced need for diabetic medication, reduced body weight, improved lipid and cardiovascular risk profile, greater self-esteem, and improved physical capacity [2]. There is an argument that exercise is a potent drug [3] which should be prescribed to all patients with DM.

The most common side effect of exercise is musculoskeletal injury [4]. Injury can become a barrier that prevents patients from achieving their ideal dose of exercise [5]. Paradoxically, this barrier to exercise is highest among patients who have the most to gain from it [1]. Many musculoskeletal conditions are more prevalent among people with diabetes [6] and there is a heightened injury risk when initiating an exercise prescription [7]. A vicious cycle of injury, physical inactivity, deteriorating metabolic parameters [8] and physical deconditioning is seen all too frequently (Fig. 19.1).

The prevalence of diabetes continues to rise, and with the disease now affecting millions of people, it has reached the scale of a pandemic [9]. Given that tendinopathies and tendon ruptures are common in the general population [10], the number of individuals with both diabetes and tendon problems is expected to continue climbing. The situation is compounded by the fact that the incidence of common tendon pathologies is higher in individuals with DM, and that recovery from injury or surgery is impaired in people with diabetes, who carry increased risk of post-operative complications including infections [11].

Types of Diabetes: Glucose and Insulin Dynamics

There are two types of diabetes. Type 1 diabetes mellitus (T1DM) results from autoimmune processes that destroy the insulin producing

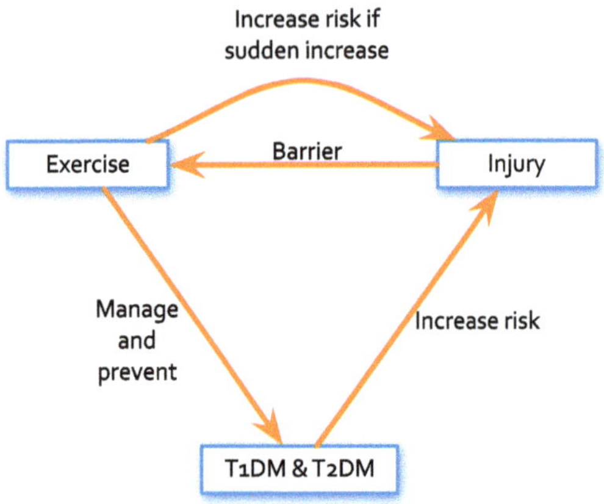

Fig. 19.1 Cyclical relationship between exercise, diabetes and injury

cells of the pancreas [12]. The loss of insulin producing cells leads to insulin deficiency. Without insulin, glucose cannot enter cells; this life-threatening situation requires life-long management with multiple daily insulin/insulin-analogue injections or infusions. Type 2 diabetes mellitus (T2DM) results from a metabolic defect called insulin resistance (IR) [13]. IR occurs when the body's cellular response to insulin becomes blunted. Insulin resistance occurs on a continuum; increasing IR manifests as impaired glucose tolerance, impaired regulation of fasting glucose levels and, at the upper end of severity, T2DM [14]. By the time T2DM is diagnosed, the body has been exposed to up to a decade of mild to moderately elevated glucose levels [15], which damages connective tissue.

How Diabetes Affects Tendon Mechanical Properties

Tendons behave as springs [16, 17]; collagen has ten times the elastic energy storage capacity of spring steel [17, 18]. Cross-linking of tendon collagen is accelerated in DM [19–21]. It has long been known that the rate of cross-linking is significantly accelerated by sustained hyperglycaemia, while more recent data show that glycaemic swings are also an important predictor [22]. Cross-linking alters tendon mechanical properties [23] by reducing sliding between collagen fibers, creating an increased reliance on the stretching of individual fibers to achieve energy storage (Fig. 19.2). There are also other potential mechanisms in tendons exposed to high glucose levels (see Chap. 18), such as changes in proteoglycan [24] and matrix metalloproteinase activity (see Chap. 17) [25], reduced synthesis of glycosaminoglycan and collagen [26], and acceleration of tenocyte senescence [27].

Does Improved Diabetes Control Translate to Accelerated Recovery from Tendon Problems?

Reasoning from first principles suggests that improved diabetes control should lead to improved recovery from a tendon injury. However, prospective studies testing this hypothesis are lacking. Massive weight loss (such as that after bariatric surgery) profoundly reduces musculoskeletal pain [28–30], and this analgesic effect both occurs in non-weight bearing locations (such as the lateral elbow, wrist and hand) and in weight-bearing locations (such as the knee, ankle and foot) [28]. Although there are many potential explanations for the reduction in symptoms (e.g. mechanical load, improved mood leading to lower pain perception, cytokines), massive weight loss unequivocally leads to improved insulin sensitivity [31]. Whether the

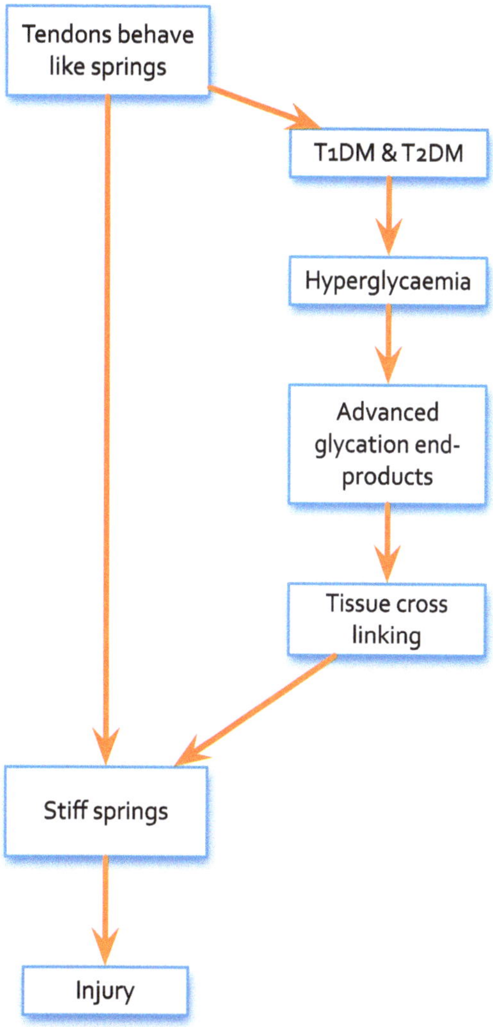

Fig. 19.2 Mechanism of diabetes increasing tendon injury risk

improved diabetes control seen with weight loss can in part explain the resolution of musculoskeletal pains, has yet to be addressed.

Considerations for Rehabilitation of Tendon Problems in People with Diabetes

This chapter will now describe the association of the more common specific tendon problems associated with DM (Table 19.1), and provide suggestions on how to tailor standard approaches to rehabilitation to be more suitable for individuals with diabetes (level V evidence).

Calcific Tendinopathy (CT)

Calcific tendinopathy (see glossary) of the rotator cuff tendons is a common occurrence in patients with DM who have shoulder pain. CT is sometimes considered part of a syndrome which involves inflammation and calcification of other periarticular shoulder structures (tendons, sheaths, joint capsule, and bursae) [32]. Use of either insulin or oral antidiabetic medications is a risk factor for rotator cuff tendinopathy (OR 1.7 and 1.8 respectively) [33]; however, differential diagnosis of shoulder pain is not standardized [34], making the true prevalence (and extent of overlap) of different causes of shoulder pain including adhesive capsulitis (frozen shoulder) difficult to measure in epidemiological studies. Over 30 % of adults with DM demonstrate CT of the shoulder, compared with 10 % of controls [35]. In one series of DM patients with shoulder CT, calcifications ranged in size from 3–22 mm, and diabetic patients were acutely symptomatic if the calcification exceeded 16 mm [36]. Those with smaller calcifications were reported to experience a milder, chronic pain condition with flare-ups and remissions; their pain was typically described as a dull ache which was worsened with shoulder active movement [36].

Table 19.1 Tendon problems with reported increased incidence with DM, or for which DM is a reported risk factor (see text for details)

Tendon problems in people with DM	Common locations
Calcific tendinopathy	Rotator cuff
Tenosynovitis	Flexor tendons of hand
Tendon rupture	Rotator cuff, Achilles
Tendinopathy	Rotator cuff, elbow, Achilles

The mechanisms of tendon calcification have not been fully explained [34, 37], but it can also occur in other diabetic tendons such as the Achilles [38]. However, the relation between calcification and symptoms is not clear in other tendons. The rehabilitation of calcific tendinopathy is typically the same as for non-calcific.

Implications for Rehabilitation – Calcific Tendinopathy of the Rotator Cuff

- Increased incidence among people with DM
  - Possibly due to lower tissue oxygenation, altered tendon structure
- Avoid exercising in painful arc range
- Encourage good shoulder posture – retraction – in all glenohumeral movements
- Be wary of decreased sensation

Tenosynovitis

Flexor tenosynovitis and De Quervain's tenosynovitis are more prevalent in adults with DM than in non-diabetic individuals, although estimates of prevalence vary [39, 40]. The presence of peripheral vascular disease is a risk factor (OR 7.26; 1.57–33.59) for the presence of tendinitis/bursitis including flexor and De Quervain's tenosynovitis [39]. In addition to the pathological changes described above, diabetes also promotes tendon thickening [41] which could increase friction between the tendon and synovium.

Diagnosis of flexor tenosynovitis is typically made by palpation of a thickened tendon nodule, with a triggering or catching sensation when moving the finger between flexion and extension.

Individuals with DM-related flexor tenosynovitis do not necessarily experience any disability related to the tenosynovitis; despite this, their grip and pinch strength may be reduced

[42]. Given that most resistance training protocols require gripping heavier weight than a person is typically exposed to in their daily life, it would be reasonable to screen for tenosynovitis and other diabetic hand complications (e.g. Dupuytren's, cheiroarthopathy), and to ensure adequate grip strength prior to strengthening – particularly if the program will involve the use of barbells or free weights.

It should be kept in mind that individuals who have tenosynovitis accompanied by DM do not experience the same clinical benefit from corticosteroid injection as do non-diabetic controls [43]. Therefore, care should be taken not to exacerbate this condition when prescribing exercise.

Implications for Rehabilitation – Tenosynovitis

- Increased incidence among people with DM
- Recommend screening for tenosynovitis and other hand complications prior to prescribing strengthening program
- May need to modify grips on weights/equipment
- Special caution if considering corticosteroid injection

Tendon Rupture

In the largest study available to date, the incidence rate ratio for tendon rupture (all locations) in patients with DM vs controls was 1.4 (95 % CI 1.10–1.87), although adjusting for alcohol and BMI suggested a less conservative estimate would be 1.84 [44]. The majority of tendon ruptures in people with DM in that study were of the rotator cuff, but cases also occurred at other sites (Achilles, distal biceps, hip/pelvis). The Achilles tendons from individuals with DM demonstrate reduced elasticity, reduced maximum load, and reduced energy at breaking point; changes which would be expected to predispose to rupture [45]. In addition to an increased risk of tendon rupture, people with DM may experience delayed healing following tendon repair (see Chap. 16) [46]. As in skin, the inflammatory and angiogenic response after tendon repair is significantly reduced in diabetic rats [47].

Because the risk of infection is higher amongst individuals with diabetes [48] and due to generally worse surgical outcomes with sports medicine procedures in people with diabetes [11], many surgeons opt for conservative management of tendon ruptures.

Exercise-based strategies to increase tendon blood flow [49] during the repair phase, such as early active ROM, may be of particular importance for diabetic patients. Healing times and progression through rehabilitation may need to be more conservative. Unlike wound healing which can be monitored visually, or bone union which can be determined from x-rays, the monitoring of tendon healing (with serial US or MRI) does not provide a very useful indicator of when it is safe to progress loading. These imaging modalities are also impractical or too costly for most rehabilitation centres. Given the magnitude of healing impairment in studies of diabetic rats [50], a conservative approach (in the absence of clinical studies) would be to delay each phase of rehabilitation by 25 %, or to start each new exercise with 25 % less loading than one would typically use. Other, more experimental, approaches to increase tendon blood flow and speed healing include intermittent pneumatic compression [51], or heat [52].

Implications for Rehabilitation – Tendon Rupture

- Increased incidence among people with DM
  - Secondary to altered tendon structure/function
- Incorporate early active ROM during repair phase
- Delay each phase and reduce intensity of rehabilitation program

Tendinopathy (Non-calcifying)

Complications of DM are wide ranging. In addition to the specific tendon problems discussed above, other musculoskeletal complications of DM include adhesive capsulitis (frozen shoulder), cheiroarthropathy (stiff hand syndrome), Dupuytren's contracture, diffuse idiopathic skeletal hyperostosis, Charcot's joints, and amyotrophy [53]. Complications in other body systems include cardiovascular and peripheral vascular disease, nephropathy, neuropathy, and retinopathy. These conditions can require rehabilitation measures for tendinopathy to be modified.

A number of non-calcifying tendinopathies have been suggested to be more common in individuals with DM. However, the associations are currently controversial. In a study of 202 individuals (100 DM and 102 non-diabetics), the prevalence of lateral (20 %) and medial (16 %) epicondylitis was higher than in controls (3.9 and 1.0 %, respectively). However, this study did not adjust for smoking [39]. Titchener et al, in a large study (4498 individuals with lateral epicondylitis) found that DM was not a risk factor, after controlling for smoking and other relevant confounders [54]. De Jonge et al found that, out of 117 incident cases of Achilles tendinopathy, 9.3 % had DM, but this was not significantly different than the general or control populations [55]. Rechardt et al found an increased risk of rotator cuff tendinopathy in men with TIDM [56], while in another small case series, Holmes and Lin found an increased risk of symptomatic AT in young men (<44 years old) [57].

Tendinopathies in individuals with DM demonstrate, on average, less blood flow on power Doppler ultrasound than tendinopathies from people without DM [58]. This finding is based on a retrospective chart review (104 diabetic and 221 controls) of sonographically diagnosed tendinopathy (Achilles, patellar, rotator cuff, elbow and wrist). The authors suggested that, as in healing wounds or ruptured tendons, diabetic tendinopathy may be characterized by reduced blood flow [58]. This could result from peripheral vascular changes affecting the area, and/or reduced angiogenesis in and around the tendon lesion. If tendon blood flow, important for healing, is on average reduced in diabetic tendons, can this situation be improved with strategies to increase blood flow? Tendon blood flow can be dynamically increased with exercise [59]. Thus, exercise which increases blood flow throughout the affected extremity without placing excessive strains on the affected tendon, may encourage healing; both isometric and isotonic exercise can be very useful in this regard. Controlled tendon loading (e.g. progressive exercise) is also thought to exert a beneficial adaptive stimulus to healing tendon and its associated muscles [60].

Common Clinical Findings in DM Patients with Tendinopathy

It is not yet known whether diabetes-induced changes in tendon properties can be reversed with exercise. Despite this, substantial gains in muscle strength and flexibility can be expected in patients with DM after several weeks of resistance and flexibility training, even in the absence of observable improvement in the tendon itself.

Table 19.2 lists some common clinical findings in DM patients with tendinopathy, with suggested clinical considerations. Of these, special mention should be given to altered foot biomechanics, and the associated risk of pressure ulcers. Altered foot biomechanics in patients with DM have been described, including Achilles tendon shortening and thickening, and premature tensioning of the plantar fascia during the gait cycle [61, 62]. Calcaneal valgus has also been noted to occur in association with diabetic neuropathy [63], the end-result of which is reduced quality of movement and poor balance [64], and increased pressure on the plantar aspect of the foot – leading to a heightened risk of diabetic pressure ulcers at this location. Flexibility of the lower extremity tendons and joint range of motion should be carefully assessed in people with DM, and a combination of calf stretching

Table 19.2 Manifestations of DM to be considered during rehabilitation of tendinopathy

Implications for rehabilitation – tendinopathy		
Potential diabetes-related finding	Potential impact on tendinopathy management	Suggested modification to standard treatment protocols
Poor glycaemic control	High blood glucose levels perpetuates poor tendon health and impairs healing [67]	Regular (daily) exercise to promote glycemic control
	Hypoglycemia can occur during or post-exercise	Measure blood sugar levels pre- and post-exercise; monitor dizziness, mood change, excessive sweating; keep sugar (e.g. juice) on hand
		N.b. hypoglycaemia may occur even several hours after exercise
Low physical activity levels	Worse prognosis/delayed progress with rehabilitation [68]	Emphasize behavioural change approach for increasing physical activity
		Provide a detailed, and relevant, and realistic exercise prescription
Reduced tendon blood flow	Might delay healing [69]	Aerobic training
		Intermittent pneumatic compression [51]
		Repetitive exercise of the specific muscle-tendon unit [70]
		Heat may be helpful [52]
Reduced muscle strength	Tendon de-adaptation	Progressive resistance training with carefully monitored loads [71]
		Emphasize importance of warm-up and cool-down for exercise program
Reduced soft-tissue flexibility	Decreased tendon length	Progressive resistance training with carefully monitored loads, progressing to eccentric activities as tolerated
		General flexibility training [72]
Reduced balance	Increased risk of injury during rehabilitation or when increasing activity levels	Evaluate balance prior to initiating treatment
		Include balance and agility exercises
Peripheral neuropathy	Numbness and tingling in extremities	Caution with prescribed exercises and other treatments
	Poor balance (see above)	Consider non-weightbearing physical activity
	Decreased exercise ability	
	Delayed healing	
	Risk of plantar pressure ulcers	
Autonomic neuropathy	Postural hypotension	Caution with prescribed exercises and other treatments
	Blunted heart rate response	
	Silent ischemia	
Peripheral vascular disease	Intermittent claudication	Consider treatment options for pressure ulcers, e.g. LLLT, ESWT [73]
	Pressure ulcers	
	Poor wound healing	
	Infections	
Altered foot biomechanics	Standard rehabilitation exercise may need to be adapted to avoid excessive/inappropriate tissue strains	Orthotics, adapted exercises, good footwear
Hypercholesterolemia	Weakens tendons; statin use may also cause myalgia and tendinopathy	Consider switching statins if a secondary tendinopathy is suspected. Consider other (non-statin) pharmacological treatment options
Potential for pressure ulcers	Rehabilitation could increase at-risk areas	Monitor for signs of skin breakdown

and progressive training prescribed using a combination of modes (isometric, concentric, eccentric), but only as far as the foot posture will allow without sacrificing a normal movement pattern, and without placing undue pressure on the plantar surface. Good footwear should also be emphasized. Although a recent randomized controlled trial suggested that orthotics do not benefit people with Achilles tendinopathy [65], this trial cannot be generalized to individuals with DM who have identified biomechanical abnormalities of the foot.

Conclusion

Clinicians are required to design physical activity programs for patients with DM, many of whom may present with a tendon problem. Patients with DM who do not have a tendon problem, are predisposed to developing one as they begin exercising more. In either situation (actual tendon problem, or increased risk), a holistic approach should be used to improve the client's physical activity level and optimize musculotendinous function, including a comprehensive exercise prescription addressing strength, flexibility, and aerobic fitness. A reasonable goal is to identify a safe form of aerobic exercise in which the client can engage naturally and with enjoyment, 30–60 min a day, at least 4 days per week, at moderate to vigourous intensity, free of tendon pain. In addition, resistance and flexibility training are an important component of the exercise prescription [66].

Glossary

Calcific tendinopathy also known as calcific periarthritis, periarticular apatite deposition, calcifying tendinitis (Ref. [37]).

References

1. Boule NG, Haddad E, Kenny GP, Wells GA, Sigal RJ (2001) Effects of exercise on glycemic control and body mass in type 2 diabetes mellitus: a meta-analysis of controlled clinical trials. JAMA 286 (10):1218–1227
2. Colberg SR, Sigal RJ, Fernhall B, Regensteiner JG, Blissmer BJ, Rubin RR et al (2010) Exercise and type 2 diabetes: the American College of Sports Medicine and the American Diabetes Association: joint position statement executive summary. Diabetes Care 33 (12):2692–2696
3. Vina J, Sanchis-Gomar F, Martinez-Bello V, Gomez-Cabrera MC (2012) Exercise acts as a drug; the pharmacological benefits of exercise. Br J Pharmacol 167 (1):1–12
4. Verhagen E, Bolling C, Finch CF (2015) Caution this drug may cause serious harm! Why we must report adverse effects of physical activity promotion. Br J Sports Med 49(1):1–2
5. Hootman JM, Macera CA, Ainsworth BE, Addy CL, Martin M, Blair SN (2002) Epidemiology of musculoskeletal injuries among sedentary and physically active adults. Med Sci Sports Exerc 34(5):838–844
6. Abate M, Schiavone C, Salini V, Andia I (2013) Occurrence of tendon pathologies in metabolic disorders. Rheumatology (Oxford) 52(4):599–608
7. Praet SF, van Rooij ES, Wijtvliet A, Boonman-de Winter LJ, Enneking T, Kuipers H et al (2008) Brisk walking compared with an individualised medical fitness programme for patients with type 2 diabetes: a randomised controlled trial. Diabetologia 51(5):736–746
8. Krogh-Madsen R, Thyfault JP, Broholm C, Mortensen OH, Olsen RH, Mounier R et al (2010) A 2-wk reduction of ambulatory activity attenuates peripheral insulin sensitivity. J Appl Physiol 108 (5):1034–1040
9. [No authors listed] (2011) The diabetes pandemic. Lancet 378(9786):99. doi:10.1016/S0140-6736(11) 61068-4. PMID: 21742159
10. Kujala UM, Sarna S, Kaprio J (2005) Cumulative incidence of achilles tendon rupture and tendinopathy in male former elite athletes. Clin J Sport Med 15 (3):133–135
11. Wolfson TS, Hamula MJ, Jazrawi LM (2013) Impact of diabetes mellitus on surgical outcomes in sports medicine. Phys Sportsmed 41(4):64–77
12. Atkinson MA, Eisenbarth GS, Michels AW (2014) Type 1 diabetes. Lancet 383(9911):69–82
13. Kahn SE (2003) The relative contributions of insulin resistance and beta-cell dysfunction to the pathophysiology of type 2 diabetes. Diabetologia 46(1):3–19
14. Guilherme A, Virbasius JV, Puri V, Czech MP (2008) Adipocyte dysfunctions linking obesity to insulin resistance and type 2 diabetes. Nat Rev Mol Cell Biol 9(5):367–377
15. Nichols GA, Glauber HS, Brown JB (2000) Type 2 diabetes: incremental medical care costs during the 8 years preceding diagnosis. Diabetes Care 23(11):1654–1659
16. Bojsen-Moller J (2005) Muscle performance during maximal isometric and dynamic contractions is influenced by the stiffness of the tendinous structures. J Appl Physiol 99(3):986–994
17. Fukashiro S, Komi PV, Jarvinen M, Miyashita M (1995) In vivo achilles tendon loading during jumping

in humans. Eur J Appl Physiol Occup Physiol 71 (5):453–458
18. Gosline J, Lillie M, Carrington E, Guerette P, Ortlepp C, Savage K (2002) Elastic proteins: biological roles and mechanical properties. Philos Trans R Soc Lond B Biol Sci 357(1418):121–132
19. Vogt BW, Schleicher ED, Wieland OH (1982) epsilon-Amino-lysine-bound glucose in human tissues obtained at autopsy. Increase in diabetes mellitus. Diabetes 31(12):1123–1127
20. Hamlin CR, Kohn RR, Luschin JH (1975) Apparent accelerated aging of human collagen in diabetes mellitus. Diabetes 24(10):902–904
21. Tanaka S, Avigad G, Brodsky B, Eikenberry EF (1988) Glycation induces expansion of the molecular packing of collagen. J Mol Biol 203(2):495–505
22. Monnier L, Mas E, Ginet C, Michel F, Villon L, Cristol JP et al (2006) Activation of oxidative stress by acute glucose fluctuations compared with sustained chronic hyperglycemia in patients with type 2 diabetes. JAMA 295(14):1681–1687
23. Li Y, Fessel G, Georgiadis M, Snedeker JG (2013) Advanced glycation end-products diminish tendon collagen fiber sliding. Matrix Biol J Int Soc Matrix Biol 32(3–4):169–177
24. Burner T, Gohr C, Mitton-Fitzgerald E, Rosenthal AK (2012) Hyperglycemia reduces proteoglycan levels in tendons. Connect Tissue Res 53(6):535–541
25. Tsai WC, Liang FC, Cheng JW, Lin LP, Chang SC, Chen HH et al (2013) High glucose concentration up-regulates the expression of matrix metalloproteinase-9 and −13 in tendon cells. BMC Musculoskelet Disord 14:255
26. Canalis EM, Dietrich JW, Maina DM, Raisz LG (1977) Hormonal control of bone collagen synthesis in vitro. Effects of insulin and glucagon. Endocrinology 100(3):668–674
27. Goldstein S, Littlefield JW, Soeldner JS (1969) Diabetes mellitus and aging: diminished planting efficiency of cultured human fibroblasts. Proc Natl Acad Sci U S A 64(1):155–160
28. Hooper MM, Stellato TA, Hallowell PT, Seitz BA, Moskowitz RW (2007) Musculoskeletal findings in obese subjects before and after weight loss following bariatric surgery. Int J Obes (Lond) 31(1):114–120
29. Iossi MF, Konstantakos EK, Teel DD 2nd, Sherwood RJ, Laughlin RT, Coffey MJ et al (2013) Musculoskeletal function following bariatric surgery. Obesity 21(6):1104–1110
30. Peltonen M, Lindroos AK, Torgerson JS (2003) Musculoskeletal pain in the obese: a comparison with a general population and long-term changes after conventional and surgical obesity treatment. Pain 104 (3):549–557
31. Mingrone G, DeGaetano A, Greco AV, Capristo E, Benedetti G, Castagneto M et al (1997) Reversibility of insulin resistance in obese diabetic patients: role of plasma lipids. Diabetologia 40(5):599–605
32. Carcia CR, Scibek JS (2013) Causation and management of calcific tendonitis and periarthritis. Curr Opin Rheumatol 25(2):204–209
33. Titchener AG, White JJ, Hinchliffe SR, Tambe AA, Hubbard RB, Clark DI (2014) Comorbidities in rotator cuff disease: a case-control study. J Shoulder Elb Surg/Am Shoulder Elb Surg [et al]
34. Carr AJ (2014) Calcification of the rotator cuff tendons and its relation to endocrine disorders. In: Via AG (ed) Metabolic diseases and tendinopathies: the missing link. Papers of Fondazione IBSA, Lugano, pp 51–55
35. Mavrikakis ME, Sfikakis PP, Kontoyannis SA, Antoniades LG, Kontoyannis DA, Moulopoulou DS (1991) Clinical and laboratory parameters in adult diabetics with and without calcific shoulder periarthritis. Calcif Tissue Int 49(4):288–291
36. Mavrikakis ME, Drimis S, Kontoyannis DA, Rasidakis A, Moulopoulou ES, Kontoyannis S (1989) Calcific shoulder periarthritis (tendinitis) in adult onset diabetes mellitus: a controlled study. Ann Rheum Dis 48(3):211–214
37. Oliva F, Via AG, Maffulli N (2012) Physiopathology of intratendinous calcific deposition. BMC Med 10:95
38. Batista F, Nery C, Pinzur M, Monteiro AC, de Souza EF, Felippe FH et al (2008) Achilles tendinopathy in diabetes mellitus. Foot Ankle Int 29(5):498–501
39. Font YM, Castro-Santana LE, Nieves-Plaza M, Maldonado M, Mayor AM, Vila LM (2014) Factors associated with regional rheumatic pain disorders in a population of Puerto Ricans with diabetes mellitus. Clin Rheumatol 33(7):995–1000
40. Leden I, Schersten B, Svensson B, Svensson M (1983) Locomotor system disorders in diabetes mellitus. Increased prevalence of palmar flexortenosynovitis. Scand J Rheumatol 12(3):260–262
41. Akturk M, Ozdemir A, Maral I, Yetkin I, Arslan M (2007) Evaluation of Achilles tendon thickening in type 2 diabetes mellitus. Exp Clin Endocrinol Diabetes 115(2):92–96
42. Savas S, Koroglu BK, Koyuncuoglu HR, Uzar E, Celik H, Tamer NM (2007) The effects of the diabetes related soft tissue hand lesions and the reduced hand strength on functional disability of hand in type 2 diabetic patients. Diabetes Res Clin Pract 77(1):77–83
43. Baumgarten KM, Gerlach D, Boyer MI (2007) Corticosteroid injection in diabetic patients with trigger finger. A prospective, randomized, controlled double-blinded study. J Bone Joint Surg Am 89 (12):2604–2611
44. Zakaria MH, Davis WA, Davis TM (2014) Incidence and predictors of hospitalization for tendon rupture in type 2 diabetes: the Fremantle diabetes study. Diabet Med J Br Diabet Assoc 31(4):425–430
45. Guney A, Vatansever F, Karaman I, Kafadar IH, Oner M, Turk CY (2015) Biomechanical properties of Achilles tendon in diabetic vs. non-diabetic patients. Exp Clin Endocrinol Diabetes 123(7):428–432
46. Abtahi AM, Granger EK, Tashjian RZ (2015) Factors affecting healing after arthroscopic rotator cuff repair. World J Orthod 6(2):211–220
47. Chbinou N, Frenette J (2004) Insulin-dependent diabetes impairs the inflammatory response and delays angiogenesis following Achilles tendon injury. Am J

Physiol Regul Integr Comp Physiol 286(5):R952–R957
48. Wukich DK, McMillen RL, Lowery NJ, Frykberg RG (2011) Surgical site infections after foot and ankle surgery: a comparison of patients with and without diabetes. Diabetes Care 34(10):2211–2213
49. Langberg H, Skovgaard D, Karamouzis M, Bulow J, Kjaer M (1999) Metabolism and inflammatory mediators in the peritendinous space measured by microdialysis during intermittent isometric exercise in humans. J Physiol 515(Pt 3):919–927
50. Ahmed AS, Schizas N, Li J, Ahmed M, Ostenson CG, Salo P et al (2012) Type 2 diabetes impairs tendon repair after injury in a rat model. J Appl Physiol 113 (11):1784–1791
51. Greve K, Domeij-Arverud E, Labruto F, Edman G, Bring D, Nilsson G et al (2012) Metabolic activity in early tendon repair can be enhanced by intermittent pneumatic compression. Scand J Med Sci Sports 22 (4):e55–e63
52. Kubo K, Yajima H, Takayama M, Ikebukuro T, Mizoguchi H, Takakura N (2010) Effects of acupuncture and heating on blood volume and oxygen saturation of human Achilles tendon in vivo. Eur J Appl Physiol 109(3):545–550
53. Smith LL, Burnet SP, McNeil JD (2003) Musculoskeletal manifestations of diabetes mellitus. Br J Sports Med 37(1):30–35
54. Titchener AG, Fakis A, Tambe AA, Smith C, Hubbard RB, Clark DI (2013) Risk factors in lateral epicondylitis (tennis elbow): a case-control study. J Hand Surg Eur Vol 38(2):159–164
55. de Jonge S, van den Berg C, de Vos RJ, van der Heide HJ, Weir A, Verhaar JA et al (2011) Incidence of midportion Achilles tendinopathy in the general population. Br J Sports Med 45(13):1026–1028
56. Rechardt M, Shiri R, Karppinen J, Jula A, Heliovaara M, Viikari-Juntura E (2010) Lifestyle and metabolic factors in relation to shoulder pain and rotator cuff tendinitis: a population-based study. BMC Musculoskelet Disord 11:165
57. Holmes GB, Lin J (2006) Etiologic factors associated with symptomatic Achilles tendinopathy. Foot Ankle Int 27(11):952–959
58. Abate M, Schiavone C, Salini S (2012) Neoangiogenesis is reduced in chronic tendinopathies of type 2 diabetic patients. Int J Immunopathol Pharmacol 25(3):757–761
59. Boushel R, Langberg H, Olesen J, Nowak M, Simonsen L, Bulow J et al (2000) Regional blood flow during exercise in humans measured by near-infrared spectroscopy and indocyanine green. J Appl Physiol 89(5):1868–1878
60. Silbernagel KG, Thomee R, Thomee P, Karlsson J (2001) Eccentric overload training for patients with chronic Achilles tendon pain – a randomised controlled study with reliability testing of the evaluation methods. Scand J Med Sci Sports 11(4):197–206
61. D'Ambrogi E, Giacomozzi C, Macellari V, Uccioli L (2005) Abnormal foot function in diabetic patients: the altered onset of Windlass mechanism. Diabet Med J Br Diabet Assoc 22(12):1713–1719
62. Grant WP, Sullivan R, Sonenshine DE, Adam M, Slusser JH, Carson KA et al (1997) Electron microscopic investigation of the effects of diabetes mellitus on the Achilles tendon. J Foot Ankle Surg 36 (4):272–278, discussion 330
63. Lazaro-Martinez JL, Aragon-Sanchez FJ, Beneit-Montesinos JV, Gonzalez-Jurado MA, Garcia Morales E, Martinez Hernandez D (2011) Foot biomechanics in patients with diabetes mellitus: doubts regarding the relationship between neuropathy, foot motion, and deformities. J Am Podiatr Med Assoc 101 (3):208–214
64. Silva P, Figueredo Borges Botelho PF, de Oliveira Guirro EC, Vaz MM, de Abreu DC (2015) Long-term benefits of somatosensory training to improve balance of elderly with diabetes mellitus. J Bodyw Mov Ther 19(3):453–457
65. Munteanu SE, Scott LA, Bonanno DR, Landorf KB, Pizzari T, Cook JL et al (2015) Effectiveness of customised foot orthoses for Achilles tendinopathy: a randomised controlled trial. Br J Sports Med 49 (15):989–994
66. Brukner P, Khan KM (2012) Clinical sports medicine, 4th edn. McGraw-Hill, Sydney
67. Wukich DK (2015) Diabetes and its negative impact on outcomes in orthopaedic surgery. World J Orthop 6 (3):331–339
68. Sayana MK, Maffulli N (2007) Eccentric calf muscle training in non-athletic patients with Achilles tendinopathy. J Sci Med Sport/Sports Med Aust 10 (1):52–58
69. Kraemer R, Lorenzen J, Rotter R, Vogt PM, Knobloch K (2009) Achilles tendon suture deteriorates tendon capillary blood flow with sustained tissue oxygen saturation – an animal study. J Orthop Surg Res 4:32
70. Kubo K, Ikebukuro T, Tsunoda N, Kanehisa H (2008) Noninvasive measures of blood volume and oxygen saturation of human Achilles tendon by red laser lights. Acta Physiol 193(3):257–264
71. Herriott MT, Colberg SR, Parson HK, Nunnold T, Vinik AI (2004) Effects of 8 weeks of flexibility and resistance training in older adults with type 2 diabetes. Diabetes Care 27(12):2988–2989
72. Tunar M, Ozen S, Goksen D, Asar G, Bediz CS, Darcan S (2012) The effects of Pilates on metabolic control and physical performance in adolescents with type 1 diabetes mellitus. J Diabetes Complications 26 (4):348–351
73. Moretti B, Notarnicola A, Maggio G, Moretti L, Pascone M, Tafuri S et al (2009) The management of neuropathic ulcers of the foot in diabetes by shock wave therapy. BMC Musculoskelet Disord 10:54

20 Inflammation in Tendon Disorders

Cathy Speed

Abstract

The role of inflammation in tendon disorders has long been a subject of considerable debate. Developments in our understanding of the basic science of inflammation have provided further insight into its potential role in specific forms of tendon disease, and the circumstances that may potentiate this. Such circumstances include excessive mechanical stresses on tendon and the presence of systemic inflammation associated with chronic diseases. In this chapter a brief review of the basic science of inflammation is provided and the influence that it may play on tendons is discussed.

Keywords

Inflammation • Tendon • Cytokines • Tendinosis • Tendinopathy • Tenosynovitis • Macrophages

Introduction

Inflammation plays a significant role in many acute and chronic musculoskeletal disorders. However its role in tendon disease has been an area of considerable debate for some years. Several texts from the early twentieth century proposed tendon disorders to be degenerative [12, 37], but nevertheless the term 'tendinitis' became the common descriptor, leading to the assumption by clinicians that inflammation is key. The term 'tendinosis' was proposed 40 years ago [39] to emphasise the lack of evidence of inflammation in chronic tendon disease, but energetic research into the role of inflammation has only been pursued in the last 15 years. Considerable insights have been gained, but the topic remains hotly discussed. Perhaps inevitably, the reader will conclude from this chapter that all parties are correct: inflammation has role to play in some tendon disorders, but not all.

C. Speed (✉)
Cambridge Centre for Health and Performance, Cambridge, UK

Fortius Clinic, London, UK

University of St Mark and St John, Plymouth, UK
e-mail: Cathy.speed@fortiusclinic.com

P.W. Ackermann and D.A. Hart (eds.), *Metabolic Influences on Risk for Tendon Disorders*, Advances in Experimental Medicine and Biology 920, DOI 10.1007/978-3-319-33943-6_20

Any discussion of the potential roles that inflammation may play in tendinopathies demands an insight into the basic physiology of inflammation and how it may have direct and indirect effects on tendons. Perhaps the two most relevant mechanisms to consider in the modern day world are mechanical load and metabolic disease and these provide excellent platforms on which to base a review of the potential role of inflammation in tendon disease. Furthermore, the reader should refer to texts on the increasing understanding of inflammation in osteoarthritis, another condition previously considered to be 'degenerative' in nature.

Inflammation

Inflammation involves a cascade of cellular and immunovascular responses to harmful stimuli [24, 30]. It has an important physiological role, and is considered to be an essential process in the resolution of healing in response to acute insult such as infection or injury. Whilst an acute controlled inflammatory response to injury can be adaptive and protective, it can be detrimental if unregulated and chronic. Furthermore, inflammation can also occur when the normal homeostasis of physiological systems are disrupted, for example by stresses, hypoxia and metabolic diseases (see Chaps. 9–18), and in this context may represent a pathological rather than protective process [36]. Uncontrolled inflammation can also be driven by chronic infections, autoimmune disease and other factors that are outside the scope of this text.

Markers/Mediators of Inflammation

Significant progress has been made in our understanding of the basic cellular and molecular processes in acute and to a lesser extent chronic inflammation. However, one of the principle reasons for the debate surrounding inflammation in tendinopathies relates to the fact that our insights into these processes in musculoskeletal disease remains limited. We do not know whether some of the processes relate to physiogical protective responses or pathological damage, and research continues on the complex interplay between cells (eg leucocytes, tenocytes) and molecular components such as pro- and anti-inflammatory cytokines, growth factors and enzymes. Whilst many mediators have been identified, their relative roles in specific tendon conditions continue to be examined. Although the response to mechanical load is the primary focus in this area, the pathological effects of metabolic dysfunction on tissue primarily as a result of systemic chronic inflammation must also be considered, in particular due to the rising incidence of obesity and other related metabolic disease.

Cellular and Molecular Mediators of Inflammation

A helpful review of the origin and physiological roles of inflammation is provided by Medzhitof [36], who describes exogenous initiators of inflammation as those that are either microbial, or non-microbial including foreign bodies, toxins, allergens and irritants. Endogenous inducers are less well defined and understood, but include signals from stressed or damaged cells (for example due to disease), products of ECM breakdown, crystals such as urate and calcium pyrophosphate (see Chaps. 10, 11 and 12), lipopolysaccharides, collagen etc.

The best understood acute inflammatory response relates to initiation through exogenous triggers such as infection or foreign bodies. Neutrophil, macrophage and mast cell activation leads to the production of mediators of inflammation. These include vasoactive amines (eg histamine, serotonin), vasoactive peptides (eg substance P) (see Chap. 5), complement products, lipid mediators (eicosanoids and platelet-activating factors), cytokines, chemokines and proteolytic enzymes. The specific mediators that are induced vary with the nature of the inflammatory trigger and the tissues involved. Some mediators have effects not only on the target tissues but also on the activity of other mediators.

In the acute phase, endothelial activation allows accumulation of leucocytes (predominantly neutrophils), which with plasma proteins form the inflammatory exudate. Neutrophils produce toxins through degranulation to kill the external agents; local tissue damage also occurs as a result [24, 30]. There then follows resolution of the inflammatory process and repair of the tissue through activated macrophages, and the production of proinflammatory prostaglandins switches to provision of anti-inflammatory lipoxins and growth factors such as TGF-β. Where for any reason there is persistence of the inflammatory response, neutrophils are replaced by macrophages and, variably, lymphocytes. Chronic inflammation through the production and activity of the inflammatory mediators ensues.

Most of our understanding of inflammatory pathways relates to the exogenous agents or macrotrauma. It is not known how closely these processes apply in the context of low-grade chronic mechanical tissue microtrauma, and in endogenous tissue stresses such as metabolic diseases and systemic inflammation. Once initiated, the typical acute inflammatory response has the purpose of adaptation to the insult and returning the tissue to its normal baseline homeostatic state. If the inflammatory stimulus persists then a maladaptive state of chronic inflammation arises.

We do know that endogenous agents induce inflammation through activating mediators of inflammatory pathways, including cytokines (eg. TNF-α, IL1-β, IL-6) and prostaglandins. For example in hyperuricaemia, the deposition of crystals in tissues such as tendon can activate macrophages that deal with the crystals as foreign bodies and involves an inflammatory response (see Chap. 11) [32, 36]. In obesity, hypertrophic adipocytes secrete chemoattractants that draw immune cells into the tissue (see Chap. 15). Secretion of pro-inflammatory mediators such as TNF-α, IL-1β, and IL-6 by adipocytes, pre-adipocytes, and infiltrating immune cells promotes pro-inflammatory macrophage and T cell activity [34]. Lipolysis results in increased levels of Free Fatty Acids (FFAs) that are in themselves pro-inflammatory [34]. This inflammatory environment negatively impacts on the insulin signaling pathway and a state of insulin resistance results. Hypertrophic adipocytes are also associated with hypoxia, which further drives a hostile environment.

Other endogenous initiators of inflammation include advanced glycation end products (AGEs) (see Chap. 18) and oxidized lipoproteins (such as high-density lipoproteins and low-density lipoproteins) [10]. AGEs are seen in ageing and in Type I and Type II Diabetes, and HDL and LDL are seen in lipid disorders. The precise mechanism by which inflammation is activated by these endogenous factors is far from clear [10, 36, 52].

Evidence for Inflammation Across the Spectrum of Tendon Disease

Inflammation plays a variable role in tendon disorders, and is influenced not only by mechanical stresses but also by underlying disease processes such as metabolic, endocrine, and autoimmune diseases, drugs, and patient characteristics (age, genetics, other comorbidities, eg. gender). It appears that all forms of the tendon can be affected, including the core tendon, tendon sheath and enthesis (Fig. 20.1).

Mechanical Effects on the Core Tendon: Tendinosis

The study of the pathophysiology of early tendinosis in humans is limited by the lack of opportunity to obtain biopsies in such cases, and much of our knowledge about acute tendon pathology in general is reliant on animal models. In chronic disease, more human

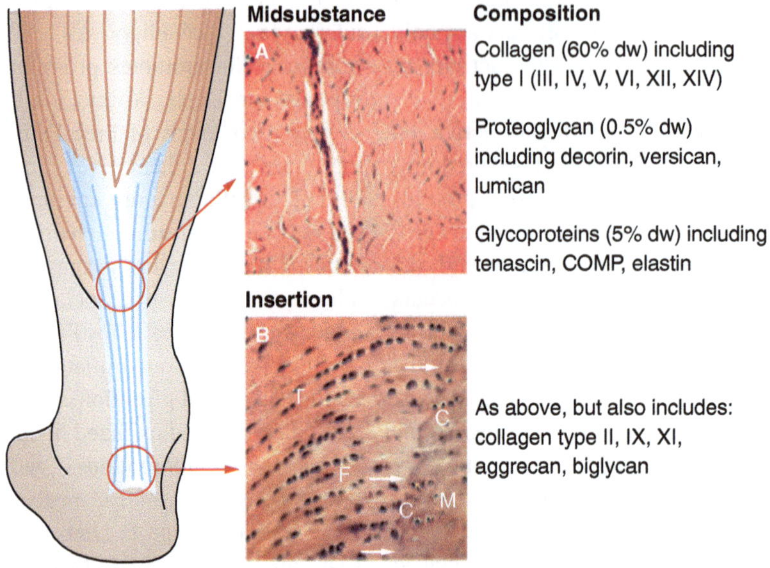

Fig. 20.1 Schematic of tendon structure and composition. The tendon midsubstance (**A**) is a dense, fibrous connective tissue of crimped fiber bundles, mostly aligned with the long axis of the tendon. The matrix consists predominantly of type I collagen, with lesser amounts of 'minor' collagens, proteoglycans and other glycoproteins. The tendon–bone insertion (enthesis) (**B**) shows more rounded cells with an Indian-file appearance and a gradual transition from tendon (T) to fibrocartilage (F) to calcified fibrocartilage (C) to mineralized bone (M). The matrix composition of the enthesis is similar to the tendon midsubstance but also contains additional 'minor' collagens and increased quantities of the proteoglycans aggrecan and biglycan. Photomicrographs (A) and (B) are H&E-stained sections of tendon midsubstance and insertion, respectively. Abbreviations: COMP, cartilage oligomeric matrix protein; dw, dry weight; H&E, hematoxylin and eosin

studies are available but are still limited in number (Fig. 20.2). Studies of humans with tendinosis have failed to show significant inflammatory cells around lesions [39, 43] leading to the dogma that inflammation plays little or no role in tendinosis [22]. Developments in immunohistochemistry and molecular biology have led to this belief to be challenged [41]. Clearly there is much more work to be done in this field but it is worth emphasising the rationale for considering inflammatory mechanisms in tendon disease.

T and B lymphocytes have been identified in chronic Achilles tendinopathy [44]. Increased levels of proinflammatory mediators including IL-1, IL-6, Cox-2 and Substance P have been found [14, 19, 20, 28, 45, 53]. Prolonged exercise is associated with increased peritenidous levels of PGE2, thromboxane, bradykinin and IL-6 in peritendinous tissue [25–27]. All are proinflammatory. Furthermore, increased levels of cyclooxygenase-2 (COX-2) are seen in patellar tendinosis [15], and repeated injection of PGE2 around tendon causes degenerative changes. These findings are not entirely straightforward, since PGE2 may also have anti-inflammatory effects [42]. Furthermore, Clearly there are other reasons for pain in tendinosis, since PGE2 levels are not correlated with pain. [5, 6].

Once initiated, by whatever trigger, tendinosis involves collagen disruption and degradation, extracellular matrix changes and tenocyte apoptosis [43]. Inflammation may act as an initial

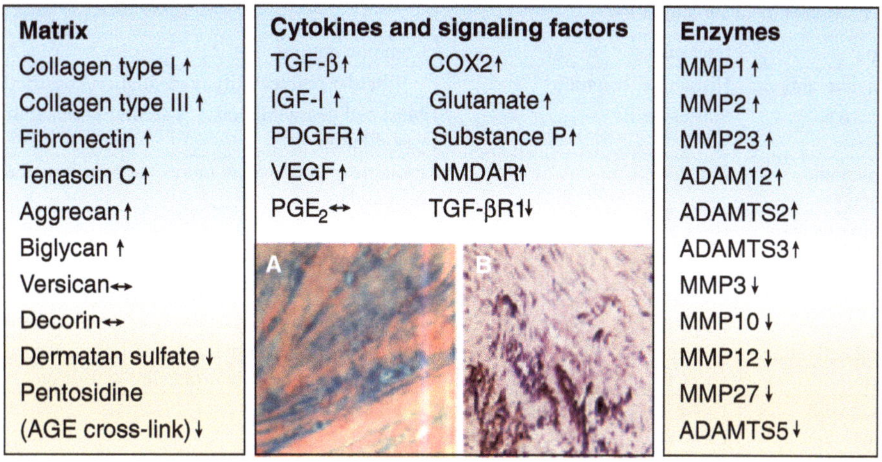

Fig. 20.2 Major structural and molecular changes in chronic tendinopathy. Typical features of tendinopathy are cell rounding and increased cell number, proteoglycan content and vascularity. Major molecular changes that have been identified by gene-expression studies, protein analysis or both are summarized. Molecules for which expression levels are increased are denoted by an upwards arrow, molecules for which expression levels are decreased are denoted by a downwards arrow and molecules for which expression levels remain unchanged are denoted by a horizontal arrow. Note that not all the changes shown have been rigorously confirmed, particularly at the protein level. Photomicrograph (**A**) is an alcian blue/H&E-stained section of supraspinatus tendinopathy, showing proteoglycans stained blue. Photomicrograph (**B**) is an H&E-stained section of Achilles tendinopathy, showing increased cellularity and proliferation of blood vessels. Abbreviations: ADAM, a disintegrin and metalloproteinase; ADAMTS, ADAM with thrombospondin motifs; AGE, advanced glycation end product; COX2, cyclooxygenase 2; H&E, hemotoxylin and eosin; IGF-I, insulin-like growth factor 1; MMP, matrix metalloproteinase; NMDAR, *N*-methyl-D-aspartate receptor; PDGFR, platelet-derived growth factor receptor; PGE_2, prostaglandin E_2; TGF-β, transforming growth factor β; TGF-βR1, TGF-β type 1 receptor; VEGF, vascular endothelial growth factor

trigger to these 'degradative' processes seen in human tendons and may still play a role in the progression of tendon disease.

The most extreme example of tendon disruption, rupture, provides the easiest model for studying the role of inflammation in tendon disease. It should be noted that this model may not reflect the processes involved in lower grade overuse injuries, but it is useful nevertheless. The inflammatory processes described earlier have been demonstrated in models of tendon rupture [44]. Cytokines and growth factors are produced not only by leucocytes but tenocytes also have the capacity for endogenous expression [21, 40, 47]. The actions of these agents on the tendon are considered to be largely proinflammatory and degradative (Tables 20.1 and 20.2). Some anti-inflammatory cytokine activity may exist.

Schulze-Tanzil and colleagues [44] reviewed the role of proinflammatory and immunoregulatory cytokines in tendon healing and rupture. They emphasise the mechanosensitivity of tendon cells, the role of proinflammatory cytokines and the influence of load on their expression. These proinflammatory cytokines are produced both by leucocytes and tenocytes in response to trauma and exercise (which can have effects not only through loading but also hypoxia). They affect extracellular matrix (ECM) homeostasis, accelerate remodeling, amplify biomechanical adaptiveness and promote tenocyte apoptosis. There are multiple interrelations between cytokines and tendon ECM synthesis, catabolic mediators such as matrix-degrading enzymes, inflammatory and angiogenic factors (COX-2, PGE2, VEGF, NO) and cytoskeleton assembly [44]. Such complex interactions may interfere with or promote healing and repair.

New vessel formation, accompanied by neural ingrowth, is commonly seen in patients with

Table 20.1 Mediators of inflammation

	Group	Examples	Common actions
1	Vasoactive amines	Histamine, serotonin	↑ Vascular permeability, vasodilation/constriction
2	Vasoactive peptides	Substance P	Mast cell degranulation, ↑ Vascular permeability, vasodilation
3	Complement fragments	C3a, C4a, C5a	Promote granulocyte & monocyte recruitment & mast cell degranulation
4	Lipid mediators		
	Ecosanoids	Prostaglandins (PgE2 & PgI2)	PgE2 and PgI2: vasodilatation
			PgE2: Principally pro-inflammatory
		Lipoxins	Anti-inflammatory, promote resolution & tissue repair
	Platelet activating factors		Recruitment of leucocytes, vasodilation, vasoconstriction, ↑ Vascular permeability, platelelt activation
5	Cytokines & chemokines	TNFα, IL-1, IL-6, others (see Table 20.2)	See Table 20.2
6	Proteolytic enzymes	Elastin, cathepsins, Matrix Metallo Proteinases (MMPs)	Diverse effects. MMPs Some promote tissue remodeling and leucocyte function

Table 20.2 Potential roles of some cytokines, growth factors and enzymes in tendon disease

Agent	Group	Pro-inflammatory activity?	Examples of sources	Comments
IL-1, IL-6, IL-10	Cytokine	✓	Macrophages, tenocytes	Eg IL-1 induces Cox-2, PgE2, MMPs, accelerating tendon degradation
TNFα	Cytokine	✓	Macrophages, tenocytes, others	Pro-inflammatory, helps collagen synthesis. May influence tenocytes to produce pro- or anti-inflammatory cytokines
IL-4	Cytokine	✗	T Lymphocytes	Anti-inflammatory: reduces TNFα, IL-1, IL-6 and promotes fibroblast proliferation
Prostaglandin E2 (PGE2)	Ecosanoid	✓	Macrphoages, Tenocytes, others	Various pro-inflmmatory effects. Can have anti-inflammatory actions
Cox-2	Isoenzyme	✓	Various: many, including tenocytes	Upregulated during inflammation
VEGF	Growth factor	✓	Macrophagesm firoblasts, tenocytes	Increased in hypoxia. Angiogenesis, can be pro-inflammatory, various effects
TGFβ	Growth factor	✗	Macrophages, tenocytes, others	Upregulates TIMP, inhibits MMPs, angiogenesis
Fibroblast growth factor (FGF)	Growth factor	✗	Various including T lymphocytes, fibroblasts, macrophages	Promototes angiogenesis, fibrosis
Platelet derived growth factor (PDGF)	Growth factor	✗	Platelets, macrophages, fibroblasts, endothelial cells,	Angiogenesis, neutrophil and fibroblast chemotaxis, fibroblast proliferation, synthesis of matrix proteins
MMPs	Enzymes	✗	Macrophages, endothelial cells, others	Degradative, catabolic potentially triggered by initial inflammatory process?
TIMP	Enzymes	✗	Various	Inhibits MMP actions

These are influenced by mechanical factors and physiological stimuli. Mechanical stress deprivation leads to over expression of pronflammatory cytokines. Note endogenous production of many by tenocytes (TNFα, il-1, IL-6, IL-10, VEGF, TGFβ). [44].

tendinosis (see Chap. 5). Its pathophysiology is still not clear. It may be driven by hypoxia, which is common in metabolic diseases and in exercise. VEGF, produced by macrophages is considered to play a role. It has a proinflammatory action but also has other mechanisms by which tendons can be damaged, including upregulation of MMPs, and down regulation of TIMP [29]. The presence of neural sprouting with angiogenesis is relevant to considering the role of inflammation. 'Neurogenic inflammation' – the release of pro-inflammatory mediators such as Substance P, Calcitonin gene related peptide and calcitonin and endothelin – may contribute to the progression and the pain of tendinopathy [29].

Inflammation in Tendinopathy in Subjects with Metabolic Diseases

Having considered the role that inflammation may have in the initiation of 'mechanical' tendinosis, it is easier to understand the mechanisms by which metabolic diseases such as obesity, diabetes, hyperlipidemia, hyperiricaemia, calcium pyrophosphate disease (CPPD) and drugs may cause tendon disease (Table 20.3).

These metabolic diseases can have negative influences on tendons through a number of mechanisms that lead to a failure of tendon homeostasis and consequential pathologies. The association between such diseases and tendon complaints is often overlooked and therefore the pathophysiology is not well understood. As will be discussed, specific potential mechanisms are associated with different metabolic conditions, but generic risks include systemic inflammation and hypoxia. Systemic inflammation causes a decline in functions and adaptive responses in many tissues through the actions of peripheral mediators (including those that are proinflammatory) and hypoxia. Hypoxia promotes the expression of proinflammatory cytokines, key apoptotic mediators and drives matrix component synthesis towards a collagen type III profile by human tenocytes. [38].

Tendon abnormalities, both of load bearing and non-load bearing tendons, are more common in individuals with higher adiposity. This is particularly the case for males and oestrogen may provide a protective benefit. Adiposity has an effect even if the subject is of normal weight [3, 17, 18, 49, 50], emphasising that although increased mechanical stresses and the effects of consequential fibre disruption undoubtedly drives some of the tendon changes seen in obese subjects, the systemic effects of adiposity play a significant role. Proinflammatory mediators including TNF-α, PGE2 PGE2, and LTB4 are seen in obesity. This is in keeping with the wider "tissue stress" effects of chronic inflammation, described earlier. Adipose tissue acts as an endocrine organ and produces active proteins and hormones, adipokines. These have the capacity to modulate cytokines, prostaglandins and MMP production.

Insulin resistance is often associated with obesity and other metabolic and systemic diseases, and it plays a potentially pivotal role in the development of tendinopathies. When blood glucose availability is increased, the production of AGE's (proinflammatory initiators, discussed earlier) is markedly elevated. These are not only associated with protein degradation, but also with nitric oxide destruction and growth factor inhibition, increasing apoptosis through oxidative stress and increased activity of pro-apoptotic and proinflammatory cytokines.

Table 20.3 Metabolic disorders associated with tendinopathies

Obesity
Diabetes
Hyperlipidaemia
Gout
CPPD
G6PD deficiency
Alkaptonuria
Hypercalcaemia
Drugs:
Fluoroquinolones
Statins
Urate lowering therapies

Dyslipidaemia also part of the metabolic syndrome, is associated with elevated total and LDL cholesterol, triglycerides and lowered HDL levels. Lipids such as LDL have proinflammatory actions, but the specific role of inflammation on tendons and indeed the association between dyslipidaemia and tendon pathologies is not clear. Dylipidaemia can be familial (Heterozygous Familial Hyperlipidaemia), where xanthomata are seen, including in tendons. Non-familial hyperlipidaemia may still be associated with tendon disease through microscopic deposition of lipid molecules which could drive inflammation; this has not been studied.

The introduction of statins in hyperlipidaemia can cause exacerbation of tendon pain. This is considered to be secondary to the rapid reduction of serum cholesterol, leading to mobilisation of lipids and reaction at least in part through a proinflammatory response [2].

The effects of hyperuricaemia are many and varied and can be increased by local hypoxia and mechanical stresses. Hyperuricameia can be part of the metabolic syndrome, and associated with obesity and dyslipidaemia or can exist as a solitary condition. Hyperuricaemia can be asymptomatic, but in the minority urate crystal deposition in joints (gout) can occur.

Intracellular uric acid levels are high in normal subjects, and extracellular levels increase in cell apoptosis, such as that seen in tendinosis. Those with hyperuricaemia would be expected to have excessive extracellular levels which would be expected to have a proinflammatory effect on tendons. Urate deposits can be found in tendon and synovium and in extreme cases may be seen on imaging, but more commonly would be expected to exist as microscopic deposits and hence would be undetectable. Enthesitis (eg patella and Achilles tendons) and tenosynovitis are commonly seen, especially in asymptomatic hyperuricaemia. Dalbeth et al. [13] studied 92 patients with tophaceous gout using dual energy CT scanning (DECT). MSU crystal deposition was observed 10.8 % of tendon/ligament sites. The Achilles tendon was the most commonly involved tendon/ligament site (39.1 % of all Achilles tendons), followed by the peroneal tendons (18.1 %). Tibialis anterior and the extensor tendons were involved less commonly (7.6–10.3 %), and the other flexor tendons, plantar fascia and deltoid ligaments were rarely involved (<5 %) ($p < 0.0001$ between sites). Involvement of the enthesis alone was more common in the Achilles tendon (OR (95 % CI) 74.5 (4.4–1264), $p < 0.0001$), as was any involvement of the enthesis (OR (95 % CI) 6.8 (3.6–13.0), $p < 0.0001$).

Crystal deposition may be enhanced by the initiation or increase of urate lowering therapy through rapid reduction of serum levels.

Calcium pyrophosphate (CPP) deposition can also cause inflammatory arthritis and potentially tendon inflammation through similar mechanisms associated with uric acid crystals. CPP tenosynovitis has been described, presumably related to crystal deposition in synovium and/or tendon [11].

Amyloidosis can present with tendon diseases such as spontaneous ruptures [33] or tenosynovitis [8]. The mechanism of action on the tendon is not known but presumed due to the effects (potentially inflammatory) of amyloid deposits in the tendon or synovial sheath.

Achilles and rotator cuff tendinopathies have been reported in patients post successful renal transplantation and seem to be related to pre-transplant hyperphosphatameia and post-transplant hypercalcaemia [7].

Surprisingly, the effects of hyercalcaemia on tendons, and the presence of inflammation, have not been documented. However idiopathic calcific tendinitis, typically occurring in the rotator cuff, is a common condition. The aetiology is unknown. In the acute phase, patients present with a very inflammatory picture of pain, stiffness, and in some inflammatory markers are raised. This indicates that inflammation plays a significant role in the condition.

Fluoroquinolone antibiotics such as ciprofloxacin inhibit fibroblast metabolism causing reduced cell proliferation and reduce collagen and matrix synthesis [51]. The role of inflammation has not been clarified.

Rare inherited disorders of metabolism can also be associated with tendinopathy. Glucose-

6-Phosphate Deficiency is associated with hyperuricaemia an can present with flexor tenosynovitis, which is likely to be driven by similar mechanisms to those described earlier.

Alkaptonuria is a deficiency in homogentisic acid oxidase, which is involved in the metabolism of homogentisic acid. Tendon rupture occurs through reduced fibril linkage but the role of inflammation is not known. [31].

Tenosynovitis in Mechanical Overuse and Metabolic Disease

The role of inflammation in disease affecting the tenosynovial sheath is highly dependent upon the underlying mechanism.

Acute inflammation of tendon sheaths can occur due to mechanical overuse. The presence of disease of the tendon proper in the acute phase is variable. Examples include tibialis posterior tenosynovitis and peroneal tenosynovitis at the ankle, and de Quervain's tenosynovitis at the wrist. There is pain, swelling, and impairment of function. Diagnostic ultrasound demonstrates tendon sheath synovial thickening, increased vascular flow on Doppler, fluid formation around the tendon. The tendon itself may be thickened with increased intratendinous blood flow.

Studies on the histopathology of the acute phase of tenosynovitis are scant. However the synovium rather than the tendon proper may be the primary initiating site of inflammation, which would explain why primary disorders of synovium such as rheumatoid arthritis frequently involve the tenosynovium and present as tenosynovitis. As has been discussed, tenosynovitis can also be the presenting feature of hyperuricaemia and calcium pyrophosphate deposition disease [16]. It is reasonable to suggest that this is due the deposition of crystals in the tendon sheath and a pro-inflammatory reaction in the synovium of the tendon sheath.

The role of inflammation in chronic tenosynovitis is less evident, particularly in overuse induced disease. Tendon thickening, tendon sheath fibrosis without the signs exhibited in the acute phase are the more common findings. The synovium may demonstrate inflammatory change or may be fibrotic. For example, chronic 'trigger finger' (flexor tendinopathy) has been attributed to tendon swelling and a thickened annular pulley. Ultrasonographic findings of increased thickness, loss of fibrillar pattern, pulley thickening and fluid collection of the tendon sheath have been reported in patients with trigger finger [23]. The histopathological features typically include hyaluronic acid-producing chondrocytoid cells originated from fibroblastic synovial B cells, and a hypocellular collagen matrix surrounding the tenosynovium. However in some patients evidence of inflammation is noted, including the presence of an inflammatory infiltrate, increased vascularity, hyperplasia of synovial lining cells, or fibrin exudation [48]. More molecular based studies are indicated.

Inflammation at the Enthesis

The tendon enthesis is best described as an entheseal organ or the synovio-entheseal complex, comprising the tendon insertion, bone and associated synovial complex [35]. The integration between synovium and tendon makes it a potential target form more classical inflammation driven through synovial inflammatory processes. The classical inflammatory enthesitis seen in autoimmune diseases is beyond the scope of this text. However inflammation may also play a role in mechanical and metabolic enthesopathies. Like the study of mid portion tendinosis, perceptions that biomechanical strain across the muscle-tendon-bone complex drives a degenerative process are based on much earlier studies showing a lack of inflammatory cell infiltrates [9]. New vessel formation at the tendon enthesis is commonly seen in 'mechanical' disease; the relevance of this in relation to hypotheses on inflammation have been discussed earlier. Local synovitis is also commonly noted [35].

As has been noted earlier, enthesopathies can be seen in those with metabolic diseases. Abate et al. [1] studied 45 patients with Achilles enthesopathy compared to 45 asymptomatic controls. An ultrasound study of the Achilles enthesis was carried out, and the presence/absence of lesions (morphologic abnormalities, calcific deposits, enthesophytes, cortical abnormalities and adjacent bursitis) was assessed. Comorbidities (osteoarthritis, diabetes, and hypertension) were recorded and body mass index (BMI), glucose, total, and HDL cholesterol were also evaluated. All symptomatic subjects showed at ultrasound evaluation at least one structural entheseal alteration; pathologic features in asymptomatic subjects were found in 6/45 (13.3 %) of cases. Symptomatic enthesopathy with detectable entheseal lesions were more likely in those with higher BMI and glucose levels. The role of inflammation in this process is not known.

Inflammation in Tendon Disease: Implications for Management

The identification of inflammation in patients with tendon disease in the clinical setting remains challenging. Whilst increased blood flow on Doppler ultrasound may indicate inflammatory activity, this needs further research. Even where inflammation may be present, the use of anti-inflammatory approaches in such patients is generally limited in effectiveness, with the exception of ultrasound guided corticosteroid injection in tenosynovitis [4]. Notably, systemic treatment with glucocorticosteroids seems to compromise the normal proliferation rate and synthetic activity of cultured tenocytes [46]. The role of rheumatological 'biologics', cytokine blocking agents (eg anti-TNFα agents), has yet to be thoroughly researched in patients with tendon disease that is unrelated to an autoimmune pathology. Optimal mechanical loading is necessary for clinical improvement in tendon disease. Defining what is optimal for a given patient at any one time is also challenging.

Management of tendon inflammation specifically in patients with metabolic diseases is reliant on the management of the underlying disease and utilisation of the same processes as used in the mechanically induced disease and in particular controlled loading strategies. In all patients, the control of all elements that may drive systemic inflammation is crucial, typically through lifestyle management and therapeutic medications where indicated.

Conclusions

It is becoming increasingly apparent that inflammation has an important role in both mechanical and metabolic related tendon disease. Further research in relation to pathways, diagnosis and management of tendon inflammation is much needed.

References

1. Abate M, Luigi Di Carlo L, Salini V, Schiavone C (2014) Metabolic syndrome associated to non-inflammatory Achilles enthesopathy. Clin Rheumatol 33(10):1517–1522
2. Abate M, Schiavone C, Vincenzo Salini V, Isabel Andia I (2013) Occurrence of tendon pathologies in metabolic disorders. Rheumatology (Oxford) 52 (4):599–608
3. Abate M (2014) How obesity modifies tendons (implications for athletic activities). Muscles Ligaments Tendons J 4(3):298–302
4. Akhtar S, Burke FD (2006) Study to outline the efficacy and illustrate techniques for steroid injection for trigger finger and thumb. Postgrad Med J 82:763–766
5. Alfredson H, Forsgren S, Thorsen K, Lorentzon R (2001) In vivo microdialysis and immunohistochemical analyses of tendon tissue demonstrated high amounts of free glutamate and glutamate NMDAR1 receptors, but no signs of inflammation, in Jumper's knee. J Orthop Res 19:881–886
6. Alfredson H, Thorsen K, Lorentzon R (1999) In situ microdialysis in tendon tissue: high levels of glutamate, but not prostaglandin E2 in chronic Achilles tendon pain. Knee Surg Sports Traumatol Arthrosc 7:378–381
7. Agarwal S, Owen R (1990) Tendinitis and tendon ruptures in successful renal transplant recipients. Clin Orthop Relat Res 252:270–275
8. Benito JR, Martinez I, Monner J, Paloma V, Castro V, Serra JM (1999) Primary amyloidosis presenting as extensor tenosynovitis. Plast Reconstr Surg 103 (2):556–558
9. Benjamin M, Toumi H, Ralphs J, Bydder G, Best T, Milz S (2006) Where tendons and ligaments meet bone: attachment sites ('entheses') in relation to exercise and/or mechanical load. J Anat 208:471–490

10. Brownlee M, Cerami A, Vlassara H (1988) Advanced glycosylation end products in tissue and the biochemical basis of diabetic complications. N Engl J Med 318:1315–1321
11. Bullocks JM, Downey CR, Gibler DP, Netscher DT (2009) Crystal deposition disease masquerading as proliferative tenosynovitis and its associated sequelae. Ann Plast Surg 62:128–133
12. Codman EA (1934) The shoulder. Thomas Todd Co., Boston
13. Dalbeth N, Kalluru R, Aati O, Horne A, Doyle A, McQueen F (2013) Tendon involvement in the feet of patients with gout: a dual-energy CT study. Ann Rheum Dis 72:1545–1548
14. Fenwick SA, Hazleman BL, Harrall RL (2000) Transforming growth factor-β isoform expression in chronic Achilles tendinopathy and their effects on tendon cell populations. Int J Exp Pathol 81:A11–A12
15. Fu SC et al (2002) Increased expression of transforming growth factor-beta1 in patellar tendinosis. Clin Orthop Relat Res 400:174–183
16. Gibler DP, Netscher DT (2009) Crystal deposition disease masquerading as proliferative tenosynovitis and its associated sequelae. Ann Plast Surg 62 (2):128–133
17. Gaida JE, Alfredson H, Kiss ZS, Bass SL, Cook JL (2010) Asymptomatic Achilles tendon pathology is associated with a central fat distribution in men and a peripheral fat distribution in women: a cross sectional study of 298 individuals. BMC Musculoskelet Disord 11:41
18. Gaida JE, Ashe MC, Bass SL, Cook JL (2009) Is adiposity an under- recognized risk factor for tendinopathy? A systematic review. Arthritis Rheum 61:840–849
19. Gotoh M, Hamada K, Yamakawa H et al (1997) Significance of granulation tissue in torn supraspinatus insertions: an immunohistochemical study with antibodies against interleukin-1 beta, cathepsin D, and matrix metalloprotease-1. J Orthop Res 15:33–39
20. Gotoh M, Hamada K, Yamakawa H (1998) Increased substance P in subacromial bursa and shoulder pain in rotator cuff diseases. J Orthop Res 16:618–621
21. John T, Lodka D, Kohl B, Ertel W, Jammrath J, Conrad C, Stoll C, Busch C, Schulze-Tanzil G (2010) Effect of pro-inflammatory and immunoregulatory cytokines on human tenocytes. J Orthop Res 28:1071–1077
22. Khan K (2002) Time to abandon the "tendinitis" myth. BMJ 324:626–627
23. Kim HR, Lee SH (2010) Ultrasonographic assessment of clinically diagnosed trigger fingers. Rheumatol Int 30(11):1455–1458
24. Kumar V, Cotran RS, Robbins SL (2003) Robbins basic pathology. Saunders
25. Langberg H, Bjørn C, Boushel R, Hellsten Y, Kjaer M (2002) Exercise-induced increase in interstitial bradykinin and adenosine concentrations in skeletal muscle and peritendinous tissue in humans. J Physiol 542:977–983
26. Langberg H, Olesen JL, Gemmer C, Kjaer M (2002) Substantial elevation of interleukin-6 concentration in peritendinous tissue, in contrast to muscle, following prolonged exercise in humans. J Physiol 542:985–990
27. Langberg H, Skovgaard D, Karamouzis M, Bülow J, Kjaer M (1999) Metabolism and inflammatory mediators in the peritendinous space measured by microdialysis during intermittent isometric exercise in humans. J Physiol 515:919–927
28. Legerlotz K, Jones ER, Screen HR (2012) Increased expression of IL-6 family members in tendon pathology. Rheumatology (Oxford) 51:1161–1165
29. Magnan B, Manuel Bondi M, Pierantoni S, Samaila E (2014) The pathogenesis of Achilles tendinopathy: a systematic review. Foot Ankle Surg 20:154–159
30. Majno G, Joris I (2004) Cells, tissues and disease. Oxford University Press, New York
31. Manoj Kumar RV, Rajasekaran S (2003) Spontaneous tendon ruptures in alkaptonuria. J Bone Joint Surg (Br) 85:883–886
32. Martinon F, Petrilli V, Mayor A, Tardivel A, Tschopp J (2006) Gout-associated uric acid crystals activate the NALP3 inflammasome. Nature 440:237–241
33. Masonis JL, Frick SL (2001) Bilateral quadriceps tendon rupture as the initial presentation of amyloidosis. Orthopaedics 24:995–996
34. McArdle MA, Finucane OM, Connaughton RM, McMorrow AM, Helen M, Roche HM (2013) Mechanisms of obesity-induced inflammation and insulin resistance: insights into the emerging role of nutritional strategies. Front Endocrinol (Lausanne) 4:52
35. McGonagle D, Benjamin M, Marzo-Ortega H, Emery P (2002) Advances in the understanding of entheseal inflammation. Curr Rheumatol Rep 4(6):500–506
36. Medzhitof R (2008) Origin and physiological roles of inflammation. Nature 454(7203):428–435
37. Meyer AW (1922) Further observations on use-destruction in joints. J Bone Joint Surg IV:491–511
38. Millar NL, Reilly JH, Kerr SC, Campbell AL, Little KJ, Leach WJ, Rooney BP, Murrell GA, McInnes IB (2012) Hypoxia a critical mediator of early human tendinopathy. Ann Rheum Dis 71(2):302–310
39. Puddu G, Ippolito E, Postacchini F (1976) A classification of Achilles tendon disease. Am J Sports Med 4:145–150
40. Pufe T, Petersen WJ, Mentlein R, Tillmann BN (2005) The role of vasculature and angiogenesis for the pathogenesis of degenerative tendons disease. Scand J Med Sci Sports 15:211–222
41. Rees JD, Stride M, Scott A (2014) Tendons – time to revisit inflammation. Br J Sports Med 48:1553–1557
42. Ricciotti E, FitzGerald GA (2011) Prostaglandins and inflammation. Arterioscler Thromb Vasc Biol 31 (5):986–1000

43. Riley G (2008) Tendinopathy—from basic science to treatment. Nat Clin Pract Rheumatol 4:82–89
44. Schulze-Tanzil G, Al-Sadi O, Wiegand E, Busch C, Kohl B, Pufe T (2011) The role of pro-inflammatory and immunoregulatory cytokines in tendon healing and rupture: new insights. Scand J Med Sci Sports 21:337–351
45. Sullo A, Maffulli N, Capasso G et al (2001) The effects of prolonged peritendinous administration of PGE1 to the rat Achilles tendon: a possible animal model of chronic Achilles tendinopathy. J Orthop Sci 6:349–357
46. Torricelli P, Giaveresi G, Nicolini A, Giardono R (2006) Effects of systemic glucocorticoid administration on tenocytes. Biomed Pharmacother 60:380–385
47. Tsuzaki M, Guyton G, Garrett W, Archambault JM, Herzog W, Almekinders L, Bynum D, Yang X, Banes AJ (2003) IL-1 beta induces COX2, MMP-1, -3 and -13, ADAMTS-4, IL-1 beta and IL-6 in human tendon cells. J Orthop Res 21:256–264
48. Uchihashi K, Tsuruta T, Mline H, Aoki S, Nishijima-Matsunobu A, Yamamoto M, Kuraoka A, Toda S (2014) Histopathology of tenosynovium in trigger fingers. Pathol Int 64(6):276–282
49. Warrender WJ, Brown OL, Abboud JA (2011) Outcomes of arthroscopic rotator cuff repairs in obese patients. J Shoulder Elb Surg 20:961–967
50. Wendelboe AM, Hegmann KT, Gren LH, Alder SC, White GL Jr, Lyon JL (2004) Associations between body-mass index and surgery for rotator cuff tendinitis. J Bone Joint Surg Am 86-A:743–747
51. Williams RJ, Attia E, Wickiewicz TL, Hannafin JA (2000) The effect of ciprofloxacin on tendon, paratenon, and capsular fibroblast metabolism. Am J Sports Med 28(3):364–369
52. Yan SF, Barile GR, D'Agati V, Du Yan S, Ramasamy R, Schmidt AM (2007) The biology of RAGE and its ligands: uncovering mechanisms at the heart of diabetes and its complications. Curr Diab Rep 7:146–153
53. Zhang J, James H-C (2010) Wang production of PGE2 increases in tendons subjected to repetitive mechanical loading and induces differentiation of tendon stem cells into non-tenocytes. J Orthop Res 28:198–203

Deep Venous Thrombosis and Tendon Healing

21

Erica Domeij-Arverud and Paul W. Ackermann

Abstract

Tendon metabolism after acute Achilles tendon rupture (ATR) is associated with major complications related to immobilization, which results in reduced circulation, high risk of deep venous thrombosis (DVT), impaired healing and functional deficits.

DVT has been demonstrated to occur in up to 50 % of the patients with ATR. Suffering from a DVT during tendon healing has been demonstrated as an independent predictive factor for impaired patient outcome at 1 year after ATR, suggesting that specific interventions are warranted to prevent DVT. Since pharmacological DVT prophylaxis has low or no effect during lower leg immobilization it is speculated whether adjuvant treatment with intermittent pneumatic compression (IPC) applied during lower limb immobilization can reduce the incidence of DVT.

IPC, which acts through mechanical, chemical and molecular mechanisms, has been demonstrated to enhance neuro-vascular ingrowth in a tendon repair model and stimulate collagen production leading to improved maximum force during healing.

Recently, a prospective randomized trial compared adjuvant IPC applied under an orthosis versus plaster cast only in ATR patients. The study found at 2 weeks post-operatively 21 % DVTs in the IPC-group compared to 37 % in the control group. Patients that received no IPC treatment exhibited an almost threefold increased odds for DVT, independently of age. Furthermore, using microdialysis technique, adjuvant IPC treatment was shown to increase the metabolic healing activity at 2 weeks post-ATR.

E. Domeij-Arverud (✉)
Department of Molecular Medicine and Surgery, Karolinska Institutet, SE-17176, Stockholm, Sweden

Department of Orthopedic Surgery, Danderyd Hospital, Stockholm SE-17176, Sweden
e-mail: erica.domeij-arverud@ds.se

P.W. Ackermann
Department of Molecular Medicine and Surgery, Karolinska Institutet, SE-17176, Stockholm, Sweden

Department of Orthopedic Surgery, Karolinska University Hospital, SE-17176, Stockholm, Sweden
e-mail: paul.ackermann@karolinska.se

P.W. Ackermann and D.A. Hart (eds.), *Metabolic Influences on Risk for Tendon Disorders*, Advances in Experimental Medicine and Biology 920, DOI 10.1007/978-3-319-33943-6_21

Tendon healing is impaired by reduced circulation and DVT. The demonstration that adjuvant IPC effectively reduced DVT incidence, and also is capable of enhancing the metabolic response suggests that IPC treatment may not only be a viable means of prophylaxis against DVT, but possibly also a method of promoting healing.

Keywords
Deep venous thrombosis • Functional outcome • Tendon healing • Tendon metabolism • Intermittent pneumatic compression • Microdialysis

List of Acronyms and Abbreviations

ACOS	Achilles Combined Outcome Score
ATR	Achilles Tendon Rupture
ATRS	Achilles tendon Total Rupture Score
BMI	Body Mass Index
CDS	Color Duplex Sonography
CDU	Compression Duplex Ultrasound
DVT	Deep Venous Thrombosis
EQ-5D™	EuroQol, a generic health-related quality of life score
IPC	Intermittent Pneumatic Compression
LSI	Limb Symmetry Index
PAS	Physical Activity Scale
RR	Relative Risk or Risk Ratio
VTE	Venous Thromboembolism

Introduction

Venous thromboembolism (VTE) is a major health problem, which is often asymptomatic, mis-diagnosed, and unrecognized. Each year 25,000 people in the UK die from VTE. This is a larger number than deaths attributable to breast cancer, AIDS and road traffic accidents combined [1, 2].

VTE often starts as a deep venous thrombosis (DVT), which can occur without any symptoms. Around 25 % of patients with a DVT also suffer from pulmonary embolism, which in 6 % of the cases are fatal [2].

VTE often arises as a serious complication following trauma and surgery requiring prolonged immobilization of the lower limb [3–5]. Other risk factors for the development of VTE consist of active cancer or cancer treatment, age, dehydration, hormone replacement therapy, thrombophilias, obesity, medical comorbidities (eg. heart disease; metabolic, endocrine or inflammatory conditions) (see Chaps. 14, 15, 16, 17, 18, 19 and 20).

VTE is thus a disease associated with reduced blood flow and metabolic disorders, which both can affect tendon metabolism. VTE resulting in a DVT is a common complication following acute Achilles tendon rupture and may have an important effect on tendon healing and functional outcome.

DVT Is an Independent Predictor of Impaired Outcome During Tendon Healing

Patients suffering from an acute Achilles tendon rupture may be at increased risk of DVT, mostly due to the immobilization in an equinus position [6]. Immobilization in an equinus position reduces the venous blood flow of the lower limb [7]. Following plaster cast treatment of acute Achilles tendon ruptures the VTE incidence verified with color duplex sonography (CDS) was 36 %, irrespective of surgery or conservative treatment [8].

The clinical significance of distal-versus proximal DVTs, especially if radiologically verified, has been debated [9]. Between 17 and 28 % of distal DVTs propagate to proximal DVTs [10, 11] but no clinical or radiological finding can predict the outcome of a single distal DVT. Since a DVT is necessary for the development of PE as well as post-thrombotic syndrome, it is reasonable to believe that even distal DVTs should be prevented and treated.

Experiencing a DVT may also have additional negative effects on outcome during tendon healing. In a recent, not yet published study by Domeij-Arverud et al. predictors of outcome in ATR patients were assessed using a prospective cohort study design. Exhibiting a DVT during the post-operative immobilization was an important independent predictor for a poor outcome after ATR together with age over 40 years and male gender [12].

Patients experiencing a DVT postoperatively exhibited a significantly lower maximum heel-rise height of 9.9 cm as compared to patients without a DVT that reached 11.1 cm ($p = 0.018$). LSI of heel-rise height was significantly different between patients without (84 %) and patients with a DVT (75 %) ($p = 0.008$). LSI of total concentric work did not reach the recommended levels of 85 % in the DVT group (61 %), nor the non-DVT group (67 %) ($p = 0.164$). ATRS was lower in the DVT group (73.2) than in the healthy group (79.2), but the difference did not reach significance ($p = 0.131$) (Table 21.1).

Intermittent Pneumatic Compression (IPC) and Tendon Healing

Intermittent pneumatic compression (IPC) for improving circulation has actually been known in the medical literature since the eighteenth century, when physicians used compressive therapy against such varied diseases as cholera and tromboangiitis. By supplying external cyclic compressions and relaxations to upper or lower extremities, IPC mimics the muscle pump. The compression pressure is chosen dependent on the type of cuff; foot compression requires 130 mmHg, while calf and thigh compression requires 40–50 mmHg, to reach a similar clinical effects [13]. IPC exerts its effect through mechanical, chemical and molecular mechanisms [14].

Mechanical compression increases venous flow with >200 %, which decreases the peripheral resistance, whereby arterial blood flow is enhanced, and thereby also the nutritive supply to the tissue (see Chaps. 4 and 5). Mechanical compression also affects the interstitial circulation, which reduces oedema and thus diffusion distance for metabolites. The stretch deformation

Table 21.1 Outcome of ATR patients with a DVT compared with patients without a DVT at 1 year post-operatively

Variable	*DVT*					
	No (n = 77)		*Yes (n = 34)*			
	Mean	*SD*	*Mean*	*SD*	*p-value*	*% Difference*
Heel-rise height injured side (cm)	11.14	2.4	9.89	2.7	**0.018**	13
Heel-rise height (LSI)	84 %	15.3	75 %	15.2	**0.008**	12
Heel-rise concentric work (LSI)	67 %	17.6	61 %	25.6	0.164	10
ATRS	79.16	18.7	73.21	18.9	0.131	8
PAS postinjury	3.84	1	3.58	1.1	0.32	7
Heel-rise concentric power (LSI)	82 %	19.7	78 %	24.1	0.105	5
Heel-rise repetitions (LSI)	82 %	17.7	78 %	22.1	0.356	5
EQ-5D	0.918	0.1	0.906	0.1	0.604	1
BMI	26.8	3.4	26.6	2.9	0.778	1
PAS preinjury	4.6	0.9	4.63	0.8	0.877	−1

LSI limb symmetry index, *PAS* physical activity scale, *ATRS* achilles tendon total rupture score, *EQ-5D* EuroQol Group's questionnaire, *BMI* body mass index, *SD* standard deviation, *n* number of patients

of tissues and cells caused by the venous flow and mechanical compression produces shear stress on the endothelial cells, which increases the production of anti-thrombotic, pro-fibrinolytic, vasodilatory and repair promoting substances (Fig. 21.1). These effects are followed by structural tissue changes and alterations in cellular gene expression [14].

The potential role of IPC in enhancing healing post-surgery has been investigated to a minor extent, mainly demonstrating that IPC can reduce postoperative oedema [15–18]. IPC has proven positive effects on tendon, wound and fracture healing in experimental studies [18–21], although the mechanisms have been mostly unexplored. However, recently IPC was demonstrated to enhance neuro-vascular ingrowth in a tendon repair model such as to increase the expression of sensory neuropeptides by up to 100 % (see Chap. 5) [15]. In the same model, IPC was able during immobilization to improve maximum force by 65 %, energy by 168 %, organized collagen diameter by 50 %, and collagen III-VI occurrence by 150 % compared with immobilization only [19]. Whether IPC can counteract the negative effects of immobilization in patients still needs to be further investigated.

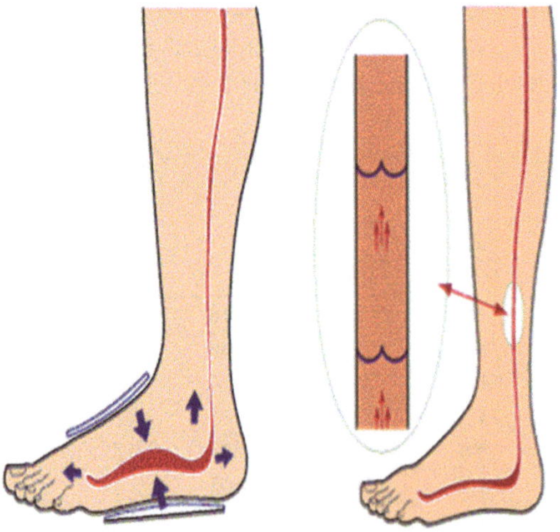

Fig. 21.1 Mechanisms of action of IPC applied to the lower limb: (1) increased venous circulation (2) increased arterial circulation (3) increased interstitial circulation with return of fluid to the veins (4) Increased venous flow produces shear stress on the endothelial cells, which (5) causes the production and release of ant-thrombotic and proliferative substances

Intermittent Pneumatic Compression (IPC) and Metabolic Activity

Adequate circulation is a healing prerequisite for sufficient supply of glucose and oxygen essential for fibroblast metabolism during ATR repair. Therefore study of the energy substrates needed for the repair process is of particular interest. Several studies have suggested that the prolonged tendon repair time until healing could be due to a low metabolic activity, which is even more restricted during limb immobilization [19, 22–25].

Essential metabolites glucose, lactate, pyruvate and glycerol and neuronal transmitters, including glutamate, have been identified in the healing tendon and are regarded to be involved in the repair process (see Chaps. 3 and 8) [26, 27]. In the study by Greve and Domeij-Arverud et al. [28], we hypothesized that vital metabolite concentrations are elevated during early human Achilles tendon repair, and that adjuvant IPC could further stimulate metabolic activity. Twenty patients were included in this prospective cohort study. The IPC group and the plaster cast group exhibited no demographical differences – except for a mean 80 h usage of adjuvant IPC in the treatment group during the first 2 weeks postoperatively.

The healing Achilles tendons exhibited at 2 weeks post-operatively significantly higher levels of glutamate, lactate, and pyruvate compared to the contralateral intact tendons. The observed findings suggest that Achilles tendon healing is associated with significantly elevated metabolic activity. Glutamate exhibited the highest (threefold) up-regulation suggesting the most central role in Achilles tendon repair. The elevation of glutamate concentrations in the healing tendons of the control group is in agreement with earlier experimental studies on Achilles tendon repair [26, 29].

Metabolite upregulations were observed in the healing Achilles tendons of the IPC-treated group as compared to the intact, contralateral tendons. Again, glutamate demonstrated the highest upregulation after IPC treatment. IPC treatment produces shear stress upon the endothelial walls of both intact and injured regenerating vessels. Vascular shear stress is giving rise to a multitude of reactions, including synthesis of nitric oxide (NO), a potent vasodilator assumed to stimulate angiogenesis, accelerated tissue healing and improved tissue quality [14].

Adjuvant compression treatment during post-operative plaster immobilization has the ability to additionally elevate the concentration of essential metabolites at the repair site. Continued studies on metabolic activity should be combined with assessments of long-term outcome of tendon repair for further understanding.

IPC Can Reduce the Risk of DVT During Lower Limb Immobilization

IPC has been shown to cause release of nitric oxide (NO), which has a vasodilating effect, and tissue factor pathway inhibitor that acts antithrombotically. IPC also cause release of tissue plasminogen activator that is pro-fibrinolytic, and has been found to up regulate mRNA for factors involved in angiogenesis and vasodilation [14].

Today, IPC is clinically used to prevent venous thromboembolism post-surgery [30]. IPC has been clearly proven to have an effect comparable to low molecular weight heparin (LMWH), but without any real side-effects and hereby offers an alternative to heparin when there is increased risk for bleeding complications [31].

Pharmacological prophylaxis with LMWH is still widely used although earlier studies have shown that LMWH does not reduce the risk of DVT during leg immobilization after ATR [32]. It has moreover been speculated that LMWH may impair tendon healing [33].

Studies also suggest that IPC in combination with LMWH can even further reduce the risk of DVT [34] The risk of poor patient compliance has sometimes been brought up as a drawback, and Eisele et al. showed that fewer than 6 h of IPC usage per day lead to more DVTs than usage exceeding 6 h per day [35]. As the devices are technically improved the conditions for adequate usage are enhanced. The American College of Chest Physicians (ACCP) has stated that further study of the efficacy of portable compression is needed [36].

In a prospective randomized trial of 150 included, operated ATR patients, the effect of adjuvant IPC during post-operative outpatient lower limb immobilization on the incidence of DVT were assessed [37]. Screening with compression duplex ultrasound (CDU) at 2 weeks post-operatively demonstrated a DVT in 23 % ($n = 16/69$) in the intention-to-treat analysis and 21 % ($n = 14/67$) in the per-protocol analysis of the patients in the IPC group and in 37 % ($n = 26/71$) of the patients in the control group (Table 21.2). The patients compliant with the IPC-treatment exhibited a significant reduction in the risk of DVT, after correction for age differences between groups (ITT analysis: $OR = 2.11$; 95 % C. I. = 0.97–4.62; $p = 0.061$ and PP analysis: $OR = 2.60$; 95 % C.I. 1.15–5.91; $p = 0.022$) (Table 21.2). No proximal DVTs and no clinical PEs were diagnosed.

The significantly reduced incidence of DVT of the IPC group indicates that IPC may prevent the development of DVT during immobilization. This study establishes patient-administered adjuvant calf IPC in an outpatient setting as a feasible method of DVT-prevention after ATR surgery.

The reduction in DVT observed at 2 weeks in the IPC group is probably related to reduced venous stasis produced by the intermittent mechanical calf compression. IPC has thus been shown to increase venous peak flow velocity by > 200 [14, 38]. The reduced incidence of DVT may also be related to effects on intrinsic fibrinolysis observed after IPC. Hence, cyclical tissue shear stress induces the production of chemical substances: antithrombotic tissue factor pathway inhibitor and pro-fibrinolytic substance tissue plasminogen activator [38]. In contrast, immobilization in equinus position and a non-

Table 21.2 Incidence of deep venous thrombosis (DVT)

	Adjuvant IPC			ITT vs. C	PP vs. C
Outcome (incidence)	ITT*	PP†	Control (C)	p-value	p-value[a]
DVTs at 2 weeks n (%)	16 *(23)*	14 *(21)*	26 *(37)*	0.083	0.042
DVTs at 6 weeks n (%)	36 *(52)*	34 *(51)*	34 *(48)*	0.612	0.737

IPC intermittent pneumatic compression, *DVT* deep vein thrombosis
a = 0.05
*ITT = Intention To Treat, i.e. all randomised patients
†PP = Per-Protocol. Two patients were withdrawn due to non-compliance regarding IPC – usage
Both patients had developed a DVT at the 2 weeks follow-up

weight-bearing regimen have been shown to reduce the venous blood flow significantly compared to full weight-bearing in a neutral cast or neutral pneumatic boot [7].

After ending the IPC intervention at week 2 and subsequent orthosis immobilization in both groups, CDS-screening at 6 weeks postoperatively showed a DVT in 52 % (n = 36/69) of the patients in the IPC group and in 48 % of (n = 34/71) the patients in the control group (OR = 0.94; 95 % C.I. 0.49–1.83). This suggests that the DVT preventive effect of the IPC therapy does not persist after cessation of treatment when continued immobilization is applied.

The correlation analysis of risk factors demonstrated that leg-immobilized patients aged >39 years exhibited an almost fivefold increased risk of DVT. Other risk factors, such as BMI, smoking, time to and time in surgery, did not significantly affect the risk of DVT in this study. Increased age is therefore a strong risk factor, as is also stated in many guidelines, e.g. by NICE, UK [39, 40]. The study therefore suggests that leg-immobilized patients aged >39 years, could be considered for prophylactic measures against DVT following ATR surgery.

Summary and Further Directions

Patients with lower limb immobilization after acute Achilles tendon rupture (ATR), represent a diverse patient population, which is affected by high incidence of DVT, and poor- and variable outcome.

Interventions to prevent the development of DVT should be of clinical focus to reduce complications and to promote the outcome after ATR. Pharmacological prophylaxis with LMWH is still widely used although earlier studies have shown that LMWH does not reduce the risk of DVT during leg immobilization after ATR [32]. Moreover, it has been speculated that LMWH may impair tendon healing [33]. Therefore the use of pharmacological prophylaxis during leg immobilization after ATR seems contradictory.

Mechanical thromboprophylaxis using IPC is an effective method of reducing the risk of DVT during leg immobilization, also in an outpatient setting. Patients with an ultrasound verified DVT exhibit an impaired outcome after ATR. Reduced blood circulation may be a common denominator to DVT and poor healing. Therefore it is our belief that increasing the blood circulation with IPC, e.g. reducing the risk of DVT, would also improve patient outcome as indicated by enhanced metabolism owing to IPC.

Glossary

Concentric muscle contraction When a muscle shortens while producing a force

Eccentric muscle contraction When a muscle lengthens while producing a force

Heel-rise The exercise in which the subject performs a plantar flexion when standing and back down again

Incidence The number of new cases of a condition/injury that develop during a specific time period

LSI The ratio of the involved limb score and the uninvolved limb score expressed in percent (involved/uninvolved x 100 = LSI)

Predictor The independent variable used to explain or to predict the outcome variable

Risk factor A variable associated with an increased risk of injury or disease

Work The product of a constant force and the distance the object is moved in the direction of the force (J)

References

1. Treasure T (2010) Venous thromboembolism: reducing the risk. In: CG92 (ed) Excellence NIfHaC. National Institute for Health and Clinical Excellence, London, p 24–8
2. Cohen AT, Agnelli G, Anderson FA et al (2007) Venous thromboembolism (VTE) in Europe. The number of VTE events and associated morbidity and mortality. Thromb Haemost 98(4):756–764
3. Nokes TJ, Keenan J (2009) Thromboprophylaxis in patients with lower limb immobilisation – review of current status. Br J Haematol 146(4):361–368
4. Bergqvist D, Lowe G (2002) Venous thromboembolism in patients undergoing laparoscopic and arthroscopic surgery and in leg casts. Arch Intern Med 162 (19):2173–2176
5. Geerts WH, Code KI, Jay RM, Chen E, Szalai JP (1994) A prospective study of venous thromboembolism after major trauma. N Engl J Med 331 (24):1601–1606
6. Karlsson JCJNvDCM N, Thermann H (2014) Achilles tendon disorders. A comprehensive overview of diagnosis and treatment. DJO Publications, London
7. Craik JD, Clark A, Hendry J, Sott AH, Hamilton PD (2015) The effect of ankle joint immobilization on lower limb venous flow. Foot Ankle Int 36(1):18–23
8. Nilsson-Helander K, Thurin A, Karlsson J, Eriksson BI (2009) High incidence of deep venous thrombosis after Achilles tendon rupture: a prospective study. Knee Surg Sports Traumatol Arthrosc 17 (10):1234–1238
9. Wille-Jorgensen P, Jorgensen LN, Crawford M (2005) Asymptomatic postoperative deep vein thrombosis and the development of postthrombotic syndrome. A systematic review and meta-analysis. Thromb Haemost 93(2):236–241
10. Lohr JM, James KV, Deshmukh RM, Hasselfeld KA, Allastair B, Karmody A (1995) Calf vein thrombi are not a benign finding. Am J Surg 170(2):86–90
11. Oishi CS, Grady-Benson JC, Otis SM, Colwell CW Jr, Walker RH (1994) The clinical course of distal deep venous thrombosis after total hip and total knee arthroplasty, as determined with duplex ultrasonography. J Bone Joint Surg Am 76(11):1658–1663
12. Domeij Arverud E (2015) Acute achilles tendon rupture: predictors and intervention to promote outcome [Doctoral PhD]. Karolinska Institutet, Stockholm
13. Khan RJK, Fick D, Keogh A, Crawford J, Brammar T, Parker M (2005) Treatment of acute achilles tendon ruptures – a meta-analysis of randomized, controlled trials. J Bone Joint Surg Am Vol 87A(10):2202–2210
14. Chen AH, Frangos SG, Kilaru S, Sumpio BE (2001) Intermittent pneumatic compression devices – physiological mechanisms of action. Eur J Vasc Endovasc Surg 21(5):383–392
15. Dahl J, Li J, Bring DK, Renstrom P, Ackermann PW (2007) Intermittent pneumatic compression enhances neurovascular ingrowth and tissue proliferation during connective tissue healing: a study in the rat. J Orthop Res 25(9):1185–1192
16. Park SH, Silva M (2003) Effect of intermittent pneumatic soft-tissue compression on fracture-healing in an animal model. J Bone Joint Surg Am Vol 85A (8):1446–1453
17. Park SH, Silva M (2008) Intermittent pneumatic soft tissue compression: changes in periosteal and medullary canal blood flow. J Orthop Res 26(4):570–577
18. Khanna A, Gougoulias N, Maffulli N (2008) Intermittent pneumatic compression in fracture and soft-tissue injuries healing. Br Med Bull 88(1):147–156
19. Schizas N, Li J, Andersson T et al (2010) Compression therapy promotes proliferative repair during rat Achilles tendon immobilization. J Orthop Res 28 (7):852–858
20. Challis MJ, Gaston P, Wilson K, Jull GA, Crawford R (2006) Cyclic pneumatic soft-tissue compression accelerates the union of distal radial osteotomies in an ovine model. J Bone Joint Surg (Br) 88(3):411–415
21. Challis MJ, Jull GJ, Stanton WR, Welsh MK (2007) Cyclic pneumatic soft-tissue compression enhances recovery following fracture of the distal radius: a randomised controlled trial. Aust J Physiother 53 (4):247–252
22. Boushel R, Langberg H, Green S, Skovgaard D, Bulow J, Kjaer M (2000) Blood flow and oxygenation in peritendinous tissue and calf muscle during dynamic exercise in humans. J Physiol 524 (Pt 1):305–313
23. Kjaer M, Langberg H, Skovgaard D et al (2000) In vivo studies of peritendinous tissue in exercise. Scand J Med Sci Sports 10(6):326–331
24. Magnusson SP, Langberg H, Kjaer M (2010) The pathogenesis of tendinopathy: balancing the response to loading. Nat Rev Rheumatol 6(5):262–268
25. Bring R, Renstrom S, Hart A (2009) Prolonged immobilization compromises up-regulation of repair genes after tendon rupture in a rat model. Scand J Med Sci Sports 20(3):411–417
26. Molloy TJ, Wang Y, Horner A, Skerry TM, Murrell GA (2006) Microarray analysis of healing rat Achilles tendon: evidence for glutamate signaling mechanisms and embryonic gene expression in healing tendon tissue. J Orthop Res 24(4):842–855
27. Ackermann PW, Salo PT, Hart DA (2009) Neuronal pathways in tendon healing. Front Biosci (Landmark edition) 14:5165–5187

28. Greve K, Domeij-Arverud E, Labruto F et al (2012) Metabolic activity in early tendon repair can be enhanced by intermittent pneumatic compression. Scand J Med Sci Sports 22(4):E55–E63
29. Schizas N, Lian O, Frihagen F, Engebretsen L, Bahr R, Ackermann PW (2010) Coexistence of up-regulated NMDA receptor 1 and glutamate on nerves, vessels and transformed tenocytes in tendinopathy. Scand J Med Sci Sports 20(2):208–215
30. Rohrer O, Eicher M (2006) Effectiveness of intermittent pneumatic compression (IPC) on thrombosis prophylaxis: a systematic literature review. Pflege 19 (3):175–187
31. Eppsteiner RW, Shin JJ, Johnson J, van Dam RM (2010) Mechanical compression versus subcutaneous heparin therapy in postoperative and posttrauma patients: a systematic review and meta-analysis. World J Surg 34(1):10–19
32. Lapidus LJ, Rosfors S, Ponzer S et al (2007) Prolonged thromboprophylaxis with dalteparin after surgical treatment of achilles tendon rupture: a randomized, placebo-controlled study. J Orthop Trauma 21(1):52–57
33. Virchenko O, Aspenberg P, Lindahl TL (2008) Low molecular weight heparin impairs tendon repair. J Bone Joint Surg (Br) 90(3):388–392
34. Kakkos SK, Caprini JA, Geroulakos G, Nicolaides AN, Stansby GP, Reddy DJ (2009) Combined intermittent pneumatic leg compression and pharmacological prophylaxis for prevention of venous thromboembolism in high-risk patients. Eur J Vasc Endovasc Surg 37(3):364–365
35. Eisele R, Kinzl L, Koelsch T (2007) Rapid-inflation intermittent pneumatic compression for prevention of deep venous thrombosis. J Bone Joint Surg Am 89 (5):1050–1056
36. Falck-Ytter Y, Francis CW, Johanson NA et al (2012) Prevention of VTE in orthopedic surgery patients: antithrombotic therapy and prevention of thrombosis, 9th ed: American College of Chest Physicians Evidence-Based Clinical Practice Guidelines. Chest 141(2 Suppl):e278S–e325S
37. Domeij-Arverud E, Labruto F, Latifi A, Nilsson G, Edman G, Ackermann PW (2015) Intermittent pneumatic compression reduces the risk of deep vein thrombosis during post-operative lower limb immobilisation: a prospective randomised trial of acute ruptures of the Achilles tendon. Bone Joint J 97-b(5):675–680
38. Domeij-Arverud E AJ, Labruto F, Ackermann P (2013) Adjuvant compression therapy in orthopaedic surgery – an evidence-based review. Eur Orthop Traumatol 49–57
39. NICE (2010) Guidelines: venous thromboembolism: reducing the risk NICE clinical guideline 92. 2010 January
40. Treasure T, Hill J (2010) NICE guidance on reducing the risk of venous thromboembolism in patients admitted to hospital. J R Soc Med 103(6):210–212

Drug-Induced Tendon Disorders

22

Karsten Knobloch

Abstract

Drug-induced tendon disorders are an often underestimated risk factor. The range from detrimental effects on the tendon include tendinopathy as well as potentially tendon rupture. As for today, four main drug classes have been reported to be associated with potentially deteriorated tendon properties: 1. Corticosteroids, 2. Chinolon antibiotics, 3. Aromatase inhbitors, 4. Statins as HMG-CoA-reductase inhibitors. Most often, the Achilles tendon is affected in terms of tendinopathy and/or subsequent tendon rupture. However, nearly every tendon of the entire body might be affected in a detrimental way by one or a combination of the aformentioned agents.

Keywords

Tendon • Corticosteroid • Tendon rupture • Statin • Chinolon • Aromatase inhibitor • Treatment • Adverse effect

Outline This chapter highlights the potential detrimental effects of four pharmacological substance families, such as corticosteroids, chinolones, aromatase inhibitors as well as statins on tendons in both, an experimental and clinical perspective.

Drug-induced tendon disorders are an often underestimated risk factor [1]. The range from detrimental effects on the tendon include tendinopathy as well as potentially tendon rupture. As for today, four main drug classes have been reported to be associated with potentially deteriorated tendon properties (Fig. 22.1):

1. Cortisone
2. Chinolone antibiotics
3. Aromatase inhibitors
4. Statins

To affect a tendon in a detrimental way, there is often times not only a single causal factor, but a multifactorial pattern. As such, patient age over

K. Knobloch (✉)
SportPraxis Prof. Knobloch, Heiligerstr. 3, Hannover 30159, Germany
e-mail: professor.knobloch@sportpraxis-knobloch.de; http://www.sportpraxis-knobloch.de

P.W. Ackermann and D.A. Hart (eds.), *Metabolic Influences on Risk for Tendon Disorders*, Advances in Experimental Medicine and Biology 920, DOI 10.1007/978-3-319-33943-6_22

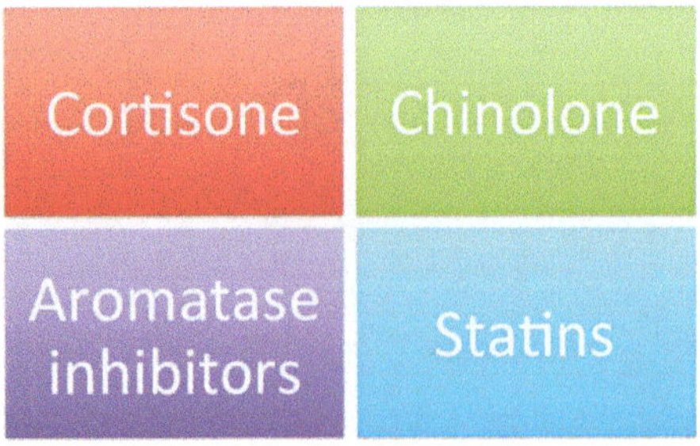

Fig. 22.1 Drugs associated with tendon disorders

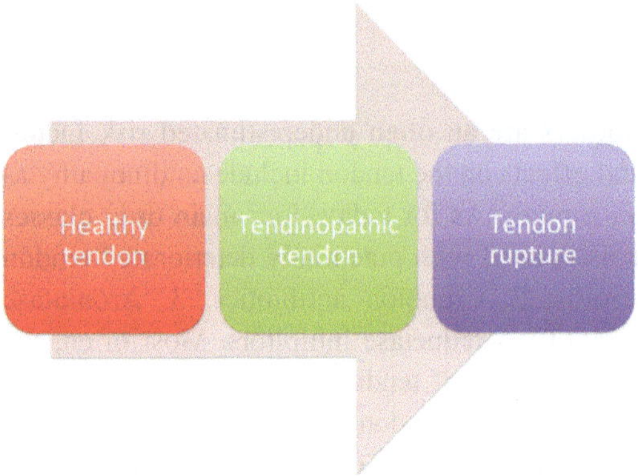

Fig. 22.2 Tendon continuum from the healthy tendon to the painful tendinopathic tendon with abnormalities in ultrasound ± MRI to the tendon rupture

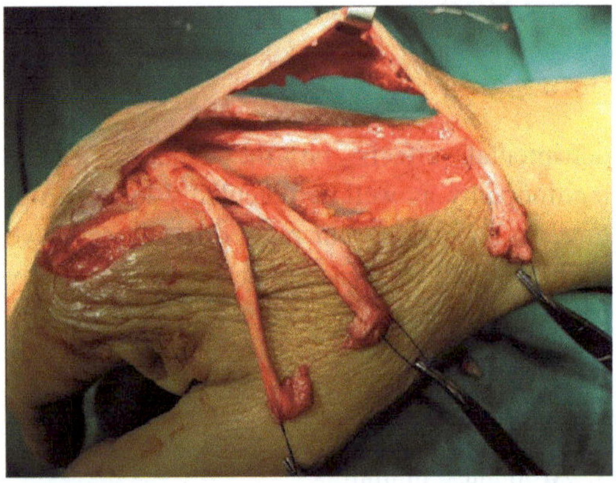

Fig. 22.3 Tendon rupture of extensor tendons of the hand following two corticosteroid injections on the extensor sheath of the wrist

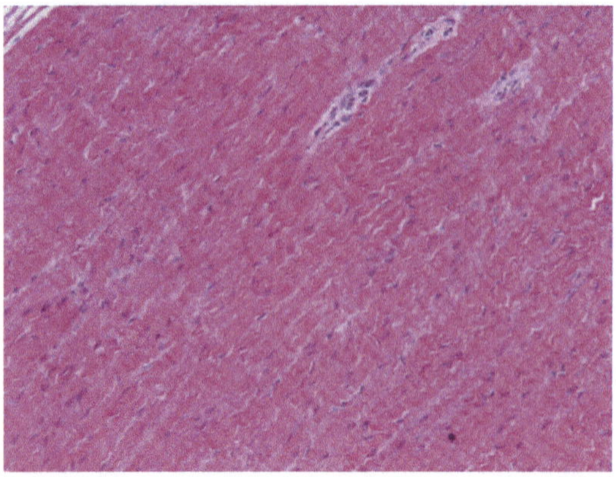

Fig. 22.4 Regular tendon staining of an extensor pollicis longus tendon

60 years, male gender, pre-existing tendon disorders or previous tendon ruptures have to be taken into account in this regard beyond the drug-related effects. A tendon might be deteriorate in a "tendon continuum" (Fig. 22.2): beginning with the healthy tendon, a number of the aforementioned risk factors might facilitate the tendinopathic stage, where affected tendons exert pain and often times show distinct alterations in grey scale ultrasound, Doppler ultrasound or MRI. The more deteriorated a tendon is, such as an Achilles tendon with a tendon diameter of >15 mm in the anterior-posterior ultrasound view, the more likely is a tendon rupture as the end of the tendon continuum (Fig. 22.3).

Cortisone

As early as 1958, a bilateral rupture of the quadriceps tendons has been reported in a patient suffering from disseminated lupus erythematosos, who was treated with cortisone [2]. In 1964 a second report highlighted another bilateral quadriceps tendon rupture in systemic lupus erythematosus was treated with cortisone medication [3]. Until now, tendon ruptures following corticosteroid medication have reported at the entire body, such as the Achilles tendons [4], triceps tendons [5], trigger finger with deep flexor tendon rupture [6], extensor pollicis longus tendon (EPL) of thumb (Figs. 22.3–22.8) [7], and many others. Repeated corticosteroid injections appear to increase the risk of a tendon rupture (see Chap. 23) [8].

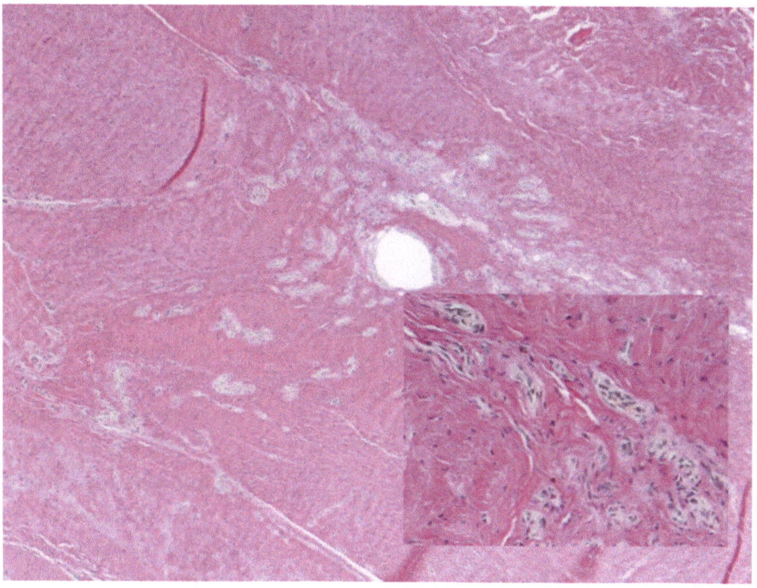

Fig. 22.5 Pathological staining with disruption of the tendon architecture with subsequent tendon rupture of an extensor pollicis longus tendon (EPL) following two corticosteroid injections

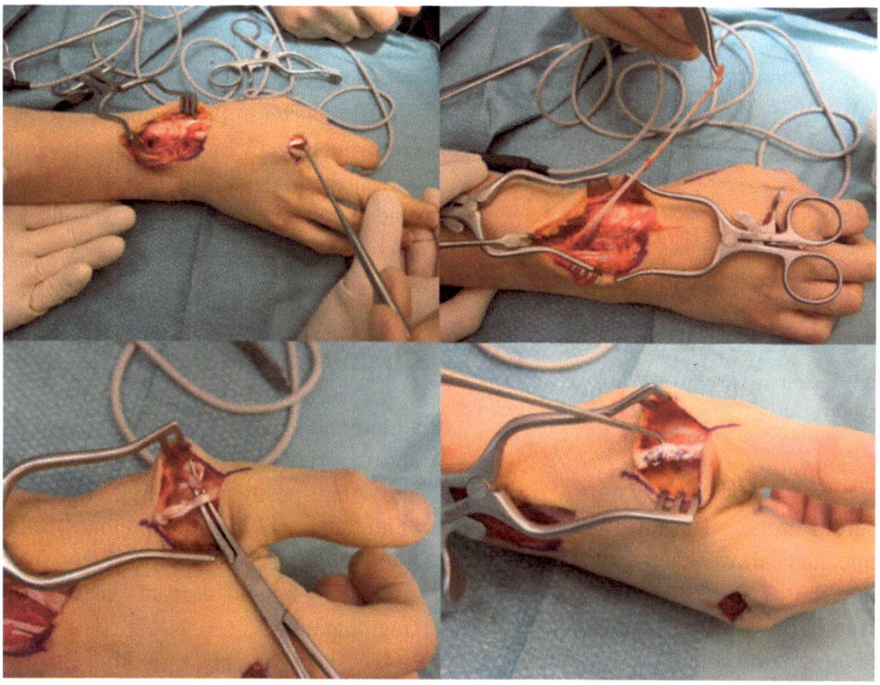

Fig. 22.6 Tendon transfer for repair of a rupture of an extensor pollicis longus tendon following two corticosteroid injections

Beyond the antiinflammatory effect of corticosteroids, tendons might substantially suffer in the following. Clinically, tendon rupture typically occur some 2–6 weeks following a corticosteroid injection, such as at the Achilles tendon level [9].

In experimental models both, triamcinolone as well as prednisolone injections have been studied in terms of subsequent tendon ruptures [10]. Histological analysis, at 1-week post-injection, showed collagen attenuation, increased expression of MMP-3 and apoptotic cells in the

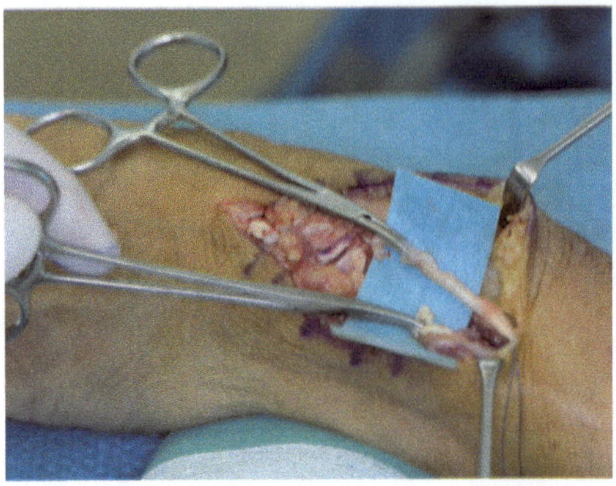

Fig. 22.7 Simultaneous rupture of the extensor indicis and the extensor digitorum communis 2 tendon following five repetitive corticosteroid injections with substantial tendon defects

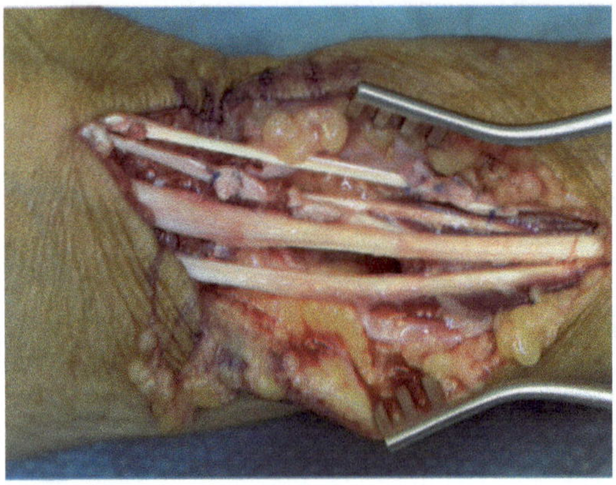

Fig. 22.8 Tendon repair by palmaris longus tendon transplantation for extensor indicis and extensor digitorum communis 2 tendon rupture following corticosteroid injections

corticosteroid groups. The histological changes and biomechanical weaknesses of the tendon were not seen at 3 weeks. These alterations appeared to be involved in tendon degeneration or rupture after corticosteroid injection.

The largest clinical metaanalysis, published 2010 in the Lancet by Coombes and coworkers [11] with 41 included randomized trials and 2672 participants showed that corticosteroid injections reduced pain in the short term compared with other interventions, but this effect was reversed at intermediate and long terms.

In a pooled analysis of treatment for lateral epicondylalgia, corticosteroid injection had a large effect on reduction of pain compared with no intervention in the short term, but no intervention was favoured at intermediate term and long term.

Of 991 participants who received corticosteroid injections in studies that reported adverse events, only one (0.1 %) had a serious adverse event (tendon rupture).

Short-term efficacy of corticosteroid injections for rotator-cuff tendinopathy is not clear. A recent experimental study in rats found s significantly lower maximal load by 20–40 % after corticosteroid injection in comparison to same volume saline injection in rotator cuffs [12].

Clinically, it appears that corticosteroid injection therapy in or besides the tendons may provoke the most detrimental effect on the tendon by local toxicity. However, to what extent a systematic corticosteroid therapy in terms of dosage and duration and to what extent a topical corticosteroid ointment therapy is detrimental for an underlying tendon is not thoroughly studied yet. From the authors point of view, corticosteroid therapy itself might harm the tendon. Thus, every corticosteroid therapy by itself should warrant a clear and substantial medical indication and a potential tendon adverse effect should be considered in my personal point of view.

Chinolone Antibiotics

Chinolone antibiotics are broad spectrum antibacterial drugs against both, gram-negative and gram-positive bacteria. The majority of chinolones are fluorochinolones. In bacteria, fluorochinolones inhibt selectively the topoisomerase II ligase domain leading to DNA fragmentation.

To date, as much as four generations of chinolones are known (table). The most prominent chinolones include 2nd generation ciprofloxacin and ofloxacin, 3rd generation levofloxacin, and 4th generation moxiflocacin (Table 22.1).

Clinically, chinolone antibiotics are predominantly used in hospital-acquired pneumonia [13] as well as urinary tract infections, both community-acquired as well as hospital-acquired including urinary tract infections associated with urinary catheters.

A German case report [14] illustrates a 45-year-old female runner with bilateral tendinopathy of the Achilles tendons after repeated ciprofloxacin treatment. One month later the right Achilles tendon ruptured without any sudden pain. The histological analysis showed cystic changes with focal necrosis. This finding differed from findings in non drug induced tendinopathies. Beyond the Achilles tendons, other tendons might deteriorate following chinolone medication, such as the gluteal tendons [15].

Recently, an Austrian working group analysed the effect of multiparametric MR imaging in healthy volunteers following ciprofloxacin medication [16]. Fourteen ankles of seven healthy men underwent 7-T MR imaging at baseline, 10 days and 5 months after ciprofloxacin administration. The sodium signal was decreased by 25 % (from 130 ± 8 AU to 98 ± 5 AU, $p < 0.05$) at day 10 and returned to baseline levels after 5 months. In line, a change in the glycosaminoglcan (GAG) imaging was noted accordingly.

A detailed animal study in rats analysed the effects of four fluoroquinolones (pefloxacin, norfloxacin, ofloxacin and ciprofloxacin) [17]. Pefloxacin mesylate dihydrate (40 mg/kg), norfloxacin (40 mg/kg), ofloxacin (20 mg/kg) and ciprofloxacin (50 mg/kg) were administered by gavage twice daily for three consecutive weeks. Six weeks after treatment, the test animals were euthanised and Achilles tendon specimens were collected. The mean elastic modulus of the control group was significantly higher than that of the norfloxacin and pefloxacin groups. The mean yield force of the control group was significantly higher than those of ciprofloxacin, norfloxacin and pefloxacin groups. The mean ultimate tensile force of the control group was significantly higher than of the ciprofloxacin, norfloxacin, and pefloxacin groups. Hyaline degeneration and fibre disarrangement were observed in the tendons of the ciprofloxacin, pefloxacin, and ofloxacin treated-groups, whereas myxomatous degeneration was observed only in the ciprofloxacin and pefloxacin groups.

Aromatase Inhibitors

Aromatase inhibitors are used to decrease circulating estrogen levels in postmenopausal women. Aromatase ist he responsible enyzme for the conversion of androgens in estrogens (Fig. 22.9). As such, aromatase inhibitors block this conversion (Fig. 22.10). This principle is

Table 22.1 First, second, third and fourth generation chinolone antibiotics which may impair tendons leading to tendinopathy and/or tendon rupture

1st generation chinolone	2nd generation chinolone	3rd generation chinolone	4th generation chinolone
Cinoxacin	Ciprofloxacin	Balofloxacin	Clinafloxcin
Nalidixic acid	Enoxacin	Grepafloxacin	Gatifloxacin
Oxolinic acid	Fleroxacin	Levofloxacin	Gemifloxacin
Piromidic acid	Lomefloxacin	Pazufloxacin	Moxiflocacin
Pipemidic acid	Nadifloxacin	Sparfloxacin	Sitafloxacin
Rosaxacin	Norfloxacin	Temafloxacin	Trovafloxacin
	Ofloxacin	Tosufloxacin	Prufiloxacin
	Pefloxacin		
	Rufloxacin		

Fig. 22.9 Conversion of testosterone to estradiol via aromatase

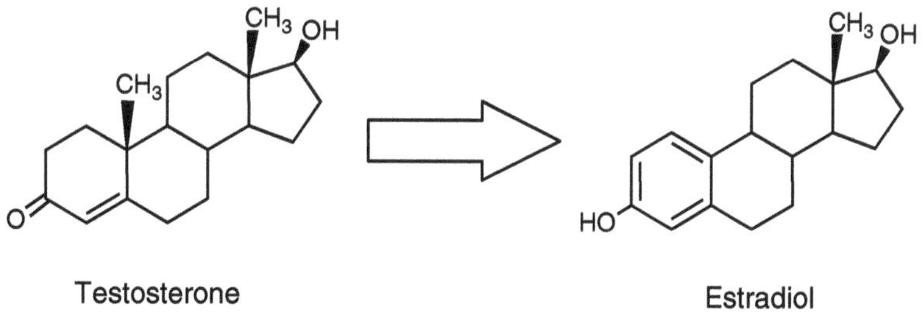

increasingly used as an adjuvant therapy in postmenopausal women with hormone receptor-positive invasive breast cancer [18, 19].

Musculoskleletal adverse effects typically arise within 3 months of therapy initiation [20]. A MRI study revealed a significant number of wrist problems with tenosynovitis among these females [21] (Table 22.2).

A prospective analysis determined the fluid accumulation and tenosynovial changes in the aromatase inhibitor-induced musculoskeletal syndrome with 2-year-follow-up data [22]. Among 33 postmenopausal breast cancer patients, 27 received aromatase inhibitors and six received tamoxifen. Between 6 and 24 months, the intraarticular fluid further increased in all aromatase inhibitor patients but not in the tamoxifen group. Grip strength further decreased in both groups. The worsened tenosynovial changes were strongly correlated with a decrease in grip strength. At 24 months, morning stiffness continued to be present in over a third of aromatase inhibitor users. The discontinuation of aromatase inhibitor treatment >1 year before assessment led to an increase in grip strength in three patients (two switched to tamoxifen and one ended endocrine therapy). Furthermore, tenosynovitis was seen on MRI in these patients after 6 months of AI therapy, and this abnormality improved at the 2-year assessment in the two patients who had switched to tamoxifen. Although the number of patients is small, these findings suggest that the AI-induced changes are reversible, at least in a subset of patients.

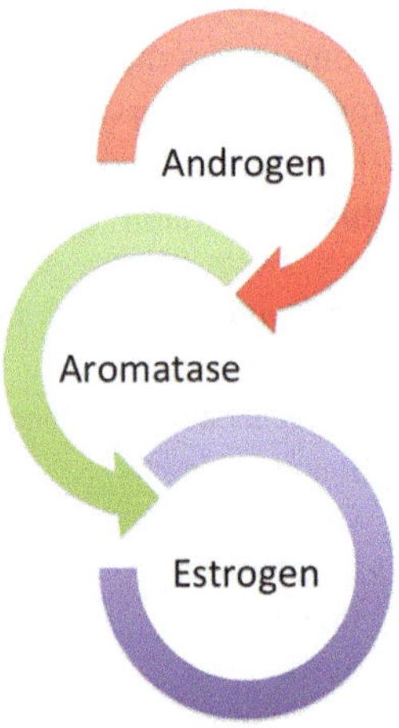

Fig. 22.10 Aromatase is the responsible enzyme for the conversion of androgen such as testosterone to estrogen

Adipose Tissue and Tendons

Fat pads, such as the Kager's fat pad [23] of the foot or the Hoffa fat pad of the knee do play a role in inflammation in tendinopathies of the associated tendons, the insertional Achilles tendon for the Kager's fat pad and the Hoffa fat pad for the patella tendon. As such, patients suffering from Achilles tendinopathy especially in insertional cases demonstrate an increased soft tissue density of the Kager's fat pad in association with an increased Achilles tendon diameter [24].

The infrapatellar fat pad (Hoffa's fat pad) may degenerate to a chronic edema with subsequent soft tissue impingement, ischemia and lipomatous tissue necrosis [25, 26]. The Hoffa fat pad may contain a number of inflammatory cells as well as is a source of adipokines, cytokines and grows factors modifying disease [27]. Adiponectin has been shown to induce matrix metallaproteinase-1 and interleukin-(IL)-6 expression in synovial fibroblasts and thus, may exert inflammatory functions [28, 29]. However, the detailed study on the Karger fat pad

Table 22.2 First, second and third generation aromatase inhibitors in clinical practice

	Nonsteroidal aromatase inhibitor	Steroidal aromatase inhibitor
1st generation	Aminoglutethimide	Testonolactone
2nd generation	Fadrozole	Formestane
3rd generation	Anastrozole	Exemestane
	Letrozole	
	Vorozole	

found a reduced Adiponectin mRNA levels in Achilles tendon patients in contrast to healthy controls, therefore Summarizing these observations, adipose tissue as such do play a role in inflammation.

Dyslipidemia has been reported to be present in Achilles tendinopathy in a matched-pair analysis (see Chap. 14) [30]. A detailed comparison of the lipid profile of 60 symptomatic patients and 60 control subjects matched for gender, age ±10 years, and body-mass-index (BMI) found higher triglyceride (TG) levels, lower %HDL-cholesterin, a higher triglyceride/HDL-c ratio, and an elevated apolioprotein B, concentration among the patients suffering from Achilles tendinopathy. The authors concluded: "If tendinopathy is confirmed to be associated with dyslipidemia and the metabolic syndrome in larger studies, it may be appropriate to redefine our concept of tendinopathy to that of a cardiovascular disease (CVD). In this case, we may be able to draw considerably on CVD research to improve our understanding of tendinopathy, and perhaps treating CVD risk factors will improve the treatment of tendinopathy."

In line, men with Achilles tendinopathy are older and have a high waist circumference [31]. Elevated adiposity is frequently associated with tendon pathology [32] in a review with a total number of 19.949 individuals. In 43 % of cases, the group with tendinopathy had significantly greater adiposity levels than the control group without tendinopathy. Two other key findings of that review are worth repeating. First, when upper-limb and lower-limb tendinopathies were compared the association with adiposity was equally strong. As the lower-limb tendons support body-weight while the upper-limb tendons support only the weight of the limb, this finding suggested that mechanical loading does not fully explain the association between adiposity and tendinopathy. Second, the longitudinal studies in the review that showed baseline adiposity predicted tendinopathy at follow-up suggested that adiposity is a risk factor for tendinopathy rather than a consequence of tendinopathy (see Chap. 15).

Statins

Statins are a cholerestrol-lowering drugs blocking the HMG-CoA-Reductase. In cardiovascular disease such as coronary artery disease, statins have been proofed to be beneficial. More recently, additional positive clinical results have been published such as in new onset atrial fibrillation [33]. Statins are believed to remodel the cytoskeletal architecture and mediate various anti-inflammatory, antioxidant, and antiproliferative effects that might limit endothelial dysfunction [34].

A number of clinical reports highlight a potential association of statin therapy and tendinopathy as well as tendon ruptures. As such, a 64-year-old man with a BMI of 31 kg/m2 under regular statin therapy suffered a bilateral quadriceps tendon rupture, where a potential associated of the statin therapy was questioned [35].

Experimental studies in rats revelaed that simvastatin, atorvastatin and rosuvastatin caused a deterioration of the biomechanical properties of the Achilles tendon (see Chap. 14) [36]. Histopathological analysis demonstrated foci of dystrophic calcification only in the statin-treated groups. However, the number and the total area of calcific deposits were similar between the statin groups in this experimental study.

Another experimental study reported a dose-related deterioration of collagen fibers and a decreased biomechanical strength of Achilles tendons follwoing atrovastatin or simvastatin treatment [37].

Recently, the effect of lovastatin was analyzed in human tendon cells [38]. The cells were cultured with different concentrations of lovastatin for up to 1 week. No changes in cell viability or morphology were observed in tenocytes incubated with therapeutic doses. Short-term exposure to lovastatin concentrations outside the therapeutic range had no effect on tenocyte viability; however, cell migration was reduced. Simvastatin and atorvastatin, two other drug family members, also reduced the migratory properties of the cells. Prolonged exposure to high concentrations of lovastatin induced changes in cytoskeleton leading to cell rounding and decreased levels of mRNA for matrix proteins, but increased BMP-2 expression. Gap junctional communication was impaired but due to cell shape change and separation rather than direct gap junction inhibition. These effects were accompanied by inhibition of prenylation of Rap1a small GTPase. Collectively, we showed that statins in a dose-dependent manner decrease migration of human tendon cells, alter their expression profile and impair the functional network, but do not inhibit gap junction function.

Potential mechanisms of statin-induced tendinopathy

- Statins inhibit the secretion of metalloproteinases (MMPs) in lung fibroblasts [39] and endothelial cells [40]
- Reduction of collagen I expression in smooth muscle [41]

Notably, on the other hand, simvastatin as a statin has been reported recently to reduce fibrosis and protects against muscle weakness in a experimental model in rats with a full-thickness supraspinatus tear [42]. In line, similar beneficial effects of atrovastatin have been underscored by Dolkart et al. in an experimental rat rotator cuff model [43]. The results indicate that atrovastatin enhances tendon healing by stimulating tenocyte proliferation, migration, and adhesion via increased COX2 activity and autocrine/paracrine PGE2 signaling. These findings also demonstrate that this effect is mediated by EP4 signaling.

> Therefore, the amount and the timing of statin therapy may well play a role whether any detrimental or potentially benefical effects of statins on tendon properties is observed. While direct detrimental effects of statins on tendons have been shown as aforementioned, the experimental data on rotator cuff healing might warrant further investigations.

References

1. Kirchgesner T, Larbi A, Omoumi P, Malghem J, Zamali N, Manelfe J, Lecouvet F, Vande Berg B, Djebbar S, Dallaudiere B (2014) Drug-induced tendinopathy: from physiology to clinical applications. Joint Bone Spine 81(6):485–492
2. Martin JR, Wilson CL, Mathews WH (1958) Bilateral rupture of the ligamenta patellae in a case of disseminated lupus erythematosus. Arthritis Rheum 1(6):548–552
3. Twinning RH, Marcus WY, Garey JL (1964) Tendon rupture in systemic lupus erythematosus. JAMA 189:377–378
4. Vallone G, Vittorio T (2014) Complete Achilles tendon rupture after local infiltration of corticosteroids in the treatment of deep retrocalcaneal bursitis. J Ultrasound 17(2):165–167
5. Celli A (2015) Triceps tendon rupture: the knowledge acquired from the anatomy to the surgical repair. Musculoskleletal Surg 2015 May 10
6. Oh J, Jo L, Lee JI (2015) Do not rush to return to sports after trigger finger injection. Am J Phys Med Rehabil 94(4):e26–e30
7. Boussakri H, Bouali A (2014) Subcutaneous rupture of the extensor pollicis longus tendon after corticosteroid injections for DeQuervain's stenosing tenovaginitis. Case Rep Orthop 2014:934384
8. Boussakri H, Bouali A (2014) Subcutaneous rupture of the extensor pollicis longus tendon after corticosteroid injections in DeQuervain stenosing tendovaginitis. Case Rep Orthop 2014:934384
9. Kearney RS, Parsons N, Metcalfe D, Costa ML (2015) Injection therapies for Achilles tendinopathy. Cochrane Database Syst Rev 5:CD010960
10. Muto T, Kokubu T, Mifune Y, Inui A, Harada Y, Yoshifumi, Takase F, Kuroda R, Kurosaka M (2014) Temporary inductions of matrix metalloprotease-3 (MMP3) expression and cell apoptosis are associated with tendon degeneration or rupture after corticosteroid injection. J Orthop Res 32(10):1297–1304
11. Coombes BK, Bisset L, Vicenzino B (2010) Efficacy and safety of corticosteroid injections and other injections for management of tendinopathy: a systematic review of randomised controlled trials. Lancet 376(9754):1751–1767. doi:10.1016/S0140-6736(10)61160-9. Epub 2010 Oct 21
12. Maman E, Yehuda C, Pritsch T, Morag G, Brosh T, Sharfman Z, Dolkart O (2015) Detrimentral effect of repeated and single subacromial corticosteroid injections on the intact and injured rotator cuff: a biomechanical and imaging study in rats. Am J Sports Med 2015 Jul 27
13. American Thoracic Society; Infectious Diseases Society of America (2005) Guidelines for the management of adults with hospital-acquired, ventilator-associated, and healthcare-associated pneumonia. Am J Respir Crit Care Med 171(4):388–416

14. Petersen W, Laprell H (1998) Insidious rupture of the Achilles tendon after ciprofloxacin-induced tendinopathy. A case report. Unfallchirurg 101 (9):731–734
15. Shimatsu K, Subramaniam S, Sim H, Aronowitz P (2014) Ciprofloxacin-induced tendinopathy of the gluteal tendons. J Gen Intern Med 29 (11):1559–1562
16. Juras V, Winhofer Y, Szomolanyi P, Vosshenrich J, Hager B, Wolf P, Weber M, Luger A, Trattnig S (2015) Multiparametric MR imaging depicts Glycosaminoglycan change in the Achilles tendon during ciprofloxacin administration in healthy men: initial observation. Radiology 275(3):763–771
17. Olcay E, Beytemur O, Kaleagasioglu F, Gulmez T, Mutlu Z, Olgac V (2011) Oral toxicity of pefloxacin, norfloxacin, ofloxacin and ciprofloxacin: comparison of biomechanical and histopathological effects on Achilles tendon in rats. J Toxicol Sci 36(3):339–345
18. Goss PE, Ingle JN, Martino S, Robert NJ, Muss HB, Piccart MJ et al (2005) Randomized trial of letrozole following tamoxifen as extended adjuvant therapy in receptor-positive breast cancer: updated findings from NCIC CTG MA. 17. J Natl Cancer Inst 97:1262–1271. [PubMed: 16145047]
19. Winer EP, Hudis C, Burstein HJ, Wolff AC, Pritchard KI, Ingle JN et al (2005) American Society of Clinical Oncology technology assessment on the use of aromatase inhibitors as adjuvant therapy for postmenopausal women with hormone receptor-positive breast cancer: status report 2004. J Clin Oncol 23:619–629. [PubMed: 15545664]
20. Henry NL, Giles JT, Ang D, Mohan M, Dadabhoy D, Robarge J et al (2008) Prospective characterization of musculoskeletal symptoms in early stage breast cancer patients treated with aromatase inhibitors. Breast Cancer Res Treat 111:365–372. [PubMed: 17922185]
21. Morales L, Pans S, Paridaens R, Westhovens R, Timmerman D, Verhaeghe J et al (2006) Debilitating musculoskeletal pain and stiffness with letrozole and exemestane: associated tenosynovial changes on magnetic resonance imaging. Breast Cancer Res Treat 104:87–91. [PubMed: 17061044]
22. Lintermans A, Laenen A, Van Calster B et al (2013) Prospective study to assess fluid accumulation and tenosynovial changes in the aromatase inhibitor-induced musculoskeletal syndrome: 2-year-follow-up data. Ann Oncol 24(2):350–355
23. Ly JQ, Bui-Mansfield LT (2004) Anatomy of and abnormalities associated with Kager's fat Pad. AJR Am J Roentgenol 182:147–154
24. Pingel J, Petersen MC, Fredberg U, Kjaer SG, Quistorff B, Langberg H, Hansen JB (2015) Inflammatory and metabolic alterations of the Kager's fat pad in chronic Achilles tendinopathy. PLoS One 10 (5):e0127811
25. Hoffa A (1904) The influence of the adipose tissue with regard to the pathology of the knee joint. JAMA 43:795–796
26. Magi M, Branca A, Bucca C, Langerame V (1991) Hoffa disease. Ital J Orthop Traumatol 17:211–216
27. Ioan-Facsinay A, Kloppenburg M (2013) An emerging player in knee osteoarthritis: the infrapatellar fat pad. Arthritis Res Ther 15:225
28. Tang CH, Chiu YC, Tan TW, Yang RS, Fu WM (2007) Adiponectin enhances IL-6 production in human synovial fibroblast via an AdipoR1 receptor, AMPK, p38, and NF-kappa B pathway. J Immunol (Baltimore, MD: 1950) 179:5483–5492
29. Gomez R, Lago F, Gomez-Reino J, Dieguez C, Gualillo O (2009) Adipokines in the skeleton: influence on cartilage function and joint degenerative diseases. J Mol Endocrinol 43:11–18. doi:10.1677/JME-08-0131
30. Gaida JE, Alfredson L, Kiss ZS, Wilson AM, Alfredson H, Cook JL (2009) Dyslipidemia in Achilles tendinopathy is characteristic of insulin resistance. Med Sci Sports Exerc 41(6):1194–1197
31. Gaida JE, Alfredson H, Kiss ZS, Bass SL, Cook JL (2010) Asymptomatic Achilles tendon pathology is associated with a central fat distribution in men and a peripheral fat distribution in women: a cross sectional study of 298 individuals. BMC Musculoskelet Disord 11:41
32. Gaida JE, Ashe MC, Bass SL, Cook JL (2009) Is adiposity an under-recognized risk factor for tendinopathy? A systematic review. Arthritis Rheum 61(6):840–849
33. Ho LT, Lin LY, Yang YH, Wu CK, Juang JJ, Wang YC, Tsai CT, Lai LP, Hwang JJ, Chiang FT, Lin JL, Chen PC (2015) Statin therapy lowers the risk of new-onset atrial fibrillation in patients with end-stage renal disease. Int J Cardiol 201:538–543
34. Bhandari S, Gupta P, Quinn P, Sandhu J, Hakimi A, Jones D, Ng L (2015) Pleiotropic effects of statins in hypercholerolaemia: a prospective observational study using a lipoproteomic based approach. Lancet 2015;385 Suppl 1:S21
35. Thomsen LL, Laursen JO (2014) Spontaneous bilateral quadriceps tendon rupture in obese patient medicated with statin. Ugeskr Laeger 176:50
36. Kaleagasioglu F, Olcay E, Olgac V. Statin-induced calcific Achilles tendinopathy in rats: comparison of biomechanical and histopathological effects of simvastatin, atorvastatin and rosuvastatin. Knee Surg Sports Traumtol Arthrosc 2015 Aug 15
37. De Oliveira LP, Vieira CP, Guerra FD, Almeida MS, Pimentel ER (2015) Structural and biomechanical changes in Achilles tendon after chronic treatment with statins. Food Chem Toxicol 77:50–57
38. Kuzma-Kuzniarska M, Cornell HR, Moneke MC, Carr AJ, Hulley PA (2015) Lovastatin-mediated changes in human tendon cells. J Cell Physiol 230 (10):2543–2551
39. Kamio K, Liu XD, Sugiura H, Togo S, Kawasaki S, Wang X, Ahn Y, Hogaboam C, Rennard SI (2010) Statins inhibit matrix

metalloproteinase release from human lung fibroblasts. Eur Respir J 35:637–646
40. Izidoro-Toledo TC, Guimarães DA, Belo VA, Gerlach RF, Tanus-Santos JE (2011) Effects of statins on matrix metalloproteinases and their endogenous inhibitors in human endothelial cells. Naunyn-Schmiedeberg's Arch Pharmacol 383:547–554
41. Schaafsma D, Dueck G, Ghavami S, Kroeker A, Mutawe MM, Hauff K, Xu FY, Mcneill KD, Unruh H, Hatch GM, Halayko AJ (2011) The mevalonate cascade as a target to suppress extracellular matrix synthesis by human airway smooth muscle. Am J Respir Cell Mol Biol 44:394–403
42. Davis ME, Korn MA, Gumucio JP, Harning JA, Saripalli AL, Bedi A, Mendias CL (2015) Simvastatin reduces fibrosis and protects against muscle weakness after massive rotator cuff tear. J Shoulder Elbow Surg 24(2):280–287
43. Dolkart O, Liron T, Chechik O, Somjen D, Brosh T, Maman E, Gabet Y (2014) Statins enhance rotator cuff healing by stimulating the COX2/PGE2/EP4 pathway: an in vivo and in vitro study. Am J Sports Med 42(12):2869–2876

The Effects of Glucocorticoid on Tendon and Tendon Derived Cells

23

Benjamin John Floyd Dean and Andrew Jonathan Carr

Abstract

Glucocorticoids are generally used to relieve pain and/or inflammation in a wide variety of musculoskeletal disorders including osteoarthritis, inflammatory arthritis, tendinopathy and degenerative spine disease. Glucocorticoids reduce tendon derived cell proliferation *in vitro* and reduce extracellular matrix synthesis both *in vitro* and *in vivo*, in particular type I collagen synthesis. Glucocorticoids also appear to result in acute deleterious changes in healthy *in vivo* tendon including collagen necrosis, collagen disorganisation and inflammatory cell infiltration; while the overall effect of glucocorticoid administration on the mechanical properties of healthy *in vivo* tendon are generally negative. Overall the existing *in vitro* and *in vivo* evidence suggests that glucocorticoids should be used with caution in treating painful tendinopathy. Certainly a real need exists to follow up the long term clinical effects of glucocorticoid in treating tendinopathy, as there is currently a paucity of evidence in this area. However in this context while the short term benefits are clear, glucocorticoids remain a useful treatment option provided they are used in the right patients in sensible moderation.

Keywords

Tendon • Glucocorticoid • Cells • Steroid • Tendinopathy

Background and Clinical Context

In the UK alone over 500 000 intra-articular glucocorticoid injections (GCIs) are administered per year in the primary care setting [1]. GCIs are generally used to relieve pain and/or inflammation in a wide variety of musculoskeletal disorders including osteoarthritis,

B.J.F. Dean (✉) • A.J. Carr
Nuffield Department of Orthopaedics, Rheumatology and Musculoskeletal Sciences, University of Oxford, Botnar Research Centre, Windmill Road, Oxford OX3 7LD, UK
e-mail: bendean1979@gmail.com

P.W. Ackermann and D.A. Hart (eds.), *Metabolic Influences on Risk for Tendon Disorders*, Advances in Experimental Medicine and Biology 920, DOI 10.1007/978-3-319-33943-6_23

inflammatory arthritis, tendinopathy and degenerative spine disease. There is an increasing evidence to suggest that chronic inflammation plays a role in the pathogenesis of tendinopathy [2–4]. The evidence regarding the clinical efficacy of GCIs is conflicting but broadly shows some short term benefits in terms of pain relief [5–8]. For example in the treatment of shoulder pain, trials have shown only short term benefits with no significant long term gains [8–10]. Emerging high quality evidence also points to poorer long term outcomes associated with GCIs in the treatment of tendinopathy [11].

GCIs are frequently applied in close proximity to tendons with common examples including the rotator cuff, the gluteus medius, the Achilles and the patellar tendons. It has been recurrently postulated that there is an increased risk of tendon rupture associated with GCI [12] but no high quality evidence exists to adequately confirm or refute this hypothesis [13, 14]. However there is strong evidence that oral corticosteroids are associated with a higher risk of tendon rupture (see Chap. 22) [15]. An increased spinal fracture risk associated with epidural GCIs has recently been reported [16]. The aim of this Chapter is first to describe the mechanisms of action of glucocorticoid, and secondly to summarise the effects of glucocorticoid on tendon cells *in vitro* and on tendon *in vivo*.

Mechanisms of Action of Glucocorticoid

The mechanisms of action of glucocorticoids are multiple, highly complex and incompletely understood [17, 18]. The most important pathway involves the activation of the cytoplasmic glucocorticoid receptor (GR) [19]. GR has a modular structure whose principal functions of transactivation, DNA binding, and ligand binding are localised to specific domains. Glucocorticoids bind to the ligand binding domain of the GR in the cytosol, thus activating the receptor by triggering a conformational change. The resulting activated form of the receptor has two principle mechanisms of action, transactivation and transrepression [19]. Transactivation involves translocation to the nucleus and binding to specific DNA responsive elements activating gene transcription. Transrepression involves the activated receptor binding with other transcription factors and repressing the activation of their relevant target genes. As well as having these well described transcriptional effects, glucocorticoids have also been increasingly demonstrated to have effects on post transcriptional, translational and post translational targets [19]. The majority of work relating to the mechanisms of action of glucocorticoids has been carried out in classically chronic inflammatory diseases such as asthma, demonstrating their anti-inflammatory effects to be mediated by both direct and indirect means. For example in asthma the reduced eosinophilia is as a result of both the direct promotion of eosinophil apoptosis, and the indirect suppression of receptor expression and production of cytokines or growth factors [20] (Fig. 23.1).

The *In Vitro* Effects of Glucocorticoid

Broadly it has been repeatedly demonstrated that glucocorticoids reduce both cell proliferation and collagen synthesis *in vitro* (Fig. 23.1) [21]. Poulsen et al. [22] demonstrated that dexamethasone reduced tendon derived cell proliferation, as well as reduced collagen and glycosaminoglycan synthesis, and that reactive oxygen species (ROS) were increased. Similar results in terms of reduced cell proliferation and matrix production have been found by several other *in vitro* studies [21, 23–26]. The mechanisms underlying these changes have not been clarified. However it has been demonstrated that glucocorticoid can an activate the forkhead proteins (FOXO1 and FOXO3a), and reduce Akt and ERK activity; it may be hypothesised that the forkhead protein activation may be at least in part mediated by the ROS production [22]. It has also been hypothesised that glucocorticoids may result in reduced collagen mRNA levels [27]. The mechanisms by which glucocorticoids may induce increased ROS production are unknown

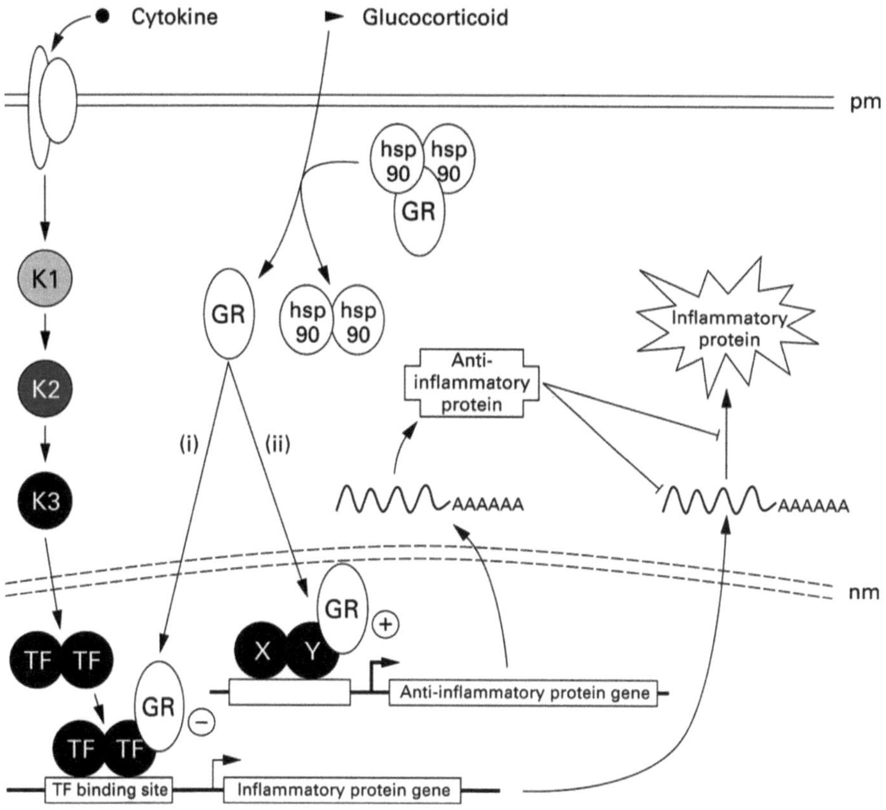

Fig. 23.1 A model for glucocorticoid dependent repression of pro-inflammatory genes. A generalised inflammatory cascade is shown. Cytokine binding to its cognate receptor localised in the plasma membrane (pm) leads to activation of a kinase cascade consisting of kinases 1, 2, and 3 (K1, K2, and K3). K3 translocates across the nuclear membrane (nm) and then phosphorylates the transcription factor (TF) which actives transcription of an inflammatory protein gene. This leads to mRNA synthesis (transcription) and protein synthesis (translation) of the inflammatory protein. Binding of glucocorticoid to the glucocorticoid receptor (GE) leads to dissociation of the heat shock proteins (hsp90) and translocation of GR to the nucleus. (i) GR may bind TF (for example NF-κB or AP-1) to repress activated transcription via tethering type interactions. (ii) Alternatively, it is hypothesised that GR interacts with other factors (X, Y) to activate gene transcription of anti-inflammatory genes. These anti-inflammatory genes are further hypothesised to promote mRNA degradation and/or repress translation of inflammatory genes. Taken with permission [19]

and is unlikely to be apoptosis related. Notably in some cell types the opposite effect has been shown in that glucocorticoids have been shown to reduce ROS production in platelets [28].

Glucocorticoids have been shown to affect the differentiation of tendon derived cells. Tempfer et al demonstrated that glucocorticoid treatment increased the number of adipocytes and chondrocyte like cells present *in vitro* [24]. It was also shown that two surrogate markers of tendon cell migration, MMP2 and MMP9 expression, were reduced after glucocorticoid treatment (see Chap. 17). An *in vitro* study using rat tendon stem cells by Chen et al showed that glucocorticoid inhibited the differentiation of these cells into tenocytes, as measured by reduced levels of tenomodulin and type I collagen [29]. Chromatin immunoprecipitation-PCR showed that dexamethasone promoted glucocorticoid receptor interaction with the TGGAAGCC sequence located between −734 and −726 base pairs upstream of the start codon of the scleraxis gene, leading the authors to conclude that glucocorticoid inhibits the differentiation of tendon stem cells to tenocytes by inhibiting the scleraxis gene.

Importantly Poulsen et al have demonstrated that glucocorticoids induce senescence in tendon derived cells *in vitro* by the p53/p21 pathway [30]. It was previously unknown whether the glucocorticoid induced reduction in cell proliferation results in cell quiescence or senescence. One of the critical features distinguishing a senescent cell from a quiescent cell is that mTOR (mammalian target of Rapamycin) remains active in the former. Poulsen et al found no change in levels of phosphorylated p70 S6kinase, a substrate of the mTORC1 enzyme, following glucocorticoid treatment of tenocytes suggesting mTOR remained active in the cells. Consistent with this finding, SA-β-gal (senescence associated beta galactosidase) activity was greater in glucocorticoid treated tendon derived cells compared with controls. Inhibition of mTOR with rapamycin markedly reduced the number of SA-β-gal-positive cells following glucocorticoid treatment. Taken together, these results confirmed that glucocorticoids induce senescence in human tendon derived cells. Poulsen et al also demonstrated that low glucose and resveratrol were protective against this glucocorticoid induced senescence and that this protection was abolishing by Sirtuin 1 (sirt1) inhibition, demonstrating that sirt1 activation was the largely responsible for the protective effect. The relevance of cellular senescence relates to its potentially key role in age related degenerative disease [31].

The *In Vivo* Effects of Glucocorticoid

Difficulties in carrying out human *in vivo* studies mean that the majority of *in vivo* work has been carried out in animals. There is certainly some evidence that glucocorticoids have significant deleterious effects on tendon *in vivo* in terms of inducing collagen bundle fragmentation and an influx of inflammatory cells [32–34]. A notable point is that these deleterious effects were only seen after relatively high doses of glucocorticoid which were repeated multiple times within a relatively short space of time, and that deleterious effects were not seen with less frequent injections [33].

In vivo studies on both the rabbit and the human have demonstrated that glucocorticoid injection results in acute localised collagen necrosis, an effect that was not seen in the saline injected controls [35, 36]. Both these studies then demonstrated a variable inflammatory response to this collagen necrosis followed by a subsequent reparative process. Whether the collagen necrosis is as a result of a direct effect of the glucocorticoid or an enzyme mediated effect is as yet unknown. A recent human *in vivo* study has demonstrated increased levels of NMDA receptor expression 6 weeks following glucocorticoid administration, it was not determined whether this change was related to an inflammatory cell infiltrate [37]. Given that it has been shown that the NMDA receptor is largely expressed by inflammatory cells within tendon [38], it is certainly eminently possible that the increase in NMDA receptor expression seen was related to an inflammatory cell infiltrate (see Chap. 5).

Not only has glucocorticoid been shown to inhibit type I collagen fibril formation *in vitro* [39], but glucocorticoids have also been shown to inhibit collagen synthesis in animal *in vivo* studies [40, 41]. Again it must be noted that a significant reduction in collagen synthesis has only been demonstrated using extremely high doses of glucocorticoid, for example Oxlund et al injected 10mhg/kg of cortisol around rat peroneal tendons every third day for 55 days [40]. A single dose of glucocorticoid has been shown to increase the ratio of type III to type I collagen expression in the rat [42]. Lee et al investigated the effect of glucocorticoid injection in a rat model of tendon injury [43]. They demonstrated that at 3 days following glucocorticoid administration both aggrecan and fibronectin gene expression was higher versus control, alongside a lower level of TNFα gene expression. They consequently hypothesised that glucocorticoid may have a short term anti-inflammatory effect which is associated with these changes in matrix gene expression.

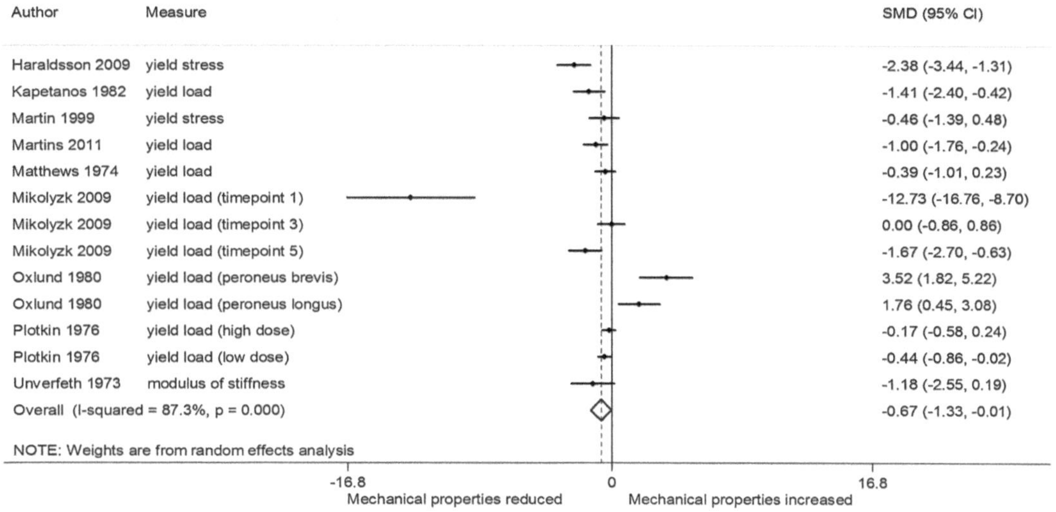

Fig. 23.2 Forest plot of studies analysing effects of glucocorticoid on the mechanical properties of tendon (X axis—effect size (standardised mean difference) and Y axis—studies included). Taken with permission [21]

Broadly the effects of glucocorticoid on the mechanical properties of tendon appear to be negative [21]. Figure 23.2 shows a forest plot demonstrating the overall negative effect of glucocorticoid on the mechanical strength of tendon. This finding is consistent the evidence that demonstrates glucocorticoid administration results in collagen necrosis and disorganisation, followed by an infiltration of inflammatory cells.

Poulsen et al investigated both the *in vitro* and *in vivo* mechanisms of glucocorticoid action on tendon derived cells and tendon respectively [30]. They found that increased levels of acetylated p53 as well as increased RNA levels of its pro-senescence effector p21 were evident in glucocorticoid-treated tendon derived cells. Knockdown of p53 or inhibition of p53 activity prevented dexamethasone-induced senescence. While immunohistochemical analysis of tendon biopsies taken before and after glucocorticoid injection revealed a significant increase in the percentage of p53-positive cells and the percentage of p21-positive cells also tended to be higher post-injection suggesting glucocorticoids activate the p53/p21 senescence inducing pathway *in vivo* as well as *in vitro*. Muto et al found that both MMP-3 expression and apoptosis were evident in the surface layers of tendons 1 week after glucocorticoid injection but that these changes had disappeared after 3 weeks [34]. Saliently the increase in MMP-3 expression and apoptosis coincided with a loss of collagen organisation after 1 week which did not persist to the 3 week stage.

Discussion

A great difficulty in terms of interpreting the clinical meaning of the observed *in vitro* and *in vivo* effects of glucocorticoid is that these experimental environments are arguably significantly different to the disease environment. It is worth considering that the effects of glucocorticoid are extremely dependent upon the cell type studied. There remains a paucity of work in terms of characterising the cell surface markers expressed by 'tendon derived cells' and it is likely that these cells are not a homogenous population. Based on previous histological studies it is a distinct possibility that cells derived from disease consist of significantly different cell groups to those derived from healthy disease-free tendon [2]. It is therefore important to consider these gaps in our knowledge before interpreting the results from studies on cells derived from different animals in different states of tendon health using different cell extraction methods at different passages. It is also worth considering that the *in vivo* models of tendinopathy are arguably not an accurate representation of human disease, while the cellular makeup of tendon is likely significantly different in a healthy *in vivo* situation when compared to real human disease.

Despite these limitations to our current understanding of the effects of glucocorticoid on tendon derived cells and tendon, several distinct themes recurrently emerge from the current scientific literature. Glucocorticoids reduce tendon derived cell proliferation *in vitro* and reduce extracellular matrix synthesis both *in vitro* and *in vivo*, in particular type I collagen synthesis. Glucocorticoids also appear to result in acute deleterious changes in healthy *in vivo* tendon including collagen necrosis, collagen disorganisation and inflammatory cell infiltration; while the overall effect of glucocorticoid administration on the mechanical properties of healthy *in vivo* tendon are generally negative.

These catabolic effects are consistent with the known effects of glucocorticoid on other connective tissue such as the skin [44]. The differing and sometimes contrasting effects of glucocorticoid on different cell types are worth considering in the context of how glucocorticoids may differentially affect healthy and normal tendon [45, 46]. The secretion of collagenase by macrophages is inhibited by glucocorticoids in a dose dependent fashion [47]. Thus one may infer that the effects of glucocorticoid on diseased tendon *in vivo*, in which macrophages are significantly more abundant, may be rather contrasting to the effects observed on healthy tendon *in vivo*. Perhaps the variability in patient response to glucocorticoid may be related to the inflammatory phenotype of their disease, and this may variability may to some degree be predictable. Overall the existing *in vitro* and *in vivo* evidence suggests that glucocorticoids should be used with caution in treating painful tendinopathy [21]. Certainly a real need exists to follow up the long term clinical effects of glucocorticoid in treating tendinopathy, as there is currently a paucity of evidence in this area. However in this context while the short term benefits are clear, glucocorticoids remain a useful treatment option provided they are used in the right patients in sensible moderation.

References

1. Authority NBS. Electronic Prescribing & Financial Information for Practices (ePFIP). http://www.nhsbsanhsuk/PrescriptionServices/963.aspx 2009
2. Millar NL, Hueber AJ, Reilly JH, Xu Y, Fazzi UG, Murrell GA et al (2010) Inflammation is present in early human tendinopathy. Am J Sports Med 38 (10):2085–2091. doi:10.1177/0363546510372613
3. Rees JD, Stride M, Scott A (2014) Tendons - time to revisit inflammation. Br J Sports Med 48 (21):533–537. doi:10.1136/bjsports-2012-091957
4. Schubert TE, Weidler C, Lerch K, Hofstadter F, Straub RH (2005) Achilles tendinosis is associated with sprouting of substance P positive nerve fibres. Ann Rheum Dis 64(7):1083–1086. doi:10.1136/ard.2004.029876
5. Arroll B, Goodyear-Smith F (2004) Corticosteroid injections for osteoarthritis of the knee: meta-analysis. BMJ 328(7444):869. doi:10.1136/bmj.38039.573970.7C
6. Arroll B, Goodyear-Smith F (2005) Corticosteroid injections for painful shoulder: a meta-analysis. Br J Gen Pract 55(512):224–228
7. Staal JB, de Bie R, de Vet HC, Hildebrandt J, Nelemans P (2008) Injection therapy for subacute and chronic low-back pain. Cochrane Database Syst Rev (Online). (3):CD001824. doi:10.1002/14651858.CD001824.pub3
8. Coombes BK, Bisset L, Vicenzino B (2010) Efficacy and safety of corticosteroid injections and other injections for management of tendinopathy: a systematic review of randomised controlled trials. Lancet 376(9754):1751–1767. doi:10.1016/s0140-6736(10)61160-9
9. Buchbinder R, Green S, Youd JM (2003) Corticosteroid injections for shoulder pain. Cochrane Database Syst Rev (Online). (1):CD004016. doi:10.1002/14651858.cd004016
10. Scott A, Docking S, Vicenzino B, Alfredson H, Zwerver J, Lundgreen K et al (2013) Sports and exercise-related tendinopathies: a review of selected topical issues by participants of the second International Scientific Tendinopathy Symposium (ISTS) Vancouver 2012. Br J Sports Med 47(9):536–544. doi:10.1136/bjsports-2013-092329
11. Coombes BK, Bisset L, Brooks P, Khan A, Vicenzino B (2013) Effect of corticosteroid injection, physiotherapy, or both on clinical outcomes in patients with unilateral lateral epicondylalgia: a randomized controlled trial. JAMA 309(5):461–469. doi:10.1001/jama.2013.129

12. Mahler F, Fritschy D (1992) Partial and complete ruptures of the Achilles tendon and local corticosteroid injections. Br J Sports Med 26(1):7–14
13. Shrier I, Matheson GO, Kohl HW 3rd (1996) Achilles tendonitis: are corticosteroid injections useful or harmful? Clin J Sport Med 6(4):245–250
14. Maffulli N, Wong J (2003) Rupture of the Achilles and patellar tendons. Clin Sports Med 22(4):761–776
15. van der Linden PD, Sturkenboom MC, Herings RM, Leufkens HM, Rowlands S, Stricker BH (2003) Increased risk of achilles tendon rupture with quinolone antibacterial use, especially in elderly patients taking oral corticosteroids. Arch Intern Med 163 (15):1801–1807. doi:10.1001/archinte.163.15.1801
16. Mandel S, Schilling J, Peterson E, Rao DS, Sanders W (2013) A retrospective analysis of vertebral body fractures following epidural steroid injections. J Bone Joint Surg 95(11):961–964. doi:10.2106/jbjs.l.00844
17. Barnes PJ (2006) Corticosteroid effects on cell signalling. Eur Respir J 27(2):413–426. doi:10.1183/09031936.06.00125404
18. Payne DN, Adcock IM (2001) Molecular mechanisms of corticosteroid actions. Paediatr Respir Rev 2 (2):145–150. doi:10.1053/prrv.2000.0122
19. Newton R (2000) Molecular mechanisms of glucocorticoid action: what is important? Thorax 55 (7):603–613
20. Giembycz MA, Lindsay MA (1999) Pharmacology of the eosinophil. Pharmacol Rev 51(2):213–340
21. Dean BJ, Lostis E, Oakley T, Rombach I, Morrey ME, Carr AJ (2014) The risks and benefits of glucocorticoid treatment for tendinopathy: a systematic review of the effects of local glucocorticoid on tendon. Semin Arthritis Rheum 43(4):570–576. doi:10.1016/j.semarthrit.2013.08.006
22. Poulsen RC, Carr AJ, Hulley PA (2011) Protection against glucocorticoid-induced damage in human tenocytes by modulation of ERK, Akt, and forkhead signaling. Endocrinology 152(2):503–514. doi:10.1210/en.2010-1087
23. Sendzik J, Shakibaei M, Schafer-Korting M, Lode H, Stahlmann R (2010) Synergistic effects of dexamethasone and quinolones on human-derived tendon cells. Int J Antimicrob Agents 35(4):366–374. doi:10.1016/j.ijantimicag.2009.10.009
24. Tempfer H, Gehwolf R, Lehner C, Wagner A, Mtsariashvili M, Bauer HC et al (2009) Effects of crystalline glucocorticoid triamcinolone acetonide on cultered human supraspinatus tendon cells. Acta Orthop 80(3):357–362. doi:10.3109/17453670902988360
25. Wong MW, Lui WT, Fu SC, Lee KM (2009) The effect of glucocorticoids on tendon cell viability in human tendon explants. Acta Orthop 80(3):363–367. doi:10.3109/17453670902988386
26. Wong MW, Tang YN, Fu SC, Lee KM, Chan KM (2004) Triamcinolone suppresses human tenocyte cellular activity and collagen synthesis. Clin Orthop Relat Res 421:277–281
27. Oikarinen J (1982) Cortisol induces (2′-5′) oligoadenylate synthetase in cultured chick embryo tendon fibroblasts. Biochem Biophys Res Commun 105(3):876–881
28. Sanner BM, Meder U, Zidek W, Tepel M (2002) Effects of glucocorticoids on generation of reactive oxygen species in platelets. Steroids 67(8):715–719
29. Chen W, Tang H, Zhou M, Hu C, Zhang J, Tang K (2015) Dexamethasone inhibits the differentiation of rat tendon stem cells into tenocytes by targeting the scleraxis gene. J Steroid Biochem Mol Biol 152:16–24. doi:10.1016/j.jsbmb.2015.04.010
30. Poulsen RC, Watts AC, Murphy RJ, Snelling SJ, Carr AJ, Hulley PA (2013) Glucocorticoids induce senescence in primary human tenocytes by inhibition of sirtuin 1 and activation of the p53/p21 pathway: in vivo and in vitro evidence. Ann Rheum Dis. doi:10.1136/annrheumdis-2012-203146
31. Heathfield SK, Le Maitre CL, Hoyland JA (2008) Caveolin-1 expression and stress-induced premature senescence in human intervertebral disc degeneration. Arthritis Res Ther 10(4):R87-R. doi:10.1186/ar2468
32. Akpinar S, Hersekli MA, Demirors H, Tandogan RN, Kayaselcuk F (2002) Effects of methylprednisolone and betamethasone injections on the rotator cuff: an experimental study in rats. Adv Ther 19(4):194–201
33. Tillander B, Franzen LE, Karlsson MH, Norlin R (1999) Effect of steroid injections on the rotator cuff: an experimental study in rats. J Shoulder Elbow Surg/Am Shoulder Elbow Surg [et al] 8 (3):271–274
34. Muto T, Kokubu T, Mifune Y, Inui A, Harada Y, Yoshifumi et al (2014) Temporary inductions of matrix metalloprotease-3 (MMP-3) expression and cell apoptosis are associated with tendon degeneration or rupture after corticosteroid injection. J Orthop Res 32(10):1297–1304. doi:10.1002/jor.22681
35. Balasubramaniam P, Prathap K (1972) The effect of injection of hydrocortisone into rabbit calcaneal tendons. J Bone Joint Surg Br Vol 54(4):729–734
36. Lee SK, Ling CM (1975) The response of human tendon to hydrocortisone injection. Singap Med J 16 (4):259–262
37. Dean BJ, Franklin SL, Murphy RJ, Javaid MK, Carr AJ (2014) Glucocorticoids induce specific ion-channel-mediated toxicity in human rotator cuff tendon: a mechanism underpinning the ultimately deleterious effect of steroid injection in tendinopathy? Br J Sports Med 48(22):1620–1626. doi:10.1136/bjsports-2013-093178
38. Dean BJF SS, Dakin S, Murphy R, Javaid MK, Carr AJ (2015) Differences in glutamate receptors and inflammatory cell numbers are associated with the resolution of pain in human rotator cuff tendinopathy. Arthritis Res Ther 17(176). doi:10.1186/s13075-015-0691-5

39. Dombi GW, Halsall HB (1986) Collagen fibril formation in the presence of dexamethasone disodium phosphate. Connect Tissue Res 15(4):257–268
40. Oxlund H (1982) Long term local cortisol treatment of tendons and the indirect effect on skin. An experimental study in rats. Scand J Plast Reconstr Surg 16 (1):61–66
41. Oikarinen A, Makela J, Vuorio T, Vuorio E (1991) Comparison on collagen gene expression in the developing chick embryo tendon and heart. Tissue and development time-dependent action of dexamethasone. Biochim Biophys Acta 1089(1):40–46
42. Wei AS, Callaci JJ, Juknelis D, Marra G, Tonino P, Freedman KB et al (2006) The effect of corticosteroid on collagen expression in injured rotator cuff tendon. J Bone Joint Surg Am 88(6):1331–1338. doi:10.2106/jbjs.e.00806
43. Lee HJ, Kim YS, Ok JH, Lee YK, Ha MY (2013) Effect of a single subacromial prednisolone injection in acute rotator cuff tears in a rat model. Knee Surg Sports Traumatol Arthrosc. doi:10.1007/s00167-013-2395-1
44. Uitto J, Teir H, Mustakallio KK (1972) Corticosteroid-induced inhibition of the biosynthesis of human skin collagen. Biochem Pharmacol 21 (16):2161–2167
45. Ahasan MM, Hardy R, Jones C, Kaur K, Nanus D, Juarez M et al (2012) Inflammatory regulation of glucocorticoid metabolism in mesenchymal stromal cells. Arthritis Rheum 64(7):2404–2413. doi:10.1002/art.34414
46. Nieuwenhuis B, Luth A, Kleuser B (2010) Dexamethasone protects human fibroblasts from apoptosis via an S1P3-receptor subtype dependent activation of PKB/Akt and Bcl XL. Pharmacol Res 61 (5):449–459. doi:10.1016/j.phrs.2009.12.005
47. Werb Z (1978) Biochemical actions of glucocorticoids on macrophages in culture. Specific inhibition of elastase, collagenase, and plasminogen activator secretion and effects on other metabolic functions. J Exp Med 147(6):1695–1712

Influence of Ageing on Tendon Homeostasis

24

Helen L. Birch, Mandy J. Peffers, and Peter D. Clegg

Abstract

Tendon functional competence and structural integrity rely on homeostasis of tendon cell metabolism and extracellular matrix macromolecules. The clear link between tendinopathies and increasing age suggests a slow change to tendon homeostasis, which increases susceptibility to damage. Despite this well evidenced association between increasing age and tendon damage, changes to tendon mechanical properties with ageing are not clear with different studies reporting conflicting results. More recent research suggests that age-related changes occur at specific sub-structure locations and may be overlooked by measuring properties of the whole tendon. In this chapter we review changes to tendon mechanical properties, structure and composition. Mechanisms speculated to contribute to tendon change with age such as cellular senescence, ageing stem cell population, reactive oxygen species and formation of advanced glycation end-product crosslinks are discussed. Understanding age-related changes to tendon homeostasis are key to understanding increased incidence of tendon injuries in the ageing population.

Keywords

Tendon • Ageing • Mechanical properties • Extracellular matrix • Collagen • Gene expression • Cellular senescence • Stem cell • Inflammageing • Advanced glycation end-product crosslink

H.L. Birch (✉)
Institute of Orthopaedics and Musculoskeletal Science, University College London, Royal National Orthopaedic Hospital, Stanmore HA7 4LP, UK
e-mail: h.birch@ucl.ac.uk

M.J. Peffers • P.D. Clegg
Department of Musculoskeletal Biology, Institute of Ageing and Chronic Disease, University of Liverpool, Leahurst Campus, Neston CH64 7TE, UK

Abbreviations

AGE	Advanced glycation end-product
CDET	Common digital extensor tendon
CSA	Cross sectional area
CSIG	Cellular senescence-inhibited gene
GAG	Glycosaminoglycan

P.W. Ackermann and D.A. Hart (eds.), *Metabolic Influences on Risk for Tendon Disorders*, Advances in Experimental Medicine and Biology 920, DOI 10.1007/978-3-319-33943-6_24

ICTP	Cross-linked carboxyterminal telopeptide of type I collagen
IL1β	Interleukin 1β
MRI	Magnetic resonance imaging
PGE_2	Prostaglandin E2
PRG4	Proteoglycan 4
ROCK1	Rho-associated coiled-coil protein kinase 1
ROS	Reactive oxygen species
SASP	Senescence-associated secretory phenotype
SDFT	Superficial digital flexor tendon
TSC	Tendon stem cell

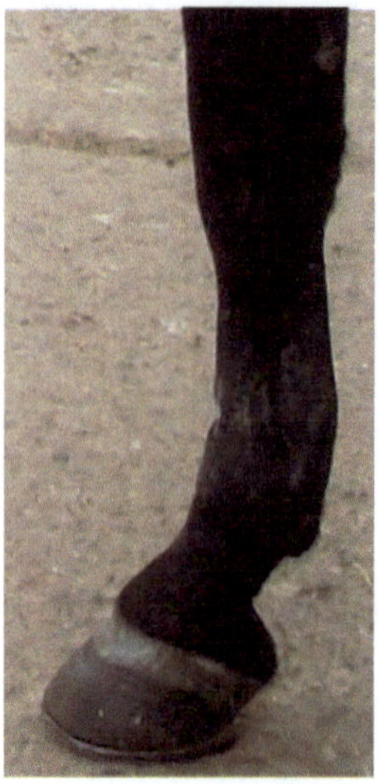
Fig. 24.1 The bowed appearance to the back of the lower limb of this horse is due to a tendon injury

Ageing and Tendon Susceptibility to Injury

Chronic tendon/ligament disorders are highly debilitating and increasingly prevalent [1], accounting for one-third of all primary-care musculoskeletal consultations in the UK [2]. Injuries to ligaments, joint capsules and tendons account for approximately 50 % of the 23 million musculoskeletal injuries that occur in the USA annually [3]. They affect sporting and sedentary individuals [1] in addition to animals such as horses [4, 5]. However, there are no effective treatments or prevention strategies for these injuries [5, 6], as a result of limited understanding of the tissues and the aetiology of injury.

The prevalence of tendon injuries is thought to be increasing due to both increasing sports participation as well as an ageing population [1]. In particular, increasing age has been demonstrated to be a risk factor for a number of different tendinopathies. A recent systematic review of rotator cuff diseases has identified a prevalence of 9.7 % in patients 20 years or younger, rising to 62 % in patients 80 years or older [7]. A separate report identified that rotator cuff tears affected 40 % of individuals older than 60 years in the USA [8]. Achilles tendinopathy is most commonly observed in the fourth and fifth decade of life [9]. Another study identified Achilles, patellar and quadriceps tendon rupture as injuries of middle age, with rotator cuff tears and biceps tendon rupture occurring more frequently in old age [10]. In the horse, a species which frequently suffers from tendinopathy (Fig. 24.1), a number of studies have demonstrated an association between increasing age and risk of tendon injury [11–13].

Age-Related Changes to Mechanical Properties

The unequivocal evidence demonstrating increased susceptibility of tendon to injury with advancing age suggests that the ability of tendon to withstand mechanical forces declines. The strength of the tendon, the degree of elongation prior to failure and ease with which the tendon deforms (stiffness) are important properties for consideration. While it might be expected that the ultimate force and strain would show a negative correlation with age following maturity, studies have not been able to demonstrate a clear link. The ultimate tensile strength of the human patellar tendon was found to show a moderate 17 % decrease between age groups of 29–50 years and 64–93 years [14] but no difference between the ages of 17–54 years [15, 16] in *in vitro* mechanical tests. The Achilles tendon demonstrated a decrease in ultimate tensile strength for ages between 36 and 100 years in embalmed specimens [17] and a lower ultimate tensile stress in fresh samples over 35 years of age compared to less than 35 years [18]. Material stiffness (modulus) might be expected to increase with ageing however some studies show no effect in human tendon [14, 16, 19] or a slight decrease in human patellar [15] and Achilles tendon [18]. In horses, despite the clear association between horse age and incidence of injury, the ultimate stress, strain and modulus of the superficial digital flexor tendon (SDFT) do not appear to change with increasing horse age [20].

Measurement of mechanical properties *in vivo* is more problematic due to difficulties in measuring both force and elongation accurately. While it is not possible to measure the failure properties of tendons *in vivo*, some studies have quantified tendon stiffness using a combination of ultrasound or MRI to track tendon elongation and joint torque to calculate muscle force generation. Measurements of this type are possible for the Achilles and patellar tendon (Fig. 24.2). The forces that tendons are subjected to decrease with advancing age from skeletal maturity through into old age as a result of a decline in muscle mass and force generation [22–25]. Some studies suggest that tendon stiffness may compensate for the decrease in the ability of the associated muscle to generate force. The tendon-aponeurosis of the vastus lateralis muscle was found to decrease in stiffness in a group of women aged from 21 to 77 years [22] although tendon strain at maximum force also decreased. In another study, the patellar tendon in a group of aged men (60–69 years) had a lower stiffness than in a young group of men (21–32 years) although there was no difference in the stiffness of the Achilles tendon between groups and no difference in the maximum strain for either tendon between young and old groups [23]. A larger study including both men and women with a broader age range (18–80 years) found that Achilles tendon stiffness decreased in the older age group [25], a finding repeated in a later study with a smaller group of women [26]. The decrease in tendon stiffness cannot be accounted for purely by a decrease in tendon size. In the study by Csapo et al. [26] the Achilles tendon showed no significant difference

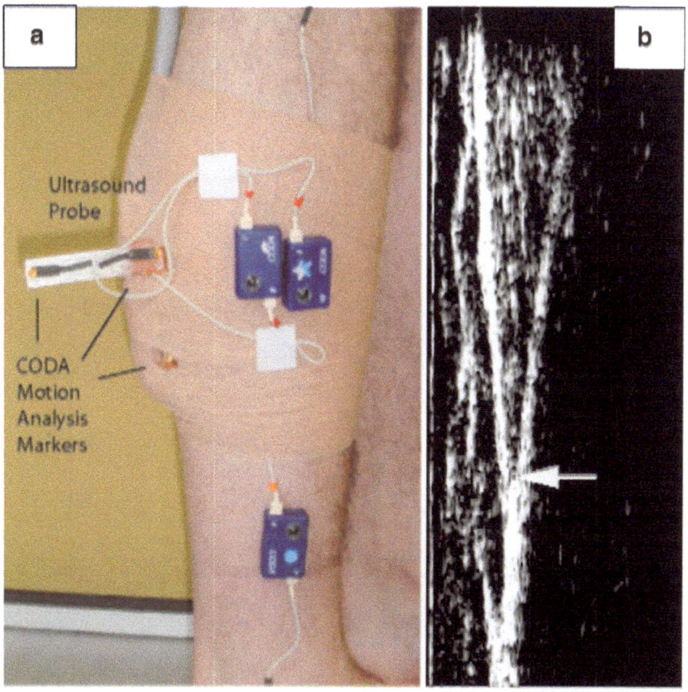

Fig. 24.2 Ultrasound probe attached to human lower limb to visualize the muscle tendon junction (**a**) and images of the Achilles tendon junction with the lateral gastrocnemius (**b**) (Adapted from [21])

in length and cross sectional area (CSA) between groups and in the study by Stenroth et al. [25] the Achilles tendon CSA increased significantly in the older group, hence both studies reported a significantly lower Young's modulus in older tendons.

Given the complex hierarchical structure of tendon, studies investigating mechanical properties of the gross tendon structure *in vitro* and *in vivo* may miss important changes occurring at sub-structure level. The work of Screen and colleagues has investigated mechanical properties of tendon fascicles and fibres and changes related to ageing. The failure properties (stress and strain) and modulus of fascicles dissected from the equine SDFT, an energy-storing tendon, showed no significant differences between a young group of horses (3–8 years) and an old group of horses (15–20 years) [20]. The stiffness of the inter-fascicular matrix however, which binds the fascicles together, increased in stiffness significantly with increasing age (Fig. 24.3) [20]. This finding suggests that the load distribution within the tendon changes with ageing such that the fascicles are loaded earlier during tendon extension in older tendons. It is interesting to note that behavior of the subunits of tendon differ between tendon types; in the equine common digital extensor tendon (CDET), a positional tendon, the inter-fascicular matrix is much stiffer [27] and does not change significantly with increasing horse age (Fig. 24.3) [20].

In addition, other mechanical properties less often considered such as hysteresis and fatigue properties are likely to be very relevant with regard to tendinopathies. These properties are particularly important in the human patellar and Achilles tendon and the equine SDFT as they are subjected to a high number of loading and unloading cycles and function as elastic energy stores. Although quantifying hysteresis *in vivo* is problematic [28], a study of the tendon-aponeurosis of the vastus lateralis showed an increase in hysteresis with ageing in a group of women between 21 and 77 years [22]. Hysteresis and post-loading recovery have been studied in fascicles from equine tendons *in vitro* and differences have been observed between tendon types and with ageing. Fascicles from the energy storing SDFT have lower hysteresis and a greater ability to recover after loading compared to those from the CDET. This difference can be explained by a difference in the extension mechanism; SDFT fascicles appear to extend by rotation of a helical structure while CDET fascicle extension is dominated by sliding of the component fibres [29]. In older horses, the ability of the fascicles from the SDFT to recover following loading is reduced and hysteresis increases [29]. Fatigue loading of SDFT fascicles *in vitro* results in changes similar to those seen in ageing; in fascicles from young horses rotation decreases and in fascicles from older horses where the helical structure is already compromised there is an increase in fibre sliding following fatigue

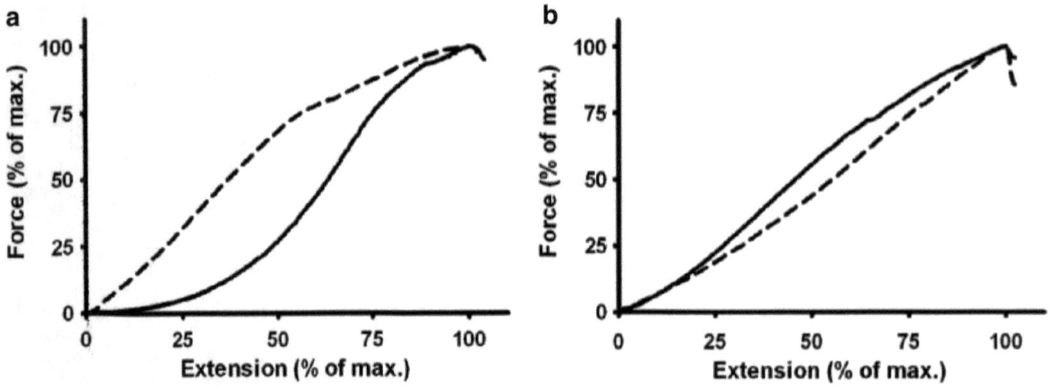

Fig. 24.3 Force extension curves for the fascicular interface in the SDFT (*solid line*) and CDET (*dashed line*) from a 3 year old horse (**a**) and a 20 year old horse (**b**) (Adapted from [20])

loading [30]. The inter-fascicular matrix in the SDFT, in addition to an increase in stiffness with ageing, also shows less ability to resist repetitive loading [31]. These studies demonstrate that the mechanical behavior of tendons is complex, as are the changes associated with ageing.

Age Related Changes to Tendon Composition

Age related changes in the response of tendon to applied force stems from a difference in the matrix structure and composition. Information on age related changes to human tendon matrix composition is limited; the majority of research in this area has been conducted on animal tissue and often short-lived species such as rats and mice. The application of these findings to human tendon should be made with care. The horse however represents a good model to study tendon ageing as this species is relatively long-lived and shows an age related decline in tendon function. Morphometric study of the SDFT in horses aged from 2 to 23 years showed that fascicles decreased in size with increasing horse age [32], however this finding was not apparent in studies where fascicles were dissected free from the tendon and CSA measured [20]. In the study by Gillis et al., [32] tendon CSA did not decrease with advancing age thus fascicle numbers appeared to increase. This would suggest a greater proportion of inter-fascicular matrix in older tendon; an interesting finding given the role that the inter-fascicular matrix plays in tendon mechanical behavior. A decrease in unit size also appears to occur at the nano-scale with increasing age, as the average collagen fibril diameter reduces with advancing age in the equine SDFT [33]. The size of the collagen fibrils has also been linked to mechanical properties, where smaller diameter collagen fibrils have been suggested to provide a more creep resistant matrix [34]. Collagen fibrils in their longitudinal course show an abrupt change of direction giving rise to a 'crimped' structure [35]; in older horses the crimp becomes less pronounced in the central core of the SDFT (Fig. 24.4) [36, 37].

In terms of molecular composition, equine tendon water content and collagen content do not change significantly with ageing [38, 39]. In contrast, tissue biopsies from the human patellar tendon were found to have a lower collagen content and higher levels of the mature crosslinks hydroxylysylpyridinoline and lysylpyridinoline in an older group of men (67 ± 3 years) compared to a young group of men (27 ± 2 years) [40]. A measure of the total sulphated glycosaminoglycan (GAG) content gives an indication of proteoglycan levels in tendon. Changes in GAG levels with age appear to depend on tendon type, with the energy storing equine SDFT showing no significant change in levels [38, 39] whereas the positional CDET showed a significant decrease in GAG levels with increasing horse age [39]. In a study of human tendons from donors ranging in age from 11 to 95 years GAG levels decreased significantly with age in the supraspinatus tendon but not the common biceps tendon [41].

Using changes in total sulphated GAG levels as an indication of changes in proteoglycan content overlooks possible disparity between

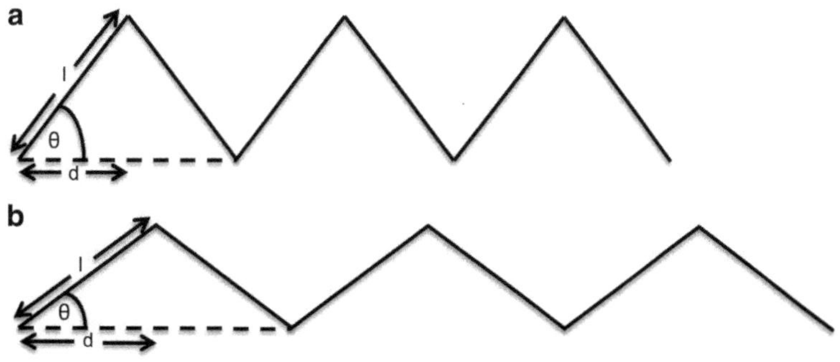

Fig. 24.4 Diagrammatic representation of collagen fibril crimp in the central of young (**a**) and old (**b**) equine SDFT Θ = crimp angle, l = crimp length, d = measured crimp length

individual proteoglycans. For example, there is some evidence to suggest that lubricin, also known as superficial zone protein and proteoglycan 4 (PRG4), increases in rabbit ligament between the ages of 1 and 3 years [42]. Lubricin is a glycoprotein that acts as a lubricant enabling gliding of cartilage surfaces [43, 44] and gliding of tendons around joints and adjacent tendons [43]. More recently, lubricin has been found within the tendon structure where it is enriched in the interfascicular matrix [45, 46]. Interestingly, the rabbit ligaments where higher levels of lubricin mRNA expression were measured showed an increased ultimate strain and decreased elastic modulus [42]; a change that may be explained by increased ability of fascicles to glide relative to each other. However, studies investigating age related changes in human Achilles tendon found that lubricin was not differentially expressed between young (19 ± 5.8 years) and old (69.4 ± 7.3 years) tendons [47]. These studies indicate the need for detailed investigation with regard to specific proteins and glycoproteins as well as sub-structure location with a clear distinction made between maturation and ageing effects.

Contrary to what might be expected, DNA levels, which provide a measure of cellularity, do not seem to decrease in ageing tendon [38, 39].

Age Related Changes to Matrix Turnover Rate

In general, protein synthesis in tissues decreases with advancing age [48], and therefore it might be expected that extracellular matrix proteins in tendon are renewed less often. It is difficult to measure the turnover rate of long-lived proteins, however measuring the ratio of D to L isomers of amino acids allows an estimate of protein half-life in tendon to be made. All amino acids are incorporated into newly synthesized proteins in their L form however over time they can spontaneously convert to the D isomer form by a process known as racemization. This is a relatively slow process but happens more quickly in aspartic acid making this amino acid convenient to measure. A comparison of the rate of accumulation of D aspartic acid with the rate that would occur if the protein were not turned over at all allows a half-life to be calculated. Measurement of the ratio of D/L forms of aspartic acid in equine SDFT and CDET tissue from horses ranging in age from 4 to 30 years suggested an average protein half-life of 7.85 years for the SDFT and 8.02 years for the CDET and half-life increased significantly with increasing horse age for both tendon types [39]. Separation of the tissue into collagenous and non-collagenous proteins showed that the collagen component has a much longer half-life than the non-collagenous proteins and this was significantly longer in the SDFT (197.53 years) compared to the CDET (34.03 years). The half-life of the collagen component increased significantly with increasing age in the SDFT but not the CDET. Conversely, the non-collagenous proteins turned over more slowly in the CDET (average half-life 3.51 years) than the SDFT (average half-life 2.18 years) and half-life increased with increasing age in the CDET but not the SDFT [39]. These studies suggest that the majority of the collagen remains in the tendon for the life-time of the horse.

A similar study has been carried out in human tendon by taking advantage of the (14)C labeling of tissues as a result of the nuclear bomb tests in 1955–1963. Levels of (14)C were measured in Achilles tendon samples and compared to known atmospheric levels. The results suggested that after the cessation of growth at about 17 years of age the turnover of tendon tissue is essentially zero [49]. Other work however has suggested a much more rapid turnover of tendon matrix. Studies using incorporation of stable isotope labeled amino acids suggested that the half-life of collagen in human patellar tendon is about 2 months and close to that of skeletal muscle proteins and higher than muscle collagen [50, 51]. Studies using microdialysis catheters to extract metabolites from the peritendinous region of human Achilles tendon have also demonstrated active turnover of collagen by measuring the pro-peptide of type I collagen as a marker of synthesis and cross-linked carboxyterminal telopeptide of type I collagen (ICTP), as a marker of degradation [52, 53]. Thus it seems likely that a component of the tendon is degraded and renewed frequently while the bulk of the tendon is relatively inert.

A decline in matrix turnover with increasing age would be most easily explained by reduced gene expression of matrix and matrix degrading enzymes by tenocytes. Expression of a range of different matrix proteins including collagens and proteoglycans and matrix degrading enzymes has been quantified in equine SDFT tissue from horses ranging in age from 3 to 30 years [54]. This study showed no significant drop in expression levels of Col1a2, Col3a1, Col5a1, or Col12a1 or proteoglycans such as decorin, biglycan and fibromodulin with increasing horse age. Furthermore, despite previous studies suggesting a slowing of collagen turnover in the SDFT with ageing, no decrease in the levels of expression of the collagenases MMP1 and MMP13 were detected or the tendon-specific transcription factor scleraxis. A full transcriptome analysis using RNA-Seq of human Achilles tendon tissue from young (19 ± 5.8 years) and old (69.4 ± 7.3 years) donors found 191 transcripts were at higher levels in the older tendon and 134 were at lower levels in the older tendon [47]. The networks identified as associated with the differentially expressed genes were cellular function and maintenance, cellular growth and proliferation, cellular cycling, and cellular development, rather than networks relating to matrix proteins, as had been identified in a transcriptome analysis of ageing cartilage [55]. Interestingly, a proteomics study comparing young and old equine SDFT tissue also found cellular proteins featured strongly in the differential analysis; with the term 'intermediate filament' identified and several cytoskeletal keratins and gap junction proteins higher in the old tendon group [56].

While data suggest that there is not a straightforward reduction in gene expression resulting in reduced matrix turnover, other age related processes may result in changes to cell behavior and response to growth factors, cytokines and mechanical signals.

Mechanisms for Age Related Decline in Matrix Turnover Rate

Cellular Senescence

Cellular senescence, the irreversibly arrest of cellular division is an intricate biological process causing alterations in the protein expression profile of the cell and resulting in replicative arrest, changes in metabolism, adhesion efficiency and secretory phenotype [57]. Several of these modifications produce beneficial tumour-suppressive effects as they diminish the proliferation capacity of mutated cells. However, senescent cells are characterised by an increase in the secretion of growth factors, inflammatory cytokines, and proteases; the 'senescence-associated secretory phenotype' (SASP), that can exert the opposite activity by creating a tumour-favoring milieu [57]. We can distinguish ageing from senescence by noting that the latter occurs at a cellular level [58].

Tendon fibroblasts from old mice exhibited low motility, a poorly organized actin cytoskeleton, and a different localization of key focal adhesion proteins as compared with young cells. Senescence associated β-galactosidase expression, a marker for senescence demonstrated that fibroblasts from old mice Achilles tendon were not senescent, but had a distinct phenotype [59] in contrast to ageing rat in which there was an increase in β-galactosidase in middle aged and old rats [60]. However, replicative senescence was demonstrated in mice Achilles tendon fibroblasts cultured for more than 50 passages [59]. Long term *in vitro* culture of cells (the Hayflick model of cellular senescence) has been used extensively to identify mechanisms of age related impairment of function [61]. This method demonstrates proliferation arrest after a number of population doublings and the associated biochemical and molecular changes. Telomere length is a further senescence marker as telomeres are known to shorten progressively during successive cell divisions [62]. In a recent study there was no decrease in cellularity or relative telomere length with increasing age in equine tendon [54].

There is a reduction in proliferation of tenocytes with age. The cellular senescence-inhibited gene (CSIG) is expressed abundantly in young tendon fibroblasts, but its expression declines during cellular senescence. In ageing tendon the reduction in proliferative capacity is associated with the down-regulation of CSIG and an increase in p27, a cell cycle inhibitor protein. CSIG modulates replicative senescence. A reduction in CSIG reduces cell growth and accelerates cellular senescence [60].

Ageing Stem Cell Population

Ageing in tissue such as muscle and brain is driven in part by an age-related reduction in regenerative potential of adult tissues related to a functional decline of their stem cell pool [63]. The number, stress resistance, and repair capacity of tissue-specific adult stem cells contributes to this [64]. The existence of tendon stem or progenitor cells (TSC) has been confirmed in a number of species [65–70].

A relationship between altered TSC properties and tissue ageing has been hypothesised. A study in human Achilles tendon demonstrated that ageing TSCs exhibited self-renewal and clonogenic insufficiencies and premature entry into senescence whilst retaining their multipotency. The group suggested that during tendon ageing the TSC pool size and functional capacity becomes exhausted [71]. It has been suggested that microRNA (miR) 135a has a role in TSC senescence through Rho-associated coiled-coil protein kinase 1 (ROCK1) whilst also promoting proliferation, migration and tenogenic differentiation [72]. Furthermore a loss of tenomodulin, a marker for the tenogenic lineage, may be a source of TSC senescence in ageing tendon. A tenomodulin knock-out mouse model demonstrated TSCs that had reduced self-renewal and demonstrated early entry into senescence [73]. A recently recognised regulator of TSC ageing is peptidyl-prolyl cis-trans isomerase NIMA-interacting 1 (Pin 1) which has a role as a post phosphorylation control in protein function regulation and participates in cellular processes including cell cycle progression, cell survival, immune response and lineage commitment. It is involved in the regulation of adult stem cells and in TSCs it affects cellular senescence possibly through miR-140-5p [74].

In ageing rat [69] and human [75] TSCs there is a reduction in both the number of TSCs, their self-renewal and differentiation potential. In rat this has been associated with a concomitant decrease in tendon lineage markers [69]. However, TSCs seem to retain their pluripotency [69, 71, 75].

Interestingly, a recent study determined the mechanical properties of ageing stem cells and identified an increase in stiffness of ageing TSCs in rat that was attributed to a dense cytoskeleton resulting in an increase in size and irregular shape of older TSCs (see Chap. 6) [76].

Inflammageing

Inflammageing is considered the age-related increase in the systemic pro-inflammatory status. The process results in the breakdown of the multi-shell cytokine network as a consequence of remodeling of the innate and acquired immune system; leading to chronic inflammatory cytokine production. Genetic, environmental and age-related factors determine susceptibility to inflammageing. Thus there is a diminished ability to modulate inflammation [77]. Dakin et al. [78] described an age-associated reduction in FPR2/ALX (a g-coupled protein receptor which binds ligands including metabolites of arachidonic acid) along with increased PGE_2 in tendon. In addition interleukin 1β (IL1β) treated tendon explants from older horses had a reduced ability to express FPR2/ALX and tenocytes from older horses had a reduced response to IL1β induced PGE_2. These results imply that inflammageing is present in ageing tendons and aged individuals exhibit a reduced capacity to resolve inflammation. Therefore ageing may contribute to deregulated tendon repair through these pathways.

In contrast, in an RNASeq study of ageing human Achilles tendon inflammatory pathways were not recognised following gene ontology of differentially expressed genes [47]. Furthermore, in a proteomics study of ageing tendon, inflammatory proteins were not differentially expressed [56].

Reactive Oxygen Species and the Free Radical Theory

One ageing theory is the 'damage-accumulating theory/free radical theory' in which there is a progressive accumulation of cell damage resulting in failure of repair and maintenance systems [79]. One of the causes of cell damage is reactive oxygen species (ROS). A free radicle is any species capable of independent existence that contains one or more unpaired electrons. Trauma, environmental and physiological stimuli may enhance ROS production and ROS are continually produced during normal cell metabolism.

There are a lack of studies on the role of ROS in age-related tendinopathy. An increase in the expression of peroxiredoxin, a thioredoxin peroxidase with antioxidant properties, in tendon degeneration suggests that oxidative stress may be involved in the pathogenesis of tendon degeneration [80], as it is in the age-related disease osteoarthritis [81].

In a recent ageing tendon proteomic study there was a reduction in catalase, α-crystalline-β chain and a number of heat shock proteins, which have protective roles in oxidative and thermal stress respectively. This could point to ageing tendon being less able to respond to increases in ROS [56]. However there was no change in oxidative stress related genes in a tendon ageing transcriptomic study [47].

There is evidence for the involvement of age-related apoptosis in tendon pathology and ageing which may be a consequence of ROS. In degenerative joint disease of the knee, an age related condition, there is an association with increased susceptibility of periarticular tenocytes to Fas ligand induced apoptosis [82].

Glycation of Matrix Proteins

Although the majority of research concerning ageing has focused on mechanisms that are cell mediated [83], an alternative explanation is that the matrix proteins themselves undergo age related changes. The long-lived nature of tendon collagen renders it susceptible to attack by reactive carbonyl groups on sugars such as glucose, in a process known as 'browning' or glycation. A series of spontaneous chemical re-arrangements occurs and further reactions with neighboring peptides results in advanced glycation end-product (AGE) crosslinks such as pentosidine. Unlike the enzyme-mediated crosslinks, which are confined to the telopeptide regions of collagen molecules, AGE crosslinks can form throughout the length of the collagen molecule. Pentosidine is relatively easy to measure as it is resistant to the convenient method of acid hydrolysis to break proteins down into constituent amino acids and in addition, pentosidine is fluorescent, so easy to detect. A number of studies have quantified levels of pentosidine in tendon tissue and shown a positive correlation with donor age. For example pentosidine levels increase with age in human patellar tendon [40], posterior tibialis tendon [84], and equine SDFT and CDET (Fig. 24.5) [39]. The levels are however relatively low (1 crosslink per 70 collagen molecules) in the study by Thorpe et al. [39] and other AGE crosslinks that are present at much higher levels are likely to be more relevant to pathophysiology. One AGE crosslink of particular interest is glucosepane; this AGE crosslink was first identified in 2002 by Biemel and colleagues [85] and later shown to be present in human skin samples at levels equivalent to those for enzyme-mediated crosslinks (see Chap. 18) [86].

Glucosepane is formed from lysine, arginine and glucose, is acid labile, non-fluorescent and present in several different stereoisomer forms making study in native tissues difficult. Initial work in our laboratory has however shown that glucosepane is present in human Achilles tendon

Fig. 24.5 Accumulation of the AGE pentosidine in the SDFT as a function of horse age (inset shows the structure of pentosidine) [39]

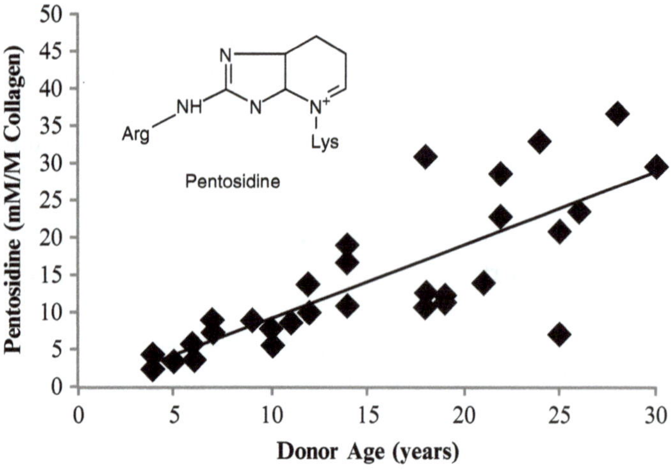

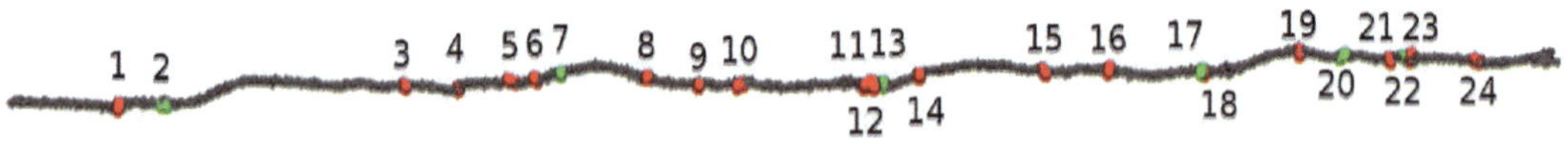

Fig. 24.6 Distribution of identified intra-molecular cross-linking sites along the length of the collagen molecule (*red* areas show energetically unfavourable sites and *green* areas show energetically favourable sites) (Taken from [87])

tissue and levels increase with increasing donor age (Birch, unpublished data). The impact that these crosslinks will have on collagen properties are determined by the sites at which glucosepane forms. Computational studies using a fully atomistic model of an entire collagen molecule in a fibrillar environment have been used to identify, based on energetics, the residues responsible for forming intra-molecular glucosepane cross-links in type I collagen [87]. Using this approach 24 sites where lysine and arginine are in close enough proximity to form glucosepane were identified and six of these sites yielded an exothermic enthalpy change on formation of the glucosepane cross-link (Fig. 24.6). The six energetically favourable sites identified all occur within regions of the collagen molecule that are involved in interactions with other matrix and bioactive molecules. Other studies have investigated the effect of AGEs on mechanical properties of collagenous tissues by incubating the tissue with sugars or other reactive carbonyl metabolites [88–91]. Incubation of rat-tail tendon with methylglyoxal resulted in increased stiffness of collagen fascicles and this seemed to result from decreased sliding between collagen fibrils rather than increased stiffness of the fibril [88]. There remains much to be discovered about AGE formation as an ageing mechanism but this will undoubtedly be very important for understanding tendon homeostasis during ageing.

Conclusion

In summary, although many characteristics of ageing tendon have been documented, there remain conflicting reports of age related effects and the cause of increased injury with age remains unresolved. We consider that age related decline in tendon function is likely to result from a combination of factors, rather than one single cause. Some of the studies discussed in this chapter suggest that a better understanding of normal tendon function and biology is required to allow the effects of ageing to be studied at more specific sub-structure levels. Understanding the factors, and their interactions, that contribute to age related changes in tendon would allow novel approaches to both prevention and better treatments for tendinopathies.

References

1. Riley G (2008) Tendinopathy – from basic science to treatment. Nat Clin Pract Rheumatol 4(2):82–89
2. Bevan S, Passmore E, Mahdon M (2007) Fit for work? Musculoskeletal disorders and labour market participation. The Work Foundation, London. http://www.fitforworkeurope.eu/Downloads/Website-Documents/44_fit_for_work_small.pdf
3. Butler DL, Juncosa N, Dressler MR (2004) Functional efficacy of tendon repair processes. Annu Rev Biomed Eng 6:303–329
4. Thorpe CT, Clegg PD, Birch HL (2010) A review of tendon injury: why is the equine superficial digital flexor tendon most at risk? Equine Vet J 42(2):174–180
5. Clegg PD (2012) Musculoskeletal disease and injury, now and in the future. Part 2: Tendon and ligament injuries. Equine Vet J 44(3):371–375
6. Kingma JJ, de Knikker R, Wittink HM, Takken T (2007) Eccentric overload training in patients with chronic Achilles tendinopathy: a systematic review. Br J Sports Med 41(6):e3
7. Teunis T, Lubberts B, Reilly BT, Ring D (2014) A systematic review and pooled analysis of the prevalence of rotator cuff disease with increasing age. J Shoulder Elb Surg 23(12):1913–1921
8. Jain NB, Higgins LD, Losina E, Collins J, Blazar PE, Katz JN (2014) Epidemiology of musculoskeletal upper extremity ambulatory surgery in the United States. BMC Musculoskelet Disord 15:4
9. Hess GW (2010) Achilles tendon rupture: a review of etiology, population, anatomy, risk factors, and injury prevention. Foot Ankle Spec 3(1):29–32
10. Clayton RA, Court-Brown CM (2008) The epidemiology of musculoskeletal tendinous and ligamentous injuries. Injury 39(12):1338–1344
11. Kasashima Y, Takahashi T, Smith RK et al (2004) Prevalence of superficial digital flexor tendonitis and suspensory desmitis in Japanese Thoroughbred flat racehorses in 1999. Equine Vet J 36(4):346–350
12. Perkins NR, Reid SW, Morris RS (2005) Risk factors for injury to the superficial digital flexor tendon and suspensory apparatus in Thoroughbred racehorses in New Zealand. N Z Vet J 53(3):184–192
13. Reardon RJ, Boden LA, Mellor DJ et al (2012) Risk factors for superficial digital flexor tendinopathy in Thoroughbred racehorses in hurdle starts in the UK (2001–2009). Equine Vet J 44(5):564–569
14. Johnson GA, Tramaglini DM, Levine RE, Ohno K, Choi NY, Woo SL (1994) Tensile and viscoelastic properties of human patellar tendon. J Orthop Res 12(6):796–803
15. Blevins FT, Hecker AT, Bigler GT, Boland AL, Hayes WC (1994) The effects of donor age and strain rate on the biomechanical properties of bone-patellar tendon-bone allografts. Am J Sports Med 22(3):328–333
16. Flahiff CM, Brooks AT, Hollis JM, Vander Schilden JL, Nicholas RW (1995) Biomechanical analysis of patellar tendon allografts as a function of donor age. Am J Sports Med 23(3):354–358
17. Lewis G, Shaw KM (1997) Tensile properties of human tendo Achillis: effect of donor age and strain rate. J Foot Ankle Surg 36(6):435–445
18. Thermann H, Frerichs O, Biewener A, Krettek C, Schandelmaier P (1995) Biomechanical studies of human Achilles tendon rupture. Unfallchirurg 98(11):570–575
19. Hubbard RP, Soutas-Little RW (1984) Mechanical properties of human tendon and their age dependence. J Biomech Eng 106(2):144–150
20. Thorpe CT, Udeze CP, Birch HL, Clegg PD, Screen HR (2013) Capacity for sliding between tendon fascicles decreases with ageing in injury prone equine tendons: a possible mechanism for age-related tendinopathy? Eur Cells Mater 25:48–60
21. Lichtwark GA, Wilson AM (2005) In vivo mechanical properties of the human Achilles tendon during one-legged hopping. J Exp Biol 208(Pt 24):4715–4725
22. Kubo K, Kanehisa H, Miyatani M, Tachi M, Fukunaga T (2003) Effect of low-load resistance training on the tendon properties in middle-aged and elderly women. Acta Physiol Scand 178(1):25–32
23. Karamanidis K, Arampatzis A (2006) Mechanical and morphological properties of human quadriceps femoris and triceps surae muscle-tendon unit in relation to aging and running. J Biomech 39(3):406–417
24. Carroll CC, Dickinson JM, Haus JM et al (2008) Influence of aging on the in vivo properties of human patellar tendon. J Appl Physiol 105(6):1907–1915
25. Stenroth L, Peltonen J, Cronin NJ, Sipila S, Finni T (2012) Age-related differences in Achilles tendon properties and triceps surae muscle architecture in vivo. J Appl Physiol 113(10):1537–1544
26. Csapo R, Malis V, Hodgson J, Sinha S (2014) Age-related greater Achilles tendon compliance is not associated with larger plantar flexor muscle fascicle strains in senior women. J Appl Physiol 116(8):961–969
27. Thorpe CT, Udeze CP, Birch HL, Clegg PD, Screen HR (2012) Specialization of tendon mechanical properties results from interfascicular differences. J R Soc Interface 9(76):3108–3117
28. Lichtwark GA, Cresswell AG, Ker RF et al (2013) Commentaries on viewpoint: On the hysteresis in the human Achilles tendon. J Appl Physiol 114(4):518–520
29. Thorpe CT, Klemt C, Riley GP, Birch HL, Clegg PD, Screen HR (2013) Helical sub-structures in energy-storing tendons provide a possible mechanism for

efficient energy storage and return. Acta Biomater 9(8):7948–7956
30. Thorpe CT, Riley GP, Birch HL, Clegg PD, Screen HR (2014) Fascicles from energy-storing tendons show an age-specific response to cyclic fatigue loading. *Journal of the Royal Society*. Interface 11(92): 1–10
31. Thorpe CT, Godinho MS, Riley GP, Birch HL, Clegg PD (2015) Screen HR. The interfascicular matrix enables fascicle sliding and recovery in tendon, and behaves more elastically in energy storing tendons. J Mech Behav Biomed Mater 52:85–94
32. Gillis C, Pool RR, Meagher DM, Stover SM, Reiser K, Willits N (1997) Effect of maturation and aging on the histomorphometric and biochemical characteristics of equine superficial digital flexor tendon. Am J Vet Res 58(4):425–430
33. Parry DA, Craig AS, Barnes GR (1978) Tendon and ligament from the horse: an ultrastructural study of collagen fibrils and elastic fibres as a function of age. Proc R Soc Lond B Biol Sci 203(1152):293–303
34. Parry DA, Barnes GR, Craig AS (1978) A comparison of the size distribution of collagen fibrils in connective tissues as a function of age and a possible relation between fibril size distribution and mechanical properties. Proc R Soc Lond B Biol Sci 203 (1152):305–321
35. Franchi M, Ottani V, Stagni R, Ruggeri A (2010) Tendon and ligament fibrillar crimps give rise to left-handed helices of collagen fibrils in both planar and helical crimps. J Anat 216(3):301–309
36. Patterson-Kane JC, Firth EC, Goodship AE, Parry DA (1997) Age-related differences in collagen crimp patterns in the superficial digital flexor tendon core region of untrained horses. Aust Vet J 75(1):39–44
37. Wilmink J, Wilson AM, Goodship AE (1992) Functional significance of the morphology and micromechanics of collagen fibres in relation to partial rupture of the superficial digital flexor tendon in racehorses. Res Vet Sci 53(3):354–359
38. Birch HL, Bailey JV, Bailey AJ, Goodship AE (1999) Age-related changes to the molecular and cellular components of equine flexor tendons. Equine Vet J 31(5):391–396
39. Thorpe CT, Streeter I, Pinchbeck GL, Goodship AE, Clegg PD, Birch HL (2010) Aspartic acid racemization and collagen degradation markers reveal an accumulation of damage in tendon collagen that is enhanced with aging. J Biol Chem 285(21):15674–15681
40. Couppe C, Hansen P, Kongsgaard M et al (2009) Mechanical properties and collagen cross-linking of the patellar tendon in old and young men. J Appl Physiol 107(3):880–886
41. Riley GP, Harrall RL, Constant CR, Chard MD, Cawston TE, Hazleman BL (1994) Glycosaminoglycans of human rotator cuff tendons: changes with age and in chronic rotator cuff tendinitis. Ann Rheum Dis 53(6):367–376
42. Thornton GM, Lemmex DB, Ono Y et al (2015) Aging affects mechanical properties and lubricin/PRG4 gene expression in normal ligaments. J Biomech 48(12):3306–3311
43. Taguchi M, Sun YL, Zhao C et al (2009) Lubricin surface modification improves tendon gliding after tendon repair in a canine model in vitro. J Orthop Res 27(2):257–263
44. Flannery CR, Hughes CE, Schumacher BL et al (1999) Articular cartilage superficial zone protein (SZP) is homologous to megakaryocyte stimulating factor precursor and Is a multifunctional proteoglycan with potential growth-promoting, cytoprotective, and lubricating properties in cartilage metabolism. Biochem Biophys Res Commun 254(3):535–541
45. Funakoshi T, Schmid T, Hsu HP, Spector M (2008) Lubricin distribution in the goat infraspinatus tendon: a basis for interfascicular lubrication. J Bone Joint Surg Am 90(4):803–814
46. Sun Y, Berger EJ, Zhao C, Jay GD, An KN, Amadio PC (2006) Expression and mapping of lubricin in canine flexor tendon. J Orthop Res 24(9):1861–1868
47. Peffers MJ, Fang Y, Cheung K, Wei TK, Clegg PD, Birch HL (2015) Transcriptome analysis of ageing in uninjured human Achilles tendon. Arthritis Res Thery 17:33
48. Tavernarakis N (2008) Ageing and the regulation of protein synthesis: a balancing act? Trends Cell Biol 18(5):228–235
49. Heinemeier KM, Schjerling P, Heinemeier J, Magnusson SP, Kjaer M (2013) Lack of tissue renewal in human adult Achilles tendon is revealed by nuclear bomb (14)C. FASEB J 27(5):2074–2079
50. Babraj JA, Cuthbertson DJ, Smith K et al (2005) Collagen synthesis in human musculoskeletal tissues and skin. Am J Physiol Endocrinol Metab 289(5):E864–E869
51. Miller BF, Olesen JL, Hansen M et al (2005) Coordinated collagen and muscle protein synthesis in human patella tendon and quadriceps muscle after exercise. J Physiol 567(Pt 3):1021–1033
52. Langberg H, Rosendal L, Kjaer M (2001) Training-induced changes in peritendinous type I collagen turnover determined by microdialysis in humans. J Physiol 534(Pt 1):297–302
53. Langberg H, Skovgaard D, Petersen LJ, Bulow J, Kjaer M (1999) Type I collagen synthesis and degradation in peritendinous tissue after exercise determined by microdialysis in humans. J Physiol 521(Pt 1):299–306
54. Thorpe CT, McDermott BT, Goodship AE, Clegg PD, Birch HL (2015) Ageing does not result in a decline in cell synthetic activity in an injury prone tendon. Scand J Med Sci Sports 26(6):684–693
55. Peffers MJ, Liu X, Clegg PD (2013) Transcriptomic signatures in cartilage ageing. Arthritis Res Ther 15(4):R98

56. Peffers MJ, Thorpe CT, Collins JA et al (2014) Proteomic analysis reveals age-related changes in tendon matrix composition, with age- and injury-specific matrix fragmentation. J Biol Chem 289 (37):25867–25878
57. Hwang ES, Yoon G, Kang HT (2009) A comparative analysis of the cell biology of senescence and aging. Cell Mol Life Sci 66(15):2503–2524
58. Sethe S, Scutt A, Stolzing A (2006) Aging of mesenchymal stem cells. Ageing Res Rev 5(1):91–116
59. Arnesen SM, Lawson MA (2006) Age-related changes in focal adhesions lead to altered cell behavior in tendon fibroblasts. Mech Aging Dev 127(9):726–732
60. Tsai WC, Chang HN, Yu TY et al (2011) Decreased proliferation of aging tenocytes is associated with down-regulation of cellular senescence-inhibited gene and up-regulation of p27. J Orthop Res 29(10):1598–1603
61. Hayflick L, Moorhead PS (1961) The serial cultivation of human diploid cell strains. Exp Cell Res 25:585–621
62. Wagner W, Horn P, Castoldi M et al (2008) Replicative senescence of mesenchymal stem cells: a continuous and organized process. PLoS One 3(5):e2213
63. Rando TA (2006) Stem cells, ageing and the quest for immortality. Nature 441(7097):1080–1086
64. Sharpless NE, DePinho RA (2007) How stem cells age and why this makes us grow old. Nat Rev Mol Cell Biol 8(9):703–713
65. Bi Y, Ehirchiou D, Kilts TM et al (2007) Identification of tendon stem/progenitor cells and the role of the extracellular matrix in their niche. Nat Med 13(10):1219–1227
66. Haasters F, Polzer H, Prall WC et al (2011) Bupivacaine, ropivacaine, and morphine: comparison of toxicity on human hamstring-derived stem/progenitor cells. Knee Surg Sports Traumatol Arthrosc 19(12):2138–2144
67. Lovati AB, Corradetti B, Lange Consiglio A et al (2011) Characterization and differentiation of equine tendon-derived progenitor cells. J Biol Regul Homeost Agents 25(2 Suppl):S75–S84
68. Tempfer H, Wagner A, Gehwolf R et al (2009) Perivascular cells of the supraspinatus tendon express both tendon- and stem cell-related markers. Histochem Cell Biol 131(6):733–741
69. Zhou Z, Akinbiyi T, Xu L et al (2010) Tendon-derived stem/progenitor cell aging: defective self-renewal and altered fate. Aging Cell 9(5):911–915
70. Williamson KA, Lee KJ, Humphreys WJ, Comerford EJ, Clegg PD, Canty-Laird EG (2015) Restricted differentiation potential of progenitor cell populations obtained from the equine superficial digital flexor tendon (SDFT). J Orthop Res 33(6):849–858
71. Kohler J, Popov C, Klotz B et al (2013) Uncovering the cellular and molecular changes in tendon stem/progenitor cells attributed to tendon aging and degeneration. Aging Cell 12(6):988–999
72. Chen L, Wang GD, Liu JP et al (2015) miR-135a modulates tendon stem/progenitor cell senescence via suppressing ROCK1. Bone 71:210–216
73. Alberton P, Dex S, Popov C, Shukunami C, Schieker M, Docheva D (2015) Loss of tenomodulin results in reduced self-renewal and augmented senescence of tendon stem/progenitor cells. Stem Cells Dev 24(5):597–609
74. Chen L, Liu J, Tao X, Wang G, Wang Q, Liu X (2015) The role of Pin1 protein in aging of human tendon stem/progenitor cells. Biochem Biophys Res Commun 464(2):487–492
75. Ruzzini L, Abbruzzese F, Rainer A et al (2014) Characterization of age-related changes of tendon stem cells from adult human tendons. Knee Surg, Sports Traumatol Arthrosc 22(11):2856–2866
76. Wu H, Zhao G, Zu H, Wang JH, Wang QM (2015) Aging-related viscoelasticity variation of tendon stem cells (TSCs) characterized by quartz thickness shear mode (TSM) resonators. Sensors Actuators 210:369–380
77. Franceschi C, Bonafe M, Valensin S et al (2000) Inflamm-aging. An evolutionary perspective on immunosenescence. Ann N Y Acad Sci 908:244–254
78. Dakin SG, Dudhia J, Werling NJ, Werling D, Abayasekara DR, Smith RK (2012) Inflamm-aging and arachadonic acid metabolite differences with stage of tendon disease. PLoS One 7(11):e48978
79. Harman D (1956) Aging: a theory based on free radical and radiation chemistry. J Gerontol 11 (3):298–300
80. Wang MX, Wei A, Yuan J et al (2001) Antioxidant enzyme peroxiredoxin 5 is upregulated in degenerative human tendon. Biochem Biophys Res Commun 284(3):667–673
81. Tiku ML, Shah R, Allison GT (2000) Evidence linking chondrocyte lipid peroxidation to cartilage matrix protein degradation. Possible role in cartilage aging and the pathogenesis of osteoarthritis. J Biol Chem 275(26):20069–20076
82. Machner A, Baier A, Wille A et al (2003) Higher susceptibility to Fas ligand induced apoptosis and altered modulation of cell death by tumor necrosis factor-alpha in periarticular tenocytes from patients with knee joint osteoarthritis. Arthritis Res Ther 5(5):R253–R261
83. Lopez-Otin C, Blasco MA, Partridge L, Serrano M, Kroemer G (2013) The hallmarks of aging. Cell 153(6):1194–1217
84. Corps AN, Robinson AH, Harrall RL et al (2012) Changes in matrix protein biochemistry and the expression of mRNA encoding matrix proteins and metalloproteinases in posterior tibialis tendinopathy. Ann Rheum Dis 71(5):746–752
85. Biemel KM, Friedl DA, Lederer MO (2002) Identification and quantification of major maillard cross-links in human serum albumin and lens protein. Evidence

for glucosepane as the dominant compound. J Biol Chem 277(28):24907–24915
86. Sell DR, Biemel KM, Reihl O, Lederer MO, Strauch CM, Monnier VM (2005) Glucosepane is a major protein cross-link of the senescent human extracellular matrix. Relationship with diabetes. J Biol Chem 280(13):12310–12315
87. Collier TA, Nash A, Birch HL, de Leeuw NH (2015) Preferential sites for intramolecular glucosepane cross-link formation in type I collagen: A thermodynamic study. Matrix Biol
88. Fessel G, Li Y, Diederich V et al (2014) Advanced glycation end-products reduce collagen molecular sliding to affect collagen fibril damage mechanisms but not stiffness. PLoS One 9(11):e110948
89. Li Y, Fessel G, Georgiadis M, Snedeker JG (2013) Advanced glycation end-products diminish tendon collagen fiber sliding. Matrix Biol 32(3–4):169–177
90. Reddy GK (2003) Glucose-mediated in vitro glycation modulates biomechanical integrity of the soft tissues but not hard tissues. J Orthop Res 21(4):738–743
91. Reddy GK, Stehno-Bittel L, Enwemeka CS (2002) Glycation-induced matrix stability in the rabbit achilles tendon. Arch Biochem Biophys 399(2):174–180

Part III

Novel Therapies that May Affect Tendon Metabolism

Does Platelet-Rich Plasma Increase Tendon Metabolism? 25

Robert-Jan de Vos

Abstract

Acute and overuse tendon disorders are frequently observed in the middle-aged active population. Tendon overuse injuries are currently designated as “tendinopathy”. Histopathological studies have shown that chronic tendinopathy is frequently characterised by degenerative changes, such as decreased organisation of collagen, altered cell distribution and neovascularisation. In the recent years, scientific research and technology in the field of regenerative medicine has provided a new perspectives on managing chronic tendinopathy. An initiation of tissue healing can be attempted by local delivery of growth factors. Nowadays, platelet-rich plasma (PRP) is a commonly applied approach to achieve this. Platelet degranulation leads to a release of various growth factors and cytokines. There is a classification system to define the different forms of PRP. In the past decade, a number of studies have been published on the effects of PRP in different basic science studies. These studies suggest that PRP modulates some aspects of tendon metabolic activity. This is one of the reasons why PRP is increasingly used by many clinicians as treatment option for tendinopathy in daily clinical practice. There is, however, evidence from the literature that it does not lead to improved outcome on imaging findings and on patient-reported outcomes. This questions the role of PRP injections as regular treatment for tendinopathy. Moreover, it results in a broader discussion on the required effects that need to occur for tendon healing and symptom relieve.

R.-J. de Vos (✉)
Department of Orthopaedics and Sports Medicine,
Erasmus MC University Medical Centre, Room
Hs-106,'s-Gravendijkwal 230, 3015 CE Rotterdam,
The Netherlands
e-mail: r.devos@erasmusmc.nl

P.W. Ackermann and D.A. Hart (eds.), *Metabolic Influences on Risk for Tendon Disorders*,
Advances in Experimental Medicine and Biology 920, DOI 10.1007/978-3-319-33943-6_25

Keywords
PRP • Growth factors • Tendinopathy • Tendon regeneration • Neovascularisation • Collagen structure

Abbreviations

PRP	platelet-rich plasma
PPP	platelet-poor plasma
P-PRP	pure platelet-rich plasma
L-PRP	leukocyte- and platelet-rich plasma
P-PRF	pure platelet-rich fibrin
L-PRF	leukocyte- and platelet-rich fibrin
PDGF	platelet-derived growth factor
TGF-β	transforming growth factor-β
VEGF	vascular-derived endothelial growth factor
EGF	epithelial growth factor
HGF	hepatocyte growth factor
IGF	insulin-like growth factor
MMPs	matrix metalloproteinases
ELISA	enzyme-linked immunology assay
US	ultrasonography
MRI	magnetic resonance imaging

Introduction

Overuse tendon disorders are frequently observed in the active middle-aged population [1]. The aetiology of these overuse injuries is largely unknown. A combination of aging and increased activity level leads to an increased risk of overuse injuries [1, 2].

The preferred terminology for tendon overuse injuries is "tendinopathy", as a descriptor of the clinical picture [3]. Tendinopathy is a clinical diagnosis that is established in the presence of pain, tendon swelling and an impaired load-bearing capacity [4]. Histopathological studies have shown that the tendon tissue of patients with chronic tendinopathy is frequently characterised by degenerative changes [5]. These changes include an abnormal collagen fibre structure and arrangement, variations in cellular distribution, rounding of tenocyte nuclei, decreased collagen and glycosaminoglycan stainability and increased vascularity. The degenerative changes are probably the prerequisites of spontaneous tendon ruptures [6] and therefore a tendon rupture can be regarded as end-stage of the degenerative process.

Based on the fact that the end-stage tendon overuse injuries have a degenerative basis, scientific research and technology in the field of regenerative medicine has provided new perspectives in this field. Initiation of tissue healing can be attempted by local delivery of growth factors. Injections with platelet-rich plasma (PRP) are nowadays commonly used with an attempt to achieve this [7]. Platelet degranulation leads to a release of various growth factors and cytokines [8]. Theoretically, local administration of PRP could enhance a regenerative response in injured tendons. Locally injected growth factors could result in an increased metabolic activity through their action on tenocytes.

The aim of this book chapter is to provide an overview of the effects of PRP on tendon tissue in basic science studies, imaging studies and from a clinical perspective.

Effects of Platelet-Rich Plasma in Basic Science Studies

In the past decade, a lot of laboratory research has been published in the field of PRP application in tendons. This enables researchers to understand how PRP could work in overuse tendon injuries. This part describes the definition of PRP and its potential working mechanisms on tendon tissue in-vitro or in-vivo.

Definition of Platelet-Rich Plasma

The main function of platelets is to contribute to haemostasis. Platelets are produced by megakaryocytes, which are formed in bone marrow. Platelet structures of interest for tissue healing are the α-granules. Degranulation of these α-granules leads to a release of many different growth factors that can support in tissue regeneration processes. Platelet-derived growth factor (PDGF), transforming growth factor-β (TGF-β), vascular-derived endothelial growth factor (VEGF), epithelial growth factor (EGF), hepatocyte growth factor (HGF) and insulin-like growth factor (IGF) are examples of such growth factors [9, 10].

PRP can be obtained through a processing technique of whole blood. Approximately 45–60 mL of autologous whole blood can be drawn from the patient. Standard cell separating devices can be used to isolate a layer which is rich of platelets. To avoid premature clotting and to preserve the integrity of the platelet membrane, a citrate anticoagulant can be added to the whole blood [11]. Centrifugation of autologous blood leads to the formation of three layers: platelet-poor plasma (PPP), platelet-rich plasma (PRP) and red blood cells (Fig. 25.1). The PRP layer can be used for injection after aspiration. Specific activating agents (e.g. calcium or thrombin) stimulate degranulation of the α-granules, but collagen type I – which is the main collagenous component of tendons – is known to have comparable stimulating effects on degranulation of the α-granules [12].

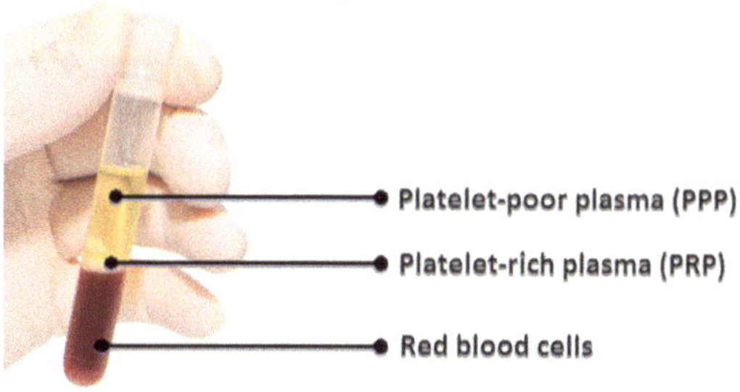

Fig. 25.1 after centrifugation of whole blood, three layers can be distinguished. Platelet-rich plasma can be aspirated and used for injection therapy

Initially, PRP was defined as "a portion of the plasma fraction of autologous blood with a platelet concentration above baseline" [13]. While this definition is still applicable, a new classification system has been proposed to improve the evaluation of PRP-based therapies. This practical consensus-based classification system divides platelet concentrates into four categories (Table 25.1) [14].

The PRP solutions have the advantage to be liquid before activation, and can therefore be used as injection. The PRF solutions only exist in a strongly activated gel form which is not applicable for injection therapy. Most studies in orthopaedic and sports medicine literature used the PRP forms.

This new classification system may be an important first step in characterisation of PRP. The simplistic approach makes it easily accessible and comprehensible for researchers and clinicians. However, due to the high number of parameters involved in the characterisation of PRP it is deemed incomplete. Other factors that may be of importance are the platelet concentration, quantity of leukocytes (if present), the detailed cell composition and preservation and

Table 25.1 Consensus-based classification system for platelet concentrates

PRP classification	Definition
Pure Platelet-Rich Plasma (P-PRP)	Preparations without leukocytes and with a low-density fibrin network after activation
Leukocyte- and Platelet-Rich Plasma (L-PRP)	Preparations with leukocytes and with a low-density fibrin network after activation
Pure Platelet-Rich Fibrin (P-PRF)	Preparations without leukocytes and with a high-density fibrin network
Leukocyte- and Platelet-Rich Fibrin (L-PRF)	Preparations with leukocytes and with a high-density fibrin network

quantity of secreted growth factors [14]. It might also be of importance to add an anti-coagulant after collection of whole blood and activate PRP after centrifugation. These variables make it challenging to structurally evaluate mechanisms and effectiveness of PRP in the scientific literature – even using a consensus based classification system.

Local Tendon Metabolism After Administration of Platelet-Rich Plasma

Mechanisms of new treatment options for tendinopathy are preferably first evaluated in the laboratory setting (see Chap. 8). From 2005, many laboratory studies have been published on the effects of PRP on musculoskeletal tissues. The proposed mechanisms of PRP on tendon tissue are discussed below.

Cell Proliferation

Anitua et al. [15] were one of the first researchers to evaluate the potential beneficial effect of growth factors released from PRP. They found that both PRP and platelet-poor plasma (PPP) within the fibrin matrices stimulated tenocyte proliferation in healthy human tendon cells. Multiple subsequent studies confirmed these results in animal or human tendon tissue [16]. In only one study, tenocytes were isolated from human rotator cuff tendons with degenerative tears [17]. These authors showed that PRP stimulates tendon cell proliferation when compared to control medium and PPP. The optimum proliferation was observed at platelet concentration of five times the baseline value. Hence, these studies implicate that affected tenocytes isolated from degenerative tissue still have the capacity to proliferate.

Growth Factor Delivery

Clotting of platelets result in secretion of platelets. Several studies also investigated whether the contents of PRP also have the ability to stimulate growth factor delivery by tenocytes. A study on the effects of a PRP clot on human tendon cells showed an increased synthesis of the angiogenetic factors VEGF and HGF after 6 days of culture. The other measured growth factors PDGF, EGF, TGF-β1 and IGF-I were, however, not increased. De Mos et al. [18] performed a cell culture study in healthy human tendon cells and administered 20 % of the platelet-rich clot releasate because they assumed that it is not realistic to consider that 100 % of the concentrate will reach the target lesion in the clinical setting. Administration of PRP clots increased concentrations of VEGF at all time-points and increased PDGF-BB after 4 days of culturing. Another placebo-controlled study in an Achilles tendon rupture model of 80 rat tendons demonstrated that there was no difference in gene expression level of PDGF-BB and TGF-β1 throughout the experiment of 6 weeks. Only at the 1-week time point, there was an increased gene-expression level of PDGFα in the PRP-treated tendons compared to the control group [19]. Most of the basic science studies identified an increased VEGF concentration after PRP administration, but this increase is not consistent for other growth factors [16].

Angiogenesis

The increased VEGF concentrations would support an angiogenetic response due to the presence of a PRP clot, which was confirmed in a study on the effects of PRP in traumatic rabbit patellar tendon defects [20]. Immunohistochemically detected neovascularisation was assessed semi-quantitatively. Neovascularisation was increased within the first 2 weeks after PRP administration and significantly decreased after 2 weeks compared to the control group. Bosch et al. performed a placebo-controlled study on the effects of PRP in surgically induced equine tendon lesions [21]. PRP resulted in a significantly increased colour Doppler signal from 1 week post-treatment until final follow-up of 24 weeks. These results could, however, not be reproduced in other placebo-controlled studies. The vascular density was deceased after administration of P-PRP in sheep Achilles tendons [22]. Additionally, vascularity was not altered after PRP treatment for a traumatic supraspinatus tendon injury model [23]. Consequently, there is conflicting evidence on the angiogenetic response after PRP administration.

Collagen Synthesis and Organisation

Theoretically, PRP has the ability to result in production of a healthy tendon matrix with aligned collagen. Indeed, PRP enhanced gene expression of markers for collagen type I and decreased collagen type III levels with a positive collagen I/III ratio in equine tendon explants [24]. PRP enhanced gene expression of markers for collagen type I and III production, without increased catabolic matrix metalloproteinases (MMPs). This implicates that PRP administration results in an anabolic response. The anabolic effects have also been established in tendon stem cells after PRP clot administration [25]. These tendon stem cells, derived from rabbit patellar tendons, differentiated into tenocytes after addition of PRP to the culture. Moreover, there was an increased cell proliferation and collagen production. Contrary, another laboratory study in healthy human tendon cells demonstrated that both PRP and PPP clots decreased gene expression of collagen type I and III [18]. PRP also resulted in an upregulation of MMP1 and MMP3 expression. This implicates that PRP might not only result in anabolic effects, but also catabolic effects on tendon tissue.

A study on the effects of PRP administration in rabbit patellar tendon defects was performed to observe histological differences [26]. The PRP-treated tendons showed better collagen stainability and more distinct oriented fibroblasts after 28 days. An in-vivo placebo-controlled study on the effects of PRP in surgically induced equine tendon lesions showed comparable results [21]. PRP resulted in increased content of collagen, glycosaminoglycans and cellularity. There was a higher strength at failure and better structural organisation of the collagen network. The above mentioned results could, however, not be reproduced in another placebo-controlled study on the effects of PRP in rat patellar tendons [27]. Comparisons between the control group and the PRP group revealed no overall differences in biomechanical testing or histopathological scores after 14 days. This may have been caused by the high variability in platelet concentration, as a subgroup analysis of successful PRP samples demonstrated improved biomechanical properties. The above mentioned studies show conflicting evidence regarding production of collagen molecules and formation of a collagen network.

Mechanical Strength and Loading

Researchers were also interested in the interaction between delivered growth factors and mechanical stimulation. The effect of a mechanical stimulus on PRP-therapy was investigated using an Achilles tendon transection model in rats [28]. PRP, in the form of liquid or gel, was derived after centrifuging whole blood of rats. After application of activated PRP, tendons were either unloaded by Botox injections into the calf muscles or mechanically stimulated in activity cages. Within the first 5 days platelets improved the mechanical properties (with tensile testing), also in the Botox-treated rats. This effect was, however, abolished after 14 days in the Botox-treated group. Without Botox, both activity and platelets resulted in improved mechanical properties. A study on the effects of PRP in rabbit patellar tendon defects confirmed the improved mechanical properties only in the early phase of PRP administration [26]. Another study on the effects of PRP and tendons stem cells in surgically-induced rat Achilles tendon lesions also demonstrated favourable effects of loading compared to unloading (see Chap. 6) [29]. Consequently, mechanical stimulation can be regarded as a prerequisite for the long-lasting effect of platelets.

Inflammation

The effects of PRP on inflammation and the role of leukocytes are a topic of debate in this field. Leukocyte- and Platelet-Rich Plasma concentrations are thought to yield bioactive catabolic cytokines. Indeed, a high leukocyte concentration in PRP contributed to the expression of inflammatory cytokines in equine tendon explants [30]. An initial pro-inflammatory response was confirmed in a study on rabbit patellar tendons, where a significant difference was found after L-PRP administration compared to P-PRP [31]. Another equine laboratory study showed that increasing the platelet concentration in leukocyte-reduced PRP leads to more anabolic growth factors and less pro-inflammatory cytokines, but decreased tendon metabolism measured with gene expression analysis for collagen type I and III production [32]. Contrary, anti-inflammatory effects of HGF in the PRP concentrate have been established in another study using a mouse Achilles tendon injury model [33]. It is therefore uncertain whether the local effects of PRP on tendon tissue are pro-inflammatory or anti-inflammatory. If a pro-inflammatory effect is present, this is probably in the early phase post-treatment using L-PRP.

Systemic Effects After Administration of Platelet-Rich Plasma

PRP might also work through chemo-attractive mechanisms. In an in-vitro study PRP was administered into the wounded area of a rat patellar tendon [34]. The numbers of fluorescent-positive cells – indicating presence of circulation-derived cells – were significantly higher after PRP administration, compared to a control group. This implicates that PRP not only acts locally, but important effects might be initiated systemically. It is also known that certain growth factors (for example IGF-1 and VEGF) are significantly elevated in the serum after an intratendinous PRP injection [35]. This suggests that the mechanism of PRP is an activation of biological pathways rather than direct delivery of growth factors. However, this hypothesis is not supported by a recent in-vivo study in rat Achilles and patellar tendons [36]. The collagenase-induced tendon lesions were treated with either a single injection of leukocyte-reduced PRP or serum and follow-up was performed using serum sample and local tendon tissue removal which was analyzed with enzyme-linked immunology assay (ELISA) techniques. This study demonstrated no difference in local or systemic markers (such as growth factors) between both groups in mid-term and long-term follow-up. Possibly, this systemic response after PRP administration is only present within the very short period after injection.

Summary of Laboratory Studies

PRP appears to have several potential effects on tendon models compared with a control treatments. Most studies showed increased cell proliferation and growth factor concentrations after PRP injection, especially the angiogenetic VEGF. An increased histological vascular network has been reported in multiple studies, but negative studies are also present and one study even showed reduced vascularity. Results are heterogeneous on collagen type I and III production and collagen organisation. The same accounts for the suggested pro-inflammatory and anti-inflammatory response. As described above, the results of PRP administration in basic science studies are inconsistent and an accurate description of the PRP concentrate is frequently lacking in the scientific literature. This was supported by a recent systematic review on this topic [16]. Another limitation of the basic science literature is that the majority of the studies analysed the effects of PRP traumatic or collagenase-induced tendon models. These models are different from the pathology observed in chronic tendinopathy. Therefore, one should be cautious with the interpretation of these findings.

Imaging Studies on the Effects of Platelet-Rich Plasma

Ultrasonography (US) and magnetic resonance imaging (MRI) are the additional diagnostics of choice in the clinical setting because of their excellent visualisation of soft tissue [37]. On US, healthy tendons display a parallel tendon structure. Disorganisation of tendon structure, hypoechoic areas and increased blood-flow, detected with the addition of Doppler techniques, are frequently observed in cases of tendinopathy

[38–40]. On MRI, a healthy tendon is displayed as a structure with dark signal due to the tendon's densely packed collagen and low intrinsic water content. Chronic tendinopathy is also characterised by tendon thickening and an enhanced intratendinous water signal as a result of the increased water-attracting proteoglycans between the fibrils [41].

Multiple ultrasonographical case series showed decreased tendon thickening and decreased number of hypoechoic areas after a PRP injection in patients with Achilles tendinopathy [42, 43]. Two case series also described an initial increase in Doppler blood-flow and a subsequent decrease after 6 months of treatment which was attributed to PRP [42, 44]. This lead to the hypothesis that the delivery of local VEGF resulted in an improved functional vascular network, which would be necessary for the restoration of healthy tendon tissue.

Improvements in gradation of MRI abnormalities after a PRP injection in patients with midportion Achilles tendinopathy have also been published [43, 45]. However, another study did not detect improvement on MRI findings after PRP administration. In a retrospective analysis after a PRP injection for patients with midportion Achilles tendinopathy, an improvement on MRI was found in only one of the six included patients [46].

One double-blind placebo-controlled trial showed no difference in change of tendon structure, measured with standardised US, within 6 months after injecting PRP for patients with chronic midportion Achilles tendinopathy [47]. Interestingly, both in the patients treated with PRP and saline, there was an initial increase in Doppler blood-flow and a decrease afterwards, but this change in was not significantly different between both treatment groups. Therefore, there is currently no evidence that the improvements on imaging – enhanced tendon structure, reduction of hypoechoic areas and improved vascularity – are induced by PRP injections. It is even questionable whether the ultrasonographic imaging modalities are useful in follow-up of patients, since there is no association between ultrasonographic abnormalities and patient-reported symptoms [48–50].

Clinical Studies on the Efficacy of Platelet-Rich Plasma

Clinical applications of PRP have become popular in the recent years. PRP injections have been proposed in the clinical setting because of the attributed improved tendon healing in basic science studies, narrative reviews and intensive marketing strategies [51, 52].

The initial clinical studies that assessed the effects of a PRP injection did not use an adequate control group, and even if a control group was present, the methodological quality was poor [8, 53, 54]. These case series all showed an improvement over time after a PRP injection in multiple tendinopathy locations [51, 55]. The first double-blind randomised placebo-controlled clinical trial in this field did not show efficacy of a single intratendinous PRP injection in 54 patients with chronic midportion Achilles tendinopathy [56]. After 24 weeks and also after 52 weeks [57], there was no improvement in pain, function and activity level compared to a local saline injection. The PRP was not classified in this study. Multiple randomised controlled trials were performed on numerous tendinopathy locations afterwards, which confirmed these results [58–62]. Positive results in randomised studies have also been published [63–65]. Some of these had poor methodological quality [66]. One high-quality study was performed on the treatment of 100 patients with chronic tendinopathy of the wrist extensors. These researchers used the same cell-separating system to obtain PRP as in the first RCT, but also in this study the PRP was not classified. The authors described that PRP significantly reduced pain and increased function, when compared to a corticosteroid injection [63]. An explanation for this finding could be the choice of the control group. Corticosteroids are less favourable than a wait-and-see policy on the mid and longer term [67]. This makes it unclear as to whether the difference observed is due to the beneficial effect of PRP or the detrimental effect of corticosteroids. This stresses the need for placebo-controlled studies. Carr et al. [68] randomised 60 patients diagnosed with rotator cuff tendinopathy to arthroscopic acromioplasty alone or an additional PRP injection in the subacromial space. Leucocyte-rich PRP with thrombin was used in this study. There were no significant differences in patient-reported outcomes up to 2 years post-treatment. Furthermore, biopsies were taken at 12 weeks after treatment. The standardised Bonar score was not significantly different between both groups, which implicates that PRP does not enhance tissue regeneration. The number of blood vessels and cell proliferation were significantly decreased and markers for apoptosis were increased in the PRP group. This suggest even a harmful effect of PRP, although complications after PRP injections are rare.

Multiple recent systematic reviews confirm that clinical application of PRP is not efficacious in symptom reduction or improvement of function in patients with tendinopathy [51, 69–71]. Frequently mentioned causes of the above mentioned results are the high variation in the handling procedures, such as the centrifuge duration, frequency and method of PRP isolation and aspiration. The preparation procedure, optimum volume of PRP injection and PRP form (liquid, light gel or solid gel material), location of the injection (intratendinous or peritendinous), the number of injections in time (single versus multiple injections), the timing of the injection (acute or chronic phase) and the post-injection rehabilitation protocol (immobilisation versus mechanical stimulation) are also variable in the literature [8]. Another explanation for the absence of efficacy is that the catabolic environment in degenerative tendons is detrimental for the PRP clot. As mentioned previously, the potential heterogeneity in these variables make it extremely difficult to evaluate and interpret the clinical efficacy of PRP. Advocators of this treatment frequently state that the negative studies did not use a proper PRP concentrate or adequate timing. This discussion should – however – not be a license to use it as standard treatment in the clinical setting.

Conclusion: Does Platelet-Rich Plasma Increase Tendon Metabolism?

PRP appears to modulate some aspects of metabolic activity of tendons in basic science studies, with increased concentrations of the angiogenetic VEGF and increased tenocyte proliferation as most robust findings. Study results are heterogeneous regarding the collagen type I and III production, biomechanical properties, collagen structural organisation, histologically assessed vascular network, (anti-)inflammatory response and systemic effects. There is currently no evidence that PRP improves findings on imaging and patient-reported symptoms. These findings do currently not support the clinical use of PRP.

Researchers in this field should also reflect which treatment effect should be required in tendinopathies. Since tendinopathy is characterised by increased metabolic activity (e.g. cell activity, matrix turnover and neovascularisation), it is at least questionable whether the proposed effects of PRP are desirable.

References

1. Maffulli N, Wong J, Almekinders LC (2003) Types and epidemiology of tendinopathy. Clin Sports Med 22(4):675–692
2. Nielsen RO, Parner ET, Nohr EA, So H, Lind M, Rasmussen S (2014) Excessive progression in weekly running distance and risk of running-related injuries: an association which varies according to type of injury. J Orthop Sports Phys Ther 44(10):739–747
3. Ackermann PW, Renstrom P (2012) Tendinopathy in sport. Sports Health 4(3):193–201
4. Alfredson H (2003) Chronic midportion Achilles tendinopathy: an update on research and treatment. Clin Sports Med 22(4):727–741
5. de Mos M, van El B, DeGroot J, Jahr H, van Schie HT, van Arkel ER et al (2007) Achilles tendinosis: changes in biochemical composition and collagen turnover rate. Am J Sports Med 35(9):1549–1556
6. Maffulli N, Barrass V, Ewen SW (2000) Light microscopic histology of achilles tendon ruptures. A comparison with unruptured tendons. Am J Sports Med 28 (6):857–863
7. Mishra A, Harmon K, Woodall J, Vieira A (2012) Sports medicine applications of platelet rich plasma. Curr Pharm Biotechnol 13(7):1185–1195

8. de Vos RJ, van Veldhoven PL, Moen MH, Weir A, Tol JL, Maffulli N (2010) Autologous growth factor injections in chronic tendinopathy: a systematic review. Br Med Bull 95:63–77
9. Sampson S, Gerhardt M, Mandelbaum B (2008) Platelet rich plasma injection grafts for musculoskeletal injuries: a review. Curr Rev Musculoskelet Med 1 (3–4):165–174
10. Eppley BL, Woodell JE, Higgins J (2004) Platelet quantification and growth factor analysis from platelet-rich plasma: implications for wound healing. Plast Reconstr Surg 114(6):1502–1508
11. Eppley BL, Pietrzak WS, Blanton M (2006) Platelet-rich plasma: a review of biology and applications in plastic surgery. Plast Reconstr Surg 118(6):147e-59e
12. Fufa D, Shealy B, Jacobson M, Kevy S, Murray MM (2008) Activation of platelet-rich plasma using soluble type I collagen. J Oral Maxillofac Surg 66 (4):684–690
13. Marx RE (2001) Platelet-rich plasma (PRP): what is PRP and what is not PRP? Implant Dent 10 (4):225–228
14. Dohan Ehrenfest DM, Andia I, Zumstein MA, Zhang CQ, Pinto NR, Bielecki T (2014) Classification of platelet concentrates (Platelet-Rich Plasma-PRP, Platelet-Rich Fibrin-PRF) for topical and infiltrative use in orthopedic and sports medicine: current consensus, clinical implications and perspectives. Muscles Ligaments Tendons J 4(1):3–9
15. Anitua E, Andia I, Sanchez M, Azofra J, del Mar Zalduendo M, de la Fuente M et al (2005) Autologous preparations rich in growth factors promote proliferation and induce VEGF and HGF production by human tendon cells in culture. J Orthop Res 23(2):281–286
16. Baksh N, Hannon CP, Murawski CD, Smyth NA, Kennedy JG (2013) Platelet-rich plasma in tendon models: a systematic review of basic science literature. Arthroscopy 29(3):596–607
17. Jo CH, Kim JE, Yoon KS, Shin S (2012) Platelet-rich plasma stimulates cell proliferation and enhances matrix gene expression and synthesis in tenocytes from human rotator cuff tendons with degenerative tears. Am J Sports Med 40(5):1035–1045
18. de Mos M, van der Windt AE, Jahr H, van Schie HT, Weinans H, Verhaar JA et al (2008) Can platelet-rich plasma enhance tendon repair? A cell culture study. Am J Sports Med 36(6):1171–1178
19. Parafioriti A, Armiraglio E, Del Bianco S, Tibalt E, Oliva F, Berardi AC (2011) Single injection of platelet-rich plasma in a rat Achilles tendon tear model. Muscles Ligaments Tendons J 1(2):41–47
20. Lyras D, Kazakos K, Verettas D, Polychronidis A, Simopoulos C, Botaitis S et al (2010) Immunohistochemical study of angiogenesis after local administration of platelet-rich plasma in a patellar tendon defect. Int Orthop 34(1):143–148
21. Bosch G, van Schie HT, de Groot MW, Cadby JA, van de Lest CH, Barneveld A et al (2010) Effects of platelet-rich plasma on the quality of repair of mechanically induced core lesions in equine superficial digital flexor tendons: a placebo-controlled experimental study. J Orthop Res 28(2):211–217
22. Fernandez-Sarmiento JA, Dominguez JM, Granados MM, Morgaz J, Navarrete R, Carrillo JM et al (2013) Histological study of the influence of plasma rich in growth factors (PRGF) on the healing of divided Achilles tendons in sheep. J Bone Joint Surg Am 95 (3):246–255
23. Dolkart O, Chechik O, Zarfati Y, Brosh T, Alhajajra F, Maman E (2014) A single dose of platelet-rich plasma improves the organization and strength of a surgically repaired rotator cuff tendon in rats. Arch Orthop Trauma Surg 134 (9):1271–1277
24. McCarrel T, Fortier L (2009) Temporal growth factor release from platelet-rich plasma, trehalose lyophilized platelets, and bone marrow aspirate and their effect on tendon and ligament gene expression. J Orthop Res 27(8):1033–1042
25. Zhang J, Wang JH (2010) Platelet-rich plasma releasate promotes differentiation of tendon stem cells into active tenocytes. Am J Sports Med 38 (12):2477–2486
26. Lyras DN, Kazakos K, Verettas D, Botaitis S, Agrogiannis G, Kokka A et al (2009) The effect of platelet-rich plasma gel in the early phase of patellar tendon healing. Arch Orthop Trauma Surg 129 (11):1577–1582
27. Spang JT, Tischer T, Salzmann GM, Winkler T, Burgkart R, Wexel G et al (2011) Platelet concentrate vs. saline in a rat patellar tendon healing model. Knee Surg Sports Traumatol Arthrosc 19(3):495–502
28. Virchenko O, Aspenberg P (2006) How can one platelet injection after tendon injury lead to a stronger tendon after 4 weeks? Interplay between early regeneration and mechanical stimulation. Acta Orthop 77 (5):806–812
29. Chen L, Dong SW, Liu JP, Tao X, Tang KL, Xu JZ (2012) Synergy of tendon stem cells and platelet-rich plasma in tendon healing. J Orthop Res 30 (6):991–997
30. McCarrel TM, Minas T, Fortier LA (2012) Optimization of leukocyte concentration in platelet-rich plasma for the treatment of tendinopathy. J Bone Joint Surg Am 94(19):e143(1–8)
31. Dragoo JL, Braun HJ, Durham JL, Ridley BA, Odegaard JI, Luong R et al (2012) Comparison of the acute inflammatory response of two commercial platelet-rich plasma systems in healthy rabbit tendons. Am J Sports Med 40(6):1274–1281
32. Boswell SG, Schnabel LV, Mohammed HO, Sundman EA, Minas T, Fortier LA (2014) Increasing platelet concentrations in leukocyte-reduced platelet-rich plasma decrease collagen gene synthesis in tendons. Am J Sports Med 42(1):42–49
33. Zhang J, Middleton KK, Fu FH, Im HJ, Wang JH (2013) HGF mediates the anti-inflammatory effects of PRP on injured tendons. PLoS One 8(6):e67303

34. Kajikawa Y, Morihara T, Sakamoto H, Matsuda K, Oshima Y, Yoshida A et al (2008) Platelet-rich plasma enhances the initial mobilization of circulation-derived cells for tendon healing. J Cell Physiol 215(3):837–845
35. Wasterlain AS, Braun HJ, Harris AH, Kim HJ, Dragoo JL (2013) The systemic effects of platelet-rich plasma injection. Am J Sports Med 41 (1):186–193
36. Dallaudiere B, Louedec L, Lenet MP, Pesquer L, Blaise E, Perozziello A et al (2015) The molecular systemic and local effects of intra-tendinous injection of Platelet Rich Plasma in tendinosis: preliminary results on a rat model with ELISA method. Muscles Ligaments Tendons J 5(2):99–105
37. de Vos RJ, d'Hooghe PPRN, de Leeuw P, Kerkhoffs GMMJ (2014) Chapter 19: Achilles Tendinopathy. In: d'Hooghe PPRN, Kerkhoffs GMMJ (eds) The ankle in football, sports and traumatology. Springer Verlag, Paris, pp 213–233
38. van Schie HT, de Vos RJ, de Jonge S, Bakker EM, Heijboer MP, Verhaar JA et al (2010) Ultrasonographic tissue characterisation of human Achilles tendons: quantification of tendon structure through a novel non-invasive approach. Br J Sports Med 44 (16):1153–1159
39. Sengkerij PM, de Vos RJ, Weir A, van Weelde BJ, Tol JL (2009) Interobserver reliability of neovascularization score using power Doppler ultrasonography in midportion achilles tendinopathy. Am J Sports Med 37(8):1627–1631
40. de Vos RJ, Weir A, Cobben LP, Tol JL (2007) The value of power Doppler ultrasonography in Achilles tendinopathy: a prospective study. Am J Sports Med 35(10):1696–1701
41. Bleakney RR, White LM (2005) Imaging of the Achilles tendon. Foot Ankle Clin 10(2):239–254
42. Gaweda K, Tarczynska M, Krzyzanowski W (2010) Treatment of Achilles tendinopathy with platelet-rich plasma. Int J Sports Med 31(8):577–583
43. Monto RR (2012) Platelet rich plasma treatment for chronic Achilles tendinosis. Foot Ankle Int 33 (5):379–385
44. Filardo G, Kon E, Di Matteo B, Pelotti P, Di Martino A, Marcacci M (2013) Platelet-rich plasma for the treatment of patellar tendinopathy: clinical and imaging findings at medium-term follow-up. Int Orthop 37(8):1583–1589
45. Oloff L, Elmi E, Nelson J, Crain J (2015) Retrospective analysis of the effectiveness of platelet-rich plasma in the treatment of Achilles tendinopathy: pretreatment and posttreatment correlation of magnetic resonance imaging and clinical assessment. Foot Ankle Spec 8:490–497
46. Owens RF Jr, Ginnetti J, Conti SF, Latona C (2011) Clinical and magnetic resonance imaging outcomes following platelet rich plasma injection for chronic midsubstance Achilles tendinopathy. Foot Ankle Int 32(11):1032–1039
47. de Vos RJ, Weir A, Tol JL, Verhaar JA, Weinans H, van Schie HT (2011) No effects of PRP on ultrasonographic tendon structure and neovascularisation in chronic midportion Achilles tendinopathy. Br J Sports Med 45(5):387–392
48. de Vos RJ (2014) Is tendon structure associated with symptoms in chronic Achilles tendinopathy? An update on tendon pain mechanisms. Aspetar Sports Med J 3(4):594–600
49. de Vos RJ, Heijboer MP, Weinans H, Verhaar JA, van Schie JT (2012) Tendon structure's lack of relation to clinical outcome after eccentric exercises in chronic midportion Achilles tendinopathy. J Sport Rehabil 21 (1):34–43
50. de Jonge S, Warnaars JL, De Vos RJ, Weir A, van Schie HT, Bierma-Zeinstra SM et al (2013) Relationship between neovascularization and clinical severity in Achilles tendinopathy in 556 paired measurements. Scand J Med Sci Sports 24:773–778
51. de Vos RJ, Windt J, Weir A (2014) Strong evidence against platelet-rich plasma injections for chronic lateral epicondylar tendinopathy: a systematic review. Br J Sports Med 48(12):952–956
52. Mishra A, Woodall J Jr, Vieira A (2009) Treatment of tendon and muscle using platelet-rich plasma. Clin Sports Med 28(1):113–125
53. Mishra A, Pavelko T (2006) Treatment of chronic elbow tendinosis with buffered platelet-rich plasma. Am J Sports Med 34(11):1774–1778
54. Kon E, Filardo G, Delcogliano M, Presti ML, Russo A, Bondi A et al (2009) Platelet-rich plasma: new clinical application: a pilot study for treatment of jumper's knee. Injury 40(6):598–603
55. Paoloni J, De Vos RJ, Hamilton B, Murrell GA, Orchard J (2011) Platelet-rich plasma treatment for ligament and tendon injuries. Clin J Sport Med 21 (1):37–45
56. de Vos RJ, Weir A, van Schie HT, Bierma-Zeinstra SM, Verhaar JA, Weinans H et al (2010) Platelet-rich plasma injection for chronic Achilles tendinopathy: a randomized controlled trial. JAMA 303(2):144–149
57. de Jonge S, de Vos RJ, Weir A, van Schie HT, Bierma-Zeinstra SM, Verhaar JA et al (2011) One-year follow-up of platelet-rich plasma treatment in chronic Achilles tendinopathy: a double-blind randomized placebo-controlled trial. Am J Sports Med 39(8):1623–1629
58. Kearney RS, Parsons N, Costa ML (2013) Achilles tendinopathy management: a pilot randomised controlled trial comparing platelet-richplasma injection with an eccentric loading programme. Bone Joint Res 2(10):227–232
59. Thanasas C, Papadimitriou G, Charalambidis C, Paraskevopoulos I, Papanikolaou A (2011) Platelet-rich plasma versus autologous whole blood for the treatment of chronic lateral elbow epicondylitis: a randomized controlled clinical trial. Am J Sports Med 39(10):2130–2134

60. Creaney L, Wallace A, Curtis M, Connell D (2011) Growth factor-based therapies provide additional benefit beyond physical therapy in resistant elbow tendinopathy: a prospective, single-blind, randomised trial of autologous blood injections versus platelet-rich plasma injections. Br J Sports Med 45 (12):966–971
61. Ruiz-Moneo P, Molano-Munoz J, Prieto E, Algorta J (2013) Plasma rich in growth factors in arthroscopic rotator cuff repair: a randomized, double-blind, controlled clinical trial. Arthroscopy 29(1):2–9
62. Rodeo SA, Delos D, Williams RJ, Adler RS, Pearle A, Warren RF (2012) The effect of platelet-rich fibrin matrix on rotator cuff tendon healing: a prospective, randomized clinical study. Am J Sports Med 40 (6):1234–1241
63. Peerbooms JC, Sluimer J, Bruijn DJ, Gosens T (2010) Positive effect of an autologous platelet concentrate in lateral epicondylitis in a double-blind randomized controlled trial: platelet-rich plasma versus corticosteroid injection with a 1-year follow-up. Am J Sports Med 38(2):255–262
64. Mishra AK, Skrepnik NV, Edwards SG, Jones GL, Sampson S, Vermillion DA et al (2014) Efficacy of platelet-rich plasma for chronic tennis elbow: a double-blind, prospective, multicenter, randomized controlled trial of 230 patients. Am J Sports Med 42 (2):463–471
65. Scarpone M, Rabago D, Snell E, Demeo P, Ruppert K, Pritchard P et al (2013) Effectiveness of platelet-rich plasma injection for Rotator Cuff tendinopathy: a prospective open-label study. Glob Adv Health Med 2(2):26–31
66. de Vos RJ, Weir A, Brasher PM, Khan KM (2014) Platelet-rich plasma for chronic tennis elbow: letters to the editor. Am J Sports Med 42(1):NP3–NP5
67. Coombes BK, Bisset L, Vicenzino B (2010) Efficacy and safety of corticosteroid injections and other injections for management of tendinopathy: a systematic review of randomised controlled trials. Lancet 376(9754):1751–1767
68. Carr AJ, Murphy R, Dakin SG, Rombach I, Wheway K, Watkins B et al (2015) Platelet-rich plasma injection with arthroscopic acromioplasty for chronic Rotator Cuff tendinopathy: a randomized controlled trial. Am J Sports Med 43(12):2891–2897
69. Moraes VY, Lenza M, Tamaoki MJ, Faloppa F, Belloti JC (2014) Platelet-rich therapies for musculoskeletal soft tissue injuries. Cochrane Database Syst Rev 4:CD010071
70. Liddle AD, Rodriguez-Merchan EC (2014) Platelet-rich plasma in the treatment of Patellar tendinopathy: a systematic review. Am J Sports Med 43:2583–2590
71. Warth RJ, Dornan GJ, James EW, Horan MP, Millett PJ (2015) Clinical and structural outcomes after arthroscopic repair of full-thickness rotator cuff tears with and without platelet-rich product supplementation: a meta-analysis and meta-regression. Arthroscopy 31(2):306–320

Can Shockwave Therapy Improve Tendon Metabolism?

26

Johannes Zwerver, Charlotte Waugh, Henk van der Worp, and Alex Scott

Abstract

Shockwave treatments are commonly used in the management of tendon injuries and there is increasing evidence for its clinical effectiveness. There is a paucity of fundamental (in vivo) studies investigating the biological action of shockwave therapy. Destruction of calcifications, pain relief and mechanotransduction-initiated tissue regeneration and remodeling of the tendon are considered to be the most important working mechanisms. The heterogeneity of systems (focussed shockwave therapy vs. radial pressurewave therapy), treatment protocols and study populations, and the fact that there seem to be responders and non-responders, continue to make it difficult to give firm recommendations with regard to the most optimal shockwave therapy approach. Specific knowledge with regard to the effects of shockwave therapy in patients with metabolic tendon disorders is not available. Further fundamental and clinical research is required to determine the value of shockwave therapy in the management of tendinopathy.

Keywords

Tendinopathy – Shockwave therapy • Pressurewave therapy • Working mechanism • Mechanotransduction • Pain relief • Tissue regeneration

J. Zwerver (✉)
Center for Sports Medicine, UMC Groningen, PO Box 30.001, 9700 RB Groningen, The Netherlands
e-mail: j.zwerver@sport.umcg.nl

C. Waugh
Centre for Hip, Health and Mobility, University of British Columbia, Vancouver, Canada

H. van der Worp
Center for Sports Medicine, University of Groningen, University Medical Center Groningen, Groningen, The Netherlands

A. Scott
Department of Physical Therapy, Centre for Hip Health and Mobility, University of British Columbia, Vancouver, BC, Canada

P.W. Ackermann and D.A. Hart (eds.), *Metabolic Influences on Risk for Tendon Disorders*, Advances in Experimental Medicine and Biology 920, DOI 10.1007/978-3-319-33943-6_26

List of Acronyms and Abbreviations

CGRP	calcitonin gene-related peptide
EFD	energy flux density
EH	electrohydraulic
EM	electromagnetic
F-SWT	focused shockwave therapy
IL	interleukin
MMP	matrix metalloproteases
PE	piezoelectric
R-PWT	radial pressure wave therapy
SWT	shockwave therapy
TGF-β1	transforming growth factor β1

Introduction

Over the past 20 years, shockwave therapy (SWT) has become a popular treatment method for chronic tendon disorders. There is growing evidence for the effectiveness of SWT when treating both upper limb tendinopathies including calcific rotator cuff tendinopathy [1, 2] and common extensor tendinopathy [3], and lower limb tendinopathies including greater trochanteric pain syndrome, patellar tendinopathy, Achilles tendinopathy and plantar fasciopathy [3–5]. No studies have been published that investigated the effects of SWT in specific subgroups of patients with metabolic tendon disorders.

Results of the SWT seem to vary between studies. This can be partly explained by the diversity of shockwave devices being used, which generate pressure waves differently. Many parameters can influence treatment outcome (Table 26.1) and there is no consensus on the most effective SWT treatment protocols [6].

Furthermore, the exact working mechanism of shockwaves on (pathologic) tendons has not been elucidated so far. There even seem to be responders and non-responders to shockwave treatment, based on different biological reactions to treatment between individuals [7], which may help to explain why shockwave treatment does not improve symptoms in all patients. These observations highlight the requirement to investigate the effects of different shockwave treatment dosage protocols as well as devices in different populations including patients with metabolic disease.

This chapter aims to provide insight in to the principles underpinning SWT and to give an update on its effect on pain and tendon metabolism.

Principles of Shockwave Therapy

Nowadays two types of 'shockwave' therapy are being used: **focused shockwave therapy** (F-SWT) and **radial pressure wave therapy** (R-PWT)

Table 26.1 Parameters which can influence 'shockwave' treatment outcome

Treatment parameter	Description
F-SWT or R-PWT	See Table 26.1, Figs. 26.1 and 26.2
Maximal positive pressure	The maximal positive pressure that is reached
Focal zone *(only with F-SWT)*	A 3- dimensional ellipsoid where the pressure is above a certain value
Energy flux density	The amount of energy per surface unit (mJ/mm^2)
Number of treatments	
Number of impulses per treatment	Energy flux * number of impulses
Total energy delivered	
Time interval between treatments	
Impulse frequency	The number shockwaves that is applied per second
Localization method	How the to-be-treated site is determined (imaging based vs feedback patient)
Anesthesia	
Concurrent treatments/Rest	

Focused Shockwave Therapy

F-SWT is called focused because a shockwave is generated that converges in the adjustable focus at selected depth in the body tissues, where the maximal pressure is reached. A **shockwave** is a special, non-linear type of pressure wave with positive peak pressures of 10–100 MPa, characterized by a short rise time and a short negative pressure phase causing cavitation at the tissue interfaces (Fig. 26.1; Table 26.2). The total duration of a shockwave is around 10 μs. There are three methods to generate focused shockwaves for F-SWT: electrohydraulic (EH), electromagnetic (EM) and piezoelectric (PE). Depending on the device the shockwaves are concentrated into small focal areas of 2–8 mm in diameter in order to optimize therapeutic effects and minimize impact on other tissues. Shockwave dosage is described as the energy flux density, (EFD, in mJ/mm^2), a term used to reflect the flow of shockwave energy in an area perpendicular to the direction of propagation. This concentrated shockwave energy per focal area is considered to be an important treatment parameter [3, 6, 8, 9]. Shockwave treatment is broadly divided in high (EFD > 0.12 mJ/mm^2) and low (EFD ≤ 0.12 mJ/mm^2) energy delivery. Low energy SWT is considered to stimulate cell regeneration and to play a role in pain therapy. High energy SWT could be applied in case of calcific tendinopathy. However, no consensus exists with regard to these high and low energy definitions and its applications.

Radial Pressure Wave Therapy

Another popular form of treatment for tendinopathies, often mistakenly described as 'radial shock wave therapy', is '**radial pressure wave therapy**' (R-PWT). Radial pressure waves are generated in a ballistic way by accelerating a projectile, using compressed air, through a tube on the end of which an applicator is placed. The projectile hits the applicator and the applicator

Table 26.2 Differences between shockwaves and pressure waves

	Shockwaves	Pressure waves
Focus	Yes	No, radial/ divergent
Propagation	Non-linear	Linear
Risetime	±10 ns	±1000 ns
Pulse duration	0.2–0.5 μs	0.2–0.5 ms
Peak positive pressure	10–100 MPa	0–10 MPa
	100–1000 bar	0–100 bar
Energy flux density	0–3 mJ/mm^2	0–0.3 mJ/mm^2
Penetration depth	Deep	Superficial (<1 cm)

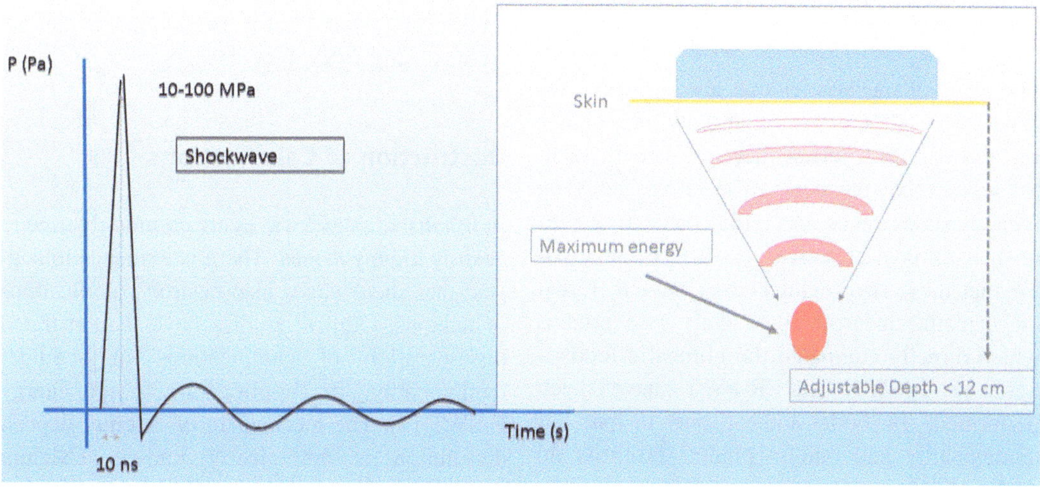

Fig. 26.1 Schematic illustration of shockwave characteristic and Focussed Shockwave Therapy

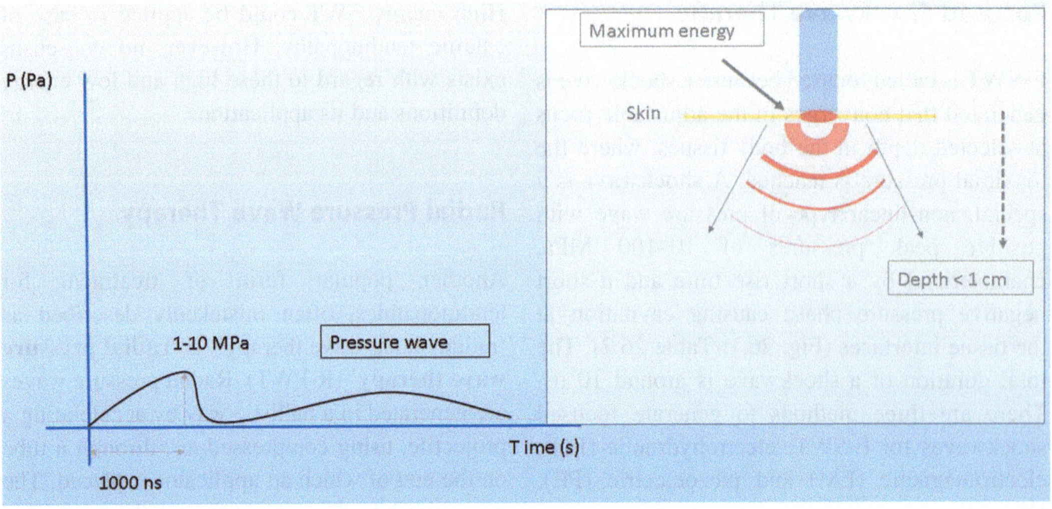

Fig. 26.2 Schematic illustration of pressurewave characteristic and Radial Pressurewave Therapy

transmits the generated pressure wave into the body. These pressure waves are not real shockwaves since they have a lower peak pressure, longer rising time and lack non-linearity (Fig. 26.2; Table 26.2). The term radial refers to the diverging pressure field of radial pressure therapy devices, which has a maximal pressure at the source of generation, not at a selected depth in the body. It has been demonstrated that radial pressure waves do not have a penetrating effect on tissue, but rather act superficially [6, 8, 10].

F-SWT vs R-PWT

The wave characteristics that are important for generating the therapeutic effects of F-SWT and R-PWT are unclear, therefore it is difficult to relate the physical differences between focused shockwaves and radial pressure waves to their clinical effectiveness. It is even possible that these two methods may have different mechanistic underpinnings. Only two studies which directly compared the clinical effectiveness of F-SWT and R-PWT have been performed. In both studies, one in patellar tendinopathy and one in plantar fasciitis, no major clinically relevant differences between the aforementioned treatment methods were found [11, 12].

Working Mechanisms of Shockwave Therapy

Research into the working mechanisms of 'shockwave therapy' has been concentrated on a few theories, which can be roughly divided in to the destruction of calcifications, pain relief and tissue regeneration. Until now most fundamental research on 'shockwave therapy' for tendinopathy has mainly been done in animal studies using F-SWT devices.

Destruction of Calcifications

In lithotripsy, shockwaves are commonly used to destroy kidney stones. There is evidence to suggest that shockwaves also destroy calcifications in tendons. Clinical studies have demonstrated disintegration of calcifications in shoulder tendinopathy after application of high energy F-SWT [2]. The mechanism of calcium deposit dissolution is not clearly known. Calcium deposits might be eliminated after shockwave therapy through a molecular mechanism of absorption associated with improved circulation at the tendon-bone junction [13].

Pain Relief

Pain relief with shockwave therapy might work by means of 'hyperstimulation analgesia'. Rompe hypothesized that overstimulation of the treated site would lead to a diminished transmission of signals to the brainstem [14]. Wess [15] described a hypothetical model that painful shockwaves might 'reset' the pain memory in the brain, which is interesting since it has recently been demonstrated that central sensitization plays a role in a subgroup of patients with chronic tendon related pain [16]. In the past decade, some molecular and cellular mechanisms have also been studied to investigate how shockwaves might mediate their pain-relieving action. Maier et al. [17] found Substance P and prostaglandin E2 were released after shockwave application to the rabbit femur. In other animal studies, shockwave application to either the distal rabbit femur or rat skin diminished the number of neurons immunoreactive for substance P and decreased calcitonin gene-related peptide (CGRP) immunoreactivity in dorsal root ganglion neurons in dorsal root ganglia [18, 19]. Hausdorf reported selective loss of unmyelinated nerve fibers after extracorporeal shockwave application to the musculoskeletal system [20]. Since neurogenic inflammation is considered to play an important role in the pathogenesis of tendinopathy, it is possible that a reduction of substance P in the target tissue in conjunction with reduced synthesis of this molecule in dorsal root ganglia cells explains the shockwave mediated long-term analgesia. Selective destruction of unmyelinated nerve fibers within the focal zone of the shock waves might also further contribute to this analgesia (see Chap. 5) [21].

Tissue Regeneration

The mechanical perturbations generated by shockwaves are likely the most important factor to explain its treatment effect. This fits within the framework of mechanotransduction [22], in which mechanical load on the cytoskeleton leads to responses of the mechanosensitive tendon cells initiating a process of tendon remodeling and repair.

In vitro studies have demonstrated that shockwaves enhance angiogenesis [23, 24], increase the synthesis of key tendon components (e.g. collagen,[25]); glycosaminoglycans & extracellular matrix proteins, [26, 27], and growth factors (transforming growth factor (TGF)-β1, [23]), and decrease the presence of inflammatory cytokines [28].

However, surprisingly little is known about the biological effect of shockwaves on tendon tissue *in vivo*. In a recent study, Waugh et al. [29] investigated the metabolic response of normal and tendinopathic tendons to shockwave treatment. Using microdialysis samples from the surrounding peri-tendon, several cytokines, growth factors and proteases were examined before and immediately after a single shockwave session. Interleukins (IL)-1β and IL-2 were detected but did not change significantly with SWT whilst IL-6 and IL-8 concentrations were elevated after shockwave treatment. Matrix metalloproteases (MMP)-2 and -9, enzymes involved in the remodeling of extra-cellular matrix, also increased after shockwave treatment. These findings suggest that shockwave treatment might aid tendon remodeling in tendinopathy by promoting the inflammatory and catabolic processes that are associated with removing damaged extra-cellular matrix components. Mani-Babu et al. [30] found that collagen synthesis was not altered in healthy Achilles tendons 72 h after a single shockwave session, however these findings are limited in their clinical application. A bigger effort is required to determine the response profile and time course of other biological indicators of matrix turnover and tendon regeneration in tendinopathy to build upon the findings of Waugh et al. [29].

Conclusion

Shockwave treatments are commonly used in the management of tendon injuries and there is increasing evidence for its clinical effectiveness. Specific knowledge with regard to the effects of SWT in patients with metabolic tendon disorders is not available. There is also a paucity of fundamental (in vivo) studies investigating the biological action of SWT. Destruction of calcifications, pain relief and mechanotransduction-initiated tissue regeneration and remodeling of the tendon are considered to be the most important working mechanisms. The – in previous studies reported – heterogeneity of systems (F-SWT vs. -R-PWT), treatment protocols and study populations, and the fact that there seem to be responders and non-responders, continue to make it difficult to give firm recommendations with regard to the most optimal SWT approach. Further research is required to determine the value of SWT in the management of tendinopathy. This research should consist of a combination of fundamental and clinical studies. Studies with clear descriptions of study populations, diagnostic criteria and treatment parameters and concurrent rehabilitation programs / tendon loading activities are necessary to advance research for this promising treatment modality.

Glossary

Energy flux density, (EFD, in mJ/mm^2) a term used to reflect the flow of shockwave energy in an area perpendicular to the direction of propagation. Shockwave dosage is broadly divided in high (EFD > 0.12 mJ/mm^2) and low (EFD ≤ 0.12 mJ/mm^2) energy delivery.

References

1. Huisstede BM, Gebremariam L, van der Sande R, Hay EM, Koes BW (2011) Evidence for effectiveness of Extracorporal Shock-Wave Therapy (ESWT) to treat calcific and non-calcific rotator cuff tendinosis – a systematic review. Man Ther 16(5):419–433
2. Ioppolo F, Tattoli M, Di Sante L, Venditto T, Tognolo L, Delicata M et al (2013) Clinical improvement and resorption of calcifications in calcific tendinitis of the shoulder after shock wave therapy at 6 months' follow-up: a systematic review and meta-analysis. Arch Phys Med Rehabil 94 (9):1699–1706
3. Speed C (2014) A systematic review of shockwave therapies in soft tissue conditions: focusing on the evidence. Br J Sports Med 48(21):1538–1542
4. Mani-Babu S, Morrissey D, Waugh C, Screen H, Barton C (2015) The effectiveness of extracorporeal shock wave therapy in lower limb tendinopathy: a systematic review. Am J Sports Med 43(3):752–761
5. Yin MC, Ye J, Yao M, Cui XJ, Xia Y, Shen QX et al (2014) Is extracorporeal shock wave therapy clinical efficacy for relief of chronic, recalcitrant plantar fasciitis? A systematic review and meta-analysis of randomized placebo or active-treatment controlled trials. Arch Phys Med Rehabil 95(8):1585–1593
6. van der Worp H, van den Akker-Scheek I, van Schie H, Zwerver J (2013) ESWT for tendinopathy: technology and clinical implications. Knee Surg Sports Traumatol Arthrosc 21(6):1451–1458
7. Waugh CMCM (2015) In vivo biological response to extracorporeal shockwave therapy in human tendinopathy. Eur Cell Mater 29:268–280
8. Gerdesmeyer L, Maier M, Haake M, Schmitz C (2002) Physical-technical principles of extracorporeal shockwave therapy (ESWT). Orthopade 31 (7):610–617
9. Cleveland RO, Chitnis PV, McClure SR (2008) Shock wave therapy: what really matters reply. Ultrasound Med Biol 34(11):1869–1870
10. Cleveland RO, Chitnis PV, McClure SR (2007) Acoustic field of a ballistic shock wave therapy device. Ultrasound Med Biol 33(8):1327–1335
11. Lohrer H, Nauck T, Dorn-Lange NV, Scholl J, Vester JC (2010) Comparison of radial versus focused extracorporeal shock waves in plantar fasciitis using functional measures. Foot Ankle Int 31(1):1–9
12. van der Worp H, Zwerver J, Hamstra M, van den Akker-Scheek I, Diercks RL (2014) No difference in effectiveness between focused and radial shockwave therapy for treating patellar tendinopathy: a randomized controlled trial. Knee Surg Sports Traumatol Arthrosc 22(9):2026–2032
13. Mariotto S, Cavalieri E, Amelio E, Ciampa AR, de Prati AC, Marlinghaus E et al (2005) Extracorporeal shock waves: from lithotripsy to anti-inflammatory action by NO production. Nitric Oxide 12(2):89–96

14. Rompe JD, Krischek O, Eysel P, Hopf C, Jage J (1998) Results of extracorporeal shock-wave application in lateral elbow tendopathy. Schmerz 12(2):105–111
15. Wess OJ (2008) A neural model for chronic pain and pain relief by extracorporeal shock wave treatment. Urol Res 36(6):327–334
16. Rio E, Moseley L, Purdam C, Samiric T, Kidgell D, Pearce AJ et al (2014) The pain of tendinopathy: physiological or pathophysiological? Sports Med 44 (1):9–23
17. Maier M, Averbeck B, Milz S, Refior H, Schmitz C (2003) Substance P and prostaglandin E-2 release after shock wave application to the rabbit femur. Clin Orthop 1(406):237–245
18. Hausdorf J, Lemmens MA, Kaplan S, Marangoz C, Milz S, Odaci E et al (2008) Extracorporeal shockwave application to the distal femur of rabbits diminishes the number of neurons immunoreactive for substance P in dorsal root ganglia L5. Brain Res 1207(1):96–101
19. Takahashi N, Wada Y, Ohtori S, Saisu T, Moriya H (2003) Application of shock waves to rat skin decreases calcitonin gene-related peptide immunoreactivity in dorsal root ganglion neurons. Auton Neurosci Basic Clin 107(2):81–84
20. Hausdorf J, Lemmens MA, Heck KD, Grolms N, Korr H, Kertschanska S et al (2008) Selective loss of unmyelinated nerve fibers after extracorporeal shockwave application to the musculoskeletal system. Neuroscience 155(1):138–144
21. Schmitz C, DePace R (2009) Pain relief by extracorporeal shockwave therapy: an update on the current understanding. Urol Res 37(4):231–234
22. Khan KM, Scott A (2009) Mechanotherapy: how physical therapists' prescription of exercise promotes tissue repair. Br J Sports Med 43(4):247–252
23. Chen Yeung-Jen YJ (2004–7) Extracorporeal shock waves promote healing of collagenase-induced Achilles tendinitis and increase TGF-beta1 and IGF-I expression. J Orthop Res 22(4):854–861
24. Wang Ching-Jen CJ (2003–11) Shock wave therapy induces neovascularization at the tendon-bone junction. A study in rabbits. J Orthop Res 21 (6):984–989
25. Vetrano Mario M (2011–12) Extracorporeal shock wave therapy promotes cell proliferation and collagen synthesis of primary cultured human tenocytes. Knee Surg Sports Traumatol Arthrosc 19(12):2159–2168
26. Bosch GG (2009–4) The effect of focused extracorporeal shock wave therapy on collagen matrix and gene expression in normal tendons and ligaments. Equine Vet J 41(4):335–341
27. Bosch GG (2007–5) Effect of extracorporeal shock wave therapy on the biochemical composition and metabolic activity of tenocytes in normal tendinous structures in ponies. Equine Vet J 39(3):226–231
28. Han SH, Lee JW, Guyton GP, Parks BG, Courneya J, Schon LC (2009) Effect of extracorporeal shock wave therapy on cultured tenocytes. Foot Ankle Int 30 (2):93–98
29. Waugh CM, Morrissey D, Jones E, Riley GP, Langberg H, Screen HR (2015) In vivo biological response to extracorporeal shockwave therapy in human tendinopathy. Eur Cell Mater 29:268–280; discussion 280
30. Mani-Babu S, Waugh CM, Screen HR, Maffulli N, Morrissey D (2013) The effects of extracorporeal shock wave therapy (ESWT) on type I collagen synthesis in the Achilles tendon: an intervention study on healthy participants. Int J Exp Pathol 94: A11–A12

Do Dietary Factors Influence Tendon Metabolism?

27

Alex Scott and Cara Nordin

Abstract

There is very little direct research to conclusively prove the relevance of diet in primary tendinopathies, however it seems prudent to ask whether our current knowledge about the impact of nutrition on collagen metabolism could be useful in assessing, preventing, or treating tendinopathy. The objective of this chapter is to discuss the potential impact (negative or positive) that nutrition may have on the metabolism of tendons by summarizing the related research. The chapter briefly discusses the roles that specific vitamins, amino acids, lipids, and antioxidants have in various processes of the body that may be directly or indirectly related to tenocyte metabolism.

Keywords

Diet • Tendon • Nutrition • Vitamin • Tendinopathy • Lipid • Amino acid • Antioxidant

Acronyms and Abbreviations

OA	Osteoarthritis
IGF-1	Insulin-like Growth Factor-1
TGF-beta	Transforming Growth Factor-Beta
FGF2	Fibroblast Growth Factor 2
COL1A1	Collagen, type 1, alpha 1
rER	Rough Endoplasmic Reticulum
MMP1	Matrix Metalloproteinase-1
PGE2	Prostaglandin E2
IL1	Interleukin-1
UV Light	Ultraviolet Light
LDL	Low-density Lipoprotein
oxLDL	Oxidized Low-density Lipoprotein
COL3A1	Collagen, type 3, alpha 1
MMP2	Matrix Metalloproteinase-2
ROS	Reactive Oxygen Species
GTE	Green Tea Extract
mRNA	Messenger RibonucleicAcid

A. Scott (✉) • C. Nordin
Department of Physical Therapy, Centre for Hip Health and Mobility, University of British Columbia, Vancouver, BC, Canada
e-mail: ascott@mail.ubc.ca

P.W. Ackermann and D.A. Hart (eds.), *Metabolic Influences on Risk for Tendon Disorders*, Advances in Experimental Medicine and Biology 920, DOI 10.1007/978-3-319-33943-6_27

Introduction

Based on recent research into the pathophysiology of tendon, a multi-factorial approach to understanding the etiology of different tendinopathies has emerged [1]. As in OA, two interacting categories of risk factors may contribute to a chronic tendon problem – biomechanical factors (volume, intensity, and quality of movement) and biological factors (sex, age, genetics, lifestyle factors, etc). In addition to these risk factors which are associated with *primary* tendinopathy, some cases of tendinopathy may be considered *secondary* (i.e. occurring mainly due to the presence of another medical condition – Table 27.1) [1].

Diet is relevant to the etiology and management of several conditions listed above that are associated with secondary tendinopathies. For example, excessive dietary intake of lipids and carbohydrates can lead to dyslipidemia and diabetes, respectively, and the impact of these conditions on tendons is discussed specifically in Chaps. 14 and 16. However, one might also ask to what extent an individual's diet may contribute to the development of *primary* tendinopathy (in the absence of other diagnosed conditions). One could hypothesize that a tendinopathy might develop in part through dietary impairment of tenocyte metabolism leading to reduced structure/function of the load-bearing tissue. Alternatively, a diet which either helps or hinders tendon healing (e.g. by influencing inflammation-repair responses) could influence the development and the prognosis of tendinopathy.

There is, unfortunately, very little direct evidence which can conclusively prove the relevance of nutrition in a typical patient presenting with a primary tendinopathy. However, given that (a) tendon is mainly composed of type I collagen [2], and (b) it has been suggested that ongoing synthesis of collagen is required to maintain a healthy extracellular matrix [3], it nevertheless seems prudent to ask whether our current knowledge about the impact of nutrition on collagen metabolism could be useful in assessing, preventing, or treating tendinopathy. This appears to be an area ripe for future research. Additionally, whereas type I collagen is relatively long-lived in the extracellular matrix, glycosaminoglycans – important for tendon hydration, regulation of collagen fibril assembly, and inter-fascicular gliding –are shorter-lived and could therefore be influenced quite rapidly by changes in diet.

The objective of this short chapter is to briefly discuss the *potential* impact that nutrition may have on extracellular matrix metabolism in tendons. Hopefully, readers with an interest in these issues may be inspired to begin conducting research to fill some of the knowledge gaps. Obviously, a major limitation of this chapter is that many of the experiments discussed have been conducted on cultured fibroblasts or animal models, as opposed to direct studies of human tenoctyes *in vivo*.

Nutrition and Tendons

Having a basic knowledge of nutrition is important for clinicians, because nutrition is vital for growth and development, and for the prevention and treatment of disease [4]. Nutrients fall into three main categories – macronutrients,

Table 27.1 Common locations associated with secondary tendinopathies [1] (With permission of JOSPT©)

Site of tendon affected	Examples of medical conditions
Mid portion	Dyslipidemias, Rheumatoid disease, tumours, infections, storage diseases, gout, pseudogout, heritable connective tissue diseases, haemachromatosis, endocrinopathies (including thyroid disease, cushings, hypogonadism, menopause), metabolic diseases including diabetes, hypercalcaemia
Enthesis	Psoriasis, gout, pseudogout, spondyloarthropathies, inflammatory bowel disease
Tendon sheath	Rheumatoid arthritis, infections, tumours

micronutrients and water [5]. Macronutrients, consisting of carbohydrates, proteins and lipids, are dietary substances which the body requires in large and regular quantities [5]. Carbohydrates are a major source of energy, of which the largest part is used to fuel muscle and brain tissue [6]. Protein is essential for building and repairing muscles, red blood cells and other tissues as well as for synthesizing hormones [6]. Dietary protein is broken down into amino acids, which are then utilized according to specific requirements [6]. Lipids are a rich source of stored energy [6]. Vitamins and minerals are considered micronutrients, and they are required in lesser amounts [5]. Vitamins act as metabolic catalysts that regulate chemical reactions in the body – e.g. see discussion of Vitamin C below [6]. Minerals are used to form structures of the body as well as help regulate body processes – e.g. matrix metallo-proteinases are zinc-dependent enzymes which play key roles in tendon metabolism and healing [6]. Many micronutrients, known as essential nutrients, cannot be made by the body therefore they need to be ingested through food or dietary supplementation.

Tenocytes, specialized fibroblasts, are the characteristic cell in tendons responsible for the secretion and maintenance of the extra-cellular matrix [7]. Tendons are composed primarily of collagen molecules, proteoglycans and glycosaminoglycans [7]. Glycosaminoglycans are responsible for the viscoelastic behavior of tendons, whereas collagen fibers act primarily to resist tension (or compression, at entheses or areas of tendon exposed to shear, such as wrap-around or other compressed areas) [7]. Type I fibril-forming collagen makes up 75 % of a tendon's dry mass and smaller amounts of other types of collagen (II, III, V, XII, XIV) also contribute to this composition [8]. Regulation of the production of collagen and glycosaminoglycans occurs through secreted growth factors (IGF-I, TGF-beta, FGF2, etc) and transcription factors (e.g. scleraxis, which promotes COL1A1 expression) [8]. Following gene expression, the messenger RNA (mRNA) brings the information from the nucleus to the ribosomes, located in the cytoplasm, to be synthesized into polypeptide precursors of the alpha chains (known as preprocollagen) [9]. The N-terminus of each preprocollagen contains a sequence of amino acids that directs it to the rough endoplasmic reticulum (rER) [9, 10]. While in the rER, a signal peptidase cleaves the signal sequence (N-terminus) to yield a precursor of collagen called the pro-alpha chain [9]. Within the lumen of the rER the proline and lysine residues of the pro-alpha chain are hydroxylated by prolyl- and lysyl- hydroxylases to form hydroxyproline and hydroxylysine, respectively [9]. Cofactors for prolylhydroxylase include ascorbic acid, ferrous ions, alpha-ketoglutarate, and oxygen therefore if these are not present in adequate amounts the enzymeis rendered nonfunctional [10]. The newly formed hydroxylysine residues are modified by glycosylation with galactose or galactosyl-glucose which are catalyzed by galactosyl and glucosyl transferases. These enzymes require manganese as a cofactor [11].

The procollagen that was formed by hydroxylation and glycosylation, now self assembles in order to form a triple helix and be prepared for secretion. Once the triple helix is formed, no further hydroxylation or glycosylation can take place and the procollagen molecules are translocated to the Golgi apparatus to be packaged into secretory vesicles and released into the extracellular space [9]. At the N- and C-terminus of the procollagen, extension peptides are cleaved off by extracellular enzymes – procollagen aminoproteinase and procollagen carboxyproteinase – resulting in the formation of collagen molecules [9]. These molecules, containing approximately 1000 amino acids per chain, now begin to spontaneously assemble into collagen fibers. The fibers are further stabilized by covalent cross-links within and between the triple helix units through the action of lysyl oxidase, a copper-dependent enzyme [9, 10]. Through lysine and hydroxylysine, this enzyme yields reactive aldehyde, which forms the stable covalent cross-links important for the tensile strength of fibers [9, 10].

Considering the many steps – both enzymatic and non-enzymatic – that are involved in collagen synthesis, errors or deficiencies can occur at multiple points throughout the process, and these could ultimately affect the quality and tensile strength of the resulting collagen molecules. The remaining paragraphs discuss how certain nutritional factors could influence the synthesis of collagen in tendons.

Vitamins

Vitamins act as catalysts of biochemical reactions [6]. Most vitamins are not produced by the human body, therefore they need to be ingested. Current research suggests that vitamins C and D may influence the metabolism and/or structure of tendons.

Vitamin C

Deficiency of vitamin C (ascorbic acid) is not common in the general population of developed countries, but it does present in individuals with poor nutrition or impaired digestion, e.g. patients who are alcoholic, critically ill, undergoing chemotherapy, etc. Advanced cases of Vitamin C deficiency can lead to scurvy, hemarthrosis, synovitis, and arthralgia. Ascorbic acid is known both for its role as an antioxidant and as a cofactor in the hydroxylation of proline and lysine to hydroxyproline and hydroxylysine needed for the synthesis of collagen [5]. This suggests that a deficiency would lead to a decrease in collagen synthesis in the musculoskeletal tissues, and clinical findings have supported this mechanism in patients with rheumatologic manifestations [12]. A study on guinea pigs found that vitamin C deficiency reduced collagen synthesis in bone, cartilage and tendon [13]. However, a recent survey of rheumatological manifestations in people with Vitamin C deficiency made no mention of tendon problems – it predominantly appears to affect more vascularized tissues such as gingiva, skin, synovium/joint and muscle [12].

Vitamin C is a potent inducer of collagen synthesis in tendon cells [14]. Following tendon injury, when collagen synthesis is maximal, vitamin C requirements to achieve optimal healing may be higher than the levels typically experienced by tenocytes [15]. Repeated injections of 150 mg of Vitamin C to injured rat tendons resulted in accelerated healing compared to controls [16].

Vitamin D

Vitamin D is also known to have a direct impact on collagen synthesis by tendon fibroblasts. The addition of Vitamin D3 and D2 metabolites to human-derived tendon fibroblasts revealed a dose-responsive, anabolic effect, with progressive increases in mRNA levels of type I collagen [17]. In these same experiments, vitamin D also reduced the levels of intracellular reactive oxygen species. These findings led the authors to suggest that vitamin D has a beneficial effect on tendon, and that vitamin D deficiency may have a negative effect on tendons by limiting type I collagen synthesis and increasing their exposure to reactive oxygen species.

Vitamin D also appears to regulate inflammation-repair mechanisms. In other closely related cell types (skin-derived fibroblasts), vitamin D has also been shown to limit MMP1 expression levels [18], an effect that would tend to increase tissue levels of type 1 collagen. Vitamin D3 was found to reduce the induction of PGE2 in IL1-treated synovial fibroblasts [19] (in other words, to have an anti-inflammatory effect). This anti-inflammatory effect has also been observed in ligament fibroblasts [20]. In ligament fibroblasts, vitamin D also inhibited osteoblastic differentiation and calcification [21]. Given that calcification can be a negative sequelae of tendon injury, this effect could be beneficial if it is found to extend to tendons.

Vitamin D deficiency resulted in impaired rotator cuff healing in a rat model [22] (reduced dietary vitamin D and restricted UV light). In one case series, over 80 % of subjects undergoing rotator cuff repair were deficient in vitamin D, but there was no correlation between serum vitamin D levels and the size or severity of the tear, or the extent of fatty infiltration in the rotator cuff muscles.

Overall, the impact of vitamin D on collagen-rich tissues such as tendon could be potentially beneficial, particularly after injury, but clinical studies are lacking.

Amino Acids

Amino acids are the building block for proteins throughout the body [6]. Leucine is an amino acid for which there is a demonstrated relation to collagen synthesis. It is an essential amino acid that requires a dietary source. In one study, experimenters placed malnourished rats on either a control diet or a leucine-rich diet. Within each dietary grouping, some rats were allowed to exercise and the remainder were sedentary [23]. The main finding of this study was that the leucine-rich diet stimulated collagen synthesis in the tendon more than a standard diet, especially when combined with exercise [23]. Leucine increased the amount of hydroxyproline [23], which is a major component of collagen, and it also plays a key role in the stability of the collagen fibres. The amino acid glycine has been demonstrated to have a direct effect on an inflamed Achilles tendon in rats. One study found that a diet containing 5 % glycine induced the synthesis of hydroxyproline and glycosaminoglycans, allowing for a faster restructuring of the collagen molecules [24]. This resulted in the tendon being more resistant to rupture, offering preliminary support for the hypothesis that dietary supplementation with glycine could be effective for individuals with injury of the Achilles tendon [24].

As discussed earlier, lysine, proline, and cysteine are other amino acids that are important factors in the synthesis of collagen. Lysine is considered essential and can be sourced through high-protein foods like meat, eggs, beans, etc or supplementation. Proline and cysteine are non-essential. A recent randomized control trial found that elderly men with sarcopenia who received collagen peptide supplementation in addition to a 12-week exercise program demonstrated a higher increase in muscle strength compared to individuals in the control group who did not use supplements [25]. Although not the purpose of the study, the difference in tendon composition of the two groups would be interesting to observe due to the inclusion of lysine, proline, leucine, hydroxyproline and hydroxylysine in the supplementation.

Lipids

Lipids are an important source of stored energy, however excessive intake of saturated fat can lead to hypercholesterolemia [6]. Most circulating cholesterol is encapsulated and transported by LDL (low density lipoprotein); the oxidized form of LDL (oxLDL) is known to be highly pathogenic. Recent studies have found that a diet rich in saturated fat causes significant tendon metabolic and structural alterations in mice [26]. The structural changes were accompanied by an increased matrix metalloproteinase-2 (MMP2) expression and accumulation of oxidized low-density protein (oxLDL) in the load-bearing extracellular matrix, and reduced biomechanical properties [26]. It appears that oxLDL caused tenocytes to adopt a proliferative degradative phenotype with reduced COL1A1 and COL3A1 (genes for collagen I and collagen III) as well as increased MMP2 expression.

The Potential Benefit of Anti-Oxidants on Tendon

Research on antioxidants has increased over the years due to its potential benefit in disease prevention and health promotion by neutralizing free radicals and limiting excessive accumulation of reactive oxygen species (ROS) in the body. ROS are produced during cellular metabolism and functional activities and have important roles in cell signaling, apoptosis, gene expression and ion transportation [27]. Although useful in adequate amounts, excessive ROS can damage DNA, RNA, proteins and lipids by inhibiting enzymes, damaging nucleic acids, oxidizing proteins, peroxidating lipids and ultimately leading to cell death [27]. Many environmental factors (UV light, pollution, cigarette smoke) are known to also cause increased ROS. Natural sources of antioxidants are available in fruits (grapes, berries), vegetables (dark greens), spices (ginger, curcumin), grains (whole grains), herbs (rosemary, oregano), and tea (green tea) [27].

The specific effect that ROS has on tendons has not been widely studied, however other type I collagen-rich areas of the body like the skin, may provide some clues. Evidence has been growing to support antioxidant dietary supplementation in reducing oxidative stress and free radical formation, thereby assisting in slowing the process of skin damage [28]. One study determined that green tea extract (GTE) appeared to delay collagen aging in adult mice due to its antioxidant capabilities; similar effects were also found with the combination of vitamin C and E [29]. Although limited, there is evidence to suggest that high concentrations of free-radical oxidants may be involved in tendon pathology [30]. One trial investigated the efficacy of polyunsaturated fatty acids and antioxidants on tendinopathies, with the findings supporting the use of dietary supplementation in its management [30].

Conclusion

Based on the discussions above there are many potential avenues by which nutrition could impact positively or negatively on tendons. Vitamin C could be a major contributor to prevention of tendinopathy and promotion of healing, due to its role as a cofactor in the hydroxylation process and as an antioxidant. Vitamin D has been shown to increase mRNA levels of type I collagen, increase tissue levels of type 1 collagen by limiting MMP1 expression levels and inhibit osteoblastic differentiation and calcification in ligament fibroblasts. Deficiency of vitamin D may also have a negative effect on tendons by limiting type 1 collagen synthesis and increasing exposure to ROS. Diets enriched leucine and glycine demonstrated positive effects on collagen synthesis in the tendons of rats and humans, respectively. Lysine is another amino acid that plays a major role in collagen synthesis, and because it is not made in the body it must be obtained through nutritional sources. A diet rich in saturated fat causes significant tendon metabolic and structural alterations in mice. Free-radical oxidants have demonstrated deleterious effects on type 1 collagen in the skin and may have similar effects on tendons, however there is apparently only one trial to date which supports this approach [30].

Knowledge Gaps

There are very few studies that address the impact of nutritional deficiency, or supplementation, on human tendon structure/function, risk of tendinopathy, or the prognosis of acute injury or chronic tendon pain. There is also a paucity of studies on the potential impact of zinc (co-factor for MMPs), manganese (cofactor for collagen glycosylation), iron (prolylhydroxylase), or copper (lysyl oxidase). In addition, as more research is performed in this area, it would be interesting to understand the time course (weeks? months? years?) over which nutritional factors may impact on the structure or function of tendons. Currently, it is not possible to make any specific evidence-based dietary recommendations for treating or preventing tendon injuries, other than to follow evidence-based guidelines to promote general health [31].

References

1. Scott A, Backman L, Speed C (2015) Tendinopathy-update on pathophysiology. J Orthop Sports Phys Ther 45(11):833–841. doi:10.2519/jospt.2015.5884
2. Birch HL, Thorpe CT, Rumian AP (2013) Specialisation of extracellular matrix for function in tendons and ligaments. Muscles Ligaments Tendons J 3(1):12–22. doi:10.11138/mltj/2013.3.1.012
3. Schechtman H, Bader DL (1997) In vitro fatigue of human tendons. J Biomech 30(8):829–835
4. Ohlhorst SD, Russell R, Bier D et al (2013) Nutrition research to affect food and a healthy lifespan. Adv Nutr 4(5):579–584. doi:10.3945/an.113.004176

5. Molnar JA, Underdown MJ, Clark WA (2014) Nutrition and chronic wounds. Adv Wound Care 3 (11):663–681. doi:10.1089/wound.2014.0530
6. Clark N (1997) Nancy Clark's sports nutrition guidebook, 2nd edn. Human Kinetics, Champaign, p 4
7. Benjamin M, Kaiser E, Milz S (2008) Structure-function relationships in tendons: a review. J Anat 212 (3):211–228. doi:10.1111/j.1469-7580.2008.00864.x
8. Kostrominova TY, Brooks SV (2013) Age-related changes in structure and extracellular matrix protein expression levels in rat tendons. Age 35 (6):2203–2214. doi:10.1007/s11357-013-9514-2
9. Chhabra N. Biochemistry for medics web site. http://www.namrata.co/collagen-synthesistypes-and-composition-part-2/. Accessed 4 Dec 2015
10. More on Collagen. http://www.uic.edu/classes/peri/peri612/persyb6.pdf. Accessed 4 Dec 2015
11. Collagen Metabolism. Medscape: multispecialty. http://www.medscape.com/viewarticle/423231_2. Accessed 4 Dec 2015
12. Mertens MT, Gertner E (2011) Rheumatic manifestations of scurvy: a report of three recent cases in a major urban center and a review. Semin Arthritis Rheum 41(2):286–290. doi:10.1016/j.semarthrit.2010.10.005
13. Kipp DE, McElvain M, Kimmel DB, Akhter MP, Robinson RG, Lukert BP (1996) Scurvy results in decreased collagen synthesis and bone density in the guinea pig animal model [abstract]. Bone 18 (3):281–288. http://www.ncbi.nlm.nih.gov/pubmed/8703585. Accessed 9 Nov 2015: PMID: 8703585
14. Kipp DE, Schwarz RI (1990) Effectiveness of isoascorbate versus ascorbate as an inducer of collagen synthesis in primary avian tendon cells. J Nutr 120(2):185–189
15. Russell JE, Manske PR (1991) Ascorbic acid requirement for optimal flexor tendon repair in vitro [abstract]. J Orthop Res 9(5):714–719, http://www.ncbi.nlm.nih.gov/pubmed/1870035. Accessed 4 Dec 2015
16. Omeroğlu S, Peker T, Türközkan N, Omeroğlu H (2009) High-dose vitamin C supplementation accelerates the Achilles tendon healing in healthy rats. Arch Orthop Trauma Surg 129(2):281–286. doi:10.1007/s00402-008-0603-0
17. Poulsen RC, Zarei A, Sabokbar A, Hulley PA (2013) Tendon, a vitamin D-responsive tissue – why the British weather may not just be bad for your bones! [abstract]. 94:A20
18. Dobak J, Grzybowski J, Liu FT, Landon B, Dobke M (1994) 1,25-dihydroxyvitamin D3 increases collagen production in dermal fibroblasts [abstact]. J Dermatol Sci 8(1):18–24. http://www.ncbi.nlm.nih.gov/pubmed/7947488. Accessed 29 Nov 2015
19. Yaron I, Meyer FA, Weisman Y, Yaron M (1993) Effect of 1,25-dihydroxyvitamin D3 on interleukin 1 beta actions and cell growth in human synovial fibroblast cultures [abstract]. J Rheumatol 20 (9):1527–1532, http://www.ncbi.nlm.nih.gov/pubmed/8164209. Accessed 29 Nov 29 2015
20. Hosokawa Y, Hosokawa I, Shindo S, Ozaki K, Matsuo T (2015) Calcitriol suppressed inflammatory reactions in IL-1β-stimulated human periodontal ligament cells. Inflammation 38(6):2252–2258. doi:10.1007/s10753-015-0209-y
21. Chen YC, Ninomiya T, Hosoya A, Hiraga T, Miyazawa H, Nakamura H (2012) 1α,25-dihydroxyvitamin D3 inhibits osteoblastic differentiation of mouse periodontal fibroblasts. Arch Oral Biol 57 (5):453–459. doi:10.1016/j.archoralbio.2011.10.005
22. Angeline ME, Ma R, Pascual-garrido C et al (2014) Effect of diet-induced vitamin D deficiency on rotator cuff healing in a rat model. Am J Sports Med 42 (1):27–34. doi:10.1177/0363546513505421
23. Barbosa AW, Benevides GP, Alferes LM, Salomao EM, Gomes-Marcondes MC, Gomes L (2012) A leucine-rich diet and exercise effect the biomechanical characteristics of the digital flexor tendon in rats after nutritional recovery. Amino Acids 42 (1):329–336. doi:10.1007/s00726-010-0810-1
24. Vieira CP, De oliveira LP, Da ré guerra F, Dos santos de almeida M, Marcondes MC, Pimentel ER (2015) Glycine improves biochemical and biomechanical properties following inflammation of the achilles tendon. Anat Rec (Hoboken) 298(3):538–545. doi:10.1002/ar.23041
25. Zdzieblik D, Oesser S, Baumstark MW, Gollhofer A, König D (2015) Collagen peptide supplementation in combination with resistance training improves body composition and increases muscle strength in elderly sarcopenic men: a randomised controlled trial. Br J Nutr 114(8):1237–1245. doi:10.1017/S0007114515002810
26. Grewal N, Thornton GM, Behzad H et al (2014) Accumulation of oxidized LDL in the tendon tissues of C57BL/6 or apolipoprotein E knock-Out mice that consume a high fat diet: potential impact on tendon health. In: Screen HR (ed) PLoS One 9(12):e114214. doi:10.1371/journal.pone.0114214
27. Lü J-M, Lin PH, Yao Q, Chen C (2010) Chemical and molecular mechanisms of antioxidants: experimental approaches and model systems. J Cell Mol Med 14 (4):840–860. doi:10.1111/j.1582-4934.2009.00897.x
28. Godic A, Poljšak B, Adamic M, Dahmane R (2014) The role of antioxidants in skin cancer prevention and treatment. Oxid Med Cell Longev 2014:860479. doi:10.1155/2014/860479
29. Rutter K, Sell DR, Fraser N et al (2003) Green tea extract suppresses the age-related increase in collagen crosslinking and fluorescent products in C57BL/6 mice. Int J Vitam Nutr Res 73(6):453–460
30. Lewis JS, Sandford FM (2009) Rotator cuff tendinopathy: is there a role for polyunsaturated fatty acids and antioxidants? J Hand Ther 22 (1):49–55. doi:10.1197/j.jht.2008.06.007
31. World Health Organization. Nutrient requirements and dietary guidelines. Available at: http://www.who.int/nutrition/publications/nutrient/en/. Accessed on 7 Dec 2015

Part IV

Summary

General Overview and Summary of Concepts Regarding Tendon Disease Topics Addressed Related to Metabolic Disorders

28

Paul W. Ackermann and David A. Hart

Abstract

Painful and non-healing musculoskeletal disorders, eg. tendinopathy, pose a tremendous burden on society and the quality of life for patients. New advances in the understanding of connective tissue disorders such as tendinopathy reveal that common health problems such as obesity, atherosclerosis, hormonal dysfunctions and diabetes mellitus are closely linked to the metabolism of components of the musculoskeletal system, particularly tendons. As tendons function as multi-component "organ systems" (Muscle-TMJ-Tendon-Enthesis to Bone), tendons can be influenced directly, or indirectly via, for instance, alterations to muscle. However, this volume/set of chapters focus mainly on the tendon.

Emerging findings in musculoskeletal research have established important new links in our understanding of tendon metabolism. Thereby, the function of the neuroendocrine/-immune axis, as well as supply of neurovascular factors, can be directly linked to the quality of tendon metabolism.

Since some conditions, eg. atherosclerosis and diabetes mellitus, are more common in individuals as they age, and aging can also affect pain and tissue repair, convergence of such complications will potentially exert an increasingly significant impact on tendons as the demographics of many societies change with expanding percentages of the populations >60–65 years of age.

Comorbidities related to metabolic dysfunction have to be identified early in patients with musculoskeletal disorders, such as acute tendon

P.W. Ackermann (✉)
Department of Molecular Medicine and Surgery, Karolinska Institutet, SE-17176, Stockholm, Sweden

Department of Orthopedic Surgery, Karolinska University Hospital, SE-17176, Stockholm, Sweden
e-mail: paul.ackermann@karolinska.se

D.A. Hart
McCaig Institute for Bone and Joint Health, University of Calgary, Calgary, AB, Canada

Centre for Hip Health and Mobility, University of British Columbia, Vancouver, BC, Canada

P.W. Ackermann and D.A. Hart (eds.), *Metabolic Influences on Risk for Tendon Disorders*, Advances in Experimental Medicine and Biology 920, DOI 10.1007/978-3-319-33943-6_28

injuries or chronic tendinopathy, for therapeutic considerations regarding both operative and non-operative treatment protocols. Necessary interactions between researchers and clinicians with different subspecialties have to be initiated in order to optimize tissue metabolism for improved healing potentials.

Keywords
Metabolism • Neuroendocrine • Healing • Neuropathy • Vasculopathy • Diabetes mellitus • Tendon disorders

Introduction

Musculoskeletal conditions affect more than 1.7 billion people worldwide and are the second greatest cause of disability, and as such have the 4th greatest impact on the overall health of the world population.

Work-related musculoskeletal disorders (MSDs) in the US are estimated at $20 billion a year in direct costs, and up to five times more in indirect costs for MSD-related workers' compensation. In addition, there is the substantial toll on affected workers who develop significant difficulties in performing simple upper extremity tasks, according to US Occupational Safety and Health Administration (OSHA 2014).

However, when it comes to current available treatment alternatives for chronic MSD, such as tendinopathy, there is a lack of proven efficient therapies partly due to poor understanding of the mechanisms leading to the development and progression of connective tissue disorders. The purpose of this integration chapter is to summarize some of the emerging new knowledge on how metabolic disturbances and associated disorders can influence tendon connective tissue homeostasis.

Our aim is to address how this new integrative knowledge will lead to novel therapeutic approaches targeting underlying metabolic deficits, and result in optimized connective tissue healing and reductions in the chronic pain of these conditions. MSD, especially with regard to tendons, will have an expanding significant impact on our society considering our more active lifestyles during aging, as well as the detrimental effects of aging on tendon connective tissue homeostasis (see Chap. 24).

The emerging knowledge in this book, and integrated in this chapter, is essential for all clinicians, researchers and those interested in musculoskeletal disorders. To advance the care of people with unmet needs, we must collaborate on all levels to address the underlying tissue pathophysiology in a targeted manner.

Basic Tendon Biology and Anatomy

Tendon anatomy and structure are optimized to fulfil their functional role in differing environments, with highly aligned and abundant collagen fibers providing the highest tensile strength required for efficient force transfer. However, in order to fully understand tendon structural relationships, the specific functional roles of each component of the tendon matrix must be determined (see Chap. 1). Presently, little is still known about the metabolically very active sheaths around the tendon, which work in concert with the tendons to provide targeted gliding, and the interfascicular matrix.

Although the metabolic activity of the tendon is generally low as compared to other tissues, the major metabolic regulatory factor is mechanical loading. Subsequent to loading or unloading, mechanotransduction and molecular anabolic/catabolic signalling result in tissue adaptations in the tendon, which particularly during youth and adolescence (growth and maturation), produce tendon size adjustments (see Chap. 2). Adaptive responses may vary in different tendon compartments or environments, and still more knowledge is needed regarding loading responses in the tendon and in surrounding structures.

Tendons, like nearly all other connective tissues, subscribe to the "use it or lose it" paradigm. Thus, tendons require consistent loading to maintain their functioning set point. Unlike ligaments that are passive structures, tendons are active connective tissues, which in many environments, work near their mechanical limit and thus are perhaps exposed to higher risks for damage (see Chaps. 1 and 2). Such risks can lead to compromised function, pain, and impaired repair mechanisms when impacted by metabolic derangements such as discussed in specific chapters. That is, metabolic disorders can exacerbate risk for tendon compromise and development and progression of certain types of MSD. The impact of metabolic disorders can be on the actual tendon proper, or even at the level of tendon regulatory elements (e.g. nerves and blood supply).

Innervation and blood supply of intact healthy tendons are localized in the surrounding interfascicular matrix of the tendon proper, i.e. paratenon, endotenon and epitenon, whereas the tendon proper is practically devoid of a neuronal supply. This anatomical finding reflects that tendon metabolism is primarily regulated from the tendon envelope or surrounding tissues (see Chaps. 4 and 5). In tendon healing and tendinopathy, however, extensive blood vessel and nerve ingrowth occur into the tendon proper, followed by a time-dependent expression of different neuronal mediators. The integrated action of the neuro-vascular system in tendon metabolism is an emerging research field, which may provide both molecular and clinical targets for future therapies. Given this scenario, disruption of tendon regulation via the "envelope" or the tendon proper by metabolic disorders could dramatically impact tendon function and development of symptoms such as pain via neuropeptides and sensory nerves.

Specific recent knowledge demonstrates how neuronal factors, such as substance P can recruit tendon stem/progenitor cells (TSCs) to sites of repair or tendinopathy. The discovery of tendon TSCs in itself is a remarkable advancement in the tendon research field (see Chap. 6). TSCs play a critical role in tendon physiology and metabolism as well as pathology such as tendinopathy. Additionally, TSCs could potentially be used for tendon tissue engineering *in vitro*, and serve as a promising source of cell-based therapies.

To better elucidate the role of TSCs, as a potential source for enhancing tendon regeneration and to engineer new tendon, tenogenic differentiation and neotissue formation has to be better understood. Thus, the structure-property relationships of embryonic tendon as well as tendon progenitor cell function during development has been studied in detail (see Chap. 6). The potential to guide tenogenic differentiation of adult mesenchymal stem cells with factors that play integral roles in tenogenic differentiation of embryonic tendon progenitor cells during normal development has been demonstrated.

Further studies are also needed to better elucidate the role of TSCs, as a potential source for enhancing endogenous repair. Emerging knowledge has shed light on fundamental tenocyte signalling pathways (see Chap. 7). Future work must also reveal pathways that can be manipulated to prevent matrix degradation, and even support functional matrix replacement with the use of TSCs to promote proper tendon metabolism (see Chap. 8).

Tendon Disorders Associated with Metabolism and Metabolic Disorders

Still today, there is a big gap in our knowledge of how basic research translates into the clinical settings dealing with lifestyle diseases, e.g. hypercholesterolemia, affecting tendon metabolism. If we aim to "solve" the riddle of musculoskeletal disorders, especially tendon problems, we need first to understand via which pathways lifestyle affects tendon metabolism in order to develop targeted means to address the connective tissue pathophysiology on a molecular level.

Let us start to look at genetic disorders by which small genetic variations in many individuals contribute to an increased susceptibility of sustaining a tendon injury. The future is already out there, online, with genetic testing, although the clinical knowledge and implications are sparse. To date, 18 genetic intervals and 32 polymorphisms have been associated with risk of tendon pathologies, which relate to collagen isoforms and variants of structural matrix homeostasis (see Chap. 9). However, these associations cannot be viewed independently and have in the future to be verified by other scientific approaches.

Other genetic disorders involve defects in genes that code for enzymes involved in metabolic pathways (see Chap. 10). The most well-known diseases are familial hypercholesterolemia leading to tendon xanthomas, alkaptonuria resulting in acid accumulation, whereby tendons get a typical ochre/yellow pigmentation (ochronosis), with ensuing inflammation, calcification and rupture, and hypophosphatasia associated with tendon deposition of hydroxyapatite crystals. However, there are likely many more subclinical diseases that may never get diagnosed, but which may provide an increased susceptibility or risk to develop subsequent tendon disorders.

One such important metabolic condition, which may not be readily diagnosed, but cause repeated tendon problems, is hyperuricemia (see Chap. 11). Hyperuricemia and monosodium urate crystal deposition can challenge tendon homeostasis because of their potential to induce inflammation in the host. Today, there is little information available regarding hyperuricemia-mediated adjuvancity in tendinopathy. More knowledge about the interactions of eg. urate with both innate immune and local cells, may help researchers and clinicians to determine if hyperuricemia is a potential target for effective treatments for a subset of tendon problems.

Other metabolic disturbances associated with specific substances that can affect tendon homeostasis include the thyroid hormones (see Chap. 12). Autoimmune thyroid diseases can lead to connective tissue disorders, as thyroid hormones can alter tendon metabolism. Other disorders of thyroid function have so far been investigated only for rotator cuff calcific tendinopathy and tears. Further research is needed to clarify the role of thyroid hormones in the onset and progression of tendinopathies.

The endocrine system holds a strong control on tendon metabolism not only via thyroid hormones, but also via sex hormones (see Chap. 13). Thus, in active young female athletes, physiological high concentrations of estrogen may lead to increased risk for injuries due to reduced fibrillar crosslinking and enhanced joint laxity. Testosterone, on the other hand, augments tendon stiffness due to an enhanced tendon collagen turnover and collagen content. Testosterone steroid injections, among many other side-effects, often result in tendon ruptures. However, the specific effects of individual hormones on tendon metabolism are not yet fully elucidated and still need further study. As well, since the natural environment contains multiple hormones and their respective receptors, analysis of individual hormone effects do not reflect the in vivo situation.

Hypercholesterolemia can exist both as hereditary dyslipidemias and as a result of lifestyle. However, associations between elevated total cholesterol and tendon problems exist in all these patients. High cholesterol environments have been demonstrated to alter tendon biomechanical properties with a few underlying mechanisms explored, showing eg.: altered protein synthesis; dysfunctional local extracellular matrix composition and turnover; and inflammatory gene expression (see Chap. 14). Future research within this area would also benefit from incorporation of additional clinically better translatable study elements.

In addition to hypercholesterolemia, obesity has been demonstrated to exert harmful effects on tendons (see Chap. 15). The pathogenesis being multi-factorial including overload, attributable to the increased body weight, and systemic factors such as bioactive peptides (e.g. chemerin, leptin, adiponectin) that contribute to a chronic, sub-clinic, low-grade inflammation (e.g. metabolic syndrome). Therefore, personalized training programs with regular check-ups are important components for the effective treatment of tendon pathology in such individuals.

Adiposity is also clearly linked to pre-diabetes and diabetes mellitus, conditions which have well documented detrimental effects on both tendon homeostasis, as well as tendon healing. Furthermore, one of the first sites of insulin resistance in obesity is in muscles, which in turn drive tendon activity. The regenerative capability of tendons is compromised in diabetes with corresponding expressional changes in collagens, matrix metalloproteinases and various inflammatory and growth mediators and their receptors (see Chap. 16 and 17). Another specific factor associated with diabetes that affects the mechanical properties of the tendon is the formation of advanced glycation end-products that lead to cross-links with collagen extracellular matrix (see Chap. 18).

So how do we treat patients with diabetes mellitus and tendon problems? Exercise is and likely will become the most important early and non-invasive intervention for these patients. However, the prescribed exercise has to be individually managed by an experienced, trained physiotherapist who is aware of all different musculoskeletal interactions associated with diabetes mellitus (see Chap. 19). A holistic approach should be used to optimize musculotendinous function, including a comprehensive exercise prescription addressing strength, flexibility, and aerobic fitness.

An important question remaining is related to whether the above mentioned metabolic disorders partly may act via inflammatory pathways to affect tendon homeostasis. Advances in our understanding of the basic science of inflammation have provided further insight into its potential role in specific forms of tendon disease, and the motive powers such as excessive mechanical stresses and systemic inflammatory diseases (see Chap. 20).

Inflammatory pathways are additionally involved in the pathogenesis of deep venous thrombosis (DVT), which may occur in up to 50 % of the patients with Achilles tendon rupture – a late stage complication of tendinopathy. DVT has recently been demonstrated as an independent predictive factor for impaired patient outcome at 1 year after Achilles tendon rupture, possibly via negative influences on tendon healing by impaired blood circulation (see Chap. 21). These findings suggest that specific interventions are warranted to prevent DVT. Thus, recently adjuvant treatments with intermittent pneumatic compression applied during lower limb immobilization was demonstrated to reduce the incidence of DVT.

An often-underestimated risk factor for tendon disorders is the influence of intake of different drugs used as prescribed medications (see Chap. 22). Thus, there can be detrimental side-effects of drugs on the tendon, including both tendinopathy and the risk for tendon ruptures. Four main drug classes have been reported to be associated with disturbed tendon metabolism: (1) Corticosteroids, (2) Chinolon antibiotics, (3) Aromatase inhbitors, and (4) Statins (HMG-CoA-reductase inhibitors). The intake of these drugs may increase tendon risk for compromise directly, or indirectly via a concurrent factor such as tendon loading, leading to detrimental effects on tendon integrity.

Corticosteroids (Glucocorticoids) are still widely used to relieve a wide variety of musculoskeletal disorders is. However, the negative influences of glucocorticoids on tendon metabolism are compelling (see Chap. 23). Glucocorticoids reduce tendon derived cell proliferation, reduce extracellular matrix synthesis, contribute to collagen disorganisation and inflammatory cell infiltration and negatively affect the mechanical properties of tendons. Therefore, glucocorticoids should be used with extreme caution in treating tendon problems.

Consider the aging population, which at the same time of life has increasing demands for an active lifestyle. This combination will lead to an increased pressure on the healthcaresystem to provide help for musculoskeletal and tendon disorders. Recent evidence clearly demonstrate increased susceptibility of tendons to injury with advancing age (see Chap. 24). These challenges suggest that we need a much better understanding of common pathways of aging and alterations to tendon homeostasis.

Novel Therapies That May Affect Tendon Metabolism

A recent highly popularized therapy for tendon disorders are injections with platelet-rich plasma (PRP). The background for PRP therapy is that platelet degranulation leads to a release of various growth factors and cytokines, which in experimental studies have shown positive effects on tendon metabolic activity. However, when it comes to clinical studies the evidence from the literature demonstrates that PRP does not lead to improved patient-reported outcomes (see Chap. 25). Therefore, currently PRP should only be used in experimental and clinical studies as to further explore if PRP has a role for treating tendon disorders.

Another increasingly used treatment for tendon disorders are shockwave treatments, which have accumulating clinical evidence for their clinical effectiveness (see Chap. 26). The few underlying mechanisms that have been explored relate to destruction of calcifications, pain relief, mechanotransduction-initiated tissue regeneration, and remodeling of the tendon. However, the heterogeneity of shockwave systems, treatment protocols and study populations, and the fact that there seems to be responders and non-responders, warrants further basic and clinical research.

Since obesity, hyperlipidemia, hyperuricemia, diabetes mellitus, and likely also inflammation in general, affect tendon metabolism negatively, it would be logical to assume that the diet could influence, at least indirectly, several tendon disorders (see Chap. 27). Today, however, the direct effects of diet on tendon metabolism, as well as long-term indirect effects on tendon disorders, are for the most part, completely unknown.

Conclusion

Today there are still no effective drug therapies for many painful and non-healing musculoskeletal disorders, eg. tendinopathy, which negatively affect the quality and functionality of life for millions of people. There is an immense need for a more holistic understanding of the underlying causes of such connective tissue problems. Recent advances in musculoskeletal research have established important new links in our understanding of how tendon disorders are linked to diseases and conditions associated with metabolic disturbances such as obesity, atherosclerosis, hormonal dysfunctions, inflammation and diabetes mellitus. However, there is still a great need for further mapping of the molecular pathways involved, as well as to characterise the extent of metabolically associated disorders in patients. Further challenging tasks are to identify biomarkers of disease and biomarkers predictive of treatment response. A prerequisite for the success of research in this area is a close interaction between researchers and clinicians with various subspecialties in order to identify underlying targets for new therapies.

The individual chapters in this book address many of the individual elements discussed here in more detail. The key to success going forward will be to both understand the individual factors and their interplay to impact subsets of people to increase risk for tendinopathies and to contribute to unique and/or common pathways for progression and resolution of these tendon disorders.

Index

P.W. Ackermann and D.A. Hart (eds.), *Metabolic Influences on Risk for Tendon Disorders*, Advances in Experimental Medicine and Biology 920, DOI 10.1007/978-3-319-33943-6

MIX
Papier aus verantwortungsvollen Quellen
Paper from responsible sources
FSC® C105338

If you have any concerns about our products,
you can contact us on
ProductSafety@springernature.com

In case Publisher is established outside the EU,
the EU authorized representative is:
Springer Nature Customer Service Center GmbH
Europaplatz 3, 69115 Heidelberg, Germany

Printed by Libri Plureos GmbH
in Hamburg, Germany